T0254711

P. Harmand D. Werner W. Werner

M-Ideals in Banach Spaces and Banach Algebras

Springer-Verlag
Berlin Heidelberg New York
London Paris Tokyo
Hong Kong Barcelona
Budapest

Authors

Peter Harmand
Fachbereich Mathematik
Universität Oldenburg
Postfach 2503
D-26111 Oldenburg, Germany

Dirk Werner
I. Mathematisches Institut
Freie Universität Berlin
Arnimallee 3
D-14195 Berlin, Germany

Wend Werner
Universität-GH-Paderbron
Fachbereich 17
Postfach 1621
D-33095 Paderborn, Germany

Mathematics Subject Classification (1991): 46BXX, 41A50, 41A65, 43A46, 46A55, 46E15, 46H10, 46J10, 46L05, 46M05, 47A12, 47D15, 52A07

ISBN 3-540-56814-X Springer-Verlag Berlin Heidelberg New York
ISBN 0-387-56814-X Springer-Verlag New York Berlin Heidelberg

Library of Congress Cataloging-in-Publication Data

Harmand, P. (Peter), 1953-
M-ideals in Banach spaces and Banach algebras / P. Harmand, D. Werner, W. Werner. p. cm. - (Lecture notes in mathematics; 1547)
Includes bibliographical references and index.
ISBN 3-540-56814-X (Berlin: pbk.: acid-free) - ISBN 0-387-56814-X (New York: pbk.: acid-free)
1. Banach spaces. 2. Ideals (Algebra) 3. Approximation theory. I. Werner, D. (Dirk), 1955- . II. Werner, W. (Wend), 1958- . III. Title. IV. Series: Lecture notes in mathemtics (Springer-Verlag); 1547.
QA3.L28 no. 1547 (QA326) 510 s-dc20 (515'.732) 93-16064 CIP

46/3140-543210 - Printed on acid-free paper

Preface

The present notes centre around the notion of an M-ideal in a Banach space, introduced by E. M. Alfsen and E. G. Effros in their fundamental article "Structure in real Banach spaces" from 1972. The key idea of their paper was to study a Banach space by means of a collection of distinguished subspaces, namely its M-ideals. (For the definition of an M-ideal see Definition 1.1 of Chapter I.) Their approach was designed to encompass structure theories for C^*-algebras, ordered Banach spaces, L^1-preduals and spaces of affine functions on compact convex sets involving ideals of various sorts. But Alfsen and Effros defined the concepts of their M-structure theory solely in terms of the norm of the Banach space, deliberately neglecting any algebraic or order theoretic structure. Of course, they thus provided both a unified treatment of previous ideal theories by means of purely geometric notions and a wider range of applicability. Around the same time, the idea of an M-ideal appeared in T. Ando's work, although in a different context.
The existence of an M-ideal Y in a Banach space X indicates that the norm of X vaguely resembles a maximum norm (hence the letter M). The fact that Y is an M-ideal in X has a strong impact on both Y and X since there are a number of important properties shared by M-ideals, but not by arbitrary subspaces. This makes M-ideals an important tool in Banach space theory and allied disciplines such as approximation theory. In recent years this impact has been investigated quite closely, and in this book we have aimed at presenting those results of M-structure theory which are of interest in the general theory of Banach spaces, along with numerous examples of M-ideals for which they apply.
Our material is organised into six chapters as follows. Chapter I contains the basic definitions, examples and results. In particular we prove the fundamental theorem of Alfsen and Effros which characterises M-ideals by an intersection property of balls. In Chapter II we deal with some of the stunning properties of M-ideals, for example their proximinality. We also show that under mild restrictions M-ideals have to be complemented subspaces, a theorem due to Ando, Choi and Effros. The last section of Chapter II is devoted to an application of M-ideal methods to the classification of L^1-preduals. In Chapter III we investigate Banach spaces X which are M-ideals in their biduals. This geometric assumption has a number of consequences for the isomorphic structure of X. For instance, a Banach space has Pełczyński's properties (u) and (V) once it is an M-

ideal in its bidual; in particular there is the following dichotomy for those spaces X: a subspace of a quotient of X is either reflexive or else contains a complemented copy of c_0. Chapter IV sets out to study the dual situation of Banach spaces which are L-summands in their biduals. The results of this chapter have some possibly unexpected applications in harmonic analysis which we present in Section IV.4. Banach algebras are the subject matter of Chapter V. Here the connections between the notions of an M-ideal and an algebraic ideal are discussed in detail. The most far-reaching results can be proved for what we call "inner" M-ideals of unital Banach algebras. These can be characterised by having a certain kind of approximation of the identity. Luckily the M-ideals which are not inner seem to be the exception rather than the rule. The final Chapter VI presents descriptions of the M-ideals in various spaces of bounded linear operators. In particular we address the problem of which Banach spaces X have the property that the space of compact operators on X is an M-ideal in the space of bounded linear operators, a problem which has aroused a lot of interest since the appearance of the Alfsen-Effros paper. We give two characterisations of those spaces X, one of them following from our work in Chapter V, the other being due to N. Kalton.

Each chapter is accompanied by a "Notes and Remarks" section where we try to give precise references and due credits for the results presented in the main body of the text. There we also discuss additional material which is related to the topics of the chapter in question, but could not be included with complete proofs because of lack of space.

Only a few prerequisites are indepensable for reading this book. Needless to say, the cornerstones of linear functional analysis such as the Hahn-Banach, Krein-Milman, Krein-Smulian and open mapping theorems are used throughout these notes, often without explicitly mentioning them. We also assume the reader to be familiar with the basics of Banach algebra theory including the Gelfand-Naimark theorem representing a commutative unital C^*-algebra in the form $C(K)$, and with various special topics such as the representation of the extreme functionals on a $C(K)$-space as multiples of Dirac measures or the principle of local reflexivity (an explicit statement of which can be found in Theorem V.1.4). Other concepts that we need but are not so well-known will be recalled as required. For our notation we refer to the list of symbols.

Several people helped us with the manuscript in one way or the other. Thanks are due in particular to G. Godefroy, N. Kalton, Å. Lima, R. Payá, T. S. S. Rao, and A. Rodríguez-Palacios. Special mention must be made of E. Behrends who introduced us to the subject and suggested that the book be written. Last but not least we are very grateful to D. Yost for reading preliminary drafts of the manuscript; his criticism has been most invaluable.

A final piece of advice to those readers who would like to relax from M-ideals, M-summands and M-structure theory: we think it's a good idea to resort to [314].

Oldenburg, Berlin, Paderborn, December 1992

Contents

CHAPTER I

Basic theory of M-ideals

I.1 Fundamental properties

In this book we shall be concerned with decompositions of Banach spaces by means of projections satisfying certain norm conditions. The essential notions are contained in the following definition. We denote the annihilator of a subspace J of a Banach space X by $J^\perp = \{x^* \in X^* \mid x^*(y) = 0 \ \forall y \in J\}$.

Definition 1.1 *Let X be a real or complex Banach space.*

(a) *A linear projection P is called an M-*projection *if*

$$\|x\| = \max\{\|Px\|, \|x - Px\|\} \qquad \textit{for all } x \in X.$$

*A linear projection P is called an L-*projection *if*

$$\|x\| = \|Px\| + \|x - Px\| \qquad \textit{for all } x \in X.$$

(b) *A closed subspace $J \subset X$ is called an M-*summand *if it is the range of an M-projection. A closed subspace $J \subset X$ is called an L-*summand *if it is the range of an L-projection.*

(c) *A closed subspace $J \subset X$ is called an M-*ideal *if $J^\perp$ is an L-summand in X^*.*

Some comments on this definition are in order. First of all, every Banach space X contains the trivial M-summands $\{0\}$ and X. All the other M-summands will be called nontrivial. (Sometimes only trivial M-summands exist as will presently be shown.) The same remark applies to L-summands and M-ideals.
There is an obvious duality between L- and M-projections:

- P is an L-projection on X iff P^* is an M-projection on X^*.
- P is an M-projection on X iff P^* is an L-projection on X^*.

This remark yields the following characterisation of M-projections which will be useful in the sequel:

A projection $P \in L(X)$ is an M-projection if and only if

$$\|Px_1 + (Id - P)x_2\| \le \max\{\|x_1\|, \|x_2\|\} \quad \textit{for all } x_1, x_2 \in X. \qquad (*)$$

In fact, $(*)$ means that the operator

$$(x_1, x_2) \mapsto Px_1 + (Id - P)x_2$$

from $X \oplus_\infty X$ to X is contractive whence its adjoint

$$x^* \mapsto (P^*x^*, (Id - P)^*x^*)$$

from X^* to $X^* \oplus_1 X^*$ is contractive. ($X \oplus_p Y$ denotes the direct sum of two Banach spaces, equipped with the ℓ^p-norm.) This means that P^* is an L-projection, and P must be an M-projection.
Turning to (b) let us note that there is only one M-projection P with $J = \operatorname{ran}(P)$ ($= \ker(Id - P)$) if J is an M-summand and only one L-projection P with $J = \operatorname{ran}(P)$ ($= \ker(Id-P)$) if J is an L-summand (cf. Proposition 1.2 below which contains a stronger statement). Consequently, there is a uniquely determined closed subspace $\widehat{J}$ such that

$$X = J \oplus_\infty \widehat{J}$$

resp.

$$X = J \oplus_1 \widehat{J}.$$

Then $\widehat{J}$ is called the complementary M- (resp. L-)summand. The duality of L- and M-projections may now be expressed as

- $X = J \oplus_\infty \widehat{J}$ iff $X^* = J^\perp \oplus_1 \widehat{J}^\perp$,
- $X = J \oplus_1 \widehat{J}$ iff $X^* = J^\perp \oplus_\infty \widehat{J}^\perp$.

It follows that M-summands are M-ideals and that the M-ideal J is an M-summand if and only if the L-summand complementary to $J^\perp$ is weak* closed. Let us note that the fact that J and $\widehat{J}$ are complementary L-summands in X means geometrically that B_X, the closed unit ball of X, is the convex hull of B_J and $B_{\widehat{J}}$.
As regards (c) the reader might wonder why we didn't introduce the notion of an "L-ideal", meaning a subspace whose annihilator is an M-summand. The reason is that such an "L-ideal" is automatically an L-summand (see Theorem 1.9 below). Note, however, that the expression "L-ideal" has occasionally appeared in the literature as a synonym of L-summand, e.g. in [11].

Proposition 1.2

(a) *If P is an M-projection on X and Q is a contractive projection on X satisfying $\operatorname{ran}(P) = \operatorname{ran}(Q)$, then $P = Q$.*

(b) *If P is an L-projection on X and Q is a contractive projection on X satisfying $\ker(P) = \ker(Q)$, then $P = Q$.*

PROOF: We first prove (b). The decisive lever for our argument is that, for an L-summand J in X, there is for a given $x \in X$ one and only one best approximant y_0 in J, that is

$$\|x - y_0\| = \inf_{y \in J} \|x - y\|,$$

namely the image of x under the L-projection onto J. (In the language of approximation theory, L-summands are Chebyshev subspaces.) We apply this remark with $J = \ker(P)$. For $x \in X$ we have $x - Px \in \ker(P) = \ker(Q)$, hence

$$\begin{aligned} \|x - (x - Qx)\| &= \|Qx\| \\ &= \|Q(x - (x - Px))\| \\ &\leq \|Q\| \cdot \|Px\| \\ &\leq \|x - (x - Px)\|. \end{aligned}$$

This means that $x - Qx \in \ker(Q) = \ker(P)$ is at least as good an approximant to x in J as $x - Px$ which is the best one. From the uniqueness of the best approximant one deduces $Qx = Px$, thus $P = Q$, as claimed.

(a) follows from (b) since $\ker(P^*) = \operatorname{ran}(P)^\perp = \ker(Q^*)$. □

Corollary 1.3 *If an M-ideal is the range of a contractive projection Q, then it is an M-summand.*

PROOF: $\ker(Q^*) = J^\perp$. Thus, the L-projection with kernel $J^\perp$ is Q^* and hence weak* continuous. □

We now discuss some examples of M-ideals and M-summands.

Example 1.4(a) Let S be a locally compact Hausdorff space. Then $J \subset C_0(S)$ is an M-ideal if and only if there is a closed subset D of S such that

$$J = J_D := \{x \in C_0(S) \mid x(s) = 0 \quad \text{for all } s \in D\}.$$

It is an M-summand if and only if D is clopen (= closed and open).

PROOF: Obviously, $\mu \mapsto \chi_D \mu$ is the L-projection from $C_0(S)^* = M(S)$ onto $J_D^\perp$ so that J_D is an M-ideal if D is closed, and J_D is an M-summand if D is clopen. Suppose now that $J \subset C_0(S)$ is an M-ideal. In order to find the set D, we make use of the following elementary lemma which we state for future reference. The set of extreme points of a convex set C is denoted by $\operatorname{ex} C$.

Lemma 1.5 *For $Z = J_1 \oplus_1 J_2$ we have (using the convention $\operatorname{ex} B_{\{0\}} = \emptyset$)*

$$\operatorname{ex} B_Z = \operatorname{ex} B_{J_1} \cup \operatorname{ex} B_{J_2}.$$

We shall apply Lemma 1.5 here with $Z = M(S)$, $J_1 = J^\perp$ and $J_2 =$ the complementary L-summand. Let

$$D = \{s \in S \mid \delta_s \in J^\perp\}.$$

Then D is closed, and by construction $J \subset J_D$. The inclusion $J_D \subset J$ follows from the Hahn-Banach and Krein-Milman theorems: By virtue of these results we only have to show $\operatorname{ex} B_{J^\perp} \subset J_D^\perp$, and this is true by Lemma 1.5 and by definition of D.
Finally, if $J = J_D$ is an M-summand, then the complementary M-summand is of the same form, $J_{\widehat{D}}$ say, consequently $D \cup \widehat{D} = S$ and D is clopen. □

In particular, c_0 is an M-ideal in $\ell^\infty = C(\beta\mathbb{N})$ (which could also be proved directly); the additional feature is that ℓ^∞ is the bidual of c_0. We shall study Banach spaces which are M-ideals in their biduals in detail in Chapter III.

Example 1.4(b) Let A be the disk algebra, that is the complex Banach space of continuous functions on the closed unit disk which are analytic in the open unit disk. It will be convenient to consider A (via boundary values) as a subspace of $C(\mathbb{T})$, where $\mathbb{T}$ is the unit circle. We claim that J is a nontrivial M-ideal in A if and only if there is a closed subset $D \neq \emptyset$ of $\mathbb{T}$ with linear Lebesgue measure 0 such that[1]

$$J = J_D \cap A = \{x \in A \mid x(t) = 0 \quad \text{for all } t \in D\}.$$

PROOF: Note first that $J_D \cap A = \{0\}$ if D is a subset of $\mathbb{T}$ having positive linear measure, cf. e.g. [545, Theorem 17.18]. To see that an M-ideal J has the form $J_D \cap A$ one proceeds exactly as in Example 1.4(a); one only has to recall that

$$\operatorname{ex} B_{A^*} = \{\lambda \cdot \delta_t|_A \mid |\lambda| = 1,\ t \in \mathbb{T}\}. \tag{$*$}$$

[This amounts to saying that every $t \in \mathbb{T}$ is in the Choquet boundary of A, and a proof of this fact is contained in [240, p. 54ff.]; for an explicit statement see e.g. [587, p. 29]. Since this example will have some importance in the sequel, we would like to sketch a direct argument: The right hand side of ($*$) is weak* closed and norming by the maximum modulus principle, therefore the Krein-Milman theorem (or rather its converse) implies "$\subset$". On the other hand, $\operatorname{ex} B_{A^*} \neq \emptyset$. So let us assume $p_0 := \lambda_0 \cdot \delta_{t_0}|_A \in \operatorname{ex} B_{A^*}$. (Here we use the fact that for $p \in \operatorname{ex} B_{X^*}$ and $X \subset Y$, p has an extension to some $q \in \operatorname{ex} B_{Y^*}$.) For $|\lambda| = 1$, $t \in \mathbb{T}$

$$(\Phi x)(s) := \frac{\lambda}{\lambda_0} \cdot x\Big(\frac{t}{t_0} s\Big)$$

defines an isometric isomorphism on A, hence $\lambda \cdot \delta_t|_A = \Phi^*(p_0) \in \operatorname{ex} B_{A^*}$.]
Now let $D \neq \emptyset$ be a closed subset of $\mathbb{T}$ of linear measure 0. We wish to find an L-projection from A^* onto $(J_D \cap A)^\perp$. A functional $p \in A^*$ may be represented as $p = \mu|_A$ for some measure $\mu \in M(\mathbb{T}) = C(\mathbb{T})^*$. Let $q = (\chi_D \mu)|_A$. Then the mapping $P : p \mapsto q$ is well-defined: If $p = \nu|_A$ is another representation, then $\nu - \mu$ annihilates A. The F. and M. Riesz theorem (cf. e.g. [545, Theorem 17.13]) implies that $\nu - \mu$ is absolutely continuous with respect to Lebesgue measure so that $\chi_D \mu = \chi_D \nu$ if D has measure 0. It

[1] Throughout, J_D will have the same meaning as in Example 1.4(a).

is easy to check that P is the required L-projection. Finally, the nontriviality follows from a theorem of Fatou [317, p. 80] which also shows that different D give rise to different M-ideals. □

We shall present a characterisation of M-ideals in a general function algebra in Theorem V.4.2. As regards this example, see also the abstract version of our approach in Corollary 1.19.

Example 1.4(c) Let K be a compact convex set in a Hausdorff locally convex topological vector space. As usual, $A(K)$ denotes the space of real-valued affine continuous functions on K. Let us recall the definition of a split face of K ([7, p. 133], [9]). A face F of K is called a split face if there is another face F' such that every $k \in K \setminus (F \cup F')$ has a *unique* representation

$$k = \lambda k_1 + (1 - \lambda) k_2 \quad \text{with } k_1 \in F,\ k_2 \in F',\ 0 < \lambda < 1.$$

It is known that every closed face of a simplex is a split face [7, p. 144].
Then J is an M-ideal in $A(K)$ if and only if there exists a closed split face F of K such that

$$J = J_F \cap A(K) = \{x \in A(K) \mid x(k) = 0 \quad \text{for all } k \in F\}.$$

The proof of this fact can be given along the lines of (b), the crucial step being the measure theoretic characterisation of closed split faces (see [7, Th. II.6.12]) which replaces the use of the F. and M. Riesz theorem.

Example 1.4(c) was one of the forerunners of the general M-ideal theory (cf. [9]). Another forerunner of the general theory is contained in the next example.

Example 1.4(d) In a C^*-algebra the M-ideals coincide with the closed two-sided ideals.

We shall present a proof of this fact in Theorem V.4.4. For the time being let us notice that in particular $K(H)$ is an M-ideal in $L(H)$ (where H denotes a Hilbert space) which was first shown by Dixmier [165] back in 1950. An independent proof of Dixmier's result will be given in Chapter VI. There we shall study the class of Banach spaces X for which $K(X)$ is an M-ideal in $L(X)$. Let us indicate at this point that this class contains all the spaces ℓ^p for $1 < p < \infty$ and c_0 as well as certain of their subspaces and quotient spaces (cf. Example VI.4.1, Corollary VI.4.20). Moreover, $K(X)$ is never an M-summand in $L(X)$ unless $\dim X < \infty$ (Proposition VI.4.3).

Example 1.4(e) Although there are certain similarities between algebraic ideals and M-ideals, there is one striking difference: Unlike the case of algebraic ideals the intersection of (infinitely many) M-ideals need not be an M-ideal. An example to this effect will be given in II.5.5. Also let us advertise now the description of M-ideals in G-spaces (Proposition II.5.2) and Theorem II.5.4 which are related to this phenomenon.

Here are some examples of L-summands.

Example 1.6(a) Consider $X = L^1(\mu)$. We assume that μ is localizable (e.g. σ-finite) so that $L^1(\mu)^* \cong L^\infty(\mu)$ in a natural fashion. Then the L-projections on $L^1(\mu)$ coincide with the characteristic projections $P_A(f) = \chi_A f$ for measurable sets A.

PROOF: Trivially, P_A is an L-projection. Conversely, for a given L-projection P the adjoint P^* is an M-projection on $L^\infty(\mu)$. Now $L^\infty(\mu)$ is a commutative unital C^*-algebra, and as such it may be represented as $C(K)$ by the Gelfand-Naimark theorem. (This representation is also possible in the real case.) Since the Gelfand transform is multiplicative, the idempotent elements in $L^\infty(\mu)$ (i.e., the measurable characteristic functions) correspond exactly to the idempotent elements in $C(K)$ (i.e., the continuous characteristic functions). Now Example 1.4(a) tells us that M-projections are characteristic projections. Thus the same is true for $L^\infty(\mu)$ so that P^*, and hence P, is a characteristic projection. □

The description of the L-projections on $L^1(\mu)$ in terms of the measure space is a bit more involved if μ is arbitrary; we refer to [66, p. 58]. However, every space $L^1(\mu)$ is order isometric to a space $L^1(m)$ where m is (even strictly) localizable [560, p. 114]. Therefore, our initial restriction concerning μ is not a severe one, since we are dealing with Banach space properties of L^1 rather than properties of the underlying measure space.
It is worthwhile adopting the point of view of Banach lattice theory in this example. Using the notions of Banach lattice theory we have established the fact that the L-projections on an L^1-space (or an (AL)-space if one prefers) coincide with the band projections and that the L-summands coincide with the (projection) bands (which are the same as the order ideals in L^1); cf. [560, p. 113 and passim] for these matters. The advantage of this description is that it avoids explicit reference to the underlying measure space.
Incidentally, we have shown that the M-projections on $L^1(\mu)^*$ are weak* continuous. This is true for every dual Banach space, see Theorem 1.9 below.

Example 1.6(b) The Lebesgue decomposition $\mu = \mu_{\mathrm{ac}} + \mu_{\mathrm{sing}}$ with respect to a given probability measure furnishes another example of an L-decomposition. Note that neither of the L-summands in

$$C[0,1]^* = M[0,1] = L^1[0,1] \oplus_1 M_{\mathrm{sing}}[0,1]$$

is weak* closed. (In fact, both of them are weak* dense.)

Example 1.6(c) In Chapter IV we shall study Banach spaces which are L-summands in their biduals. Prominent examples will be the L^1-spaces as well as their "noncommutative" counterparts, i.e., the preduals of von Neumann algebras.

The preceding examples have shown a variety of situations where M-ideals or L-summands arise in a natural way. Sometimes, however, it will also be of interest to have a criterion at hand in order to show that a given Banach space does not contain any nontrivial M-ideal or L-summand.

Proposition 1.7 *If X is smooth or strictly convex, then X contains no nontrivial M-ideal and no nontrivial L-summand.*

PROOF: Since points of norm one which are contained in a nontrivial L-summand never have a unique supporting hyperplane, we conclude that a smooth space cannot contain a nontrivial L-summand. On the other hand, if X is smooth and $x^* \in S_{X^*}$ attains its norm, then x^* is an extreme functional. Now the Bishop-Phelps theorem (see Theorem VI.1.9) yields that ex B_{X^*} is norm dense in S_{X^*}. By Lemma 1.5, X cannot contain a nontrivial M-ideal either. (Actually, we have shown the (by virtue of Proposition V.4.6) stronger statement that all the L-summands in the dual of a smooth space are trivial. However, we shall eventually encounter the dual of a smooth space which contains nontrivial M-ideals, namely L^∞/H^∞ (Remark IV.1.17).)
As for the case of strictly convex spaces, all the L-summands are trivial by Lemma 1.5, and the triviality of the M-ideals is a consequence of Corollary II.1.5 below. □

Proposition 1.7 applies in particular to the L^p-spaces for $1 < p < \infty$.
The following theorem describes a dichotomy concerning the existence of L-summands and M-ideals.

Theorem 1.8 *A complex Banach space or a real Banach space which is not isometric to $\ell^\infty(2)$ $(= (\mathbb{R}^2, \|\cdot\|_\infty))$ cannot contain nontrivial M-ideals and nontrivial L-summands simultaneously.*

PROOF: Suppose that the real Banach space X admits nontrivial decompositions

$$\begin{aligned} X &= J \oplus_\infty \widehat{J} \\ &= Y \oplus_1 \widehat{Y}. \end{aligned}$$

Our aim is to show that under this assumption Y must be one-dimensional. Once this is achieved we conclude by symmetry that $\widehat{Y}$ must be one-dimensional as well so that X is isometric to $\ell^1(2)$ $(\cong \ell^\infty(2))$. If X is a complex Banach space, then the above reasoning applies to the underlying real space and yields that X is $\mathbb{R}$-isometric to $\ell^\infty_{\mathbb{R}}(2)$ which of course is impossible.
Let us now present the details of the proof that $\dim(Y) = 1$. We denote by P the M-projection onto J and by π the L-projection onto Y. We first claim:

$$J \cap Y = \{0\} \tag{1}$$

Assume to the contrary that there exists some $u \in J \cap Y$ with $\|u\| = 1$. Let $x \in \widehat{J}$, $\|x\| = 1$. We shall show $x \in Y$: Since $\|u \pm x\| = 1$ we have

$$\begin{aligned} 2 &= \|u + x\| + \|u - x\| \\ &= \|u + \pi(x)\| + \|x - \pi(x)\| + \|u - \pi(x)\| + \|\pi(x) - x\| \\ &\geq 2 \cdot \|u\| + 2 \cdot \|x - \pi(x)\| \end{aligned}$$

so that

$$x = \pi(x) \in Y.$$

This shows $\widehat{J} \subset Y$ and in particular $\widehat{J} \cap Y \neq \{0\}$. Thus we may repeat the same argument with the duo $\widehat{J}$ & Y to obtain $J \subset Y$ as well. Consequently $Y = X$ in contrast to our assumption that the L-decomposition is nontrivial. Therefore (1) holds.

To show that Y is one-dimensional we again argue by contradiction. Suppose there exists a two-dimensional subspace Y_0 of Y. By (1), $P_{|Y_0}$ is injective so that $J_0 := P(Y_0)$ is two-dimensional, too. By Mazur's theorem (e.g. [318, p. 171]) (or since every convex function on $\mathbb{R}$ has a point of differentiability) J_0 contains a smooth point z, i.e., $\|z\| = 1$ and

$$\ell(x) := \lim_{h\to 0} \frac{1}{h}\Big(\|z + hx\| - 1\Big)$$

exists for all $x \in J_0$. Since J_0 is supposed to be two-dimensional, we can find some $x \in J_0$, $\|x\| = 1$, such that

$$\lim_{h\to 0} \frac{1}{h}\Big(\|z+hx\| - 1\Big) = 0. \tag{2}$$

Let us write $z = \dfrac{Py}{\|Py\|}$ for some $y \in Y_0$, $\|y\| = 1$. We next claim:

$$\lim_{h\to 0} \frac{1}{h}\Big(\|y + hx\| - 1\Big) = 0 \tag{3}$$

To prove this we note

$$\|y + hx\| = \max\{\|Py + hx\|, \|y - Py\|\}$$

since P is an M-projection and $x \in J$. Also note

$$1 = \|y\| = \max\{\|Py\|, \|y - Py\|\}.$$

Thus, if $\|Py\| < 1$, we have for sufficiently small h

$$\|y + hx\| = \|y - Py\| = 1,$$

and (3) follows immediately. If $\|Py\| = 1$ (hence $z = Py$) and $\|y - Py\| < 1$, we have for sufficiently small h

$$\|y + hx\| = \|Py + hx\| = \|z + hx\|,$$

and (3) follows from (2). It is left to consider the case where $1 = \|Py\| = \|y - Py\|$. In this case we have

$$\|y + hx\| = \max\{\|z + hx\|, 1\},$$

and again (3) follows from (2).
We can now conclude the proof as follows. For $h > 0$ we have

$$\|y + hx\| = \|y + h\pi(x)\| + h\|x - \pi(x)\|,$$

$$\|y - hx\| = \|y - h\pi(x)\| + h\|x - \pi(x)\|.$$

Observing

$$\|y + h\pi(x)\| + \|y - h\pi(x)\| \geq 2 \cdot \|y\| = 2$$

we obtain from the above equations

$$\frac{1}{h}\Big(\|y + hx\| - 1\Big) + \frac{1}{h}\Big(\|y - hx\| - 1\Big) \geq 2 \cdot \|x - \pi(x)\|,$$

which shows $x = \pi(x) \in Y$ by virtue of (3). Since $0 \neq x \in J$ by choice of x we have arrived at a contradiction to (1). Therefore, Y must be one-dimensional. □

The real space $\ell^\infty(2)$ has to be excluded because it contains the nontrivial M-summand $\{(s,t) \mid s = 0\}$ and the nontrivial L-summand $\{(s,t) \mid s + t = 0\}$.
Theorem 1.8 implies that an L^1-space does not contain any nontrivial M-ideal whereas a $C(K)$-space does not contain any nontrivial L-summand unless the scalars are real and the space is two-dimensional.

Next we are going to collect several results on M-ideals and L-summands which will frequently be used.

Theorem 1.9 *An M-summand in a dual space X^* is weak* closed and hence of the form $J^\perp$ for some L-summand J in X. Consequently, an M-projection on a dual space is weak* continuous.*

PROOF: Given a decomposition $X^* = J_1 \oplus_\infty J_2$ let us assume to the contrary that J_1 is not weak* closed. In this case there exists a net (x_i^*) in the unit ball of J_1 which converges in the weak* sense to some $x^* \in J_2$, $x^* \neq 0$. (Here the Krein-Smulian theorem enters.) Then $y_i^* = x_i^* + x^*/\|x^*\|$ defines a net in the unit ball of X^* whose weak* limit has norm $1 + \|x^*\| > 1$: a contradiction.
It is easily checked that a projection whose range and kernel are weak* closed is weak* continuous so that the remaining assertion follows. □

REMARKS: (a) Theorem 1.9 can be thought of as a "localized" version of Grothendieck's result that L^1 is the only predual of L^∞ [282].

(b) We will eventually prove in Corollary II.3.6 that a weak* closed M-ideal in a dual space is an M-summand and hence the annihilator of an L-summand.

(c) In Proposition V.4.6 we present an example of a space X without nontrivial M-ideals such that X^* has infinitely many L-summands; see also Example IV.1.8.

Theorem 1.10

(a) *Two L- (resp. M-)projections commute.*

(b) *The set $\mathbb{P}_L(X)$ of all L-projections on X forms a complete Boolean algebra under the operations*

$$P \wedge Q = PQ,\ P \vee Q = P + Q - PQ,\ P^c = Id - P.$$

(c) *The set $\mathbb{P}_M(X)$ of all M-projections on X forms a (generally not complete) Boolean algebra. However, for a dual space X^* $\mathbb{P}_M(X^*)$ is isomorphic to $\mathbb{P}_L(X)$ and hence complete.*

PROOF: (a) Let P and Q be L-projections. Then

$$\begin{aligned}
\|Qx\| &= \|PQx\| + \|(Id - P)Qx\| \\
&= \|QPQx\| + \|(Id - Q)PQx\| + \|Q(Qx - PQx)\| + \|(Id - Q)(Qx - PQx)\| \\
&= \|QPQx\| + 2 \cdot \|PQx - QPQx\| + \|Qx - QPQx\| \\
&\geq \|Qx\| + 2 \cdot \|PQx - QPQx\|
\end{aligned}$$

so that $PQ = QPQ$. Likewise we obtain $P(Id - Q) = (Id - Q)P(Id - Q)$ which is equivalent to $QP = QPQ$, hence the result.

(b) PQ is a projection by (a), and it is an L-projection since

$$\begin{aligned} \|x\| &= \|Qx\| + \|(Id - Q)x\| \\ &= \|PQx\| + \|Qx - PQx\| + \|x - Qx\| \\ &\geq \|PQx\| + \|x - PQx\| \\ &\geq \|x\|, \end{aligned}$$

and so is $P + Q - PQ = Id - (Id - P)(Id - Q)$. It is routine to verify that $\mathbb{P}_L$ has the structure of a Boolean algebra.
Let $\mathbb{M}$ be a family of L-projections. To prove that $\mathbb{P}_L$ is complete we have to show that the supremum of $\mathbb{M}$ (for the induced order: $P \leq Q$ if and only if $PQ = P$) exists. Upon replacing $\mathbb{M}$ by its family of finite suprema one may assume without loss of generality that $\mathbb{M}$ is upward directed. Since

$$\begin{aligned} \|Qx\| &= \|PQx\| + \|Qx - PQx\| \\ &= \|Px\| + \|Qx - Px\| \\ &\geq \|Px\| \end{aligned}$$

for $P \leq Q$ and since $\{\|Px\| \mid P \in \mathbb{M}\}$ is bounded we see that $(Px)_{P \in \mathbb{M}}$ is a Cauchy net for every x: If $\|P_0 x\| \geq \sup_{P \in \mathbb{M}} \|Px\| - \varepsilon$, then $\|Q_1 x - Q_2 x\| \leq 2\varepsilon$ for $Q_1, Q_2 \geq P_0$. It is not hard to verify that its pointwise limit $Sx = \lim_{P \in \mathbb{M}} Px$ is an L-projection, and that $Q \geq S$ if and only if $Q \geq P$ for all $P \in \mathbb{M}$ which means that S is the supremum of $\mathbb{M}$.

(c) $\mathbb{P}_M$ is a Boolean algebra by the duality of L- and M-projections. That $\mathbb{P}_M$ need not be complete can be seen from Example 1.4(a): For $X = c = C(\alpha\mathrm{N})$ the family of M-projections onto the M-summands J_D, $D = \{2\}, \{2,4\}, \{2,4,6\}, \ldots$, has no infimum. For dual spaces combine Theorem 1.9 and (b). □

Proposition 1.11 *Let J_i $(i \in I)$, J_1, J_2 be M-ideals in X.*

(a) $\overline{\mathrm{lin}}\bigcup_i J_i$ *is an M-ideal in X. (Equivalently, the intersection of weak* closed L-summands is a weak* closed L-summand.)*

(b) $J_1 \cap J_2$ *is an M-ideal in X.*

(c) $J_1 + J_2$ *is closed and hence an M-ideal in X. Moreover*

$$J_1/(J_1 \cap J_2) \cong (J_1 + J_2)/J_2.$$

PROOF: (a) and (b) are consequences of Theorem 1.10(b) (cf. [51, p. 37] for details). (c) is a special case of Proposition 1.16 below. □

We again stress that in general an arbitrary intersection of M-ideals need not be an M-ideal; see II.5.5 for a counterexample.
The following proposition is very important, though its proof is immediate from the definition of an M-ideal. It states that M-ideals are "Hahn-Banach smooth".

Proposition 1.12 *Let J be an M-ideal in X. Then every $y^* \in J^*$ has a unique norm preserving extension to a functional $x^* \in X^*$.*

PROOF: By assumption, $J^\perp$ is an L-summand so that there is a decomposition

$$X^* = J^\perp \oplus_1 J^\#. \qquad (*)$$

But $J^\#$ can explicitly be described, since there are canonical isometric isomorphisms

$$J^* \cong X^*/J^\perp \cong J^\#$$

so that

$$J^\# = \{x^* \in X^* \mid \|x^*\| = \|x^*|_J\|\}.$$

The result now follows. □

Remark 1.13 Proposition 1.12 enables us to view J^* as a subspace of X^*. Accordingly we shall replace (*) above by

$$X^* = J^\perp \oplus_1 J^* \qquad (**)$$

in the sequel. Thus, it makes sense to consider the (generally non-Hausdorff) topology $\sigma(X, J^*)$, which we shall occasionally meet. (In the special case where J is an M-ideal in J^{**} (see Chapter III) this is nothing but the weak* topology on J^{**}.)
The Hahn-Banach theorem implies that B_J is $\sigma(X, J^*)$-dense in B_X. (Note that this holds even in the case when $\sigma(X, J^*)$ is not Hausdorff.) This simple observation has important consequences when applied to operator spaces (Proposition VI.4.10).

Next we wish to consider the question whether an M-ideal in a Banach space induces an M-ideal in a subspace or a quotient space. We first collect some general facts in this direction.

Lemma 1.14 *Let J and Y be closed subspaces in a Banach space X.*

(a) *$J+Y$ is closed in X if and only if $J^\perp+Y^\perp$ is closed in X^* if and only if $J^\perp+Y^\perp$ is weak* closed in X^*. In this case $J^\perp + Y^\perp = (J \cap Y)^\perp$, and $(J+Y)/J$ is isomorphic to $Y/(J \cap Y)$, $(J^\perp + Y^\perp)/Y^\perp$ is isomorphic to $J^\perp/(J^\perp \cap Y^\perp)$.*

(b) *Suppose $J^\perp$ is the range of a projection P such that $P(Y^\perp) \subset Y^\perp$. Then the assertions of* (a) *hold. If P is contractive we even have*

$$(J+Y)/J \cong Y/(J \cap Y),$$
$$(J^\perp + Y^\perp)/Y^\perp \cong J^\perp/(J^\perp \cap Y^\perp).$$

If $Id - P$ is contractive we have

$$(J+Y)/Y \cong J/(J \cap Y).$$

PROOF: (a) To prove the asserted equivalence apply the closed range theorem to the operator $y \mapsto y + J$ from Y to X/J (cf. [522] for details). The sum $J^\perp + Y^\perp$ is always weak* dense in $(J \cap Y)^\perp$ by the Hahn-Banach theorem so that the next assertion obtains. Finally, the operator

$$T : Y/(J \cap Y) \longrightarrow (J+Y)/J, \quad T(y + (J \cap Y)) = y + J$$

is a well-defined bijective contractive operator between Banach spaces, hence an isomorphism. Similarly,

$$S: J^\perp/(J^\perp \cap Y^\perp) \longrightarrow (J^\perp + Y^\perp)/Y^\perp, \quad S(z^* + (J^\perp \cap Y^\perp)) = z^* + Y^\perp$$

is an isomorphism.

(b) Let us prove that $J^\perp + Y^\perp$ is closed. Suppose

$$J^\perp + Y^\perp \ni z_n^* + \xi_n^* \to x^* \in X^*$$

as $n \to \infty$. Then

$$\xi_n^* - P\xi_n^* = (Id - P)(z_n^* + \xi_n^*) \to x^* - Px^*$$

which is seen to lie in $Y^\perp$ by assumption on P. Hence

$$x^* = Px^* + (x^* - Px^*) \in J^\perp + Y^\perp.$$

It is left to prove that T^{-1} and S^{-1} are contractive. The latter follows from the estimate (valid for $z^* \in J^\perp$)

$$\begin{aligned} &\inf\{\|z^* - \xi^*\| \mid \xi^* \in J^\perp \cap Y^\perp\} \\ &\quad = \inf\{\|Pz^* - P\xi^*\| \mid \xi^* \in X^*,\ P\xi^* \in Y^\perp\} \\ &\quad \le \inf\{\|z^* - \xi^*\| \mid \xi^* \in X^*,\ P\xi^* \in Y^\perp\} \\ &\quad \le \inf\{\|z^* - \xi^*\| \mid \xi^* \in Y^\perp\} \end{aligned}$$

(where we used $\|P\| \le 1$ and $P(Y^\perp) \subset Y^\perp$); for the former observe that

$$\begin{aligned} \|y + J\cap Y\| &= d(y, J\cap Y) \\ &= \sup\{|\langle x^*, y\rangle| \mid \|x^*\| \le 1,\ x^* \in (J\cap Y)^\perp = J^\perp + Y^\perp\}, \\ \|y + J\| &= d(y, J) \\ &= \sup\{|\langle x^*, y\rangle| \mid \|x^*\| \le 1,\ x^* \in J^\perp\}. \end{aligned}$$

Now for $x^* = z^* + \xi^* \in J^\perp + Y^\perp$ we have $x^* - Px^* = \xi^* - P\xi^* \in Y^\perp$, hence $\langle x^*, y\rangle = \langle Px^*, y\rangle$. Since $\|Px^*\| \le \|x^*\|$ the result follows.
The remaining isometry is established in the same way. □

Lemma 1.15 *Let P be a projection on X and let $Y \subset X$ be a closed subspace. We suppose $P(Y) \subset Y$ so that*

$$P_{|Y} : Y \longrightarrow Y, \quad y \mapsto Py$$

and

$$P/Y : X/Y \longrightarrow X/Y, \quad x + Y \mapsto Px + Y$$

are well-defined projections. We have

$$\begin{aligned} \operatorname{ran}(P_{|Y}) &= \operatorname{ran}(P) \cap Y, \\ \operatorname{ran}(P/Y) &= (\operatorname{ran}(P) + Y)/Y, \end{aligned}$$

and

$$\mathrm{ran}(P/Y) \cong \mathrm{ran}(P)/(\mathrm{ran}(P) \cap Y)$$

if P is contractive. Moreover, $P|_Y$ and P/Y are L- (resp. M-) projections if P is.

PROOF: All the assertions concerning $P|_Y$ are obvious. Note that $P^*(Y^\perp) \subset Y^\perp$ and thus $(P/Y)^* = P^*|_{Y^\perp}$. Taking into account the duality of L- and M-projections and Lemma 1.14 we infer the assertions concerning P/Y. □

Proposition 1.16 *Suppose J is an M-ideal in X with corresponding L-projection P from X^* onto $J^\perp$. Suppose in addition that Y is a closed subspace of X such that*

(a) $$P(Y^\perp) \subset Y^\perp.$$

Then $J \cap Y$ is an M-ideal in Y, $J + Y$ is closed, and $(J+Y)/Y$ is isometric with $J/(J \cap Y)$ and an M-ideal in X/Y.
Moreover, (a) *is equivalent to either of the conditions*
(b) $\quad J \cap Y$ *is $\sigma(X, J^*)$-dense in Y,*
(b_1) $\quad B_{J\cap Y}$ *is $\sigma(X, J^*)$-dense in B_Y.*

PROOF: If one identifies Y^* with $X^*/Y^\perp$ then $P/Y^\perp$ defines an L-projection from Y^* onto the annihilator of $J \cap Y$ in Y^* by Lemma 1.15 so that $J \cap Y$ is an M-ideal in Y. To show that $(J+Y)/Y$ is an M-ideal in X/Y employ the L-projection $P|_{Y^\perp}$. The other assertions are special cases of Lemma 1.14. Now that $J \cap Y$ is an M-ideal in Y under assumption (a), we see from Remark 1.13 that (b_1) holds, and (b_1) trivially implies (b). Next assume (b). For $\xi^* \in Y^\perp$ we have $\xi^* - P\xi^* \in \ker(P) = J^*$. Now $\xi^* \in Y^\perp \subset (J \cap Y)^\perp$ and $P\xi^* \in J^\perp \subset (J \cap Y)^\perp$ imply that $\xi^* - P\xi^*$ annihilates $\overline{J \cap Y}^{\sigma(X,J^*)}$ so that by (b) $\xi^* - P\xi^* \in Y^\perp$. Hence (a) follows. □

REMARKS: (a) The reader should have no difficulty in recovering Example 1.4(b) from 1.4(a) using condition (a) of Proposition 1.16. This is the abstract argument behind our reasoning in Example 1.4(b), and it will be expressively mentioned in Corollary 1.19 below.

(b) We remark that (a) is not necessary to ensure that $J \cap Y$ is an M-ideal. For example, let $Y = \ell^\infty(2)$, $K = B_{Y^*} (= \{(s,t) \in \mathbb{K}^2 \mid |s| + |t| \le 1\})$ and $X = C(K)$ so that $Y \subset X$ in a canonical fashion. Let $U = \{(0,t) \mid t \in \mathbb{K}\}$ and

$$D = \{y^* \in K \mid y^*|_U = 0\} = \{(s,0) \mid |s| \le 1\}.$$

Then U is the intersection of the M-ideal J_D with Y and also itself an M-ideal in Y (in fact, an M-summand). Nevertheless, assumption (a) in Proposition 1.16 is violated: If $\mu \in M(K)$ is the discrete measure

$$\mu = \delta_{(1/2,1/2)} + \delta_{(1/2,-1/2)} + \delta_{(-1,0)}$$

then $\mu \in Y^\perp$, but

$$P(\mu) = \chi_D \mu = \delta_{(-1,0)} \notin Y^\perp.$$

For a positive result involving maximal measures we refer to Theorem 2.4.

(c) Condition (a) in 1.16 is fulfilled if Y, too, is an M-ideal, since L-projections commute (1.10(a)). This provides a proof of Proposition 1.11(c).

Proposition 1.17

(a) *If J_1 and J_2 are M-ideals in X, then the canonical images of J_1 and J_2 in $(J_1 + J_2)/(J_1 \cap J_2)$ are complementary M-summands.*

(b) *Let J be an M-ideal in X.*

- *$Y \subset J$ is an M-ideal in X if and only if it is an M-ideal in J.*
- *$Y \subset X/J$ is an M-ideal if and only if it is the image of an M-ideal in X under the quotient map.*
- *If $Y \subset J$, then J/Y is an M-ideal in X/Y.*
- *If $J \subset Y \subset X$, then J is an M-ideal in Y.*

PROOF: These assertions are easily verified; for details cf. [51, p. 39f.]. □

We recall from the fundamental Example 1.4(a) that the M-ideals in $C(K)$ coincide with the ideals J_D. The following proposition says that the J_D are in fact the ancestors of all M-ideals, since every Banach space is a subspace of some $C(K)$.

Proposition 1.18 *Let X be a closed subspace of $C(K)$, and let J be an M-ideal in X. Then there is a closed subset D of K such that $J = J_D \cap X$.*

PROOF: Let $D = \{k \in K \mid \delta_{k|_X} \in J^\perp\}$. Then D is a closed set, and $J \subset J_D \cap X$ by construction. If the inclusion were proper, we could separate a certain $x_0 \in J_D \cap X$ from J by a functional $p \in J^\perp$. We may even assume $p \in \operatorname{ex} B_{J^\perp}$ by the Krein-Milman theorem and thus (Lemma 1.5) $p \in \operatorname{ex} B_{X^*}$. Such a p is of the form $p(x) = \lambda \cdot x(k)$ for some $k \in K$, $|\lambda| = 1$. Since $p \in J^\perp$ we must have $k \in D$ and hence $x_0(k) = 0$. On the other hand $p(x_0) \neq 0$ since p is a separating functional: a contradiction. □

We turn to the converse of this proposition, which, of course, will generally not hold. The following is a special case of Proposition 1.16.

Corollary 1.19 *Let X be a closed subspace of $C(K)$. Then $J_D \cap X$ is an M-ideal in X if the L-projection $\mu \mapsto \chi_D \mu$ from $C(K)^*$ onto $J_D^\perp$ leaves $X^\perp$, the annihilator of X in $C(K)^*$, invariant.*

Let us formulate an important example where this occurs. Suppose X is a closed subspace of $C(K)$ and suppose $D \subset K$ is closed. We let $X|_D$ be the space of all restrictions $\{x|_D \mid x \in X\}$. Following [445] one says that $(X|_D, X)$ has the *bounded extension property* if there exists a constant C such that, given $\xi \in X|_D$, $\varepsilon > 0$ and an open set $U \supset D$, there is some $x \in X$ such that

$$\begin{aligned} x|_D &= \xi, \\ \|x\| &\leq C \cdot \|\xi\|, \\ \|x(k)\| &\leq \varepsilon \qquad \text{for } k \notin U. \end{aligned}$$

(Note that $X_{|D}$ is closed under this assumption.) For example, if $X \subset C(K)$ is a subalgebra and $D \subset K$ is a subset of the form $f^{-1}(\{1\})$ for some $f \in B_X$ (a "peak set"), then the pair $(X_{|D}, X)$ has the bounded extension property: Let $\xi \in X_{|D}$ and $g \in X$ such that $g_{|D} = \xi$. First of all we remark that replacing f by $(1+f)/2$ permits us to assume that in addition $|f(k)| < 1$ if and only if $k \notin D$. Then $g \cdot f^n$ meets the requirements of the above definition if n is large enough. (The argument easily extends to intersections of peak sets, the so-called p-sets.)
We now have:

Proposition 1.20 *If $(X_{|D}, X)$ has the bounded extension property, then $J_D \cap X$ is an M-ideal in X.*

Instead of proving this proposition now with the help of Corollary 1.19 we prefer to provide a proof using the methods of the next section, see p. 24. In fact, not only could one prove that under the assumption of the bounded extension property the condition of Corollary 1.19, namely $\chi_D \mu \in X^\perp$ for all $\mu \in X^\perp$, is fulfilled; it is even true that the two conditions are equivalent. This follows from [239]. Let us add that Hirsberg [312] has obtained necessary and sufficient conditions on a closed subset D to ensure that $J_D \cap X$ is an M-ideal in X, where $X \subset C(K)$ is assumed to be a closed subspace separating points and containing the constants.

As a counterpart to the preceding results we now discuss the L-structure of subspaces of L^1-spaces. (In contrast to the situation presented in Proposition 1.18 and Corollary 1.19, not every Banach space is a subspace of some L^1-space.) Recall from Example 1.6(a) that the L-projections on L^1 coincide with the band projections.

Proposition 1.21 *Let X be a closed subspace of some L^1-space and let $P : X \longrightarrow X$ be an L-projection. Then P can be extended to an L-projection on L^1. More precisely, if P_B denotes the L-projection from L^1 onto the band B generated by $P(X)$, then X is an invariant subspace of P_B and $P_{B|X} = P$.*

PROOF: We shall use the slightly unusual notation ($A \subset L^1$ a given subset)

$$A^s = \{y \in L^1 \mid |x| \wedge |y| = 0 \quad \text{for all } x \in A\}.$$

(Here s stands for singular. The usual notation $A^\perp$ collides with our symbol for the annihilator of A.) Then ([559, p. 210]) $B = P(X)^{ss}$ and $B^s = P(X)^s$. We first claim:

(a) For $x \in X$ there is a measurable set E (depending on x) such that $Px = \chi_E \cdot x$.

In fact, for $f, g \in L^1$ with $\|f \pm g\| = \|f\| + \|g\|$ we have $|f| \wedge |g| = 0$, since $|f(\omega) \pm g(\omega)| = |f(\omega)| + |g(\omega)|$ almost everywhere which implies $f(\omega) \cdot g(\omega) = 0$ almost everywhere. In particular $|Px| \wedge |x - Px| = 0$, and our claim obtains with $E = \{Px \neq 0\}$.
We now show

(b) $$(Id - P)(X) = X \cap B^s,$$

(c) $$P(X) = X \cap B,$$

which will prove Proposition 1.21.

ad (b): For $x \in X \cap B^s$ we have $|x| \wedge |Px| = 0$. But $Px = \chi_E \cdot x$ for a certain E by (a) so that $Px = 0$ and $x \in (Id - P)(X)$. Conversely, let $Px = 0$. We wish to prove $|x| \wedge |y| = 0$ for all $y \in P(X)$. Now choose E such that $P(x+y) = \chi_E \cdot (x+y)$. It follows $y = P(x+y) = \chi_E \cdot (x+y)$ and $\chi_{\complement E} \cdot y = \chi_E \cdot x$. This implies $|x| \wedge |y| = 0$.

ad (c): "$\subset$" is trivial. Let $x \in X \cap B$. Since $Px \in B$ we have $x - Px \in B$. On the other hand $x - Px \in B^s$ by (b). Thus $x = Px \in P(X)$. □

Corollary 1.22 *Let X and Y be closed subspaces of $C(K)$ with $Y \subset X$. Let B be the band in $M(K)$ generated by $X^\perp$, the annihilator of X in $M(K)$, and let P_B be the band projection onto B. Then X/Y is an M-ideal in $C(K)/Y$ if and only if*

$$P_B(Y^\perp) \subset Y^\perp \tag{1}$$

$$B \cap Y^\perp = X^\perp. \tag{2}$$

PROOF: By standard duality, X/Y is an M-ideal if and only if there is an L-projection from $Y^\perp$ onto $X^\perp$. Now, if (1) and (2) hold, the restriction of P_B to $Y^\perp$ is such a projection. The converse follows from Proposition 1.21. □

The final part of this section is devoted to the possible dependence of results on the choice of the scalar field. Up to now we have not encountered the necessity of treating real and complex spaces separately. This is not accidental.

Proposition 1.23 *Let X be a complex Banach space, and denote by $X_{\mathbb{R}}$ the same space, considered as a real Banach space. If P is an L- (resp. M-) projection on $X_{\mathbb{R}}$, then P is complex linear and thus an L- (resp. M-) projection on X. Consequently, X and $X_{\mathbb{R}}$ have the same L-summands, M-summands and M-ideals.*

PROOF: Let P be an $\mathbb{R}$-linear L-projection. It is enough to prove

$$z \in \operatorname{ran}(P) \implies iz \in \operatorname{ran}(P) \tag{$*$}$$

since then $iPx \in \operatorname{ran}(P)$ and, by symmetry, $i(Id - P)x \in \ker(P)$ for every $x \in X$ which shows

$$iPx = P(iPx) = P(ix - i(Id - P)x) = P(ix).$$

For the proof of $(*)$ let $Qx = i^{-1}P(ix)$. This is an $\mathbb{R}$-linear L-projection because multiplication by i is an isometry. Consequently, P and Q commute (1.10(a)). Now we have for $z \in \operatorname{ran}(P)$

$$z - Qz = Pz - QPz = P(z - Qz) \in \operatorname{ran}(P)$$

and

$$i(z - Qz) = iz - P(iz) \in \ker(P)$$

so that

$$\begin{aligned} \sqrt{2} \cdot \|z - Qz\| &= \|(z - Qz) + i(z - Qz)\| \\ &= \|z - Qz\| + \|z - Qz\| \\ &= 2 \cdot \|z - Qz\|. \end{aligned}$$

Hence $z = Qz$, i.e. $iz = P(iz) \in \operatorname{ran}(P)$. □

There are, however, several phenomena particular to complex spaces. We now discuss one of them. Suppose X is a complex Banach space and $T \in L(X)$. We let

$$\Pi(X) = \{(x^*, x) \in B_{X^*} \times B_X \mid \langle x^*, x\rangle = 1\}.$$

Recall that T is called *hermitian* if

$$V(T) := \{\langle x^*, Tx\rangle \mid (x^*, x) \in \Pi(X)\} \subset \mathbb{R}.$$

The set $V(T)$ is called the spatial numerical range of T. Equivalently, T is hermitian if and only if $\|\exp(itT)\| = 1$ for all $t \in \mathbb{R}$. We refer to [84] for a proof of this fact and related information. As an application, one easily shows that L- and M-projections are hermitian.

We wish to prove that M-ideals are invariant subspaces of hermitian operators on complex Banach spaces. This will be obtained as a corollary to the following technical proposition, which will be used again in Lemma V.6.7.

Proposition 1.24 *Suppose X is a complex Banach space, and let $J \subset X$ be an M-ideal. Moreover, let $T \in L(X)$ be an operator whose numerical range $V(T)$ is contained in the strip $R(\varepsilon) := \mathbb{R} \times [-\varepsilon, \varepsilon]i$ of the complex plane. Then for $x \in J$*

$$d(Tx, J) \le \varepsilon \cdot \|x\|.$$

PROOF: Let us assume $\|x\| = 1$. Since

$$d(Tx, J) = \|Tx + J\| = \sup\{|\langle x^*, Tx\rangle| \mid x^* \in B_{J^\perp}\}$$

we have to show

$$|\langle x^*, Tx\rangle| \le \varepsilon \quad \text{for all } x^* \in S_{J^\perp}.$$

So let $x^* \in S_{J^\perp}$. Taking into account the canonical duality $(J^\perp)^* = (J^*)^\perp$ (recall Remark 1.13), we may choose $x^{**} \in B_{(J^*)^\perp}$ such that $\langle x^{**}, x^*\rangle = 1$, i.e. $(x^{**}, x^*) \in \Pi(X^*)$. Then

$$\langle \lambda x^*, x + \overline{\lambda} x^{**}\rangle = 1$$

and

$$\|x + \overline{\lambda} x^{**}\| = \max\{\|x\|, \|x^{**}\|\} = 1$$

if $|\lambda| = 1$. Consequently $(x + \overline{\lambda} x^{**}, \lambda x^*) \in \Pi(X^*)$ so that

$$\lambda\langle x^*, Tx\rangle + \langle x^*, T^{**}x^{**}\rangle \in V(T^*) \overset{(*)}{\subset} \overline{V(T)} \subset R(\varepsilon)$$

if $|\lambda| = 1$ (cf. [86, p. 11] for the inclusion marked $(*)$). But this implies $|\langle x^*, Tx\rangle| \le \varepsilon$, as requested. □

Letting $\varepsilon \to 0$ in this proposition we obtain:

Corollary 1.25 *A hermitian operator on a complex Banach space leaves M-ideals invariant.*

I.2 Characterisation theorems

The definition of an M-ideal was given in terms of projections on the dual space. The first aim of the present section is to derive an equivalent condition for J to be an M-ideal which avoids mentioning the dual space. We start with an easy lemma.

Lemma 2.1 *Suppose J is an M-summand in X, and suppose $(B(x_i,r_i))_{i\in I}$ is a family of closed balls satisfying*

$$B(x_i,r_i)\cap J\neq\emptyset \quad \text{for all } i\in I \tag{1}$$

and

$$\bigcap_i B(x_i,r_i)\neq\emptyset. \tag{2}$$

Then

$$\bigcap_i B(x_i,r_i)\cap J\neq\emptyset.$$

PROOF: Let P be the M-projection onto J, and let $x\in\bigcap_i B(x_i,r_i)$. We claim $Px\in\bigcap_i B(x_i,r_i)$.
In fact, if $y_i\in B(x_i,r_i)\cap J$ then

$$\begin{aligned} r_i &\geq \|x_i-y_i\| \\ &= \|(Px_i-y_i)+(x_i-Px_i)\| \\ &= \max\{\|Px_i-y_i\|,\ \|x_i-Px_i\|\} \\ &\geq \|x_i-Px_i\| \end{aligned}$$

so that

$$\|x_i-Px\|=\max\{\|Px_i-Px\|,\ \|x_i-Px_i\|\}\leq r_i. \qquad \square$$

It will be shown in Proposition II.3.4 that the intersection condition of Lemma 2.1 actually characterises M-summands. Let us now characterise M-ideals by means of an intersection condition which turns out to be a powerful tool for detecting M-ideals.

Theorem 2.2 *For a closed subspace J of a Banach space X, the following assertions are equivalent:*

(i) *J is an M-ideal in X.*

(ii) (The n-ball property)
For all $n\in\mathbb{N}$ and all families $(B(x_i,r_i))_{i=1,\dots,n}$ of n closed balls satisfying

$$B(x_i,r_i)\cap J\neq\emptyset \quad \text{for all } i=1,\dots,n \tag{1}$$

and

$$\bigcap_{i=1}^{n} B(x_i,r_i)\neq\emptyset \tag{2}$$

the conclusion

$$\bigcap_{i=1}^{n} B(x_i, r_i + \varepsilon) \cap J \neq \emptyset \quad \textit{for all } \varepsilon > 0$$

obtains.

(iii) *Same as* (ii) *with* $n = 3$.

(iv) (The [restricted] 3-ball property)
For all $y_1, y_2, y_3 \in B_J$, *all* $x \in B_X$ *and all* $\varepsilon > 0$ *there is* $y \in J$ *satisfying*

$$\|x + y_i - y\| \leq 1 + \varepsilon \qquad (i = 1, 2, 3).$$

(v) (The strict n-ball property)
For all $n \in \mathbb{N}$ *and all families* $(B(x_i, r_i))_{i=1,\dots,n}$ *of* n *closed balls satisfying*

$$B(x_i, r_i) \cap J \neq \emptyset \quad \textit{for all } i = 1, \dots, n \tag{1}$$

and

$$\operatorname{int} \bigcap_{i=1}^{n} B(x_i, r_i) \neq \emptyset \tag{2'}$$

the conclusion

$$\bigcap_{i=1}^{n} B(x_i, r_i) \cap J \neq \emptyset$$

obtains.

PROOF: (i) $\Rightarrow$ (ii): We consider X as a subspace of X^{**} and

$$B_{X^{**}}(x_i, r_i) = \{x^{**} \in X^{**} \mid \|x^{**} - x_i\| \leq r_i\}.$$

Now $J^{\perp\perp}$ is an M-summand in X^{**}, hence by Lemma 2.1 we find some

$$x_0^{**} \in \bigcap_i B_{X^{**}}(x_i, r_i) \cap J^{\perp\perp}.$$

If (ii) were false, we could even assume that, for some $\varepsilon > 0$, $D := \bigcap_i B(x_i, r_i + \varepsilon)$ and J have positive distance and hence can be separated strictly. Thus, there is $x_0^* \in J^\perp$ with $\operatorname{Re} x_0^*|_D \geq 1$. Let $E := \operatorname{lin}\{x_0^{**}, x_1, \dots, x_n\} \subset X^{**}$ and $\delta = \min \varepsilon / r_i$.
The principle of local reflexivity assures the existence of an operator $T \in L(E, X)$ satisfying

- $\|T\| \leq 1 + \delta$
- $Tx_i = x_i \qquad (i = 1, \dots, n)$
- $\langle Tx_0^{**}, x_0^* \rangle = \langle x_0^{**}, x_0^* \rangle$.

It follows that

$$\|Tx_0^{**} - x_i\| \;\leq\; \|T\| \cdot \|x_0^{**} - x_i\| \;\leq\; (1+\delta)r_i \;\leq\; r_i + \varepsilon,$$

i.e.

$$Tx_0^{**} \in D,$$

thus

$$1 \;\leq\; \mathrm{Re}\,\langle Tx_0^{**}, x_0^*\rangle = \mathrm{Re}\,\langle x_0^{**}, x_0^*\rangle \;=\; 0,$$

a contradiction. (Variants of this proof will be discussed in the Notes and Remarks.)

(ii) $\Rightarrow$ (iii): This is more than obvious.

(iii) $\Rightarrow$ (iv): (iv) is just the special case $x_i = x + y_i$, $r_i = 1$.

(iv) $\Rightarrow$ (i): We put

$$J^\# = \{x^* \in X^* \mid \|x^*\| = \|x^*|_J\|\}.$$

STEP 1: Each $x^* \in X^*$ can be written as

$$x^* = x_1^* + x_2^*$$

with $x_1^* \in J^\perp$, $x_2^* \in J^\#$.

[PROOF: Let x_2^* be a Hahn-Banach extension of $x^*|_J$ and put $x_1^* = x^* - x_2^*$.]

STEP 2: $\|x_1^* + x_2^*\| = \|x_1^*\| + \|x_2^*\|$ for all $x_1^* \in J^\perp$, $x_2^* \in J^\#$.

[PROOF: Given $\varepsilon > 0$, choose $x \in B_X$ and $z \in B_J$ such that $\langle x_1^*, x\rangle$ and $\langle x_2^*, z\rangle$ are real and

$$\begin{aligned}\langle x_1^*, x\rangle &\;\geq\; \|x_1^*\| - \varepsilon,\\ \langle x_2^*, z\rangle &\;\geq\; \|x_2^*\| - \varepsilon.\end{aligned}$$

An application of (iv) with $y_1 = y_2 = z$, $y_3 = -z$ yields

$$\|x \pm z - y\| \leq 1 + \varepsilon$$

for some $y \in J$. Hence

$$\begin{aligned}&(1+\varepsilon)(\|x_1^* + x_2^*\| + \|x_1^* - x_2^*\|)\\ &\qquad\geq\; |\langle x_1^* + x_2^*, x + z - y\rangle + \langle x_1^* - x_2^*, x - z - y\rangle|\\ &\qquad=\; 2\,|\langle x_1^*, x\rangle + \langle x_2^*, z\rangle|\\ &\qquad\geq\; 2\|x_1^*\| + 2\|x_2^*\| - 4\varepsilon\\ &\qquad\geq\; \|x_1^*\| + \|x_2^*\| + \|x_1^* - x_2^*\| - 4\varepsilon\end{aligned}$$

from which Step 2 follows.]

STEP 3: The decomposition in Step 1 is unique.

[PROOF: If $x_1^* + x_2^* = y_1^* + y_2^*$, then $x_2^* = (y_1^* - x_1^*) + y_2^* \in J^\perp + J^\#$. Since $x_2^*|_J = y_2^*|_J$, Step 2 shows $\|y_1^* - x_1^*\| = 0$.]

Thus, $P : x^* \mapsto x_1^*$ is a well-defined idempotent map onto $J^\perp$ which fulfills the norm condition of L-projections. To finish the proof we must show the linearity of P which easily follows from the

STEP 4: $J^\#$ is a linear subspace of X^*.

[PROOF: Obviously $J^\#$ is a cone. Now let $x^*, y^* \in J^\#$. Then we have a unique decomposition

$$x^* + y^* = x_1^* + x_2^* \in J^\perp + J^\#.$$

We wish to show $x_1^* = 0$. To this end, let $x \in B_X$. Given $\varepsilon > 0$, choose $y_0, y_1, y_2 \in B_J$ satisfying

$$\begin{aligned} \mathbb{R} &\ni \langle x^*, y_0\rangle &\geq& \|x^*\| - \varepsilon \\ \mathbb{R} &\ni \langle y^*, y_1\rangle &\geq& \|y^*\| - \varepsilon \\ \mathbb{R} &\ni -\langle x_2^*, y_2\rangle &\geq& \|x_2^*\| - \varepsilon. \end{aligned}$$

Use (iv) to obtain $y \in J$ such that

$$\|x + y_i - y\| \leq 1 + \varepsilon \qquad (i = 0, 1, 2).$$

We then have

$$\begin{aligned} &(1+\varepsilon)(\|x^*\| + \|y^*\| + \|x_2^*\|) \\ \geq\ & |\langle x^*, x + y_0 - y\rangle + \langle y^*, x + y_1 - y\rangle - \langle x_2^*, x + y_2 - y\rangle| \\ \geq\ & \mathrm{Re}\,\langle x_1^*, x\rangle + \|x^*\| + \|y^*\| + \|x_2^*\| - 3\varepsilon. \end{aligned}$$

Consequently

$$\mathrm{Re}\,\langle x_1^*, x\rangle \leq 0 \quad \text{for all } x \in B_X$$

so that $x_1^* = 0$.]

(ii) $\iff$ (v): Since (v) is apparently stronger than (ii), the proof will be completed if we can show the following for a fixed number n:

- If (ii) holds for collections of $n+1$ balls, then (v) holds for collections of n balls.

So let us assume the balls $B(x_i, r_i)$, $i = 1, \ldots, n$, satisfy (1) and (2′). By (2′) there exist $y_0 \in X$ and $\delta > 0$ such that

$$\|y_0 - x_i\| \leq r_i - \delta, \quad i = 1, \ldots, n.$$

Let $r = \min r_i$. Next we wish to construct a sequence $y_1, y_2, \ldots$ in J such that for $k \in \mathbb{N}$

$$\begin{aligned} \|y_k - y_{k-1}\| &\leq 2^{-k} \cdot 4r, \\ \|y_k - x_i\| &\leq r_i + 2^{-k}\delta \qquad (i = 1, \ldots, n) \end{aligned}$$

hold. (Note that we are not claiming $y_0 \in J$.) Once this is achieved we see that the limit y of the Cauchy sequence (y_k) belongs to $\bigcap_{i=1}^n B(x_i, r_i) \cap J$.
To construct y_1 consider the balls $B(y_0, 2r - \delta)$ and $B(x_i, r_i)$ for $i = 1, \dots, n$. These $n+1$ balls fulfill the assumptions of (ii) by the definition of y_0, hence there is some

$$y_1 \in B(y_0, 2r) \cap \bigcap_{i=1}^n B(x_i, r_i + \delta/2) \cap J,$$

which is what we are looking for.
Now suppose $y_1, \dots, y_k$ have already been constructed. To find y_{k+1} consider the balls $B(y_k, (2^{-(k+1)} - 2^{-(2k+1)}) \cdot 4r)$ and $B(x_i, r_i + (2^{-(k+1)} - 2^{-(2k+1)})\delta)$ for $i = 1, \dots, n$. Obviously each of them intersects J (note $y_k \in J$ for $k \geq 1$). Let us observe that

$$z_k := 2^{-(k+1)} y_0 + (1 - 2^{-(k+1)}) y_k$$

lies in the intersection of these $n+1$ balls:

$$\begin{aligned}
\|z_k - y_k\| &= 2^{-(k+1)} \|y_0 - y_k\| \\
&\leq 2^{-(k+1)} (\|y_0 - y_1\| + \dots + \|y_{k-1} - y_k\|) \\
&\leq 2^{-(k+1)} \sum_{i=1}^k 2^{-i} \cdot 4r \\
&= (2^{-(k+1)} - 2^{-(2k+1)}) \cdot 4r, \\
\|z_k - x_i\| &\leq 2^{-(k+1)} \|y_0 - x_i\| + (1 - 2^{-(k+1)}) \|y_k - x_i\| \\
&\leq 2^{-(k+1)} (r_i - \delta) + (1 - 2^{-(k+1)})(r_i + 2^{-k}\delta) \\
&= r_i + (2^{-(k+1)} - 2^{-(2k+1)})\delta.
\end{aligned}$$

An application of (ii) yields some

$$y_{k+1} \in B(y_k, 2^{-(k+1)} \cdot 4r) \cap \bigcap_{i=1}^n B(x_i, r_i + 2^{-(k+1)}\delta) \cap J,$$

as requested. □

Remarks 2.3 (a) Two balls do not suffice in the *real* case. For example, in the real space $X = L^1(\mu)$, μ a positive measure, the subspace J of X consisting of those functions whose integral vanishes satisfies (ii) in Theorem 2.2 with $n = 2$, but fails to be an M-ideal if $\dim X > 2$. (We remark that only recently such an example was found in the complex case, see the Notes and Remarks.)
[Proof: To prove that J satisfies the 2-ball property suppose that the balls $B(f_1, r_1)$ and $B(f_2, r_2)$ have nonvoid intersection (i.e., $\|f_1 - f_2\| \leq r_1 + r_2$) and both of them meet J (i.e., $\|g_i - f_i\| \leq r_i$ for some $g_i \in J$). If $B(f_1, r_1 + \varepsilon) \cap B(f_2, r_2 + \varepsilon) \cap J = \emptyset$ for some

$\varepsilon > 0$, we could separate the set $D = \{(x_1, x_2) \mid \|x_i - f_i\| \le r_i\} \subset L^1 \oplus L^1$ strictly from $\{(y, y) \mid y \in J\}$ by means of some functional $(F_1, F_2) \in L^\infty \oplus L^\infty$. Thus we would have

$$F_1 + F_2 \in J^\perp = \operatorname{lin}\{1\}, \tag{1}$$

and

$$\sup_{(x_1,x_2)\in D} \int (F_1 x_1 + F_2 x_2)\, d\mu < 0. \tag{2}$$

On the other hand, each L^∞-function F may be decomposed as $F = c1 + G$ with $c = (\operatorname{ess\,sup} F + \operatorname{ess\,inf} F)/2$ so that

$$\|F\| = |c| + \|G\|.$$

By (1) above we may then write

$$F_1 = c_1 1 + G_1, \qquad \|F_1\| = |c_1| + \|G_1\|,$$

$$F_2 = c_2 1 - G_1, \qquad \|F_2\| = |c_2| + \|G_1\|.$$

Hence we obtain

$$\begin{aligned}
&\sup_{(x_1,x_2)\in D} \int (F_1 x_1 + F_2 x_2)\, d\mu \\
&= \sup_{(x_1,x_2)\in D} \int \Big(F_1(x_1 - f_1) + F_1 f_1 + F_2(x_2 - f_2) + F_2 f_2 \Big)\, d\mu \\
&= \|F_1\| r_1 + \|F_2\| r_2 + c_1 \int f_1\, d\mu + \int G_1(f_1 - f_2)\, d\mu + c_2 \int f_2\, d\mu \\
&= \|F_1\| r_1 + \|F_2\| r_2 + c_1 \int (f_1 - g_1)\, d\mu + \int G_1(f_1 - f_2)\, d\mu + c_2 \int (f_2 - g_2)\, d\mu \\
&\ge \|F_1\| r_1 + \|F_2\| r_2 - \Big(|c_1| r_1 + \|G_1\|(r_1 + r_2) + |c_2| r_2 \Big) \\
&= 0,
\end{aligned}$$

a contradiction to (2).

Clearly, J is not an M-ideal by Theorem 1.8 if $\dim(X) > 2$.]

An examination of the proof of Theorem 2.2 reveals that the 2-ball property yields a nonlinear "L-projection" onto $J^\perp$; in order to establish linearity we needed an intersection property involving three balls. We refer to the Notes and Remarks section for some information on spaces with the 2-ball property.

(b) One can even guarantee $\|y\| \le 1 + \varepsilon$ in (iv). This is a consequence of the 4-ball property applied to the balls $B(x + y_i, 1)$ $(i = 1, 2, 3)$ and B_X.

(c) One may also consider infinite collections of balls in (ii) provided the centres form a relatively compact set and the radii depend (lower semi-) continuously on the centres. (The proof is straightforward.)

(d) $\varepsilon = 0$ is not admissible in (ii). By way of example, let A be the disk algebra, and let

$$J = \{x \in A \mid x(1) = 0\}.$$

By Example 1.4(b) J is an M-ideal in A. Consider now the Möbius transform $w(z) = \frac{i-z}{1-iz}$ and the function $f(z) = w(\sqrt{w(z)})$. Then, by elementary complex analysis, f maps the open unit disk $\mathbb{D}$ onto $\{z \in \mathbb{D} \mid \operatorname{Re} z < 0\}$ and the unit circle $\mathbb{T}$ onto the boundary of that region. Moreover, we have $f \in A$ and $f(1) = 0$, hence $f \in J$; and since $|f(z)| = 1$ on $\{z \in \mathbb{T} \mid \operatorname{Re} z \leq 0\}$ we conclude $f \in \operatorname{ex} B_A$ [317, p. 139]. This means $B(1+f,1) \cap B(1-f,1) = \{1\}$, and therefore

$$J \cap B(1+f,1) \cap B(1-f,1) = \emptyset.$$

(It may seem counterintuitive that an M-ideal contains extreme points of the ambient space, since this is impossible for an M-summand.)

(e) The presence of some $\varepsilon > 0$ in (ii) is due to the fact that some openness assumptions are needed in order to apply the Hahn-Banach theorem. This assumption is shifted to condition (2′) in (v).

One can see Theorem 2.2 at work at various places of this book, e.g., II.2.3, II.5.2, III.1.4, III.3.4, V.3.2, VI.2.1, VI.4.1, VI.5.3. Let us now give a first application of the 3-ball property in that we provide the still missing

Proof of Proposition 1.20:

To begin with we observe that the constant C appearing in the definition of the bounded extension property may be chosen as close to 1 as we wish. To see this let $U \supset D$ be an open set and let $\varepsilon > 0$. Applying the bounded extension property with $U_1 = U$ and ε yields an extension x_1 of a given $\xi \in X|_D$ (w.l.o.g. $\|\xi\| = 1$) such that $\|x_1\| \leq C$ and $|x_1| \leq \varepsilon$ off U_1. Then we repeat this procedure with $U_2 = \{k \mid |x_1(k)| < 1 + \varepsilon/2\} \cap U_1$ and obtain an extension x_2 such that $\|x_2\| \leq C$ and $|x_2| \leq \varepsilon$ off U_2. In the third step one applies the bounded extension property with $U_3 = \{k \mid |x_2(k)| < 1 + \varepsilon/2\} \cap U_2$ etc. This yields a sequence of extensions (x_n) with $\|x_n\| \leq C$ and $|x_n| \leq \varepsilon$ off U_n. Let $x = \frac{1}{N}\sum_{n=1}^{N} x_n$. Obviously we have $x|_D = \xi$ and $|x| \leq \varepsilon$ off U. Finally, if $k \in U$, then by construction $|x_n(k)|$ is big (but $\leq C$) for at most one n, and $|x_n(k)| \leq 1 + \varepsilon/2$ otherwise. It follows

$$|x(k)| \leq \frac{1}{N}\Big(C + (N-1)(1+\varepsilon/2)\Big) \leq 1 + \varepsilon$$

for sufficiently large N. (As a consequence of this one may remark that $X/(J_D \cap X) \cong X|_D$.)

Now we can prove that $J_D \cap X$ is an M-ideal employing the 3-ball property. Thus, let y_1, y_2, y_3 in the unit ball of $J_D \cap X$, $x \in B_X$ and $\varepsilon > 0$. With the help of the $(1+\varepsilon)$-bounded extension property we may find $\widehat{x} \in X$ such that

$$\widehat{x}|_D = x|_D,$$

$$\|\widehat{x}\| \leq (1+\varepsilon)\|x|_D\| \leq 1 + \varepsilon,$$

$$|\widehat{x}(k)| \leq \varepsilon \quad \text{if } \max_i |y_i(k)| \geq \varepsilon.$$

Let $y = x - \widehat{x}$ so that $y \in J_D \cap X$. Distinguishing whether or not $\max_i |y_i(k)| \geq \varepsilon$ one can immediately verify that

$$|(x + y_i - y)(k)| = |y_i(k) + \widehat{x}(k)| \leq 1 + 2\varepsilon$$

for all $k \in K$, i.e.

$$\|x + y_i - y\| \leq 1 + 2\varepsilon \qquad (i = 1, 2, 3). \qquad \square$$

We now turn to a characterisation of M-ideals with a different, namely measure theoretic flavour. In the following we shall make use of the basic notions of integral representation theory, for the weak* compact convex set B_{X^*}. By a measure on B_{X^*} we shall understand a regular Borel measure on this compact set. A measure μ is called maximal (or boundary measure) if its variation is maximal with respect to Choquet's ordering. A measure μ has a unique barycentre (or resultant), denoted by $r(\mu)$, which is defined by

$$\langle r(\mu), x\rangle = \int_{B_{X^*}} p(x)\, d\mu(p) \quad \text{for all } x \in X.$$

Note $r(\mu) \in B_{X^*}$ for probability measures μ. Conversely, every $x^* \in X^*$ may be represented by a maximal measure μ on B_{X^*} in that $r(\mu) = x^*$. (For a detailed discussion see [7] and [496].)

Let J be a closed subspace of X. For $\mu \in M(B_{X^*})$ we define the restricted measure $\mu|_{J^\perp} \in M(B_{X^*})$ by $\mu|_{J^\perp}(E) = \mu(E \cap J^\perp)$. Note that $\mu|_{J^\perp}$ is maximal if μ is.

Now the characterisation theorem reads as follows:

Theorem 2.4 *$J \subset X$ is an M-ideal if and only if the following requirements are fulfilled:*

(1) $\|p\| + \|q - p\| = \|q\|$ *and* $q \in J^\perp$ *imply* $p \in J^\perp$.

(2) *If* $\mu \in M(B_{X^*})$ *is maximal and* $r(\mu) = 0$, *then also* $r(\mu|_{J^\perp}) = 0$.

PROOF OF THE "IF" PART:
Given $x^* \in X^*$, choose a maximal measure μ with $r(\mu) = x^*$. By (2), $r(\mu|_{J^\perp})$ depends only on x^* and not on the particular choice of μ so that $P(x^*) = r(\mu|_{J^\perp})$ gives rise to a well-defined mapping $P : X^* \to X^*$. By construction P is linear. We wish to show that P is an L-projection onto $J^\perp$. Since we have $\|r(\nu)\| \leq \|\nu\|$ for every measure $\nu \in M(B_{X^*})$ we conclude

$$\begin{aligned}
\|x^*\| &\leq \|Px^*\| + \|x^* - Px^*\| \\
&= \|r(\mu|_{J^\perp})\| + \|r(\mu - \mu|_{J^\perp})\| \\
&\leq |\mu|(B_{X^*} \cap J^\perp) + |\mu|(B_{X^*} \setminus J^\perp) \\
&= \|\mu\|.
\end{aligned}$$

Since one may choose such a μ with $\|\mu\| = \|x^*\|$, one obtains

$$\|x^*\| = \|Px^*\| + \|x^* - Px^*\|.$$

Also, P is a projection by the remark preceding Theorem 2.4, and $\operatorname{ran}(P) \subset J^\perp$ is clear. Now let $q \in J^\perp$ with $\|q\| = 1$, say. Represent q by a maximal probability measure μ. Then there is a net of discrete measures (ν_α) with $r(\nu_\alpha) = q$ and weak*-lim $\nu_\alpha = \mu$ [7, Prop. I.2.3]. Let

$$\nu_\alpha = \sum \lambda_i \delta_{p_i}$$

be such a measure. Then

$$\|q\| = \left\| \sum \lambda_i p_i \right\| \le \sum \lambda_i \|p_i\| \le \sum \lambda_i = 1 = \|q\|$$

so that $\lambda_i = 0$ or $p_i \in J^\perp$ as a consequence of (1). Therefore, the support of μ must be contained in $J^\perp$, and it follows $\mu = \mu|_{J^\perp}$ so that $q = P(q)$. (□)

The proof of the "only if" part will be given as a consequence of several lemmas which are of independent interest. We fix some notation. In the following we assume that J is an M-ideal in the *real* Banach space X with $P : X^* \to X^*$ denoting the corresponding L-projection onto $J^\perp$. (This assumption is made merely for simplicity of notation. The modifications to be made in the complex case will be sketched at the end of the proof of Theorem 2.4.) Furthermore we shall write

$$K = B_{X^*}, D = K \cap J^\perp, D' = K \cap \ker(P).$$

Given $x \in X$ we define $h_x : K \to \mathbb{R}$ by

$$h_x(x^*) = \begin{cases} x^*(x) & \text{if } x^* \in D \text{ and } x^*(x) \ge 0 \\ 0 & \text{otherwise.} \end{cases}$$

In other words, $h_x = \chi_D \cdot x \vee 0$.
Also, we recall the definition of the upper envelope $\widehat{h_x}$ of h_x :

$$\widehat{h_x}(x^*) = \inf a(x^*)$$

where a runs through the set of those functions $a \ge h_x$ which are affine and weak* continuous on K. Then $\widehat{h_x}$ is concave and weak* upper semicontinuous.

Lemma 2.5

$$\langle P^*x, x^* \rangle = \langle x, Px^* \rangle = \widehat{h_x}(x^*) - \widehat{h_x}(-x^*) \quad \textit{for all } x \in X,\ x^* \in K.$$

(Here the adjoint P^* of P is considered as a map from X to X^{**}.)

Proof: We first note

$$\begin{aligned} \mathcal{S}_x &:= \{(x^*, r) \in K \times \mathbb{R} \mid 0 \le r \le \widehat{h_x}(x^*)\} \\ &= \overline{\text{co}}^{w^*}\{(x^*, r) \mid 0 \le r \le h_x(x^*)\} \end{aligned}$$

which results from the Hahn-Banach theorem. ($\mathcal{S}_x$ is the truncated subgraph of $\widehat{h_x}$.) Also

$$\{(x^*, r) \mid 0 \le r \le h_x(x^*)\} = \{(x^*, r) \mid x^* \in D,\ 0 \le r \le x^*(x)\} \cup (K \times \{0\}).$$

Since the latter two sets are weak* compact and convex, so is the convex hull of their union. Therefore, taking into account

$$K = \mathrm{co}\,(D \cup D')$$

we have

$$S_x = \mathrm{co}\,(\{(x^*,r) \mid x^* \in D,\ 0 \le r \le x^*(x)\} \cup (D \times \{0\}) \cup (D' \times \{0\})).$$

Now let $x^* \in K$. Then $(x^*, \widehat{h_x}(x^*)) \in S_x$. Accordingly there is a convex combination

$$x^* = \lambda_1 x_1^* + \lambda_2 x_2^* + \lambda_3 x_3^* \tag{1}$$

with

$$x_1^*, x_2^* \in D,\ x_3^* \in D' \tag{2}$$

$$x_1^*(x) \ge 0 \tag{3}$$

$$\widehat{h_x}(x^*) = \lambda_1 x_1^*(x). \tag{4}$$

By concavity of $\widehat{h_x}$

$$\begin{aligned} \widehat{h_x}(x^*) &\ge \sum \lambda_i \widehat{h_x}(x_i^*) \\ &\ge \sum \lambda_i h_x(x_i^*) \\ &\ge \lambda_1 x_1^*(x) + \lambda_2 x_2^*(x) \end{aligned}$$

so that

$$0 \ge \lambda_2 x_2^*(x).$$

In case $\lambda_2 = 0$ we could have chosen $x_2^* = 0$, hence we may conclude $x_2^*(x) \le 0$ and thus

$$h_x(x_2^*) = 0, \qquad h_x(-x_2^*) = -x_2^*(x).$$

We employ the concavity of $\widehat{h_x}$ once again to obtain

$$\begin{aligned} \widehat{h_x}(-x^*) &\ge \sum \lambda_i h_x(-x_i^*) \\ &\ge -\lambda_2 x_2^*(x). \end{aligned}$$

Together with (4) this implies

$$\begin{aligned} \widehat{h_x}(x^*) - \widehat{h_x}(-x^*) &\le (\lambda_1 x_1^* + \lambda_2 x_2^*)(x) \\ &= \langle Px^*, x\rangle \qquad \text{(by (1) and (2))}. \end{aligned}$$

Since both sides of this inequality are odd functions of x^*, the inequality is in fact an equality, and the lemma is proved. □

From Lemma 2.5 one concludes immediately that $P^*x_{|K}$, being a difference of weak* upper semicontinuous functions, is weak* Borel. Even more: the points of continuity of an upper semicontinuous function on a compact space form a dense G_δ-set, cf. e.g. [205, p. 87]. By Baire's theorem this property is shared by linear combinations of such functions, in particular by P^*x restricted to K or any compact subset. A result of Choquet's [7, p. 16] now yields:

Lemma 2.6 *$P^*x|_K$ is a bounded affine weak* Borel function which satisfies the barycentric calculus, i.e.*

$$\langle P^*x, r(\mu)\rangle = \int_K P^*x \; d\mu$$

for all $\mu \in M(K)$.

Of course, Lemma 2.6 is completely trivial if J is an M-summand, since in this case $P^*x \in X$, i.e., P^*x is weak* continuous. Now Lemma 2.6 implies that P^*x is not very discontinuous in the general case, either. In fact, if X is separable then $P^*x|_K$ is of the first Baire class by what we have seen and Baire's classification theorem [158, p. 67]. One should mention that the weak* closedness of $\operatorname{ran}(P)$ is crucial for Lemma 2.6: for example, consider the L-projection P defined on $C[0,1]^*$ by $P(\nu)$ = atomic part of ν. Then $P^*\mathbf{1}$ is of the second Baire class (as a function on the dual unit ball), yet fails the barycentric calculus [7, Ex. I.2.10].

Lemma 2.7 *$r(\chi_D\mu) = P(r(\mu))$ for every maximal measure $\mu \in M(K)$.*

PROOF: With the help of Lemmas 2.5 and 2.6 one obtains

$$\begin{aligned} \langle P(r(\mu)), x\rangle &= \langle r(\mu), P^*x\rangle \\ &= \int_K P^*x \; d\mu \\ &= \int_K (\widehat{h_x}(x^*) - \widehat{h_x}(-x^*)) \; d\mu(x^*) \\ &= \int_D (h_x(x^*) - h_x(-x^*)) \; d\mu(x^*) \\ &= \int_D x^*(x) \; d\mu(x^*) \\ &= \langle r(\chi_D\mu), x\rangle. \end{aligned}$$

(In the fourth line we used the fact that a maximal measure is concentrated on $\{h = \widehat{h}\}$ for all upper semicontinuous functions h [7, p. 35].) □

Now it is easy to give the

PROOF OF THE "ONLY IF" PART OF THEOREM 2.4:

(1) is an elementary calculation:

Since

$$\|p\| = \|P(p)\| + \|p - P(p)\|$$

and

$$\|q - p\| = \|q - P(p)\| + \|p - P(p)\|$$

(because $q \in J^\perp$) we have

$$\begin{aligned}\|q\| &= \|p\| + \|q - p\| \\ &= \|P(p)\| + \|q - P(p)\| + 2\|p - P(p)\| \\ &\geq \|q\| + 2\|p - P(p)\|,\end{aligned}$$

hence $p = P(p) \in J^\perp$.

(2) is a consequence of Lemma 2.7:

If $r(\mu) = 0$, then we conclude in the case of real Banach spaces $r(\mu^+) = r(\mu^-)$ $(\mu = \mu^+ - \mu^-$ denoting the Hahn decomposition; note that μ^+ and μ^- are maximal), hence

$$\begin{aligned}r(\chi_D\mu) &= r(\chi_D\mu^+) - r(\chi_D\mu^-) \\ &= P(r(\mu^+)) - P(r(\mu^-)) \\ &= 0.\end{aligned}$$

The proof of the complex case can be reduced to the real case by considering real and imaginary parts. (We remark that the above lemmas remain valid, too; h_x should be defined as $(\mathrm{Re}\,\chi_D x) \vee 0$ in the complex case.) □

We finish this section with a theorem stating that every vector in a Banach space X has a kind of unconditional expansion into a series of elements from a separable M-ideal J, with the convergence being taken with respect to the topology $\sigma(X, J^*)$. (Recall from Remark 1.13 that J^* can naturally be identified with a subspace of X^*.) The proof of that result (Theorem 2.10) relies on the above Lemma 2.5 as well as the following two lemmas which are of independent interest.

Lemma 2.8 *Let H and K be compact Hausdorff spaces and $\rho : K \to H$ a continuous surjection. Suppose $\sigma : H \to \mathbb{R}$ is a function such that $\tau = \sigma \circ \rho$ is a difference of two positive lower semicontinuous functions $\tau = g_1 - g_2$ with the additional property that $g_1(t) + g_2(t) \leq 1$ for all $t \in K$. Then there are positive lower semicontinuous functions $f_1, f_2 : H \to \mathbb{R}$ such that $\sigma = f_1 - f_2$ and $f_1(s) + f_2(s) \leq 1$ for all $s \in H$.*

PROOF: We let

$$f_i(s) := \inf\{g_i(t) | \ \rho(t) = s\} = \min\{g_i(t) | \ \rho(t) = s\}.$$

The infimum defining f_i is actually a minimum, as lower semicontinuous functions on compact sets attain their infimum. Clearly the f_i are well-defined and positive.

To show that f_i ($i = 1$ or 2) is lower semicontinuous we pick $s_0 \in H$ and $\alpha \in \mathbb{R}$ such that $f_i(s_0) > \alpha$. Then, by definition of f_i, $\rho^{-1}(\{s_0\}) \subset V := \{g_i > \alpha\}$, which is an open set by assumption on g_i. Now $\bigcap_W \rho^{-1}(W) \cap \complement V = \emptyset$, where the intersection is taken over all compact neighbourhoods of s_0. Then the compactness of K produces a finite void intersection which in turn yields a compact neighbourhood W_0 of s_0 such that $\rho^{-1}(W_0) \subset V$. This says $g_i(t) > \alpha$ whenever $\rho(t) \in W_0$, and as a result $f_i(s) > \alpha$ whenever $s \in W_0$. Thus the lower semicontinuity is proved.

Next we show that $\sigma = f_1 - f_2$. Given $s \in H$ there are $t_1, t_2 \in \rho^{-1}(\{s\})$ such that $f_i(s) = g_i(t_i)$ and therefore $g_1(t_1) \le g_1(t_2)$, $g_2(t_2) \le g_2(t_1)$. Hence

$$f_1(s) - f_2(s) = g_1(t_1) - g_2(t_2) \le g_1(t_2) - g_2(t_2) = \tau(t_2) = \sigma(s)$$

and likewise $f_1(s) - f_2(s) \ge \sigma(s)$.
Finally, for s and t_i as above,

$$f_1(s) + f_2(s) = g_1(t_1) + g_2(t_2) \le g_1(t_2) + g_2(t_2) \le 1,$$

which completes the proof of Lemma 2.8. □

Lemma 2.9 *Let E be a Banach space, $F \subset E$ a (not necessarily closed) subspace and $x^{**} \in F^{\perp\perp}$ such that there is a sequence (x_n) in E and a constant $C > 0$ satisfying*

$$x^{**} = \text{weak}^*\text{-}\sum_{n=1}^{\infty} x_n,$$

$$\sup_{|\varepsilon_n| \le 1} \left\| \sum_{n=1}^{N} \varepsilon_n x_n \right\| \le C\|x^{**}\| \qquad \forall N \in \mathbb{N}.$$

Then there is, for every $\varepsilon > 0$, a sequence (y_n) in F satisfying

$$x^{**} = \text{weak}^*\text{-}\sum_{n=1}^{\infty} y_n,$$

$$\sup_{|\varepsilon_n| \le 1} \left\| \sum_{n=1}^{N} \varepsilon_n y_n \right\| \le (C + \varepsilon)\|x^{**}\| \qquad \forall N \in \mathbb{N}.$$

PROOF: We may assume $\|x^{**}\| = 1$. To begin with we show that x^{**} is a weak* limit of some sequence (z_m) from F. First of all one concludes from the Hahn-Banach theorem that the distance between F and $\text{co}\{\sum_{n=1}^{m} x_n, \sum_{n=1}^{m+1} x_n, \ldots\}$ vanishes for all $m \in \mathbb{N}$. Hence there are $z_m \in F$ and $\xi_m \in \text{co}\{\sum_{n=1}^{m} x_n, \sum_{n=1}^{m+1} x_n, \ldots\}$ such that $\|z_m - \xi_m\| \to 0$, and we obtain

$$\text{weak}^*\text{-}\lim z_m = \text{weak}^*\text{-}\lim \xi_m = \text{weak}^*\text{-}\sum_{n=1}^{\infty} x_n = x^{**}.$$

Now $v_n = z_n - \sum_{i=1}^{n} x_i$ defines a weakly null sequence so that a sequence of convex combinations tends to 0 strongly. More precisely, we can define a strictly increasing sequence of integers $0 = p_0 < p_1 < \ldots$, a sequence of real numbers $\lambda_n \ge 0$ with $\sum_{i=p_{n-1}+1}^{p_n} \lambda_i = 1$ and $u_n = \sum_{i=p_{n-1}+1}^{p_n} \lambda_i v_i$ to obtain $\|u_n\| \le \varepsilon/2^{n+1}$ for all n. Setting now

$$w_n = \sum_{i=p_{n-1}+1}^{p_n} \lambda_i z_i$$

and

$$y_1 = w_1, \qquad y_{n+1} = w_{n+1} - w_n$$

for $n \geq 1$, we have

$$y_1 = \sum_{j=1}^{p_1} \mu_j^0 x_j + u_1,$$

$$y_{n+1} = \sum_{j=p_{n-1}+1}^{p_{n+1}} \mu_j^n x_j + u_{n+1} - u_n$$

with

$$\mu_j^0 = \sum_{i=j}^{p_1} \lambda_i,$$

$$\mu_j^n = \begin{cases} 1 - \sum_{i=j}^{p_n} \lambda_i & \text{if } p_{n-1}+1 \leq j \leq p_n, \\ \sum_{i=j}^{p_{n+1}} \lambda_i & \text{if } p_n + 1 \leq j \leq p_{n+1}. \end{cases}$$

Since $0 \leq \mu_j^n \leq 1$ for all j and n, we obtain

$$\left\| \sum_{n=1}^{N} \varepsilon_n y_n \right\| \leq \left\| \sum_{n=1}^{N-1} \sum_{j=p_{n-1}+1}^{p_n} (\varepsilon_n \mu_j^{n-1} + \varepsilon_{n+1}\mu_j^n) x_j + \sum_{j=p_{N-1}+1}^{p_N} \varepsilon_N \mu_j^{N-1} x_j \right\| + \varepsilon$$

$$\leq C + \varepsilon$$

because $\mu_j^{n-1} + \mu_j^n = 1$ for $p_{n-1}+1 \leq j \leq p_n$ and hence $|\varepsilon_n \mu_j^{n-1} + \varepsilon_{n+1}\mu_j^n| \leq 1$. □

We now state the promised "unconditional expansion".

Theorem 2.10 *Let X be a Banach space, and suppose J is a separable M-ideal in X. Then there is, for each $x \in X$ and each $\varepsilon > 0$, a sequence (y_n) in J such that*

$$x = \sum_{n=1}^{\infty} y_n$$

with respect to the topology $\sigma(X, J^*)$ and

$$\sup_{|\varepsilon_n| \leq 1} \left\| \sum_{n=1}^{N} \varepsilon_n y_n \right\| \leq (1+\varepsilon)\|x\| \qquad \textit{for all } N \in \mathbb{N}.$$

In fact, such y_n can be picked from any given dense subspace of J.

PROOF: It is clearly sufficient to prove this for real Banach spaces and for $x \in B_X$. We will retain the notation $K = B_{X^*}$, $P : X^* \to J^\perp$ the L-projection and h_x from Lemma 2.5. Putting

$$\tau : K \to \mathbb{R}, \quad \tau(x^*) = \langle x^* - Px^*, x \rangle,$$

we obtain from that lemma the representation

$$\tau(x^*) = \left[\frac{1 + x^*(x)}{2} - \widehat{h_x}(x^*) \right] - \left[\frac{1 - x^*(x)}{2} - \widehat{h_x}(-x^*) \right]$$

of τ as a difference of two lower semicontinuous functions $\tau = g_1 - g_2$, where in addition $g_1 \geq 0$, $g_2 \geq 0$ and $g_1 + g_2 \leq 1$ as is easily checked. Now, if $H = B_{J^*}$ (equipped with the topology $\sigma(J^*, J)$) and $\rho : K \to H$ is the restriction map, then there is, by Lemma 2.8, a corresponding representation of the function $\sigma : H \to \mathbb{R}$, $\sigma(y^*) = \langle y^*, x \rangle$ as a difference $\sigma = f_1 - f_2$ of positive lower semicontinuous functions on H whose sum does not exceed 1. (To define the duality between y^* and x we have identified J^* with $\ker(P)$, as in Remark 1.13. Note also $\sigma \circ \rho = \tau$.)
The assumption that J is separable, i.e. H is metrizable and $C(H)$ is separable, permits us to write f_1 as a pointwise converging series of positive continuous functions on H. [In fact, since f_1 is lower semicontinuous and positive, we have

$$f_1(y^*) = \sup\{\varphi(y^*) | \ 0 \leq \varphi \leq f_1 \text{ and } \varphi \in C(H)\}$$

for each $y^* \in H$. If (φ_n) is a uniformly dense sequence in $\{\varphi \in C(H) | \ 0 \leq \varphi \leq f_1\}$ and $\varphi_1 = 0$, then clearly $f_1 = \sup \varphi_n$, and the functions $h_{1,n} = (\varphi_1 \vee \cdots \vee \varphi_{n+1}) - (\varphi_1 \vee \cdots \vee \varphi_n)$ have the required properties.] Likewise $f_2 = \sum_{n=1}^\infty h_{2,n}$ for some positive continuous functions $h_{2,n}$. Hence, using the notation $h_n = h_{1,n} - h_{2,n}$, we obtain a pointwise converging series of continuous functions $\sigma = \sum_{n=1}^\infty h_n$ such that

$$\begin{aligned}
\left\| \sum_{n=1}^N \varepsilon_n h_n \right\| &\leq \sup_{y^*} \left(\left| \sum_{n=1}^N \varepsilon_n h_{1,n}(y^*) \right| + \left| \sum_{n=1}^N \varepsilon_n h_{2,n}(y^*) \right| \right) \\
&\leq \sup_{y^*} \left(\sum_{n=1}^N h_{1,n}(y^*) + \sum_{n=1}^N h_{2,n}(y^*) \right) \\
&\leq \sup_{y^*} (f_1(y^*) + f_2(y^*)) \\
&\leq 1
\end{aligned}$$

whenever $|\varepsilon_n| \leq 1$ and $N \in \mathbb{N}$, and

$$\langle y^*, x \rangle = \sigma(y^*) = \sum_{n=1}^\infty h_n(y^*)$$

for each $y^* \in H$.
We would like to replace $\sum h_n$ by a series of functions from $A_0(H)$, the space of affine continuous functions on H vanishing at 0. To achieve this we employ Lemma 2.9. Clearly

σ is a bounded affine function on the compact convex set H, $\sigma(0) = 0$ and moreover, by what we have just proved, it is of the first Baire class. Hence σ is continuous on a dense G_δ-set as is every restriction $\sigma|_{H'}$ of σ to a compact subset H'. Choquet's theorem referred to in the paragraph preceding Lemma 2.6 now yields that σ satisfies the barycentric calculus. The upshot of this argument is that $\sigma \in A_0(H)^{\perp\perp} \subset C(H)^{**}$: In fact, since σ is measurable and bounded we have that $\sigma \in C(H)^{**}$ in a natural way. If $\mu \in C(H)^*$ annihilates $A_0(H)$, then the resultant $r(\mu)$ is 0; consequently $\int_H \sigma \, d\mu = \sigma(0) = 0$.
Of course, $A_0(H)$ is canonically isometrically isomorphic with J by the Krein-Smulian theorem; and the dominated convergence theorem yields $\sigma = \sum_{n=1}^{\infty} h_n$ in the topology $\sigma(C(H)^{**}, C(H)^*)$. Therefore an application of Lemma 2.9 with $E = C(H)$ and $F = A_0(H)$ shows how to find such a sequence (y_n) in J. If a dense subspace of J is considered, then we only have to apply Lemma 2.9 with the corresponding dense subspace of $A_0(H)$. □

REMARK: Another way to see that $\sigma \in A_0(H)^{\perp\perp}$ in the above proof is to use a result from [458] (see also [158, p. 235]) stating that a bounded affine function which is a pointwise limit of a sequence of continuous functions is even a pointwise limit of a sequence of continuous affine functions.

We shall apply Theorem 2.10 in Theorem III.3.8 and Theorem VI.4.21.

I.3 The centralizer of a Banach space

In this section we shall briefly discuss an algebra of operators on a Banach space X which has close relations to the M-ideal structure of X. In the same way as the M-ideals of a Banach space correspond to the closed ideals of a C^*-algebra, this operator algebra, called the centralizer of X, corresponds to the centre of a C^*-algebra (cf. the revisited Example 3.4(h)). It must be stressed that there are some differences between the real and complex case, here; cf. the Notes and Remarks.
For the most part, proofs of the results presented in this section have already appeared in E. Behrends' monograph [51]. Instead of repeating these arguments we prefer to give due references.
We start with the basic notion.

Definition 3.1 *$T \in L(X)$ is called a* multiplier *if every $p \in \operatorname{ex} B_{X^*}$ is an eigenvector of T^*, with eigenvalue $a_T(p)$ say, i.e.*

$$T^*p = a_T(p)p \qquad \text{for all } \; p \in \operatorname{ex} B_{X^*}. \qquad (*)$$

The collection of all multipliers is called the multiplier algebra *and denoted by* Mult(X). Mult(X) *is said to be trivial if* Mult$(X) = \mathbb{K} \cdot Id$.

It is an immediate consequence of the Krein-Milman theorem that Mult(X) is a closed commutative unital subalgebra of $L(X)$.

The following lemma is sometimes quite helpful for deciding whether a given operator belongs to $\mathrm{Mult}(X)$. We let

$$Z_X = \overline{\mathrm{ex}}^{w^*} B_{X^*} \setminus \{0\}.$$

This weak* locally compact space will turn out to be of importance later; see Theorem II.5.9.

Lemma 3.2 *Suppose that E is a weak* dense subset of Z_X and that $T \in L(X)$. Then $T \in \mathrm{Mult}(X)$ if and only if for all $p \in E$ there is a number $a_T(p)$ such that*

$$T^* p = a_T(p)p.$$

In this case, a_T may be extended to a weak continuous function on Z_X, and each $p \in Z_X$ is an eigenvector of T^* with eigenvalue $a_T(p)$.*

This lemma is a direct consequence of the definition. Next we will collect a few easy-to-prove properties of the functions a_T appearing in $(*)$ which we will understand to be defined on Z_X.

Lemma 3.3 *Let $T \in \mathrm{Mult}(X)$.*

(a) *a_T is bounded and, moreover, $\|a_T\|_\infty = \|T\|$.*

(b) *$a_T(p) = a_T(-p)$ for all $p \in Z_X$.*

(c) *$T \mapsto a_T$ is an algebra homomorphism.*

By these properties, $\mathrm{Mult}(X)$ is isometric to a closed subalgebra of $C^b(Z_X)$, where Z_X is equipped with the weak* topology. Consequently, $\mathrm{Mult}(X)$ is a function algebra.

Examples 3.4

(a) Let S be locally compact. Then $\mathrm{Mult}(C_0(S))$ consists exactly of the multiplication operators

$$M_z : x \mapsto x \cdot z$$

with bounded continuous functions z as can easily be proved [51, p. 55]. Hence

$$\mathrm{Mult}(C_0(S)) \cong C^b(S) \cong C(\beta S)$$

as algebras. If S is compact, then $\mathrm{Mult}(C(S)) \cong C(S)$.

(b) Let $A \subset C(\mathbb{T})$ be the disk algebra. Since the extreme functionals are given by

$$x \mapsto \lambda \cdot x(t), \quad |\lambda| = 1, \ t \in \mathbb{T}$$

(cf. p. 4) we conclude that a multiplier must be of the form M_z as above, for some function z. Since $z = M_z(\mathbf{1}) \in A$,

$$\mathrm{Mult}(A) = \{M_z | z \in A\} \cong A.$$

(c) The same reasoning as in (b) applies to any complex function algebra A. [If $S \in \mathrm{Mult}(A)$, then $Sx(t) = (S\mathbf{1})(t) \cdot x(t)$ for all t in the Choquet boundary of A, hence $S = M_{S\mathbf{1}}$.] This example will be of importance in Chapter V.

(d) Now consider $A_{\mathbb{R}}$, that is the disk algebra as a real Banach space. For $T \in \text{Mult}(A_{\mathbb{R}})$, T^* is subject to having real eigenvalues so that $T = M_z$ for some real-valued analytic function on the unit disk. Hence $\text{Mult}(A_{\mathbb{R}}) = \mathbb{R}\cdot Id$. Consequently, the multiplier algebra depends on the choice of the scalar field.

(e) Every M-projection P belongs to $\text{Mult}(X)$, the eigenvalues of P^* being 0 or 1. This is immediate from Lemma 1.5.

(f) For a compact convex set K we let $A(K)$ be the space of real-valued affine continuous functions on K. Then T is a multiplier on $A(K)$ if and only if T is order bounded. This is proved in [7, II.7.10].

(g) In a sense example (a) comprises the most general case. Given X, consider the locally compact space Z_X as above. Then there is a canonical isometric embedding of X into $C_0(Z_X)$, and by definition $T \in L(X)$ is a multiplier if and only if T is the restriction of a multiplication operator on $C_0(Z_X)$ which leaves X invariant.

We shall need the following lemma in Chapter VI (Lemma VI.2.2).

Lemma 3.5 *Let $T \in \text{Mult}(X)$.*
(a) $\|x+y\| = \max\{\|x\|, \|y\|\}$ *if* $x \in \ker(T)$, $y \in \text{ran}(T)$.
(b) $T(J) \subset J$ *if J is an M-ideal in X.*

PROOF: (a) We have $y = Tz$ for some z and $Tx = 0$ so that $0 = p(Tx) = a_T(p)p(x)$ for all $p \in \text{ex}\, B_{X^*}$ and, as a result, $a_T(p) = 0$ or $p(x) = 0$. It follows

$$\begin{aligned}\|x+y\| &= \sup\{\text{Re}\, p(x+y) \mid p \in \text{ex}\, B_{X^*}\} \\ &= \sup\{\text{Re}\, p(x) + \text{Re}\, a_T(p)p(z) \mid p \in \text{ex}\, B_{X^*}\} \\ &= \max\{\|x\|, \|y\|\}.\end{aligned}$$

(b) follows from $T^*(\text{ex}\, B_{J^\perp}) \subset J^\perp$, the Krein-Milman theorem and the Hahn-Banach theorem. (Recall $\text{ex}\, B_{J^\perp} \subset \text{ex}\, B_{X^*}$ from Lemma 1.5.) □

The following theorem characterises multipliers without recourse to the dual space.

Theorem 3.6 *For a real or complex Banach space X and $T \in L(X)$, the following assertions are equivalent:*
(i) $T \in \text{Mult}(X)$.
(ii) *For all $x \in X$, Tx is contained in every closed ball which contains*

$$\{\lambda x \mid \lambda \in \mathbb{K},\ |\lambda| \le \|T\|\}.$$

PROOF: [11, p. 151] and [51, p. 57] in the complex case. □

In the real case, $\text{Mult}(X) \cong \{a_T \mid T \in \text{Mult}(X)\}$ is the self-adjoint part of the C^*-subalgebra $\{a_S + ia_T \mid S, T \in \text{Mult}(X)\}$ of $C^b_{\mathbb{C}}(\text{ex}\, B_{X^*})$ and thus $\text{Mult}(X) \cong C_{\mathbb{R}}(K_X)$ for a (uniquely determined) compact Hausdorff space K_X by the Gelfand-Naimark theorem. In the complex case, $\text{Mult}(X)$ need not be self-adjoint (cf. Example 3.4(b)). This is the motivation for the next definition.

Definition 3.7 *The centralizer of X, $Z(X)$, consists of all those $T \in \text{Mult}(X)$ for which there exists $\overline{T} \in \text{Mult}(X)$ such that $a_{\overline{T}}(p) = \overline{a_T(p)}$ for all $p \in \text{ex}\, B_{X^*}$. We let*

$$Z_{\mathbb{R}}(X) = \{T \in Z(X) \mid a_T \text{ real valued}\}$$

and

$$Z_{0,1}(X) = \{T \in Z_{\mathbb{R}}(X) \mid 0 \le a_T \le 1\}.$$

$Z(X)$ is called trivial if $Z(X) = \mathbb{K} \cdot Id$.

Obviously, $Z(X) = \text{Mult}(X)$ for real Banach spaces and $Z(X) = Z_{\mathbb{R}}(X) + iZ_{\mathbb{R}}(X)$ in the complex case. By construction $Z(X)$ is a commutative unital C^*-algebra, thus $Z(X) \cong C(K_X)$, with K_X the Gelfand space of this C^*-algebra. One can prove that X has a representation as a Banach space of sections in a bundle with base space K_X such that the section of norms is upper semicontinuous [139], [51, Chap. IV]. We will not pursue this idea, but invite the reader to consult the above mentioned literature and the Notes and Remarks section.
For the sake of easy reference we remark in addition:

Lemma 3.8 *Every $T \in Z_{\mathbb{R}}(X)$ is hermitian.*

PROOF: $\|e^{itT}\| = \|e^{ita_T}\|_\infty = 1$ for all $t \in \mathbb{R}$. □

Examples 3.4 (revisited)

(a) Here we have $Z(C_0(S)) = \text{Mult}(C_0(S)) \cong C^b(S)$.

(b) If $T = M_z \in Z(A)$ has an "adjoint" $\overline{T} = M_{z^*}$ for some $z^* \in A$, then $z^* = \overline{z}$ (the complex conjugate function) so that z is constant. Consequently $Z(A) = \mathbb{C} \cdot Id$.

(c) Here we have $Z(A) = \{x \in A \mid \overline{x} \in A\}$.

(e) Since a_P is real-valued, we have $P \in Z(X)$. In fact, a projection belongs to $Z(X)$ if and only if it is an M-projection. This is a consequence of (∗) from p. 2 and Proposition 3.9 below.

(g) If X is canonically embedded in $C_0(Z_X)$, then the operators in $Z(X)$ are exactly the restrictions of multiplication operators $M_h : f \to f \cdot h$ on $C_0(Z_X)$ such that M_h and $M_{\overline{h}}$ leave X invariant.

(h) For a unital C^*-algebra A, $Z(A) = \{M_z \mid z \in \text{centre}(A)\}$. (This will be proved in Theorem V.4.7.)

The following proposition should be compared with (∗) on p. 2.

Proposition 3.9 *$T \in L(X)$ belongs to $Z_{0,1}(X)$ if and only if*

$$\|Tx_1 + (Id - T)x_2\| \le \max\{\|x_1\|, \|x_2\|\}$$

for all $x_1, x_2 \in X$.

PROOF: It is elementary to verify that $T \in Z_{0,1}(X)$ has the announced property. Now suppose that T fulfills the above norm condition. Then we have for $p \in \operatorname{ex} B_{X^*}$ (assuming w.l.o.g. $T^*p \neq 0$ and $T^*p \neq p$)

$$p = \|T^*p\| \frac{T^*p}{\|T^*p\|} + \|p - T^*p\| \frac{p - T^*p}{\|p - T^*p\|}.$$

But this is a convex combination since there are, given $\varepsilon > 0$, $x_1, x_2 \in B_X$ such that

$$\begin{aligned} \|T^*p\| + \|p - T^*p\| - 2\varepsilon &\leq p(Tx_1) + p(x_2 - Tx_2) \\ &\leq \|Tx_1 + (x_2 - Tx_2)\| \\ &\leq 1 \end{aligned}$$

by assumption on T. The extremality of p now yields

$$T^*p = \|T^*p\|\, p.$$

□

We are now going to discuss the problem of characterising those functions a on $\operatorname{ex} B_{X^*}$ (or Z_X if you prefer) which give rise to operators in $Z(X)$. In view of the (revisited) Example 3.4(g) this is the case if and only if multiplication by a and $\overline{a}$ (defined on $C_0(Z_X)$) leaves X invariant. Eventually we shall state a characterisation of those functions a in terms of a topological condition involving M-ideals (Theorem 3.12).
First, however, we shall present a result of Stone-Weierstraß type which gives a necessary and sufficient condition for $g \in C_0(Z_X)$ to be in X. It will be convenient to take a slightly broader view. So let us fix some notation. Let L be a locally compact space and suppose that X is a closed subspace of $C_0(L)$. Put

$$Z(X, C_0(L)) = \{f \in C^b(L) \mid f \cdot X \subset X \text{ and } \overline{f} \cdot X \subset X\}.$$

We further denote by $\mathcal{F}(X, C_0(L))$ the set of equivalence classes which are obtained from the equivalence relation

$$s \sim t \quad \Longleftrightarrow \quad f(s) = f(t) \quad \forall f \in Z(X, C_0(L))$$

on L. Finally, for a subset $F \subset L$, we let

$$X|_F = \{x|_F \mid x \in X\}.$$

With this notation, we may state:

Theorem 3.10 *Let X be a closed subspace of $C_0(L)$. Then $g \in C_0(L)$ belongs to X if and only if*

$$g|_F \in X|_F \qquad \textit{for all } F \in \mathcal{F}(X, C_0(L)).$$

PROOF: The "only if" part is trivial. For the "if" part, we start by showing that the support of each measure $\mu \in \operatorname{ex} B_{X^\perp}$ is contained in some $F_0 \in \mathcal{F}(X, C_0(L))$. To this end let $f \in Z(X, C_0(L))$ be given. We wish to show

$$f|_{\operatorname{supp}(\mu)} = \text{const.}$$

Since Re $f \in Z(X, C_0(L))$, too, we may assume that f is real-valued, and a simple scaling procedure allows us to assume

$$0 < f(s) < 1 \qquad \text{for all } s \in L.$$

We then have

$$\|f\mu\| + \|(1-f)\mu\| = \int_L f\,d|\mu| + \int_L (1-f)\,d|\mu| = \|\mu\| = 1.$$

Hence μ can be represented as a convex combination

$$\mu = \|f\mu\| \frac{f\mu}{\|f\mu\|} + \|(1-f)\mu\| \frac{(1-f)\mu}{\|(1-f)\mu\|}$$

so that by extremality (note that by the very choice of f we have $f\mu \in X^\perp$)

$$\|f\mu\|\,\mu = f\mu,$$

i.e.

$$\|f\mu\| = f \qquad \mu\text{-a.e.},$$

and the claim follows.
Now consider $g \in C_0(L)$ with

$$g|_F \in X|_F \quad \forall F \in \mathcal{F}(X, C_0(L)).$$

If $g \notin X$, there would be some $\mu \in \operatorname{ex} B_{X^\perp}$ with $\int_L g\,d\mu \neq 0$. However, if $\operatorname{supp}(\mu) \subset F_0$, $F_0 \in \mathcal{F}(X, C_0(L))$, then, since $g|_{F_0} = x|_{F_0}$ for some $x \in X$,

$$\int_L g\,d\mu = \int_{F_0} g\,d\mu = \int_{F_0} x\,d\mu = \int_L x\,d\mu = 0.$$

This contradiction proves $g \in X$, as claimed. □

We remark that for a subalgebra X, $\mathcal{F}(X, C_0(L))$ is the well-known maximal X-antisymmetric decomposition of L, and Theorem 3.10 reduces to Bishop's generalisation of the Stone-Weierstraß theorem. We hasten to add that our proof of Theorem 3.10 is nothing but a minor modification (if any) of the de Branges-Glicksberg proof of Bishop's theorem, cf. [240, p. 60].
An application of Theorem 3.10 will be given in Theorem II.5.9; for others see the Notes and Remarks section.

To indicate the links between the centralizer of X and the M-ideals of X, which will enable us to characterise those functions on ex B_{X^*} which arise as a_T for some $T \in Z(X)$, we need the notion of the structure topology.

Definition 3.11 *The collection of sets* ex $B_{X^*} \cap J^\perp$*, where J runs through the family of M-ideals in X, is the collection of closed sets for some topology on* ex B_{X^*} *(this follows from Proposition 1.11), called the* structure topology.

This topology is never Hausdorff, since p and $-p$ cannot be separated. For this reason we will also consider the quotient space $\operatorname{ex} B_{X^*}/\!\sim$ (where $p \sim q$ iff p and q are linearly dependent) equipped with the corresponding quotient topology. This topological space will be denoted by E_X.
The richness of the M-ideal structure is reflected by topological qualities of the structure topology. For example, if X has no nontrivial M-ideals, then E_X has no nontrivial structurally open sets. On the other hand, if E_X satisfies the T_1-separation axiom, then $\ker p$ is an M-ideal in X for all $p \in \operatorname{ex} B_{X^*}$ so that X has in fact many M-ideals. Surely, E_X is a T_1-space if X^* is isometric to an L^1-space, but this condition is not necessary, as the example of the disk algebra shows. We will consider certain strengthenings of the T_1-axiom for E_X and will study the Banach spaces with such a structure space in Section II.5.
We can now state the announced connection between M-ideals and the centralizer in the following theorem of Dauns-Hofmann type. These authors obtained a corresponding result in the setting of C^*-algebras [146].

Theorem 3.12 *For a bounded function $a : \operatorname{ex} B_{X^*} \to \mathbb{K}$, the following assertions are equivalent:*
(i) *a is structurally continuous.*
(ii) *$a = a_T$ for some $T \in Z(X)$.*

PROOF: [11, p. 153], [202] or [51, Th. 3.13(ii)], where in addition the complex case is treated. □

As a corollary one obtains: If $Z(X)$ is nontrivial, then X contains a nontrivial M-ideal. We shall later encounter Banach spaces which show that the converse is not true, see p. 88 or the discussion following Theorem III.2.3. Another example is the disk algebra; but note that here the multiplier algebra is nontrivial.

Finally, we would like to briefly sketch the dual situation. We give one more definition.

Definition 3.13 *The closed linear span of the set of L-projections on X is called the* Cunningham algebra *of X, denoted by* $\operatorname{Cun}(X)$.
$\operatorname{Cun}(X)$ is a commutative algebra of operators by virtue of Theorem 1.10(a). For instance, by Example 1.6(a) the Cunningham algebra of $L^1(\mu)$ coincides with the algebra of multiplication operators with L^∞-functions.
The connection between Definitions 3.13 and 3.7 is expressed in the following theorem.

Theorem 3.14
(a) *$T \in Z(X)$ if and only if $T^* \in \operatorname{Cun}(X^*)$.*
(b) *$T^* \in Z(X^*)$ if and only if $T \in \operatorname{Cun}(X)$.*
(c) *Every operator $T \in Z(X^*)$ is weak* continuous so that*

$$\operatorname{Cun}(X) \cong Z(X^*) \cong C(K_{X^*})$$

where K_{X^*} is hyperstonean, and

$$Z(X^*) = \overline{\operatorname{lin}}\,\{P \mid P \text{ is an } M\text{-projection on } X^*\}.$$

Moreover, $\operatorname{Mult}(X^*)$ *and* $Z(X^*)$ *are dual Banach spaces.*

(d) *If X is a complex Banach space or a real Banach space not isometric to $\ell^\infty(2)$, then either $Z(X)$ or* $\mathrm{Cun}(X)$ *is trivial.*

PROOF: (a) [11, p. 151] or [51, Th. 3.12] (real scalars) and [56] (complex scalars).

(b) and (c) Let us first note that $L(X^*)$ is a dual Banach space with canonical predual $X^*\widehat{\otimes}_\pi X$ and that the corresponding weak* topology (henceforth called "the" weak* topology) on $L(X^*)$ coincides with the weak* operator topology on bounded sets, i.e.

$$T_i \xrightarrow{w*} T \quad\Longleftrightarrow\quad \langle T_i x^*, x\rangle \to \langle Tx^*, x\rangle \quad \forall x^* \in X^*,\ x \in X$$

if $\sup_i \|T_i\| < \infty$. (Cf. e.g. [159, Chap. VIII].)
Now we show that $\mathrm{Mult}(X^*)$ is weak* closed in $L(X^*)$. Let (T_i) be a bounded net in $\mathrm{Mult}(X^*)$ and $T \in L(X^*)$ such that $T_i \xrightarrow{w*} T$. There is no loss of generality in assuming $\|T\| = \|T_i\| = 1$ throughout. We shall verify the condition of Theorem 3.6(ii). So suppose

$$\|x_0^* - \lambda x^*\| \le r \quad \text{for all } |\lambda| \le 1.$$

Since the T_i are multipliers we obtain

$$\|x_0^* - T_i x^*\| \le r \quad \text{for all } i.$$

Passing to the limit entails by weak* lower semicontinuity of the norm

$$\|x_0^* - Tx^*\| \le r$$

so that our assertion follows from Theorem 3.6 and the Krein-Smulian theorem.
Hence, if the scalars are real, $Z(X^*) = \mathrm{Mult}(X^*)$ is a weak* closed subspace of $L(X^*)$, and in the complex case $Z_{\mathbb{R}}(X^*)$, which is $\mathbb{R}$-isometric with $\mathrm{Mult}(X^*_{\mathbb{R}})$, is weak* closed in $L(X^*_{\mathbb{R}})$. Consequently they are isometric to dual Banach spaces. The point of these considerations is that, regardless of the scalar field, $Z(X^*)$ is isometric to a dual $C(K)$-space. By a theorem due to Grothendieck [386, p. 96] we conclude that K_{X^*} is hyperstonean; in particular, the algebra $C(K_{X^*})$ is generated by its idempotents, and the same must be true for $Z(X^*)$. Taking into account that the idempotents in the centralizer are exactly the M-projections (cf. the revisited Example 3.4(e)) we obtain the assertions made in (b) and (c).
For a different line of reasoning, see [51, Th. 5.9] along with [51, Prop. 1.16] and [66, p. 24ff.].
(d) is a consequence of (a) – (c) and Theorem 1.8, applied to X^*. □

Formally, Theorem 3.14 generalises both Theorem 1.8 and Theorem 1.9, but its proof relies on these results.

Corollary 3.15

(a) *If $T \in Z(X)$, then also $T^{**} \in Z(X^{**})$.*

(b) *If X is a complex Banach space, then $T \in Z(X)$ commutes with every hermitian operator.*

PROOF: (a) is an immediate consequence of Theorem 3.14.
(b) Since the adjoint of an hermitian operator is hermitian, we may – by passing to second adjoints – assume from the outset that T is an M-projection (by (a) and 3.14(c)). In this case the assertion to be proved is equivalent with saying that a hermitian operator leaves M-summands invariant, which we already know from Corollary 1.25. □

We remark that the assertions of Corollary 3.15 extend to $T \in \mathrm{Mult}(X)$; see [56] and [68] for a proof.

I.4 Notes and remarks

GENERAL REMARKS. In 1972 the seminal paper "Structure in real Banach spaces" [11] by Erik M. Alfsen and Edward G. Effros appeared. There they initiated a study of general Banach spaces based on the notion of an M-ideal, proceeding in much the same way as in C^*-algebra theory where the notion of a closed two-sided ideal is instrumental. Effros [187] proved in 1963 that the closed left ideals in a C^*-algebra correspond in a one-to-one fashion to the closed faces in the state space, thus paving a way towards a geometric approach to structural questions. (The result was independently obtained by Prosser [510, Th. 5.11], as well.) Then it was proved in a series of articles ([187], [585], [9]) that the closed two-sided ideals are in one-to-one correspondence with the closed split faces of the state space. Similar developments of general ideal theories were encountered in the theory of ordered Banach spaces ([10], [492], [638]), L^1-predual spaces ([190], [215]; cf. Section II.5) and compact convex sets ([9], [188], [189]; cf. Example 1.4(c)). In order to provide a unified treatment, Alfsen and Effros defined the concept of an M-ideal; around the same time the same notion arises in Ando's work [19], however under a different name ("splittable convex set") and for a different purpose.
The key result of part I of the Alfsen-Effros paper is the characterisation of M-ideals by means of an intersection property of balls; more precisely they proved (i) $\iff$ (ii) $\iff$ (iii) $\iff$ (v) of Theorem 2.2 for real Banach spaces. Their proof was anything but simple and used a theorem on "dominated extensions" [11, Th. 5.4], based on Theorem 2.4 which is also due to them [11, Th. 4.5], as a decisive step. (Alfsen adapted this approach to complex spaces in [8].) Simpler proofs, covering the case of complex scalars, too, were devised by Lima [400], who added the useful condition (iv), Behrends [51, Chap. 2D] and Yost [650]. The latter paper also contains the counterexample of Remark 2.3(d); further counterexamples appear in [11, p. 126], [649] and [651]. The example of Remark 2.3(a) is due to Lima [400]; a three-dimensional version can already be found in [11].
The proof of the implication (iv) $\Rightarrow$ (i) in Theorem 2.2 follows [400], and for (ii) $\iff$ (v) we follow [651]. Our proof of the implication (i) $\Rightarrow$ (ii) using the principle of local reflexivity, however natural this approach may be, seems to be new. One could even shortcut the proof if one knew in advance that a local reflexivity operator T can always be found with the additional property that $T(J^{\perp\perp} \cap E) \subset J$ for a given M-ideal J and a finite dimensional subspace $E \subset X^{**}$. As a matter of fact this is the case as was recently shown in [62], see also [69] and [168]. However we decided to present the slightly longer proof, since it employs well known tools only. Lima's idea in [400] to show (i) $\Rightarrow$ (ii) is to

use the Hahn-Banach theorem in the space $X \oplus_\infty \cdots \oplus_\infty X$. If (ii) were false one could separate the "diagonal" $\{(y,\dots,y) \mid y \in J\}$ from $B(x_1,r_1) \times \cdots \times B(x_n,r_n)$, from which Lima derives a contradiction. Yost's approach in [650] uses the $1\frac{1}{2}$-ball property to be discussed in the Notes and Remarks to Chapter II.

We remark that Proposition 1.2 and Corollary 1.3 first appeared in [293]. Theorem 1.8 is due to Behrends whose original proof is quite tedious albeit elementary (cf. [51, p. 24ff.]). We have followed the line of reasoning by Payá and Rodríguez [480]. Theorem 1.9 comes from [142], and Theorem 1.10 from [138]. (For the importance of part (b) of this theorem see below.) Proposition 1.16 was observed in [148] in order to obtain best approximation results in the setting of C^*-algebras. For another account of Proposition 1.20 we refer to [642, Chap. III.D]. Proposition 1.21 and Corollary 1.22 can be found in [241] where a different proof is given. Proposition 1.23 is due to Effros and was published in [312], see [477] for another proof. Corollary 1.25 made its first appearance in [474] and, independently, [477] and [478]; our proof and Proposition 1.24 are taken from [631]. Theorem 2.10 represents a general version of the main result of Godefroy's and Li's paper [265] where the case $X = J^{**}$ is treated and also draws on the related paper [397]. Lemma 2.9 is in essence a classical result due to Pełczyński [485] and can be found for example in [423, p. 32] or [573, p. 446ff.].

Part II of the Alfsen-Effros paper centres around the notions of Section I.3, notably the centralizer, the Cunningham algebra and the structure topology. These notions are patterned after similar concepts in C^*-algebra theory, the analogue of the structure topology being the so-called hull-kernel topology. The main result describes their connection in the Dauns-Hofmann type theorem 3.12. However, as the title of [11] suggests, only real Banach spaces are considered in that paper. The wish to cover complex spaces as well necessitates distinguishing between the centralizer and the multiplier algebra. This was first done by Behrends in [51]. Let us remark that M-ideals in complex Banach spaces were first studied in [312]. With the help of Proposition 1.23 one may conclude that most of the results on M-ideals in real Banach spaces extend – mutatis mutandis – to the complex case. For the multiplier algebra things are not so easy. For example it is not known if a complex Banach space such that $\mathrm{Mult}(X)$ is nontrivial must contain a nontrivial M-ideal. (For $Z(X)$ this follows from 3.12, though a direct proof is possible, too.) Wodinski [639] has shown that this is so if in addition the existence of a nontrivial $T \in \mathrm{Mult}(X)$ such that T^* attains its norm is supposed. Also, it is not hard to show that a strictly convex space has a trivial centralizer. The corresponding result for the multiplier algebra holds, too, but is considerably more difficult to prove [339, Th. 12.7], [639].

Most of the material of Section I.3 appears in [11] and [51], cf. the references given in the text. Proposition 3.9 and Theorem 3.10 come from [628] and [630]. There it is also pointed out that 3.10 implies that a unital C^*-algebra is commutative if and only if its centre separates the weak* closure of its pure states (which represents a special case of Théorème 11.3.1 in [167]) and that a compact convex set K is a Bauer simplex if and only if the order bounded operators on $A(K)$ separate the closure of $\mathrm{ex}\, K$.

For a detailed discussion of results related to Section I.3 we refer to the monographs [35] and [51]. In addition we mention the papers [53], [56], [162], [163], [340], [453], [454], [455].

SEMI M-IDEALS. We have pointed out in Remark 2.3(a) that the 2-ball property and the 3-ball property are not equivalent, whereas it follows from Theorem 2.2 that the n-ball property and the 3-ball property are in fact equivalent if $n \geq 3$. Let us call a closed subspace a *semi M-ideal* if it satisfies the 2-ball property. (By the way, the 2-ball property and the strict 2-ball property coincide [650].) The arguments of Theorem 2.2, (iv) $\Rightarrow$ (i), and of Remark 2.3(a) yield that J is a semi M-ideal in X if and only if there is a (nonlinear) projection P from X^* onto $J^\perp$ such that

$$\begin{aligned} P(\lambda x^* + Py^*) &= \lambda Px^* + Py^*, \\ \|x^*\| &= \|Px^*\| + \|x^* - Px^*\| \end{aligned}$$

for all $x^*, y^* \in X^*$, $\lambda \in \mathbb{K}$. Such a projection is called a *semi L-projection* and its range a *semi L-summand*. Thus the above result (due to Lima [400]) says that J is a semi M-ideal if and only if $J^\perp$ is a semi L-summand. (Actually, Lima defines a semi M-ideal in terms of semi L-projections, and the characterisation by the 2-ball property is a theorem of his.)

The example given in 2.3(a) seems to be one of the few natural occurrences of (proper) semi M-ideals resp. semi L-summands. It shows that M-ideals and semi L-summands can appear simultaneously, in contrast to the situation described in Theorem 1.8. However, in this case it is necessary that the semi L-summand is one-dimensional and that the ambient space is isometric to an $A(K)$-space. This is proved by Payá and Rodríguez [480]. A lot more examples of semi M-ideals in real Banach spaces which are not M-ideals are constructed in [482] and [413]. As a matter of fact, these authors show that every real Banach space can be represented as a quotient space X/J where J is a semi M-ideal, but not an M-ideal. On the other hand, occasionally a semi M-ideal is automatically forced to be an M-ideal. For example, if X is a semi M-ideal in X^{**} then it is already an M-ideal ([552], [405]). This is due to the fact that a natural contractive linear projection with kernel $J^\perp$ is available. In fact, the proof of Proposition 1.2 extends to show: If P is a semi L-projection and Q is a contractive linear projection with $\mathrm{ran}(P) = \ker(Q)$, then $P = Id - Q$ so that P is linear, too. Applying this to $P =$ the semi L-projection onto the annihilator $X^\perp$ of X in X^{***} and $Q =$ the natural projection from X^{***} onto X^* (see III.1.2) we obtain that P is actually an L-projection. The same holds for the subspace $K(X)$ of compact operators in $L(X)$ provided X has the metric compact approximation property. (The natural linear projection will be described in the Notes and Remarks to Chapter VI.) Also, semi M-ideals in $C(K)$-spaces and, more generally, in C^*-algebras are ideals and hence M-ideals ([400], [440]).

Semi M-ideals share a number of properties with M-ideals; for example, the sum of two semi M-ideals is closed and a semi M-ideal [400], hermitian operators leave semi M-ideals invariant [440], semi M-ideals are Hahn-Banach smooth and proximinal (see Section II.1 for M-ideals) etc. [400]. A striking difference from the theory of M-ideals is that the intersection of two semi M-ideals need not be a semi M-ideal and that there are weak* closed semi M-ideals in a dual space which are not M-summands [400]; for M-ideals this is not possible as will be shown in II.3.6(b).

It is worth mentioning that the annihilator $J^\perp$ of a semi M-ideal J is always the *range* of a contractive linear projection which, however, in general does not enjoy decent norm properties. This follows from Lindenstrauss' work [415, p. 270] (see also [489, p. 62] for a simpler proof) since the semi L-projection onto $J^\perp$ can be shown to be 1-Lipschitz [648]. The most intriguing problem about semi M-ideals is to find examples of semi M-ideals in a complex Banach space which are not M-ideals. Only recently the existence of such examples was established by Behrends [63], thus refuting the long-standing conjecture that complex M-ideals might be characterised by the 2-ball property. (We shall discuss in the Notes and Remarks to Chapter II which results led to this conjecture.) Previously Yost [653] proved the equivalence of the following conjectures:

- There exists a semi M-ideal in some complex Banach space which is not an M-ideal.
- There exists a compact convex set $\Delta \subset \mathbb{C}^2$ such that $\ell(\Delta)$ is a disk for all linear maps $\ell : \mathbb{C}^2 \to \mathbb{C}$ which is not symmetric.

Further reformulations appear in [655] and [413], see also [657]. Using a very delicate and involved construction Behrends produces a set Δ as above and hence shows the existence of proper complex semi M-ideals. Let us remark that the equivalence of the two conjectures holds for real Banach spaces as well (here "disk" has of course to be replaced by "interval"); in this case, however, both of them were already known to be true. (For the former consult Remark 2.3(a), for the latter take Δ to be a triangle.) Another relevant paper on this topic is [64].

UNIQUE HAHN-BANACH EXTENSIONS. It was shown in Proposition 1.12 that a functional which is defined on an M-ideal can be extended to a functional on the whole space preserving its norm in a unique manner. Following [574] and [588] we shall refer to this property of a closed subspace as *Hahn-Banach smoothness.* The study of the phenomenon of unique Hahn-Banach extensions was initiated by Phelps in [495] who proved for example that the subspace J is Hahn-Banach smooth if and only if $J^\perp$ is a Chebyshev subspace of X^* (cf. Section II.1 for the definition of this notion); in Phelps' terminology Hahn-Banach smoothness is called property (U).
Obviously, being an M-ideal is far stronger than being Hahn-Banach smooth. For this reason Hennefeld [305] has proposed a weakening of the notion of an M-ideal in that he requires that there be a projection P on X^* whose kernel is $J^\perp$ such that

$$\|Px^*\| < \|x^*\| \quad \text{if } x^* \neq Px^* \tag{1}$$

$$\|x^* - Px^*\| \leq \|x^*\|. \tag{2}$$

In this case J is called an *HB-subspace.* It is known that, for all Banach spaces X and for $1 < p < \infty$, $K(X, \ell^p)$ is an *HB*-subspace of $L(X, \ell^p)$ without generally being an M-ideal [462], cf. also [305], [306].
It is easy to see that an *HB*-subspace is in fact Hahn-Banach smooth, but actually inequality (2), which may contain welcome information, has nothing to do with the problem of unique Hahn-Banach extensions. Let us say that a subspace J is *strongly Hahn-Banach smooth* if there is a projection on X^* with kernel $J^\perp$ which satisfies solely (1). In [405] Lima characterises the difference between strongly Hahn-Banach smooth subspaces and Hahn-Banach smooth subspaces by an intersection property akin to the

n-ball property of Theorem 2.2. More precisely, he has proved that a Hahn-Banach smooth subspace J is strongly Hahn-Banach smooth if and only if for $\varepsilon > 0$, all $n \in \mathbb{N}$ (equivalently: for $n = 3$), all $x_1, \ldots, x_n \in J$ the implication

$$\bigcap_{i=1}^{n} B(x_i, r_i) \neq \emptyset \quad \Longrightarrow \quad \bigcap_{i=1}^{n} B(x_i, r_i) \cap J \neq \emptyset$$

holds. (Note that one requires here all the centres to lie in J!) Incidentally, Lima seems to have overlooked inequality (2) since he claims to provide a characterisation of *HB*-subspaces. This was pointed out by E. Oja in [461]. She also presents examples of strongly Hahn-Banach smooth subspaces which are not *HB*-subspaces and of Hahn-Banach smooth subspaces which are not strongly Hahn-Banach smooth. One way of distinguishing between Hahn-Banach smoothness and its strong version is as follows: J is Hahn-Banach smooth if and only if there is only one norm preserving extension function $\Phi : J^* \to X^*$. The subspace J is strongly Hahn-Banach smooth if in addition Φ is linear!

MORE GENERAL NORM DECOMPOSITIONS. A natural generalisation of the concept of L- and M-projections is the concept of an L^p*-projection*, defined by the norm condition

$$\|x\|^p = \|Px\|^p + \|x - Px\|^p.$$

L^p-summands are defined in the obvious way. This notion is discussed in detail in [66]. The nontrivial fact about these projections is that two L^p-projections commute if $p \neq 2$. This was first proved by Behrends [47], an easier proof can be found in [110], and for complex scalars there is a simple argument in [477]. If we agree to call an M-projection an L^∞-projection, then it is also true that L^p- and L^r-projections cannot exist simultaneously on a given Banach space if $p \neq r$ (except for $\ell^\infty_{\mathbb{R}}(2)$). This is proved in [47] and [110], too. Moreover, the set of all L^p-projections forms a complete Boolean algebra, and L^p-projections on dual spaces are weak* continuous if $p > 1$. Therefore, there is a complete duality between L^p-projections in X and L^q-projections in X^* (with q denoting the index conjugate to p); in particular there is no need for "L^p-ideals". This fact seems to be responsible for the extreme lack of nontrivial examples of L^p-projections apart from the characteristic projections in L^p-spaces, including L^p-spaces of vector valued functions (Bochner L^p-spaces). In fact, one can characterise L^p-spaces by a richness condition on the set of L^p-projections ([138] for $p = 1$, [66, p. 54]). Let us mention one other instance where L^p-projections arise naturally: the 1-complemented subspaces of the Schatten classes c_p for $1 < p < \infty$, $p \neq 2$, were recently characterised as those subspaces which can be decomposed into a sequence of L^p-summands of a special nature (Cartan factors) ([29], [30]). (The boundary cases $p = 1, \infty$ were treated in [28].) However, it follows from the results of [581] that the Schatten class c_p itself does not admit any nontrivial L^p-projections.

L^p-projections also turn out to be of interest in the discussion of ergodic theorems [283] and isometries for Bochner L^p-spaces [278] [279]. Fortunately, it is the lack of L^p-projections of the range space which is of importance for the description of the isometric isomorphisms of Bochner L^p-spaces. For example, Greim describes in [278] the surjective isometries of $L^p(\mu, V)$ if V is separable and has no nontrivial L^p-summands.

Yet more general norm decompositions are discussed in [439] and the subsequent papers [440], [441]. Let F be a norm on $\mathbb{R}^2$ such that

$$\begin{aligned} F(0,1) &= F(1,0) = 1 \\ F(s,t) &= F(|s|,|t|) \qquad \forall (s,t) \in \mathbb{R}^2. \end{aligned}$$

An *F-projection* is defined by the equation

$$\|x\| = F(\|Px\|, \|x - Px\|) \qquad \forall x \in X,$$

and the range of an F-projection is called an F-summand. The dual norm F^* is defined by

$$F^*(a,b) = \sup_{F(s,t)\le 1} |at + bs|,$$

and an F-ideal of X is a closed subspace J for which $J^\perp$ is an F^*-summand. One can show that the existence of F-ideals which are not F-summands depends decisively on the geometry of the two-dimensional unit ball $B_F = \{(s,t) | \ F(s,t) \le 1\}$. In fact, if $(0,1) \in \operatorname{ex} B_F$, then every F-ideal is actually an F-summand. [One way to prove this is to show by duality that a G-summand in X^* is weak* closed if G is differentiable at (1,0), i.e., $\lim_{h\to 0}(G(1,h) - 1)/h = 0$. This can be achieved by an argument similar to the one leading to Theorem 1.9.]
The authors of [439] also consider the corresponding "semi" notions, where the additivity of the projections involved is replaced by the weaker quasiadditivity condition

$$P(x + Py) = Px + Py \qquad \forall x, y \in X.$$

At this level of abstraction another dualisation procedure leads to a new concept: If $J^\perp$ is a semi F^*-ideal, then J is called a semi F-idealoid. Luckily, further dualisation has no effect, since J can be shown to be a semi F-ideal if $J^\perp$ is a semi F^*-idealoid. Again, one can prove that the geometry of B_F and of B_{F^*} governs the problem whether several of these notions coincide for a given F. For details we refer to [439].

BOOLEAN ALGEBRAS OF PROJECTIONS. In Theorem 1.10, which is due to Cunningham [138], we proved the completeness of the Boolean algebra $\mathbb{P}_L$ of all L-projections. (Actually, not only is $\mathbb{P}_L$ complete as an abstract Boolean algebra, but the supremum of an upwards filtrating family is its limit in the strong operator topology. This refined version of completeness is known as Bade completeness and of importance in operator theory.) By the Stone representation theorem (cf. e.g. [284, p. 78]) $\mathbb{P}_L$ is isomorphic to the Boolean algebra of clopen sets of some compact Hausdorff space Ω. Since $\mathbb{P}_L$ is complete, Ω is extremally disconnected, meaning that the closure of each open set is open. But even more is true: as Cunningham shows there is a Borel measure m on Ω with the following properties.

- m is regular on sets of finite measure.
- $m(A) = 0$ if A is nowhere dense.
- $m(C) > 0$ if $\emptyset \neq C$ is clopen.
- Every Borel set contains a Borel set of finite measure.

[Idea of construction: Let $C \mapsto E_C$ be the Boolean isomorphism between the clopen sets of Ω and $\mathbb{P}_L$. For $x \in X$ and closed $B \subset \Omega$ let $m_x(B) = \inf\{\|E_C x\| \mid C \supset B \text{ clopen}\}$. One can show that m_x extends to a uniquely determined finite regular Borel measure on Ω. Let $\{m_{x_i} | \, i \in I\}$ be a maximal family of pairwise singular measures of this type (which exists by Zorn's lemma) and put $m(B) = \sum_i m_{x_i}(B)$. (Details can be found in [66, p. 24ff.].)]
An extremally disconnected compact space admitting such a measure is called hyperstonean; equivalently, the order continuous functionals on $C(\Omega)$ ("normal measures") separate the points of $C(\Omega)$. (This is the usual definition to be found e.g. in [386].) It is now possible to represent X as a Banach space of "integrable sections" in a product $\prod_{\omega \in \Omega} X_\omega$ such that $\|x\| = \int_\Omega \|x(\omega)\|_{X_\omega} \, dm(\omega)$. If X is represented in this way the operators in the Cunningham algebra (Definition 3.13) correspond exactly to the multiplication operators $Tx(\cdot) = f(\cdot)x(\cdot)$ for $f \in C(\Omega)$. This representation, called the integral module representation, is proved in [66].
Furthermore one can show that the operators in the Cunningham algebra behave like normal operators on Hilbert space; more precisely they are spectral operators of scalar type in the sense of Dunford and Schwartz [180], see [48], [66].
These theorems extend to the setting of L^p-projections for $1 < p < \infty$, $p \neq 2$ [66]. Also, Evans proves in [207] that a Boolean algebra of F-projections necessarily consists solely of L^p-projections for some p under some natural additional assumption on F.
The case $p = \infty$ differs from the previous ones in that the Boolean algebra $\mathbb{P}_M$ need not be complete. As a matter of fact, there need not exist any nontrivial M-projections on a $C(K)$-space (e.g. $K = [0, 1]$) so that such a space cannot be "resolved" in terms of M-projections although it certainly should have a fairly large "M-structure". The idea here is to have the role of the algebra generated by $\mathbb{P}_M$ taken over by the centralizer, and now a representation analogous to the integral module representation is possible. This was shown – up to the terminology – by Cunningham in [139]. More specifically, let K_X be the Gelfand space of the commutative C^*-algebra $Z(X)$. Then X is isometric to a subspace of a product $\prod_{k \in K_X} X_k$ in such a way that the numerical function $k \mapsto \|x(k)\|_{X_k}$ is upper semicontinuous on K_X and $\|x\| = \sup \|x(k)\|$. Moreover, the operators $T \in Z(X)$ correspond exactly to the multiplication operators $Tx(\cdot) = f(\cdot)x(\cdot)$ for $f \in C(K_X)$. In particular, the copy of X in $\prod X_k$ has the structure of a $C(K)$-module. The above representation is called the maximal function module representation in [51, Chap. 4]. The reader should observe that this procedure is reminiscent of von Neumann's reduction theory of operator algebras [166, Chap. II]. We refer to [139], [51] or [245] for exhaustive information, moreover we mention the papers [162], [246], [378] and [563]. A general view of integral and function module representation theorems is given in Evans' paper [208], see also [209].

CHAPTER II

Geometric properties of M-ideals

II.1 M-ideals and best approximation

In this section we shall establish several approximation theoretic properties of M-ideals. Also, some preparatory work for Section II.3 will be done where we shall study the difference between M-ideals and M-summands in some detail.

Let us first recall some notions from approximation theory. A subspace J of a Banach space X is called *proximinal* if

$$\forall\, x \in X \quad \exists\, y \in J \quad \|x - y\| = d(x, J).$$

The set of all such y is called the set of best approximants and denoted by $P_J(x)$. Thus J is proximinal if and only if $P_J(x) \neq \emptyset$ for all $x \in X$. The set-valued map P_J is called the *metric projection*. J is called a *Chebyshev* subspace if $P_J(x)$ is a singleton for each $x \in X$. Finally, the *metric complement* J^θ is defined as

$$J^\theta = \{x \in X \mid \|x\| = d(x, J)\} = \{x \in X \mid 0 \in P_J(x)\}.$$

To begin with, let us study these concepts for an M-summand J in X. P denotes the M-projection with range J.
First of all, J is proximinal, and $P_J(x)$ is a ball with radius $d(x, J) = \|x - Px\|$ and centre Px. To see this note that for any $y \in J$

$$\|y - x\| = \max\{\|y - Px\|, \|x - Px\|\}.$$

As a consequence, $\operatorname{lin} P_J(x) = J$ for each $x \in X \setminus J$. So J is highly non-Chebyshev. Another consequence is the existence of a continuous linear selection for P_J, namely P.

(A *selection* f of a set-valued map F is a function such that $f(x) \in F(x)$ for all x.) Furthermore,

$$\{x \in X \mid \|Px\| < \|x\|\}$$

is an open subset of J^θ.

We wish to examine the validity of these results for M-ideals J. Naturally, one has to face certain limitations. For example, the M-ideal c_0 in ℓ^∞ is easily seen to be proximinal (this is not accidental, cf. Proposition 1.1 below), but

$$P_{c_0}(1) = \{(s_n) \in c_0 \mid |s_n - 1| \le 1 \text{ for all } n\}$$

is not a ball (though still "big"), there is no continuous linear selection for P_{c_0} (since there is no continuous linear operator whatsoever from ℓ^∞ onto c_0), and the metric complement $c_0^\theta = \{(s_n) \in \ell^\infty \mid \|(s_n)\| = \limsup |s_n|\}$ has empty interior.

Proposition 1.1 *M-ideals are proximinal.*

PROOF: Let $J \subset X$ be an M-ideal, and let $x \in X$ with $d = d(x, J) > 0$. We shall inductively construct a sequence (y_n) in J with

$$\|y_{n+1} - y_n\| \le \left(\frac{3}{4}\right)^n \tag{1}$$

$$\|y_n - x\| \le d + \left(\frac{3}{4}\right)^{n-1} \tag{2}$$

for $n \ge 1$. Once this is achieved, (1) yields that (y_n) is a Cauchy sequence, while (2) gives $\|\lim y_n - x\| \le d$, whence $P_J(x) \ne \emptyset$. To show how the induction proceeds let $\varepsilon > 0$. First, choose $y \in J$ satisfying $\|y - x\| \le d + \varepsilon$. Then consider the balls $B(x, d + \varepsilon/2)$ and $B(y, \varepsilon/2)$. Since the distance of the centres does not exceed the sum of the radii, they have a point in common, and both balls meet J. Now J has the 2-ball property (Theorem I.2.2), so there is

$$z \in J \cap B(x, d + 3\varepsilon/4) \cap B(y, 3\varepsilon/4).$$

Applying this procedure with the sequence $(\varepsilon_n) = ((\frac{3}{4})^{n-1})$ we obtain the desired y_n. □

Next, we are going to discuss the largeness of the set of best approximants. We shall meet the following notion again in Section II.3.

Definition 1.2 *A closed convex bounded subset B of a Banach space Y is called a pseudoball of radius r if its diameter is $2r > 0$ and if for each finite collection $y_1, \dots, y_n$ of points with $\|y_i\| < r$ there is $y \in B$ such that*

$$y + y_i \in B \text{ for } i = 1, \dots, n.$$

If $\rho = \sup\{s \ge 0 \mid B$ contains a ball of radius $s\}$, then the grade of B is defined to be

$$g(B) = 1 - \frac{\rho}{r}.$$

Singletons are considered as pseudoballs with radius 0 and grade 0.

Equivalently B is a pseudoball of radius $r > 0$ if and only if

$$\bigcap_i (y_i + B) \neq \emptyset$$

for each finite family $y_1, \ldots, y_n$ satisfying $\|y_i\| < r$.
Note that B is a closed ball of radius $r > 0$ if and only if

$$\bigcap_{\|y\|<r} (y + B) \neq \emptyset,$$

in which case the intersection consists of the centre of the ball. In a pseudoball there is only a "centre" for any finite set of directions. Also, note that $g(B) = 0$ if and only if B is a ball. Thus, the larger the grade of B is the more proper a pseudoball is B. The idea of properness for M-ideals will be studied in Section II.3.

To see a simple example, let $0 \leq s \leq 1$. It is easy to check that

$$B_s = \{(s_n) \in c_0 \mid |s_n - s| \leq 1 \text{ for all } n \in \mathbb{N}\}$$

is a pseudoball in c_0 with radius 1 and $g(B_s) = s$.

The following result explains our interest in pseudoballs here.

Proposition 1.3

(a) *Let J be an M-ideal in X, and let $x \in X$. Then $P_J(x)$ is a pseudoball in J with radius $d(x, J)$.*

(b) *If $P_J(x)$ is a pseudoball of radius $d(x, J)$ for all $x \in X$, then J is an M-ideal.*

PROOF: (a) Obviously, $P_J(x)$ is a closed convex set with diameter $\leq 2 \cdot d(x, J)$. There is no loss in generality in assuming $d(x, J) = 1$. So, let $y_1, \ldots, y_n \in \operatorname{int} B_J$ be given. Then we have

$$P_J(x + y_i) = J \cap B(x + y_i, 1) \neq \emptyset \tag{1}$$

since J is proximinal (Proposition 1.1), and

$$\operatorname{int} \bigcap_i B(x + y_i, 1) \neq \emptyset \tag{2'}$$

since $\|y_i\| < 1$. By Theorem I.2.2(v) one concludes

$$\bigcap_i (y_i + P_J(x)) = J \cap \bigcap_i B(x + y_i, 1) \neq \emptyset.$$

Hence $P_J(x)$ is a pseudoball of radius $d(x, J)$.

(b) We shall verify the 3-ball property (Th. I.2.2(iv)). Let x, y_i, ε be given as in I.2.2(iv). Let $r < d(x, J) < r + \varepsilon$. Then there is $y \in J$ such that

$$-ry_i + y \in P_J(x), \quad i = 1, 2, 3.$$

Hence

$$\|x + y_i - y\| \leq d(x, J) + (1-r)\|y_i\| \leq 1 + \varepsilon. \qquad \square$$

We shall prove in Theorem 3.10 that every pseudoball arises in this way. This and the preceding proof show that B is a pseudoball if there is a "centre" for any 3 directions.

Proposition 1.4 *Let $B \subset Y$ be a pseudoball of radius r. Then*

$$\operatorname{int} B(0,r) \subset \frac{1}{2}(B - B) \subset B(0,r).$$

PROOF: Let $y \in Y$, $\|y\| < r$. Choose

$$z \in (-y + B) \cap (y + B).$$

Then

$$y = \frac{1}{2}((y+z) - (-y+z)) \in \frac{1}{2}(B - B).$$

The other inclusion is clear. $\square$

Combining Propositions 1.3 and 1.4 we arrive at the following stunning approximation properties of M-ideals.

Corollary 1.5 *If J is an M-ideal in X and $x \in X \backslash J$, then every element in J can be represented as a linear combination of two points from $P_J(x)$. More precisely,*

$$\operatorname{int} B_J(0, d(x,J)) \subset \frac{1}{2}(P_J(x) - P_J(x)) \subset B_J(0, d(x,J)).$$

In particular, $P_J(x)$ is compact if and only if J is finite dimensional.

Thus M-ideals are far from being Chebyshev subspaces. As a result, no strictly convex space can contain a nontrivial M-ideal (otherwise the corresponding metric projection would have to be single-valued).

Before we continue investigating the metric projection in more detail we want to give a characterisation of pseudoballs in terms of the bidual space.

Theorem 1.6 *For a closed convex bounded subset $B \subset Y$ and $r \geq 0$, equivalence between* (i) *and* (ii) *holds:*

(i) *B is a pseudoball of radius r.*

(ii) *The weak* closure of B in Y^{**} is a ball with radius r (and centre y_B^{**}, say).*

In this case

$$r \cdot g(B) = d(y_B^{**}, Y). \qquad (*)$$

PROOF: We may assume $r = 1$.

(i) $\Rightarrow$ (ii): Let $K = \overline{B}^{w*}$. It is easy to check that for any $y^* \in Y^*$ the set $y^*(K) = \overline{y^*(B)}$ is a pseudoball in $\mathbb{K}$ with radius $\|y^*\|$, and it is easy to prove, too, that pseudoballs in $\mathbb{K}$ are actually balls. For $y^* \in Y^*$ let $\ell(y^*)$ denote the centre of the ball $y^*(K)$.

CLAIM: ℓ is a continuous linear functional on Y^*.

Obviously ℓ is homogeneous. To prove additivity choose, given $\varepsilon > 0$ and $y_1^*, y_2^* \in Y^*$, $y_1, y_2, y_3 \in Y$ such that

$$\begin{array}{lllcl} \|y_1\| < 1, & \mathbb{R} \ni & \langle y_1^*, y_1\rangle & \geq & (1-\varepsilon^2/2)\ \|y_1^*\| \\ \|y_2\| < 1, & \mathbb{R} \ni & \langle y_2^*, y_2\rangle & \geq & (1-\varepsilon^2/2)\ \|y_2^*\| \\ \|y_3\| < 1, & \mathbb{R} \ni & \langle y_1^*+y_2^*, y_3\rangle & \geq & (1-\varepsilon^2/2)\ \|y_1^*+y_2^*\|. \end{array}$$

We infer

$$|\langle y_1^*, y\rangle - \ell(y_1^*)| \ \leq\ \varepsilon\|y_1^*\|$$

and in the same way

$$\begin{aligned} |\langle y_2^*, y\rangle - \ell(y_2^*)| &\leq \varepsilon\|y_2^*\| \\ |\langle y_1^*+y_2^*, y\rangle - \ell(y_1^*+y_2^*)| &\leq \varepsilon\|y_1^*+y_2^*\|, \end{aligned}$$

from which the additivity of ℓ follows. The continuity of ℓ is clear.
Let us write y_B^{**} instead of ℓ. It is left to show that

$$K = B_{Y^{**}}(y_B^{**}, 1).$$

This is an immediate consequence of the Hahn-Banach theorem, since

$$y^*(K) \ =\ B_{\mathbb{K}}(\langle y_B^{**}, y^*\rangle, \|y^*\|) \ =\ y^*(B_{Y^{**}}(y_B^*, 1))$$

for all $y^* \in Y^*$.

(ii) $\Rightarrow$ (i): A moment's reflection shows that we have to prove

$$\operatorname{int}(B_Y \times \cdots \times B_Y) \subset B \times \cdots \times B - \Delta_B$$

where

$$\Delta_B = \{(y, \ldots, y) \mid y \in B\} \subset Y^n$$

and $n \in \mathbb{N}$ is arbitrary. By a separation theorem due to Tukey [605] (see also [179, p. 461] or [337, Cor. 22.5]) this will follow from

$$B_Y \times \cdots \times B_Y \ \subset\ \overline{B \times \cdots \times B - \Delta_B}. \tag{1}$$

To prove this inclusion we first note

$$\overline{B}^{w*} \times \cdots \times \overline{B}^{w*} - \Delta_{\overline{B}^{w*}} \ \subset\ \overline{B \times \cdots \times B - \Delta_B}^{w*} \tag{2}$$

thanks to the weak* continuity of $(y_1^{**}, \ldots, y_n^{**}, y^{**}) \mapsto (y_1^{**} - y^{**}, \ldots, y_n^{**} - y^{**})$. Now $\overline{B}^{w*}$ is a ball of radius 1 so that the left hand side of (2) contains $B_{Y^{**}} \times \cdots \times B_{Y^{**}}$ and a fortiori $B_Y \times \cdots \times B_Y$.
The desired inclusion easily follows from this and the Hahn-Banach theorem. It remains to prove formula $(*)$. We first observe

$$B_{Y^{**}}(y_B^{**}, 1) \cap Y = B.$$

This and the triangle inequality give

$$B(y, 1 - \|y_B^{**} - y\|) \subset B$$

and hence

$$g(B) \le \|y_B^{**} - y\|$$

for any $y \in Y$. This shows "$\le$" in $(*)$.
Conversely, if a ball $B(y,s)$ is contained in B, then

$$B_{Y^{**}}(y,s) \subset B_{Y^{**}}(y_B^{**}, 1).$$

This and the triangle inequality give

$$\|y_B^{**} - y\| \le 1 - s$$

and hence $d(y_B^{**}, Y) \le g(B)$. □

Corollary 1.7 *Let J be an M-ideal in X, and let P be the associated M-projection from X^{**} onto $J^{\perp\perp}$. Consider the pseudoball $B = P_J(x)$ for some $x \in X$. Then $Px = y_B^{**}$, the centre of $\overline{B}^{w*}$. More precisely: the canonical isometry i_J^{**} from J^{**} onto $J^{\perp\perp}$ maps y_B^{**} onto Px. Moreover (if $d(x,J) = 1$)*

$$g(P_J(x)) = d(Px, J) = \sup\{|\langle y^*, x\rangle| \mid y^* \in B_{J^*}\}.$$

PROOF: We consider J^* via unique Hahn-Banach extensions as a subspace of X^* (Remark I.1.13). We have for $y^* \in J^*$

$$y^*(B_{J^{**}}(y_B^{**}, 1)) = B_{\mathbb{K}}(\langle y_B^{**}, y^*\rangle, \|y^*\|).$$

On the other hand we know from Theorem 1.6 (w.l.o.g. $d(x,J) = 1$)

$$\begin{aligned} B_{\mathbb{K}}(\langle y_B^{**}, y^*\rangle, \|y^*\|) &= \overline{y^*(B)} \\ &\subset \overline{y^*(B_X(x,1))} \\ &= B_{\mathbb{K}}(\langle y^*, x\rangle, \|y^*\|). \end{aligned}$$

Hence $\langle y_B^{**}, y^*\rangle = \langle x, y^*\rangle$ for all $y^* \in J^*$, which gives $y_B^{**} = Px$. □

We now return to best approximation and investigate the problem of whether there is a continuous (possibly linear) selection for the metric projection. We recall the definition of the Hausdorff metric d_H on the set of closed subsets of X:

$$d_H(A,B) = \sup(\{d(a,B) \mid a \in A\} \cup \{d(b,A) \mid b \in B\})$$

Proposition 1.8 *The metric projection associated to an M-ideal J in X is Lipschitz continuous with respect to the Hausdorff metric. More precisely*

$$d_H(P_J(x_1), P_J(x_2)) \le 2 \cdot \|x_1 - x_2\| \qquad \text{for all } x_1, x_2 \in X.$$

PROOF: The assertion is clearly true if either of x_1 or x_2 is in J. For the general case we have to show: given $x_1, x_2 \in X$, $\varepsilon > 0$, and $y_2 \in P_J(x_2)$, there is $y_1 \in P_J(x_1) = J \cap B(x_1, d(x_1, J))$ such that

$$\|y_1 - y_2\| \leq 2 \cdot \|x_1 - x_2\| + \varepsilon.$$

By the strict 2-ball property (Theorem I.2.2(v)) this will be assured by

$$\|x_1 - y_2\| < 2 \cdot \|x_1 - x_2\| + \varepsilon + d(x_1, J) \qquad (*)$$

(since int $(B(a_1, r_1) \cap B(a_2, r_2)) \neq \emptyset$ if $\|a_1 - a_2\| < r_1 + r_2$ and $r_1, r_2 > 0$). Finally, $(*)$ is a consequence of

$$\begin{aligned} \|x_1 - y_2\| &\leq \|x_1 - x_2\| + \|x_2 - y_2\| \\ &= \|x_1 - x_2\| + d(x_2, J) \\ &\leq \|x_1 - x_2\| + d(x_2 - x_1, J) + d(x_1, J) \\ &< 2 \cdot \|x_1 - x_2\| + d(x_1, J) + \varepsilon. \end{aligned}$$

□

REMARKS: (a) The constant 2 is optimal here. In fact, let $X = \ell^\infty$, $J = c_0$, $x_1 = (0,0,0,\dots)$, $x_2 = (1,1,1,\dots)$, $y_2 = (2,0,0,\dots)$. Then

$$d_H(P_J(x_1),\ P_J(x_2)) \geq d(y_2, P_J(x_1)) = \|y_2 - x_1\| = 2.$$

(b) A similar argument shows

$$d_H(x_1 - P_J(x_1), x_2 - P_J(x_2)) \leq d_H(x_1 - x_2, J).$$

Theorem 1.9 *Let J be an M-ideal in X.*

(a) *There exists a continuous homogeneous mapping $\pi : X \to J$ with $\pi(x) \in P_J(x)$ for all $x \in X$ (i.e. a continuous selection for P_J) which is quasiadditive, meaning $\pi(x + y) = \pi(x) + y$ for $x \in X$, $y \in J$.*

(b) *There exists a continuous homogeneous mapping $f : X/J \to X$ with $f(x+J) \in x + J$ and $\|f(x + J)\| = \|x + J\|$ for all $x \in X$ (i.e. a continuous norm preserving lifting for the quotient map).*

PROOF: Consider the set-valued homogeneous mapping Σ on X/J defined by

$$\Sigma\,(x + J) = \{x\} - P_J(x).$$

It is well-defined, and by Proposition 1.8 Σ is continuous for the Hausdorff metric, a fortiori it is lower semicontinuous (that is to say for an open set U, the set $\{x+J \mid \Sigma(x+J) \cap U \neq \emptyset\}$ is open in X/J). Furthermore the $\Sigma(x + J)$ are closed, convex and bounded. Of course, the same is true for the restriction Σ_0 of Σ to the unit sphere in X/J so that Michael's selection theorem [443, Theorem 3.2″] applies to yield a continuous selection

σ_0 for Σ_0. The technique described in [443, p. 376] permits the extension of σ_0 to a homogeneous selection σ for Σ on X/J.
Now, (a) is achieved by letting $\pi(x) = x - \sigma(x + J)$, and (b) by $f = \sigma$. □

Note that (a) and (b) above are in complete duality: given π as in (a), $f(x+J) = x - \pi(x)$ fulfills the requirements of (b), and given f as in (b), $\pi(x) = x - f(x + J)$ fulfills (a).

The question of whether the mappings π and f above can be chosen to be Lipschitzian (or even linear) arises immediately. Let us first address the question of Lipschitz projections. This problem remains open. It is pointed out in [511] and [656] that a Lipschitz set-valued map need not have a Lipschitz selection (see the Notes and Remarks); on the other hand, there does exist a 2-Lipschitz projection from $\ell^\infty_{\mathbb{R}}$ onto c_0 [415]. Here we give a lower estimate for the Lipschitz constants of such projections in terms of the grade of certain pseudoballs.

Proposition 1.10 *Suppose J is an M-ideal in X. Let*

$$g^*(J, X) := \sup\{g(P_J(x)) \mid d(x, J) = 1\}.$$

If there exists a Lipschitz projection π from X onto J, then its Lipschitz constant L is at least $2 \cdot g^*(J, X)$.

The number $g^*(J, X)$ defined above will be put into its proper milieu in the Notes and Remarks section. It will be remarked there that $g^*(J, X) = 1$ in a number of cases, e.g. for the familiar M-ideals J_D in $C(K)$ if D is not clopen, for $J = K(H)$ in $X = L(H)$ or, more generally, in the case $X = J^{**}$. On the other hand, clearly $g^*(J, X) = 0$ if J is an M-summand.

PROOF: Let $x \in X$, $d(x, J) = 1$, and consider $B = P_J(x) = B_X(x, 1) \cap J$. Then

$$B = \pi(B) \subset \pi(B_X(x, 1)) \subset B_J(\pi(x), L).$$

Now we take $\sigma(J^{**}, J^*)$-closures and use Proposition 1.3, Theorem 1.6, and Corollary 1.7 to obtain

$$B_{J^{**}}(Px, 1) \subset B_{J^{**}}(\pi(x), L)$$

where P is the M-projection from X^{**} onto $J^{\perp\perp} \cong J^{**}$. Hence

$$2Px - y \in B_{J^{**}}(\pi(x), L) \quad \text{for all } y \in B$$

so that by Theorem 1.6 and Corollary 1.7

$$\begin{aligned} L &\geq 2 \cdot \|Px - (y + \pi(x))/2\| \\ &\geq 2 \cdot d(Px, J) \\ &= 2 \cdot g(B). \end{aligned}$$

□

The grade of the pseudoballs $P_J(x)$ is the appropriate tool for investigating the (possibly empty) interior of the metric complement J^θ.

Proposition 1.11 *Let J be an M-ideal in X, and let*

$$D = \{x \in X \mid \|x\| = d(x, J) = 1\}.$$

(a) *If $B_J(y_0, 2r) \subset P_J(x_0)$ for some $x_0 \in D$, then $B_X(x_0 - y_0, r) \subset J^\theta$.*

(b) *If $B_X(x_0, r) \subset J^\theta$ for some $x_0 \in D$, then $P_J(x_0)$ contains a ball of radius $2r - \varepsilon$ for whatever $\varepsilon > 0$.*

Therefore, J^θ has empty interior if and only if

$$g_*(J, X) := \inf\{g(P_J(x)) \mid d(x, J) = 1\} = 1.$$

PROOF: (a) We may assume $y_0 = 0$. Define

$$|x| = \sup\{|\langle x, y^* \rangle| \mid y^* \in B_{J^*}\}$$

(cf. Remark I.1.13). This is a seminorm, and from Lemma I.1.5 we get

$$\|x\| = \max\{|x|, d(x, J)\}. \tag{1}$$

Now let $z \in X$, $\|z - x_0\| < r$. We wish to show $|z| < d(z, J)$ (so that $z \in J^\theta$), then (a) will follow from the closedness of J^θ.
First note $|z - x_0| < r$ and $d(z - x_0, J) < r$ by (1). Secondly,

$$|x_0| \le 1 - 2r. \tag{2}$$

Proof of (2): Given $\varepsilon > 0$ choose $y^* \in S_{J^*}$ such that

$$|x_0| \le \langle y^*, x_0 \rangle + \varepsilon.$$

Next choose $y \in B_J$ such that

$$\langle y^*, y \rangle \le -1 + \varepsilon.$$

Note $2ry \in P_J(x_0)$, i.e. $\|x_0 - 2ry\| = 1$. The estimate

$$\begin{aligned} |x_0| &\le \langle y^*, x_0 \rangle + \varepsilon \\ &= \langle y^*, x_0 - 2ry \rangle + 2r\langle y^*, y \rangle + \varepsilon \\ &\le 1 + 2r(-1 + \varepsilon) + \varepsilon \end{aligned}$$

now proves (2).
Altogether one obtains

$$|z| < |x_0| + r \le 1 - r = d(x_0, J) - r < d(z, J)$$

as desired.

(b) Let P be the M-projection from X^{**} onto $J^{\perp\perp} \cong J^{**}$. In view of Theorem 1.6 and Corollary 1.7 we have to show $d(Px_0, J) \le 1 - 2r$. This is equivalent to

$$|\langle y^*, x_0 \rangle| \le 1 - 2r \quad \text{for all } y^* \in B_{J^*}.$$

To prove this for some given $y^* \in S_{J^*}$, fix $\varepsilon > 0$ and choose $y \in \operatorname{int} B_J$ such that $\langle y^*, y\rangle$ is real and so close to 1 that

$$|\langle y^*, y\rangle \pm \alpha| \leq 1 \text{ only if } |\alpha| \leq \varepsilon.$$

In particular $\langle y^*, y\rangle \geq 1 - \varepsilon$.
Since $P_J(x_0)$ is a pseudoball (Proposition 1.3), we may find $\overline{y} \in P_J(x_0)$ with $\overline{y} \pm y \in P_J(x_0)$. Let us first observe

$$|\langle y^*, \overline{y}\rangle - \langle y^*, x_0\rangle| \leq \varepsilon. \tag{3}$$

In fact,

$$\langle y^*, \overline{y}\rangle \pm \langle y^*, y\rangle \in y^*(P_J(x_0)) \subset B_{\mathbb{K}}(\langle y^*, x_0\rangle, 1)$$

(cf. Theorem 1.6) so that (3) follows. On the other hand, for $z = (1-r)x_0 + r\overline{y}$ we have

$$y^*(z) = y^*(z + ry) - ry^*(y) \leq (1-r) - r(1-\varepsilon) \tag{4}$$

where we used $\|x_0 - (z + ry)\| \leq r$ (since $\overline{y} + y \in P_J(x_0)$), consequently

$$\|z + ry\| = d(z + ry, J) = d((1-r)x_0, J) = 1 - r.$$

(3) and (4) together give our claim. □

With the help of Remark 3.8(d) below we conclude from Proposition 1.11 for example that J^θ has empty interior if J is an M-ideal in J^{**}. Elementary calculations show that $J_D^\theta \subset C(K)$ has empty interior if and only if D has empty interior.

II.2 Linear projections onto M-ideals

In this section we shall investigate whether there are *linear* mappings π and f in the setting of Theorem 1.9 above. Since there are noncomplemented M-ideals (e.g. c_0 in ℓ^∞; an entirely elementary argument for this fact is contained in [634]) we cannot generally expect such a linear projection or lifting to exist. We next present a sufficient condition. It involves Grothendieck's approximation property which we now recall.

Let X be a Banach space. Then X has the *approximation property (AP)* if $Id \in \overline{F(X)}$, the closure being taken with respect to the topology τ of uniform convergence on compact sets. We say that X has the *λ-approximation property (λ-AP)* where $\lambda \geq 1$ if $Id \in \overline{B_{F(X)}(0,\lambda)}^{\tau}$. Usually the 1-AP is called *metric approximation property (MAP)*. Finally X has the *bounded approximation property (BAP)* if it has the λ-AP for some λ.
If one requires only approximation by compact operators instead of finite rank operators, X is said to have the *compact approximation property (CAP)* resp. *λ-compact approximation property (λ-CAP)* etc.

We remark that the closure with respect to τ in the definition of the λ-AP and λ-CAP may be replaced by the closure in the strong operator topology or even the weak operator topology (note that the latter two topologies yield the same dual space [179, p. 477]).

For a detailed exposition of the approximation properties see [159], [281], [422], [423]. We mention that the classical $L^p(\mu)$- and $C(K)$-spaces enjoy the MAP. However, each of the spaces ℓ^p ($1 \le p < \infty$, $p \neq 2$) and c_0 has a closed subspace failing the MCAP (even the CAP) [423, p. 107 and passim], [355].
The main result of this section, due to Ando and, independently, Choi and Effros, is as follows.

Theorem 2.1 *Suppose J is an M-ideal in the Banach space X, Y is a separable Banach space, and $T \in L(Y, X/J)$ with $\|T\| = 1$. Assume further*

(a) *Y has the BAP*

or

(b) *J is an L^1-predual.*

Then there is a continuous linear lifting L for T, i.e. $L \in L(Y,X)$ such that $qL = T$ where $q: X \to X/J$ denotes the quotient map. More precisely, we obtain a lifting with

$$\|L\| \le \lambda \qquad \text{if } Y \text{ has the } \lambda\text{-AP},$$

resp.

$$\|L\| = 1 \qquad \text{under assumption (b).}$$

Before giving the proof we would like to make some remarks.

Remarks 2.2 (a) Instead of the separability of Y and (a) or (b) one may of course use the weaker assumptions

(a) *T factors through a separable space with the BAP*

resp.

(b′) *J is an L^1-predual, and T has a separable range.*

(b) If Theorem 2.1 is applicable to $Y = X/J$ and L is a lifting for Id_Y, then Lq is a continuous linear projection whose kernel is J so that J is a complemented subspace of X.

(c) One might wonder to what extent the additional assumptions in Theorem 2.1 are really needed. As for the separability one just has to activate the standard example of the noncomplemented M-ideal c_0 in ℓ^∞. (In view of the examples in [137] or [418, Prop. 3.5] Theorem 2.1 does not even extend to weakly compactly generated spaces Y.) Also, it is not enough to assume that $J^\perp$ is norm one complemented by some projection as is shown by the example of a subspace J of ℓ^1 such that $\ell^1/J = L^1[0,1]$. As for the approximation property we first note an easy proposition.

Proposition 2.3 *For every separable Banach space Y there are a separable Banach space X enjoying the MAP and an M-ideal J in X such that $Y \cong X/J$.*

Applying Proposition 2.3 to a space Y without the BAP we infer that there can be no continuous linear lifting for the quotient map, because otherwise Y would be isomorphic to a complemented subspace of X and hence would have the BAP.

PROOF OF PROPOSITION 2.3: We let (E_n) be an increasing sequence of finite dimensional subspaces of Y such that $\bigcup E_n$ is dense and define

$$X = \{(x_n) \mid x_n \in E_n, \ \lim x_n \text{ exists}\},$$
$$J = \{(x_n) \mid x_n \in E_n, \ \lim x_n = 0\}.$$

(These spaces are, of course, equipped with the sup norm.)
It is quickly verified that Y is isometric to X/J. Moreover, J is an M-ideal. To prove this we use the 3-ball property (Th. I.2.2(iv)). In fact, given normed vectors $\xi = (x_n) \in X$, $\eta_i = (y_n^i) \in J$ $(i = 1,2,3)$ and $\varepsilon > 0$, choose N such that

$$\|y_n^i\| \le \varepsilon \qquad \text{for } n > N \text{ and } i = 1,2,3.$$

If $\eta = (y_n)$ with $y_n = x_n$ $(n \le N)$, $y_n = 0$ $(n > N)$, then

$$\|\xi + \eta_i - \eta\| \le 1 + \varepsilon.$$

It remains to notice that X has the MAP since the contractive finite rank projections $P_m : (x_n) \mapsto (x_1, \dots, x_m, x_m, x_m, \dots)$ converge strongly to the identity. □

We now proceed to the proof of Theorem 2.1. The essential part of the argument is purely finite dimensional, and we shall present it in the following lemma.

Lemma 2.4 *Let X be a Banach space, $J \subset X$ an M-ideal and $q : X \to X/J$ the quotient map. Suppose further that E is a finite dimensional space and $F \subset E$ is a subspace. Let $T \in L(E, X/J)$ with $\|T\| = 1$. We assume*

(a) *there exists a contractive projection π from E onto F*

or

(b) *J is an L^1-predual.*

Then, given a contractive linear lifting $L_F : F \to X$ for $T_{|F}$ and $\varepsilon > 0$, there exists a contractive linear lifting $L_E : E \to X$ for T such that $\|L_{E|F} - L_F\| \le \varepsilon$.

For the time being, let us take this lemma for granted and give the

PROOF OF THEOREM 2.1:
We will first assume (b). Let (E_n) be an increasing sequence of finite dimensional subspaces of Y such that its union is dense. Let $E_0 = \{0\}$ and $L_0 = 0$. Using Lemma 2.4 one may inductively define a sequence $L_n : E_n \to X$ of contractions such that

$$qL_n = T_{|E_n} \qquad \text{and} \qquad \|L_{n|E_{n-1}} - L_{n-1}\| \le 2^{-n}$$

for all $n \in \mathbb{N}$. For $y \in \bigcup E_n$ the sequence $(L_n y)$ is eventually defined and Cauchy. Hence

$$Ly := \lim_{n\to\infty} L_n y \tag{$*$}$$

exists for these y, and $(*)$ defines a linear operator which can be extended to a contraction (still called L) from Y to X. It is clear that $qL = T$, i.e. L is a lifting.

The proof using assumption (a) is more delicate. We let $c(X)$, $c(X/J)$, ... be the sup-normed spaces of convergent sequences of vectors in X, X/J, ... Then T induces a contraction $c(T): c(Y) \to c(X/J)$ defined by

$$c(T)((y_n)_n) = (Ty_n)_n.$$

Further note that $c(q): c(X) \to c(X/J)$, defined in the same vein, is a quotient map with kernel $c(J)$ so that

$$c(X/J) \cong c(X)/c(J).$$

Finally we remark that $c(J)$ is an M-ideal in $c(X)$ as can easily be verified either by means of the 3-ball property (Theorem I.2.2) or by directly checking the definition. (This is a special case of Proposition VI.3.1.)
Now that we know that Y is a separable space with the λ-AP there is a sequence of finite rank operators (S_n) converging strongly to Id_Y with $\|S_n\| \leq \lambda$. Next we define an auxiliary subspace $H \subset c(Y)$ as the closed linear span of the sequences

$$(S_1y, \ldots, S_{m-1}y, S_my, S_my, \ldots)$$

where $m \in \mathbb{N}$, $y \in Y$. Note $(S_ny)_n \in H$ for all $y \in Y$. For $m \in \mathbb{N}$ and $(y_n)_n \in H$ we define

$$\pi_m((y_n)_n) = (y_1, \ldots, y_{m-1}, y_m, y_m, y_m, \ldots).$$

Obviously, the π_m form an increasing sequence of contractive finite rank projections on H converging strongly to Id_H. If we let $E_m = \operatorname{ran}(\pi_m)$ and use the same technique as in the first part (based on Lemma 2.4(a) applied to the M-ideal $c(J)$ in $c(X)$) we obtain a contractive linear lifting

$$\Lambda: H \to c(X)$$

for $c(T)|_H$, i.e.

$$c(q)\Lambda = c(T)|_H.$$

To get the desired lifting for T we define

$$Ly = \text{limit } \Lambda((S_ny)_n).$$

Then

$$\|L\| \leq \|\Lambda\| \cdot \sup_n \|S_n\| \leq \lambda$$

and

$$\begin{aligned} q(Ly) &= \text{limit } c(q)(\Lambda((S_ny)_n)) \\ &= \text{limit } (c(q)\Lambda)((S_ny)_n) \\ &= \lim_n T(S_ny) \\ &= Ty. \end{aligned}$$

This concludes the proof of Theorem 2.1. □

PROOF OF LEMMA 2.4: We shall work with the Banach space $L(E,X)$. There is a canonical isometry (since $\dim E < \infty$)

$$L(E,X)^{**} \cong L(E,X^{**}) \qquad (*)$$

which we shall employ in the sequel. ($(*)$ is a special case of Grothendieck's duality theory of tensor products [281], but a direct argument is available in [152].) Let us consider the following subspaces of $L(E,X)$:

$$\begin{aligned} W &= \{S \in L(E,X) \mid \operatorname{ran}(S) \subset J\} \quad (\cong L(E,J)) \\ V &= \{S \in W \mid \ker(S) \supset F\} \end{aligned}$$

Using $(*)$ one checks

$$\begin{aligned} W^{\perp\perp} &= \{S \in L(E,X^{**}) \mid \operatorname{ran}(S) \subset J^{\perp\perp}\} \quad (\cong L(E,J^{\perp\perp})) \\ V^{\perp\perp} &= \{S \in W^{\perp\perp} \mid \ker(S) \supset F\}. \end{aligned}$$

At this point we observe that W is an M-ideal in $L(E,X)$: in fact, if P denotes the M-projection from X^{**} onto $J^{\perp\perp}$, then $S \mapsto PS$ is the M-projection from $L(E,X^{**})$ onto $W^{\perp\perp}$. (We refer to Chapter VI for more general results.)
Now let $L_1 \in L(E,X)$ be any extension of L_F such that $qL_1 = T$ (this is possible since $\dim E < \infty$). Abbreviating $B_{L(E,X)}$ by B we claim

$$L_1 \in \overline{B+V}. \qquad (**)$$

To prove this, we think of L_1 as an element of $L(E,X^{**})$ and will show

$$L_1 \in \overline{B+V}^{w*}. \qquad (***)$$

In fact, let P be as above. We will first show how assumption (a) entails $(***)$. We decompose L_1 as

$$L_1 = ((Id-P)L_1 + PL_1\pi) + PL_1(Id-\pi)$$

and note

- $PL_1(Id-\pi) \in V^{\perp\perp}$
- $\|PL_1\pi\| \le \|P\| \cdot \|L_F\| \cdot \|\pi\| \le 1$
- $\|(Id-P)L_1\| = \|T\| = 1$

(since $\operatorname{ran}(Id-P) \cong (X/J)^{**}$ and the diagram

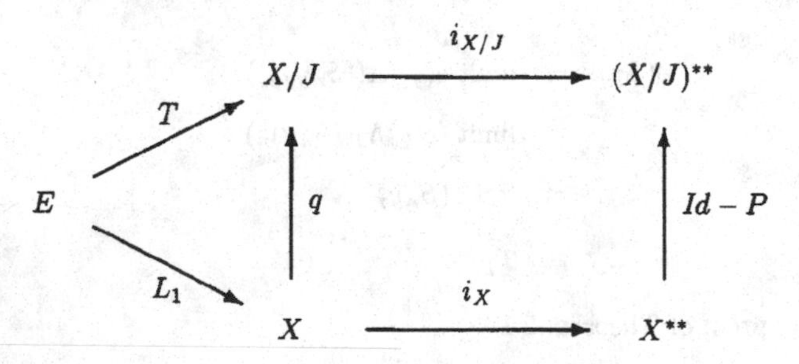

commutes) so that

- $\|(Id-P)L_1+PL_1\pi\| = \max\{\|(Id-P)L_1\|, \|PL_1\pi\|\} \leq 1$

(cf. Lemma VI.1.1(c)).
It follows that

$$L_1 \in B_{L(E,X^{**})} + V^{\perp\perp} = \overline{B}^{w*} + \overline{V}^{w*} = \overline{B+V}^{w*}.$$

This proves (***) under assumption (a). If (b) is assumed we know that $J^{\perp\perp}$ is isometric to an L^∞-space and hence enjoys the Hahn-Banach extension property (cf. e.g. [386, p. 86ff.] for that matter). In particular, there is a contractive extension $\Lambda: E \to J^{\perp\perp}$ of $PL_F: F \to J^{\perp\perp}$. We now decompose L_1 as

$$L_1 = ((Id-P)L_1 + \Lambda) + (PL_1 - \Lambda)$$

and deduce $L_1 \in \overline{B+V}^{w*}$ as above.
Given (**) we conclude the proof of the lemma as follows. By (**) there are $S_1 \in B$, $S_2 \in V$ such that

$$\|L_1 - (S_1 + S_2)\| \leq \varepsilon/2.$$

Let $L_2 = L_1 - S_2$. Then L_2 is a lifting for T extending L_F which is nearly a contraction ($\|L_2\| \leq 1+\varepsilon/2$). We wish to disturb L_2 so as to obtain a lifting which nearly extends L_F but *is* a contraction. The decisive tool to achieve this is an intersection property of M-ideals. To wit:

$$\begin{aligned} L_2 &\in (L_1+V)\cap(1+\varepsilon/2)B \\ &\subset \overline{B+V}\cap(1+\varepsilon/2)B \qquad \text{by } (**) \\ &\subset \overline{B+W}\cap(1+\varepsilon/2)B \\ &\subset B + \varepsilon(B\cap W) \end{aligned}$$

where we used Lemma 2.5 below in the last line. Thus there is a contraction L_E with

$$\|L_E - L_2\| \leq \varepsilon \quad \text{and} \quad \operatorname{ran}(L_E - L_2) \subset J.$$

It follows

$$\|L_E|_F - L_F\| \leq \varepsilon \quad \text{and} \quad qL_E = T,$$

as desired. □

It remains to prove:

Lemma 2.5 *For an M-ideal J in a Banach space X and $\varepsilon > 0$*

$$\overline{B_X + J}\cap(1+\varepsilon/2)B_X \subset B_X + \varepsilon B_J$$

holds.

PROOF: For $x \in \overline{B_X + J}$ certainly $d(x, J) \leq 1$ holds. Since J is proximinal (1.1) we have $B(x,1) \cap J \neq \emptyset$, and, if $\|x\| \leq 1 + \varepsilon/2$, $B(0,\varepsilon) \cap B(x,1)$ has nonempty interior. Therefore there is $y \in B(0,\varepsilon) \cap B(x,1) \cap J$ by the strict 2-ball property (Theorem I.2.2(v)), and $x = (x-y) + y \in B_X + \varepsilon B_J$, as requested. □

Theorem 2.1 contains several well-known results on the existence of linear extension operators as a special case. We give a sample.

Corollary 2.6 (Borsuk-Dugundji)
Let K be a compact Hausdorff space and let $D \subset K$ be a closed metrizable subset. Then there is a linear extension operator $T : C(D) \to C(K)$ with $\|T\| = 1$, i.e. $(Tx)(t) = x(t)$ for $x \in C(D)$ and $t \in D$.

PROOF: Consider the M-ideal $J_D = \{x \in C(K) \mid x|_D = 0\}$ (Example I.1.4(a)). Then $C(D) \cong C(K)/J_D$ meets the requirements of Theorem 2.1. (As a matter of fact, both (a) and (b) are fulfilled: $C(D)$ has the MAP, and J_D is an L^1-predual.) □

Corollary 2.7 (Pełczyński)
Let A be the disk algebra, and suppose D is a subset of the unit circle with Lebesgue measure 0. Then there is a contractive linear extension operator from $C(D)$ to A.

PROOF: Consider the M-ideal $J = \{x \in A \mid x|_D = 0\}$ (Example I.1.4(b)). By the Rudin-Carleson theorem [240, p. 58] we have $C(D) \cong A/J$, hence the result. □

More general corollaries can be formulated along the same lines on the basis of the Glicksberg peak interpolation theorem [240, p. 58, Th. 12.5 and Th. 12.7] and Theorem V.4.2 below.

Corollary 2.8 (Michael and Pełczyński, Ryll-Nardzewski)
Suppose $X \subset C(K)$ and $D \subset K$ is closed such that the pair $(X|_D, X)$ has the bounded extension property. If $X|_D$ is separable and has the MAP, then there is a contractive linear extension operator from $X|_D$ into X.

PROOF: This follows from Corollary I.1.20. □

A slightly different type of corollary is the following.

Corollary 2.9 (Sobczyk)
If X is a separable Banach space and $Y \subset X$ is a closed subspace isometric to c_0, then there is a continuous linear projection π from X onto Y with $\|\pi\| \leq 2$.

PROOF: $Y^{\perp\perp}$, which is canonically isometric to Y^{**}, is isometric to ℓ^∞, and Y (more precisely $i_X(Y)$) is an M-ideal in $Y^{\perp\perp}$, since c_0 is an M-ideal in ℓ^∞. Since $Y^{\perp\perp}$ is isometric to ℓ^∞ there is a contractive projection P from X^{**} onto $Y^{\perp\perp}$. Hence $Y^\perp$ is the kernel of a contractive projection on X^*, viz. $Q = i_X^* P^* i_{X^*}$. (To see this check $\operatorname{ran}(Id - Q) \subset Y^\perp \subset \ker Q$ which shows $Q(Id - Q) = 0$ and $\ker Q = Y^\perp$.) We now renorm X^* so that Q becomes an L-projection:

$$|x^*| := \|Qx^*\| + \|x^* - Qx^*\|$$

i.e.

$$(X^*, |\,.\,|) = \operatorname{ran}(Q) \oplus_1 Y^\perp.$$

Unfortunately $|\,.\,|$ need not be a dual norm, therefore we cannot conclude directly that Y is an M-ideal in some renorming of X. However, we have with respect to the dual norm

$$(X^{**}, |\,.\,|) = \ker(Q^*) \oplus_\infty Y^{\perp\perp}$$

so that $Y^{\perp\perp}$ is an M-summand in $(X^{**}, |\,.\,|)$. Note that $|\,.\,|$ and $\|\,.\,\|$ coincide on $Y^{\perp\perp}$; it follows that Y is an M-ideal in $(X^{**}, |\,.\,|)$ (Prop. I.1.17(b)), a fortiori Y is an M-ideal in the intermediate space $(X, |\,.\,|)$. By Theorem 2.1 there is a contractive (with respect to $|\,.\,|$) linear lifting L of the identity map $Id_{X/Y}$ with respect to quotient map $q: X \to X/Y$ (Y is an L^1-predual!), consequently $\pi = Id - Lq$ is a linear projection onto Y. It remains to estimate the norm:

$$\|\pi(x)\| = |\pi(x)| \le |\pi| \cdot |x| \le 2 \cdot \|x\|$$

since $\|x^*\| \le |x^*|$ for all $x^* \in X^*$ whence $|x^{**}| \le \|x^{**}\|$ for all $x^{**} \in X^{**}$. □

We hasten to add that the above argument is probably the most complicated proof of Sobczyk's theorem that has appeared in the literature; a very simple one can be found in [422, Th. 2.f.5]. Reading our proof from the bottom to the top will, however, yield an interesting result in Chapter III (Theorem III.3.11).

We finish this section with a proposition which is contained in the proof of Corollary 2.9 and worth stating explicitly.

Proposition 2.10 *Let X be a Banach space and $Y \subset X$ a subspace isometric to c_0. Then there is an equivalent norm on X which agrees with the original norm on Y so that Y becomes an M-ideal.*

II.3 Proper M-ideals

This section in devoted to the study of the distinction between M-ideals and M-summands.

Definition 3.1 *An M-ideal which is not an M-summand is called a* proper M-ideal. *A Banach space Y can be a* proper M-ideal *if it is isometric to a proper M-ideal in a suitable superspace X.*

Trivially, a reflexive space Y cannot be a proper M-ideal, since in the decomposition

$$X^* = Y^\perp \oplus_1 Y^*$$

the complementary L-summand Y^* is reflexive, hence weak* closed. On the other hand, if Y is an M-ideal in Y^{**} and if Y is not reflexive, then it is a proper M-ideal. (If Y were an M-summand, then Y would be a weak* closed subspace of Y^{**} by Theorem I.1.9. Goldstine's theorem then yields $Y = Y^{**}$.)

The approximation theoretic results of Section II.1 imply the following useful criterion.

Proposition 3.2 *For a closed subspace $J \subset X$, the following assertions are equivalent:*
(i) *J is an M-summand.*
(ii) *$P_J(x)$ is a ball of radius $d(x,J)$ for all $x \in X$.*

PROOF: (i) $\Rightarrow$ (ii) is elementary (compare p. 49).
(ii) $\Rightarrow$ (i): First of all J is an M-ideal by Proposition 1.3. Let P denote the M-projection from X^{**} onto $J^{\perp\perp}$. By Theorem 1.6 and Corollary 1.7 we obtain $Px \in J$ for all $x \in X$ from (ii). This means that P maps X onto J. □

Proposition 3.2 has an interesting consequence. The 3-ball property of Theorem I.2.2(iv) implies:

- J is an M-ideal in X iff J is an M-ideal in $\text{lin}\,(J \cup \{x\})$ for all $x \in X$.

Proposition 3.2 shows the same equivalence for M-summands! Thus:

Corollary 3.3 *Y can be a proper M-ideal if and only if there is a Banach space X containing Y as a proper M-ideal and $\dim X/Y = 1$.*

Another consequence is a characterisation of M-summands by means of an intersection property.

Proposition 3.4 *For a subspace J in X, the following assertions are equivalent:*
(i) *J is an M-summand.*
(ii) *For all families $(B(x_i,r_i))_{i\in I}$ of closed balls satisfying*

$$B(x_i,r_i) \cap J \neq \emptyset \quad \text{for all } i \in I \tag{1}$$

and

$$\bigcap_i B(x_i,r_i) \neq \emptyset \tag{2}$$

the conclusion

$$\bigcap_i B(x_i,r_i) \cap J \neq \emptyset$$

obtains.

PROOF: (i) $\Rightarrow$ (ii) is Lemma I.2.1.
(ii) $\Rightarrow$ (i): We show that $P_J(x)$ is a ball of radius 1 if $d(x,J) = 1$. This is the case if and only if

$$\bigcap_{\|y\|<1} (P_J(x) + y) \neq \emptyset. \tag{$*$}$$

This set equals

$$\bigcap_{\substack{\|y\|<1 \\ \varepsilon>0}} B_X(x+y, 1+\varepsilon) \cap J,$$

and $(*)$ follows since the above collection of balls satisfies (1) and (2). □

We next describe a class of Banach spaces which cannot be proper M-ideals.

Proposition 3.5 *Suppose Y is norm one complemented in Y^{**}. Then Y cannot be a proper M-ideal.*

PROOF: Suppose Y is an M-ideal in X. Let P denote the M-projection from X^{**} onto $Y^{\perp\perp}$ and Q a contractive projection from $Y^{**} \cong Y^{\perp\perp}$ onto Y. Then $QP|_X$ is a contractive projection from X onto Y, hence Y is an M-summand (Corollary I.1.3). □

We remark that the assumption of Proposition 3.5 is equivalent to saying that Y is a 1-complemented subspace of a dual; this is well known.

Corollary 3.6

(a) *If Y is isometric to a dual Banach space (or merely a 1-complemented subspace of a dual Banach space) then Y cannot be a proper M-ideal.*

(b) *A weak* closed M-ideal in a dual Banach space X^* is an M-summand which is the annihilator of an L-summand in X.*

PROOF: (a) follows from Proposition 3.5, and (b) follows from (a), since the complementary M-summand is weak* closed by Theorem I.1.9. □

REMARKS: (a) Proposition 3.5 and Corollary 3.6 are isometric results; the corresponding isomorphic versions are false in general. For example, consider $\ell^\infty = C(\beta\mathbb{N})$ and $t \in \beta\mathbb{N}\setminus\mathbb{N}$. Then the proper M-ideal $J_{\{t\}}$ is isomorphic to ℓ^∞ as is easily seen. (Compare, however, Corollary III.3.7(e).)

(b) Another argument for Corollary 3.6 appears in the Notes and Remarks section.

We are going to characterise Banach space which can be proper M-ideals intrinsically. It will be useful to introduce some notation. We first recall the notion of the characteristic $r(V, X^*)$ of a subspace V of a dual space X^*, which was introduced by Dixmier [164],

$$r(V, X^*) = \max\{r \geq 0 \mid rB_{X^*} \subset \overline{B_V}^{w^*}\}.$$

Obviously, $0 \leq r(Y, X^*) \leq 1$. Also $r(V, V^{**}) = 1$ by Goldstine's theorem, and an application of the bipolar theorem shows

$$r(V, X^*) = \inf_{x \in S_X} \sup_{x^* \in B_V} |\langle x^*, x\rangle|.$$

We refer to [175, p. 42ff.] for detailed information.

We now define:

Definition 3.7

(a) *If J is a one-codimensional M-ideal in X then the* grade of J in X, $g(J, X)$, *is the number $r(J^*, X^*)$ in case J is proper and $g(J, X) = 0$ otherwise.*

(b) *For a Banach space Y we define the* grade of Y *by*

$$g(Y) = \sup g(Y, X)$$

where the sup is taken over all superspaces X of Y containing Y as a one-codimensional M-ideal. If there is such an X with $g(Y, X) = 1$ we call Y an extreme M-ideal.

Remarks 3.8 (a) Corollary 3.3 suggests studying one-codimensional M-ideals in order to find intrinsic characterisations.

(b) Note that $r(J^*, X^*)$ is well defined for proper one-codimensional M-ideals, since J^* is either weak* closed or weak* dense if $\dim(X/J) = 1$. (Recall from Remark I.1.13 that J^* is a subspace of X^* for M-ideals.) Also, $g(J, X) = 0$ only if J is an M-summand: $r(J^*, X^*) = 0$ implies that B_{J^*} and consequently J^* is $\sigma(X^*, X)$-closed since J^* has codimension one. (In general, there do exist weak* dense subspaces V with $r(V, X^*) = 0$, e.g. for $X = c_0$ [164], [175, p. 45]. As a matter of fact, this characterises non-quasireflexivity of X [150], [175, p. 99].) Thus we obtain the equivalence

$$Y \text{ can be a proper } M\text{-ideal} \iff g(Y) > 0,$$

and the bigger $g(Y)$ is, the more proper an M-ideal can Y be.

(c) As an example consider the space X_r of convergent sequences equipped with the norm

$$\|(s_n)\|_r = \max\left\{ \|(s_n)\|_\infty , \frac{1}{r} \lim |s_n| \right\}$$

where $0 < r \le 1$ and $\|\cdot\|_\infty$ is the usual sup-norm. An application of the 3-ball property (Theorem I.2.2(iv)) reveals that c_0 is an M-ideal in X_r (note that $\|\cdot\|_r = \|\cdot\|_\infty$ on c_0). Also, an easy computation shows $g(c_0, X_r) = r$. (Some readers might find the following representation of X_r as a G-space helpful:

$$X_r \cong \{x \in C(\{0\} \cup \alpha\mathbb{N}) \mid r \cdot x(0) = x(\infty)\}$$

where the latter space carries the sup-norm.)

(d) Suppose there exists $y^{**} \in Y^{**}$ such that Y is an M-ideal in $X = \operatorname{lin}(Y \cup \{y^{**}\})$. Goldstine's theorem implies $g(Y, X) = 1$. Moreover, Y is an extreme M-ideal. The converse is also true: the canonical operator $I_{X,Y^*} : X \to Y^{**}$ is isometric if $g(Y, X) = 1$, see the proof of Proposition 3.9. We note as a particular case

$$g(Y) = 1 \text{ if } Y \text{ is an } M\text{-ideal in } Y^{**}.$$

We now link the grade of an M-ideal Y with the grade of certain pseudoballs in Y.

Proposition 3.9 *Let Y be an M-ideal in X, $\dim X/Y = 1$. Then*

$$g(Y, X) = g(P_Y(x)) \quad \forall x \in X \backslash Y.$$

PROOF: Since Y is an M-summand if and only if either of the two numbers equals 0 (Proposition 3.2 and Remark 3.8(b)) we may suppose that Y is a proper M-ideal, i.e. $g(Y, X) > 0$. A look at the definition of the characteristic reveals that the canonical operator (recall $Y^* \subset X^*$)

$$I_{X,Y^*} : X \to Y^{**},\ \langle I_{X,Y^*}(x), y^* \rangle = \langle x, y^* \rangle$$

is an (into-) isomorphism with $\|I_{X,Y^*}^{-1}\| = g(Y,X)^{-1}$. On the other hand, we note that $g(P_Y(x))$ is actually independent of x (because codim $(Y) = 1$) and equals (cf. Cor. 1.7, including the notation used there), in the case $d(x,Y) = 1$,

$$g(P_Y(x)) = d(Px, Y) = \sup\{|\langle x, y^*\rangle| \mid y^* \in B_{Y^*}\}.$$

These two observations together prove Proposition 3.9. □

Theorem 3.10 *A Banach space Y can be a proper M-ideal if and only if Y contains a pseudoball B which is not a ball. In the latter case, there exists a Banach space X containing Y such that $\dim(X/Y) = 1$, Y is a proper M-ideal in X, and $g(Y,X) = g(B)$.*

PROOF: The "only if" part is a consequence of Proposition 1.3, Corollary 3.3 and Proposition 3.9. We now turn to the "if" part. Let $B \subset Y$ be a pseudoball which is not a ball. We consider the vector space $X = Y \oplus \mathbb{K}$. We will find a norm $|\,.\,|$ on X such that

a) the map $y \mapsto (y,0)$ from Y into X is an isometry,

b) $P_{Y\oplus\{0\}}(x) = B \times \{0\}$ for some x,

c) $Y \cong Y \oplus \{0\}$ is an M-ideal in X.

The inequality $g(B) > 0$ then yields that $Y \oplus \{0\}$ is a proper M-ideal with grade $g(B)$ by Proposition 3.9.
In the sequel we assume w.l.o.g. that B has diameter 2 and that $0 \in B$. We put

$$\begin{aligned} K &:= \overline{\text{co}}\,\{(\theta y, \theta) \mid \theta \in \mathbb{K},\ |\theta| = 1,\ y \in B\} \\ &= \overline{\text{aco}}\, B \times \{1\} \end{aligned}$$

(the closures being taken with respect to the product topology) and let $|\,.\,|$ be the associated Minkowski functional. (To see that K is absorbing, use Proposition 1.4 to obtain

$$\text{int}\, B_X \times \{0\} \subset \frac{1}{2}(B \times \{1\} - B \times \{-1\}) \subset K.) \tag{$*$}$$

Since by $(*)$

$$|(y,r)| \le \|y\| + |r| \le 4 \cdot |(y,r)|,$$

$|\,.\,|$ is a norm which generates the product topology. Thus X is a Banach space.
ad a): It remains to prove (look at $(*)$)

$$K \cap (Y \oplus \{0\}) \subset B_Y \times \{0\}. \tag{$**$}$$

So let $(y,0) \in K$. We may write

$$\begin{aligned} (y,0) &= \lim(y_n, r_n) \\ r_n &= \sum_k \mu_{n,k}, \qquad \sum_k |\mu_{n,k}| \le 1 \\ y_n &= \sum_k \mu_{n,k} y_{n,k}, \qquad y_{n,k} \in B. \end{aligned}$$

If y_B^{**} denotes the centre of $\overline{B}^{w*}$ in Y^{**}, then (cf. Theorem 1.6)

$$\begin{aligned} \|y_n\| &\leq \left\| \sum_k \mu_{n,k}(y_{n,k} - y_B^{**}) \right\| + \left| \sum_k \mu_{n,k} \right| \|y_B^{**}\| \\ &\leq 1 + |r_n| \, \|y_B^{**}\|. \end{aligned}$$

Since $r_n \to 0$ and $\|\,.\,\|$ and $|\,.\,|$ are equivalent on Y we get

$$\|y\| = \lim \|y_n\| \leq 1,$$

i.e. $y \in B_Y$.
ad b): Consider $x = (0, -1)$. A moment's reflection reveals $d(x, Y \oplus \{0\}) = 1$ and $P_{Y \oplus \{0\}}(x) = \{(y, 0) \mid (y, 1) \in K\}$. (For the proof take into account that $|r| \leq |(y, r)|$ for all $(y, r) \in X$.) Therefore, $B \times \{0\} \subset P_{Y \oplus \{0\}}(x)$. For the converse inclusion consider $y \in Y$ such that $(y, 1) \in K$. We write

$$(y, 1) = \lim (y_n, r_n)$$

with

$$\begin{aligned} r_n &= \sum_k \lambda_{n,k} \theta_{n,k}, \qquad \lambda_{n,k} \geq 0, \ \sum_k \lambda_{n,k} = 1, \ |\theta_{n,k}| = 1 \\ y_n &= \sum_k \lambda_{n,k} \theta_{n,k} y_{n,k}, \qquad y_{n,k} \in B. \end{aligned}$$

We have

$$\hat{y}_n := y_n + \sum_k \lambda_{n,k}(1 - \theta_{n,k}) y_{n,k} \in B$$

and

$$\hat{y}_n \to y$$

thanks to the following lemma. This proves $y \in B$ as desired.
ad c): We remark that $P_{Y \oplus \{0\}}((y, r)) = -r(B - y) \times \{0\}$, hence it is a pseudoball in $Y \oplus \{0\}$ with radius $|r| = d((y, r), Y \oplus \{0\})$. Thus c) is a consequence of Proposition 1.3(b). □

So, the proof of Theorem 3.10 will be completed as soon as the following lemma is proved.

Lemma 3.11 *For* $m \in \mathbb{N}$, $\lambda_k \geq 0$, $\sum_{k=1}^m \lambda_k = 1$, *and* $\theta_k \in \mathbb{C}$ *with* $|\theta_k| = 1$ *we have*

$$\sum_{k=1}^m \lambda_k |1 - \theta_k| \leq \sqrt{2\sqrt{\varepsilon}} + 2\sqrt{\varepsilon}$$

provided that $|\sum \lambda_k (1 - \theta_k)| \leq \varepsilon$.

PROOF: We define

$$\begin{aligned} N_1 &:= \{k \in \{1,\dots,m\} \mid \operatorname{Re}\theta_k > 1-\sqrt{\varepsilon}\}, \\ N_2 &:= \{k \in \{1,\dots,m\} \mid \operatorname{Re}\theta_k \le 1-\sqrt{\varepsilon}\}, \\ \lambda &:= \sum_{k\in N_1} \lambda_k, \end{aligned}$$

and we claim that

$$1-\lambda \le \sqrt{\varepsilon}, \tag{1}$$

$$|1-\theta_k| \le \sqrt{2\sqrt{\varepsilon}} \quad \text{for every } k \in N_1. \tag{2}$$

ad (1): By assumption we have $|1-z| \le \varepsilon$ and therefore $\operatorname{Re} z \ge 1-\varepsilon$, where $z := \sum \lambda_k\theta_k$. Let $z_1 := \sum_{k\in N_1}(\lambda_k/\lambda)\theta_k$ and $z_2 := \sum_{k\in N_2}(\lambda_k/(1-\lambda))\theta_k$. The convexity of $\{w \in \mathbb{C} \mid |w| \le 1,\ \operatorname{Re} w \le 1-\sqrt{\varepsilon}\}$ implies that $\operatorname{Re} z_2 \le 1-\sqrt{\varepsilon}$. Hence

$$1-\varepsilon \le \operatorname{Re} z = \operatorname{Re}(\lambda z_1 + (1-\lambda)z_2) \le \lambda + (1-\lambda)(1-\sqrt{\varepsilon}),$$

and this gives $1-\lambda \le \sqrt{\varepsilon}$.
ad (2): For $k \in N_1$ we have $1-\operatorname{Re}\theta_k < \sqrt{\varepsilon}$ and thus

$$|1-\theta_k|^2 = 2(1-\operatorname{Re}\theta_k) < 2\sqrt{\varepsilon}.$$

The lemma follows by combining (1) and (2):

$$\begin{aligned} \sum_{k=1}^m \lambda_k|1-\theta_k| &= \sum_{k\in N_1}\lambda_k|1-\theta_k| + \sum_{k\in N_2}\lambda_k|1-\theta_k| \\ &\le \lambda\sqrt{2\sqrt{\varepsilon}} + (1-\lambda)\cdot 2 \\ &\le \sqrt{2\sqrt{\varepsilon}} + 2\sqrt{\varepsilon}. \end{aligned}$$

□

We next aim at giving an example of a Banach space which can be a proper M-ideal, yet not an extreme M-ideal. (A geometric characterisation of extreme M-ideals is contained in the Notes and Remarks section.)
First we present a result on finite dimensional subspaces of proper M-ideals which will eventually lead to a necessary condition for an M-ideal to be extreme.

Proposition 3.12 *Suppose Y is a proper M-ideal in X with $\dim X/Y = 1$ and $\alpha := g(Y,X) > 0$. Then, given $\varepsilon > 0$ and a finite dimensional subspace $E \subset Y$, one may find $y_0 \in Y$ satisfying*

$$\frac{\alpha}{1+\varepsilon}\max\{\|e\|,|\lambda|\} \le \|e+\lambda y_0\| \le (1+\varepsilon)\max\{\|e\|,|\lambda|\}$$

for all $e \in E$, $\lambda \in \mathbb{K}$.

PROOF: We recall that under our present assumptions the canonical operator

$$I_{X,Y^*}: X \to Y^{**}, \ \langle I_{X,Y^*}(x), y^* \rangle = \langle x, y^* \rangle$$

is an (into-) isomorphism whose inverse has norm α^{-1}. To get started, consider the decomposition

$$X^{**} = Y^{\perp\perp} \oplus_\infty \operatorname{lin}\{x_0^{**}\}$$

in which we assume $\|x_0^{**}\| = 1$. Let $\delta > 0$ and choose with the help of the principle of local reflexivity an injective operator

$$T: \ \operatorname{lin}(E \cup \{x_0^{**}\}) \to X$$

such that

$$\|T\| \cdot \|T^{-1}\| \leq 1 + \delta$$

and

$$Te = e \ \text{ for all } \ e \in E.$$

For $x_0 := Tx_0^{**}$ we obtain

$$\frac{1}{1+\delta} \max\{\|e\|, |\lambda|\} \ \leq \ \|e + \lambda x_0\| \ \leq \ (1+\delta) \max\{\|e\|, |\lambda|\}$$

for all $e \in E$, $\lambda \in \mathbb{K}$.
We have to "push" x_0 into Y. As a first step, we push it into Y^{**} by means of $y_0^{**} := I_{X,Y^*}(x_0)$. It follows easily

$$\frac{\alpha}{1+\delta} \max\{\|e\|, |\lambda|\} \leq \|e + \lambda y_0^{**}\| \leq (1+\delta) \max\{\|e\|, |\lambda|\}$$

for all $e \in E$, $\lambda \in \mathbb{K}$. As a second step, we again apply the principle of local reflexivity to obtain an injection

$$S: \ \operatorname{lin}(E \cup \{y_0^{**}\}) \to Y$$

with $\|S\| \cdot \|S^{-1}\| \leq 1 + \delta$ which extends the identity on E.
With an appropriate choice of δ and $y_0 := S(y_0^{**})$ we finally achieve the desired result. □

Proposition 3.12 claims that there are uniformly M-orthogonal directions in a proper M-ideal Y for every finite dimensional subspace with the uniformity constant depending only on the grade. If Y is even an extreme M-ideal (i.e. $\alpha = 1$), then an obvious induction process shows that Y contains arbitrarily good copies of $\ell^\infty(n)$ "everywhere", i.e. the induction process can be started with an arbitrary $y \in Y$. Moreover, these subspaces are even nested so that Y contains $(1+\varepsilon)$-copies of c_0, for every $\varepsilon > 0$. This result extends to all proper M-ideals as will be shown using a different approach in Theorem 4.7.

The following corollary supports the point of view that elements of (proper) M-ideals should – loosely speaking – vanish at infinity.

Corollary 3.13 *If Y is an extreme M-ideal, then*

$$0 \in \overline{\mathrm{ex}}^{w^*} B_{Y^*}.$$

PROOF: Let $y_1, \dots, y_n \in Y$, $\|y_i\| = 1$, and $\delta > 0$. We have to produce some $p \in \mathrm{ex}\, B_{Y^*}$ such that

$$|p(y_i)| \le \delta \quad \text{for all } i.$$

To this end consider $E := \mathrm{lin}\,\{y_1, \cdots, y_n\}$ and $\varepsilon = \delta/3 > 0$. Choose y_0 according to 3.12. In particular (since $\alpha = 1$ here) $\|y_0\| \ge 1/(1+\varepsilon)$ so that there exists $p \in \mathrm{ex}\, B_{Y^*}$ with $p(y_0) \in \mathbb{R}$ and $p(y_0) > 1/(1+\varepsilon) - \varepsilon$. Hence for $i = 1, \dots, n$ and suitable scalars θ_i with modulus 1

$$\begin{aligned} |p(y_i)| &= p(\theta_i y_i + y_0) - p(y_0) \\ &\le \|\theta_i y_i + y_0\| - p(y_0) \\ &\le 1 + \varepsilon - \left(\frac{1}{1+\varepsilon} - \varepsilon\right) \\ &\le \delta. \end{aligned}$$

□

The preceding corollary enables us to prove the existence of nonextreme M-ideals. (The example of Remark 3.8(c) does not give this.)

Example 3.14 *Consider the sup-normed spaces*

$$\begin{aligned} X &= \{(s_n)_{n \ge 0} \in c \mid s_0 = s_1 + 2s_\infty\} \\ Y &= \{(s_n) \in X \mid s_0 = 0\}. \end{aligned}$$

(Here, $s_\infty := \lim s_n$.) Then

(a) *Y can be a proper M-ideal - in fact Y is a proper M-ideal in X - but*
(b) *Y cannot be an extreme M-ideal in any superspace.*

PROOF: A straightforward application of the 3-ball property (Theorem I.2.2) shows that Y is an M-ideal in X. To show that it is proper, let $\delta_m : (s_n) \mapsto s_m$ be the evaluation functional on X. It is quickly seen that $2\delta_n + \delta_1$ attains its norm on Y so that $2\delta_n + \delta_1 \in Y^*$ ($\subset X^*$, cf. Remark I.1.13). But $(2\delta_n + \delta_1)$ tends to $\delta_0 \in Y^\perp$ with respect to $\sigma(X^*, X)$, hence Y^* is not weak* closed in X^*. This proves (a). To prove (b) we use Corollary 3.13. We have

$$\mathrm{ex}\, B_{Y^*} \subset \{\lambda \cdot \delta_n|_Y \mid n = 1, 2, \dots, \infty,\ |\lambda| = 1\}$$

by the Krein-Milman theorem (more precisely Milman's converse to it) since the latter set is norming. But it is also $\sigma(Y^*, Y)$-closed, therefore $0 \notin \overline{\mathrm{ex}}^{w^*} B_{Y^*}$. □

II.4 The intersection property IP

In Theorem 3.10 we gave an internal characterisation of Banach spaces which can (or cannot) appear as proper M-ideals. In this section we are going to discuss an easy to check intersection property whose presence in a Banach space X entails that X fails to be a proper M-ideal.

Definition 4.1 *A Banach space X has the intersection property (IP for short) if, given $\varepsilon > 0$, there are a finite family $\{x_i \mid i = 1, \dots, n\} \subset S_X$ and $\delta > 0$ such that*

$$\bigcap_{i=1}^{n} B(\pm x_i, 1 + \delta) \subset B(0, \varepsilon).$$

An equivalent formulation, which saves one existence quantifier, is of course

$$\forall \varepsilon > 0 \ \exists n \in \mathbb{N} \ \exists x_1, \dots, x_n \in \operatorname{int} B_X \ \ \forall x \in X \ (\forall_i \|x \pm x_i\| \le 1 \Rightarrow \|x\| \le \varepsilon).$$

It is easily verified that c_0 fails the IP while $C(K)$ has it. (For the latter, take $n = 1$ and $x_1 = \mathbf{1}$, independently of ε.) The $C(K)$-proof admits immediate generalisation. Recall that $x \in B_X$ is called a strong extreme point of B_X if

$$x_k \in X, \ \ \|x \pm x_k\| \to 1 \ \Rightarrow \ x_k \to 0.$$

Taking $n = 1$ and x_1 a strong extreme point in Definition 4.1 we conclude that Banach spaces X for which B_X contains a strong extreme point enjoy the IP. It is, however, worth remarking at this point that we shall eventually encounter a strictly convex space which fails the IP (see the remarks following Example 4.6).
Let us collect more examples of spaces with IP.

Proposition 4.2 *Each of the following properties is sufficient for a Banach space X to have the IP:*

(a) *B_X contains a strong extreme point,*
(b) *X is a (real or complex) unital Banach algebra (e.g. $X = L(E)$),*
(c) *there exists $x_0 \in X$ such that $|p(x_0)| = 1$ for all $p \in \operatorname{ex} B_{X^*}$,*
(d) *$\overline{\operatorname{ex}}^{w^*} B_{X^*} \subset S_{X^*}$ (e.g. X is a C_Σ-space),*
(e) *B_X is dentable (e.g. B_X contains a strongly exposed point),*
(f) *X has the Radon-Nikodým property (e.g. X is reflexive or a separable dual space),*
(g) *X contains a nontrivial L-summand (e.g. X is an L^1-space),*
(h) *X is the space $C_\Sigma(S^m)$.*

The definition of a C_Σ-space will be recalled in the next section (p. 83); for the notions employed in (e) and (f) we refer to the monograph [159].
PROOF: (a) was observed above.

(b) The complex case follows from Theorem 5, p. 38 in [84], where it is shown that the unit of a Banach algebra is a strong extreme point of the unit ball, and (a). The real case can be reduced to the complex case by a complexification procedure (p. 68ff. in [85]).

(c) is a consequence of (a), since such an x_0 is a strong extreme point.

(d) We first claim:

For each $\varepsilon > 0$ there is a finite family $\{x_i \mid i \leq n\} \in \operatorname{int} B_X$ such that

$$(*) \qquad \max_i |p(x_i)| > 1 - \varepsilon \qquad \text{for all } p \in \operatorname{ex} B_{X^*}.$$

If this were false, we could find a net (p_F) in $\operatorname{ex} B_{X^*}$, indexed by the finite subsets of $\operatorname{int} B_X$, such that

$$\max_{x \in F} |p_F(x)| \leq 1 - \varepsilon_0,$$

for a suitable $\varepsilon_0 > 0$. Then also

$$|x^*(x)| \leq 1 - \varepsilon_0 \qquad \text{for all } x \in \operatorname{int} B_X,$$

where x^* is any weak*-accumulation point of the net (p_F). Hence $x^* \in (\overline{\operatorname{ex}}^{w^*} B_{X^*}) \backslash S_{X^*}$, which contradicts (d). (By the way, it is not hard to show that conversely $(*)$ implies the condition in (d).) Now let $\varepsilon > 0$ and (x_i) as in $(*)$. We wish to show

$$\|x \pm x_i\| \leq 1 \text{ (all } i) \Rightarrow \|x\| \leq \sqrt{2\varepsilon},$$

which will yield the IP.

Let $p \in \operatorname{ex} B_{X^*}$. Then by assumption

$$|p(x) \pm p(x_i)| \leq 1 \qquad \text{(all } i).$$

But $|p(x_i)| > 1 - \varepsilon$ for a certain i, so that

$$|p(x)| \leq \sqrt{2\varepsilon}$$

and, consequently,

$$\|x\| = \sup_{p \in \operatorname{ex} B_{X^*}} |p(x)| \leq \sqrt{2\varepsilon}.$$

That C_Σ-spaces have the property in (d) is contained in [386, p. 72f.]; as a matter of fact, for C_Σ-spaces X, $\operatorname{ex} B_{X^*}$ is weak* closed.

(e) Suppose X fails the IP. We will show that there exists $\varepsilon > 0$ with

$$x \in \overline{\operatorname{co}}\,(B_X \backslash B(x, \varepsilon/2)) \text{ for all } x \in S_X,$$

i.e., the unit ball is not dentable.

By assumption there exists $\varepsilon > 0$ such that for all $x \in S_X$ and $n \in \mathbb{N}$ there is $z_n \in X$, $\|z_n\| > \varepsilon$, with $\|x \pm z_n\| \leq 1 + \frac{1}{n} =: r_n$.

Then

$$\frac{1}{2}\Big((x + z_n)r_n + (x - z_n)r_n\Big) \to x$$

and

$$\|(x \pm z_n)r_n - x\| > \varepsilon/2$$

for large n.

(f) follows from (e), by well-known results ([159, Chap. V]).

(g) Suppose $X = X_1 \oplus_1 X_2$ and consider any $x_i \in X_i$, $\|x_i\| = 1$. We claim

$$B(\pm x_1, 1+\varepsilon) \cap B(\pm x_2, 1+\varepsilon) \subset B(0, 2\varepsilon).$$

So let y be a member of the left hand side and decompose

$$y = y_1 + y_2 \in X_1 \oplus_1 X_2.$$

Then

$$\|y_1 \pm x_1\| + \|y_2\| \leq 1 + \varepsilon,$$

hence

$$\|x_1\| + \|y_2\| \leq 1 + \varepsilon,$$

so that

$$\|y_2\| \leq \varepsilon.$$

Analogously, $\|y_1\| \leq \varepsilon$ and so

$$\|y\| = \|y_1\| + \|y_2\| \leq 2\varepsilon.$$

(h) As usual, S^m is the Euclidean sphere in $\mathbb{R}^{m+1}$ and

$$C_\Sigma(S^m) = \{f \in C(S^m) \mid f(s) = -f(-s) \quad \forall\, s \in S^m\}.$$

This is a C_Σ-space and consequently enjoys the IP by (d). The upshot of this example is, however, that the number n appearing in Definition 4.1 must be larger than m for $X = C_\Sigma(S^m)$. So let us prove:

(*) For $f_1, \dots, f_m$ in the open unit ball of $C_\Sigma(S^m)$ there is $g \in C_\Sigma(S^m)$ with

$$\|g \pm f_i\| \leq 1 \text{ (all } i\text{), but } \|g\| > 1/3.$$

The decisive tool to prove this assertion is a corollary to the Borsuk-Ulam theorem [6, p. 485, Satz VIII] according to which any m functions in $C_\Sigma(S^m)$ have a common zero, say $f_1(s_0) = \ldots = f_m(s_0) = 0$. To construct g choose a neighbourhood U of s_0 such that

$$U \cap -U = \emptyset \text{ and } |f_i(s)| < 1/2 \text{ for } s \in U,\ i \leq m.$$

Let $h : S^m \to [0,1]$ be a continuous function vanishing outside U with $h(s_0) = 1$. Put $g(s) = \frac{1}{2}(h(s) - h(-s))$ and $V = -U \cup U$. Then $g \in C_\Sigma(S^m)$, $\|g\| = |g(s_0)| = 1/2$, and g vanishes off V. It follows easily $\|g \pm f_i\| \leq 1$. □

One cannot expect many permanence properties for the IP. For example, $c_0 \oplus_1 c_0$ has the IP by Proposition 4.2(g), but the one-complemented subspace c_0 fails it. Also, ℓ^2 has the IP while the injective tensor product $\ell^2 \widehat{\otimes}_\varepsilon \ell^2$ fails it. This follows from Theorem 4.4 below and Example I.1.4(d). We stress, however, that in connection with M-ideals positive results can be achieved.

Proposition 4.3 *Suppose X has the IP and J is an M-ideal in X. Then X/J has the IP. In particular, the IP is inherited by M-summands.*

PROOF: For every $\varepsilon > 0$, we find by assumption $x_1, \ldots, x_n \in \operatorname{int} B_X$ such that

$$\|z \pm x_i\| \leq 1 \quad (\text{all } i) \quad \Rightarrow \quad \|z\| \leq \varepsilon.$$

We claim that $q(x_1), \ldots, q(x_n)$ will do for X/J, where $q : X \to X/J$ is the quotient map. So let

$$\|q(x) \pm q(x_i)\| \leq 1 \qquad (\text{all } i).$$

Then all the balls $B(x \pm x_i, 1)$ meet J since J is proximinal (Proposition 1.1), and their intersection contains a small ball around x, since $\|x_i\| < 1$. Apply Theorem I.2.2(v) to obtain $y \in J$ such that

$$\|x \pm x_i - y\| \leq 1 \qquad (\text{all } i).$$

By assumption we have $\|x - y\| \leq \varepsilon$, i.e. $\|q(x)\| \leq \varepsilon$. □

The following theorem furnishes a large class of Banach spaces which fail the IP. Note that by Proposition 4.2 the failure of the IP is connected to some kind of "vanishing at infinity" (especially (c) and (d) suggest this point of view). So we think that Theorem 4.4 is quite a natural result.

Theorem 4.4 *A proper M-ideal X fails the IP.*

PROOF: We know from Theorem 3.10 (or Corollary 1.7 if you prefer) that X contains a pseudoball B with radius 1 which is not a ball so that

$$\alpha := d(x^{**}, X) > 0$$

where x^{**} is the centre of $\overline{B}^{w*}$ in X^{**}. Let $\varepsilon < \alpha$ and $x_1, \ldots, x_n \in \operatorname{int} B_X$. We wish to produce $z \in X$ with $\|z\| \geq \varepsilon$ and $\|z \pm x_i\| \leq 1$ for all i, thus showing that X fails the IP. First of all choose $\eta > 0$ such that

$$(1 + \eta) \max \|x_i\| < 1 \quad \text{and} \quad \varepsilon < \alpha/(1 + \eta)^2.$$

The defining property of pseudoballs provides us with an $x \in X$ such that $x \pm (1+\eta)x_i \in B$ and thus $\|x \pm (1 + \eta)x_i - x^{**}\| \leq 1 \quad (i = 1, \ldots, n)$. We now use the principle of local reflexivity to obtain an injection

$$T\colon\ E := \operatorname{lin}\{x^{**}, x, x_1, \ldots, x_n\} \to X$$

with $\|T\|, \|T^{-1}\| \leq 1 + \eta$ which is the identity on $E \cap X$. Hence

$$\|x - Tx^{**} \pm (1 + \eta)x_i\| \leq 1 + \eta$$

and

$$\|x - Tx^{**}\| \geq \|x - x^{**}\|/(1 + \eta) \geq \alpha/(1 + \eta).$$

Therefore $z = (x - Tx^{**})/(1 + \eta)$ has the required properties. □

The converse of Theorem 4.4 does not hold. To prepare a counterexample we prove:

Lemma 4.5 *Suppose $X_1, X_2, \dots$ are Banach spaces and put*

$$X = X_1 \oplus_\infty X_2 \oplus_\infty \cdots = \{(x_m) \mid x_m \in X_m, \|(x_m)\| := \sup_m \|x_m\| < \infty\}.$$

Then:

(a) *If no X_m can be a proper M-ideal, then neither can X.*

(b) *If X has the IP, then each X_m has the IP. Moreover, the number $n = n(\varepsilon)$ in the definition of the IP can be chosen to be the same for X and each X_m.*

PROOF: (a) Let P_m be the M-projection from X onto X_m (we consider X_m as naturally embedded in X). Suppose X is an M-ideal in some superspace Z. Then X_m is an M-ideal in Z (Proposition I.1.17(b)) so that by assumption on X_m there is an M-projection Q_m from Z onto X_m. Proposition I.1.2(a) shows $P_m = Q_m|_X$.
Define $Q : Z \to Z$ by $Qz = (Q_m z)_{m\in\mathbb{N}}$. We have shown that Q is a contractive projection onto the M-ideal X, hence X cannot be a proper M-ideal by Corollary I.1.3.
(b) can immediately be verified. □

Example 4.6 *There is a Banach space which fails the IP, yet cannot be a proper M-ideal.*

PROOF: Let $X_m = C_\Sigma(S^m)$ (cf. 4.2(h)) and consider the ℓ^∞-sum $X = X_1 \oplus_\infty X_2 \oplus_\infty \cdots$. Then X_m has the IP (Proposition 4.2(h)), thus cannot be a proper M-ideal (Theorem 4.4). Therefore X cannot be a proper M-ideal either (Lemma 4.5(a)).
On the other hand, to verify the IP for X_m we need at least $m+1$ vectors in Definition 4.1 as was pointed out in the proof of Proposition 4.2(h). Therefore X fails the IP (Lemma 4.5(b)). □

As a consequence of Theorem 4.4 and Proposition 4.2(a) one obtains that the extreme point structure of the unit ball of a proper M-ideal is quite weak; at any rate there are no strong extreme points, let alone strongly exposed points. However, it is shown in [304] that $B_X = \overline{\mathrm{co}}\, \mathrm{ex}\, B_X$ for $X = K(\ell^p)$, $1 < p < \infty$, $p \neq 2$. Since $K(\ell^p)$ is a proper M-ideal in $L(\ell^p)$ for these p (this will be proved in Example VI.4.1), none of these extreme points can be strongly extreme. Even better: The quotient space $C(\mathbb{T})/A$ ($\mathbb{T}$ = unit circle, A = disk algebra) is strictly convex and an M-ideal in its bidual (Remark IV.1.17). Though all the points with norm one are extreme, none of them is strongly extreme.
As another consequence of 4.4 and 4.2 we observe that a proper M-ideal must fail the RNP. This also follows from the following result which was already announced in the previous section, where a special case was treated (see the discussion following Proposition 3.12).

Theorem 4.7 *A Banach space X which fails the IP contains a subspace isomorphic to c_0. In particular, every proper M-ideal contains c_0.*

PROOF: A by now classical theorem due to Bessaga and Pełczynski [422, Prop. 2.e.4] asserts that X contains a copy of c_0 if and only if there is a weakly unconditionally

Cauchy series which does not converge. So, our aim is to produce a sequence (x_i) such that

$$\left\| \sum_{i=1}^{n} \varepsilon_i x_i \right\| < 1 \qquad \text{for all } n \in \mathbb{N} \text{ and all } \varepsilon_i = \pm 1,$$

yet

$$\inf \|x_i\| > 0.$$

Let $\alpha > 0$ be such that for all finite families $\{z_1, \ldots, z_n\} \subset \operatorname{int} B_X$ there is $z \in X$ with $\alpha < \|z\| < 1$ and $\|z \pm z_i\| < 1$ for all i. Such an α exists by assumption on X. (The inequality $\|z\| < 1$ can be obtained upon adding 0 to $\{z_1, \ldots, z_n\}$.) Now the desired sequence can easily be defined inductively: Start with an arbitrary x_1 satisfying $\alpha < \|x_1\| < 1$. Suppose $x_1, \ldots, x_n$ have been found such that

$$\left\| \sum_{i=1}^{n} \varepsilon_i x_i \right\| < 1 \text{ for all choices of signs } \varepsilon_i = \pm 1, \tag{1}$$

$$\|x_i\| > \alpha \text{ for } i = 1, \ldots, n. \tag{2}$$

Apply the above version of non-IP to the finite family $\{\sum_1^n \varepsilon_i x_i \mid \varepsilon_i = \pm 1\}$ to obtain x_{n+1} extending (1) and (2) to $n+1$.
This completes the proof of Theorem 4.7. □

James' distortion theorem (e.g. [422, Prop. 2.e.3]) permits us to infer that there are even $(1+\varepsilon)$-isomorphic copies of c_0, for every $\varepsilon > 0$. One might wonder if Theorem 4.7 can be strengthened so as to even yield isometric copies of c_0; after all, the assumptions of 4.7 are of isometric type as well. This, however, is impossible; the strictly convex space $C(\mathbb{T})/A$, which is an M-ideal in its bidual, serves as a counterexample (Remark IV.1.17). Another counterexample will be presented in Corollary III.2.12 where we will exhibit a smooth space X which is an M-ideal in X^{**}. Another enlightening counterexample concerning $(1+\varepsilon)$-isomorphisms will be presented in Proposition III.2.13.

In the last part of this section we will consider a question of isomorphic nature, namely what conditions imply that a Banach space can be renormed so as to become a proper M-ideal. It will turn out that in the isomorphic setting an example such as 4.6 does not exist. The key idea is contained in the following notion.

Definition 4.8 *A closed bounded convex subset S of a Banach space is called a* quasiball *if*

(a) $\operatorname{int}(S - S) \neq \emptyset$,
(b) $\bigcap_{i=1}^{n} (x_i + S) \neq \emptyset$ *for any finite family of vectors* $x_1, \ldots, x_n \in \frac{1}{2}(S - S)$.

A quasiball is called proper *if it is not symmetric.*

Proposition 1.4 shows that every pseudoball is a quasiball. But unlike the definition of a pseudoball, the notion of a quasiball is only dependent on the topology of X and not on the particular choice of an equivalent norm. On the other hand, a quasiball S will be a pseudoball for the equivalent norm $|\,.\,|$ on X whose unit ball is $\frac{1}{2}\overline{S - S}$. (By the way, the remark following Proposition 1.3 thus implies that it would be enough to require (b) of Definition 4.8 for $n = 3$.) Moreover, S is proper if and only if it is not a ball for $|\,.\,|$. (This follows for example from Theorem 1.6.)

Theorem 4.9 *For a Banach space X, the following assertions are equivalent:*

(i) *X can be renormed to become a proper M-ideal.*

(ii) *X can be renormed to fail the IP.*

(iii) *X contains an isomorphic copy of c_0.*

(iv) *X contains a proper quasiball.*

PROOF: (iv) $\Rightarrow$ (i): This follows from the above considerations together with Theorem 3.10.

(i) $\Rightarrow$ (ii) is Theorem 4.4.

(ii) $\Rightarrow$ (iii) is Theorem 4.7.

(iii) $\Rightarrow$ (iv): Let Y be subspace of X isomorphic to c_0. Then Y contains a proper quasiball S_0 since c_0 does. We let $S = \overline{S_0 + B_X}$. Obviously, S is closed, convex, bounded, and $\operatorname{int}(S - S) \neq \emptyset$. Moreover, S is not symmetric. In fact, an easy application of the Hahn-Banach theorem reveals

$$y + B_X \subset S \;\Rightarrow\; y \in S_0. \tag{$*$}$$

Thus, should S be symmetric, w.l.o.g. with centre 0, $(*)$ implies that S_0 is symmetric around 0, too.

It is left to establish (b) from Definition 4.8. We first observe

$$\begin{aligned} \operatorname{int}\frac{1}{2}(S-S) &\subset \operatorname{int}\overline{\frac{1}{2}(S_0+B_X)-(S_0+B_X)} \\ &= \operatorname{int}\overline{\left(\frac{1}{2}(S_0-S_0)+B_X\right)} \\ &= \operatorname{int}_Y\frac{1}{2}(S_0-S_0) + \operatorname{int} B_X. \end{aligned}$$

[Proof of this equality: Let us abbreviate $\frac{1}{2}(S_0 - S_0)$ by V. Then obviously

$$C := \operatorname{int}_Y V + \operatorname{int} B_X \subset \operatorname{int}\overline{V + B_X} =: C'$$

and also $C' \subset \overline{C}$ since

$$V + B_X = \overline{\operatorname{int}_Y V} + \overline{\operatorname{int} B_X} \subset \overline{C}.$$

Thus C and C' are open convex sets satisfying

$$C \subset C' \subset \overline{C}.$$

An application of the Hahn-Banach theorem now shows $C = C'$.]

Now let $x_i \in \operatorname{int}\frac{1}{2}(S-S)$ $(i = 1, \dots, n)$. Write

$$x_i = y_i + z_i \in \operatorname{int}_Y\frac{1}{2}(S_0 - S_0) + \operatorname{int} B_X.$$

Then, since $-z_i \in B_X$,

$$\begin{aligned}\bigcap_{i=1}^{n}(x_i + S) &\supset \bigcap_{i=1}^{n}(y_i + z_i) + (S_0 + B_X)\\ &\supset \bigcap_{i=1}^{n}(y_i + S_0)\\ &\neq \emptyset.\end{aligned}$$

This completes the proof of Theorem 4.9. □

II.5 L^1-preduals and M-ideals

This final section is devoted to the study of L^1-predual spaces (i.e. Banach spaces X whose duals are isometrically isomorphic to spaces of type $L^1(\mu)$) and certain of their subclasses by means of M-structure methods. We shall restrict ourselves to *real* Banach spaces, although several of the results presented here have direct analogues for complex spaces, owing to the refinements in complex integral representation theory made in the 70s.

Our first result connects the M-ideals in an L^1-predual space with certain faces in the unit ball of L^1-spaces. Let us call a subset H of B_Y, Y a Banach space, a *biface* if $H = \mathrm{co}\,(F \cup -F)$ for some face F in B_Y.

Proposition 5.1 *In an L^1-predual space X, there is a one-to-one correspondence between M-ideals in X and weak* closed bifaces in B_{X^*}: $J \subset X$ is an M-ideal if and only if J is the annihilator of a weak* closed biface in B_{X^*}.*

PROOF: Let H be a weak* closed biface in B_{X^*}. It is proved in [386, p. 220] that

(a) H equals the unit ball of its linear span $\mathrm{lin}\, H$.

(Note that (a) does not hold for bifaces in arbitrary Banach spaces, as shown by two-dimensional examples.) Using similar arguments we wish to show:

(b) If $p \in H$ and $\|p\| = \|p - q\| + \|q\|$, then $q \in H$.

To prove (b) we may assume that $H \neq B_{X^*}$ and that X^* is isometric to some L^1-space in such a way that F corresponds to a subface of

$$\{u \in L^1 \mid \|u\| = 1,\ u \geq 0\}$$

where $\geq$ is the usual order of L^1 (see [386, p. 171]).

CLAIM: If $0 \leq u \leq v$, $u \neq 0$ and $\dfrac{v}{\|v\|} \in F$, then $\dfrac{u}{\|u\|} \in F$.

In fact, writing $v = u + w$ with $w \geq 0$ we obtain from the additivity of the L^1-norm on positive elements

$$\|v\| = \|u\| + \|w\|$$

and

$$\frac{v}{\|v\|} = \frac{\|u\|}{\|v\|}\frac{u}{\|u\|} + \frac{\|w\|}{\|v\|}\frac{w}{\|w\|} \in F,$$

hence the claim.

Now the norm equality in (b) is equivalent to

$$|p| = |p-q| + |q| \qquad \text{a.e.}$$

which in turn means

$$p^+ \geq q^+, \qquad p^- \geq q^-.$$

(p^+ is the positive part of the L^1-function p.) By assumption on p we may write

$$p = \lambda u_1 - (1-\lambda)u_2$$

with some $0 \leq \lambda \leq 1$, $u_i \in F$. Hence

$$\lambda u_1 \geq p^+ \geq q^+,$$

and the claim applies to show that $q^+ = 0$ or else $\frac{q^+}{\|q^+\|} \in F$. An analogous statement applies to q^- so that

$$\begin{aligned} q &= \|q^+\|\frac{q^+}{\|q^+\|} - \|q^-\|\frac{q^-}{\|q^-\|} + (1-\|q\|)\cdot 0 \\ &\in \operatorname{co}(F \cup -F \cup \{0\}) \\ &= \operatorname{co}(F \cup -F) \\ &= H. \end{aligned}$$

(a) and (b) together yield:

(c) $\operatorname{lin} H$ is a weak* closed order ideal in L^1 and hence an L-summand.

Weak* closedness follows from (a) and the Krein-Smulian theorem. If $|q| \leq |p|$ and $p \in H$, then by (b) $p^+ \in H$ and $p^- \in H$, hence $|p| \in H$. By the same argument q^+ and q^- are in H, hence $q \in \operatorname{lin} H$. For the equivalence of order ideals and L-summands cf. Example I.1.6(a). Thus, we have shown that

$$J := \{x \in X \mid p(x) = 0 \text{ for all } p \in H\}$$

is an M-ideal in X.

Conversely let J be an M-ideal in X. Representing X^* as $L^1(\mu)$ we see that there is a measurable set A such that (Example I.1.6(a) again)

$$J^\perp = \{p \in L^1 \mid p|_A = 0\}.$$

Now

$$F = \left\{p \in L^1 \;\middle|\; \|p\| \leq 1,\ \int_{\complement A} p\, d\mu = 1\right\}$$

defines a face in B_{L^1} such that

$$J^\perp = \text{lin}\,(\text{co}\,(F \cup -F)) \;\; (= \text{lin}\, F). \qquad \square$$

We wish to consider the following subclasses of L^1-predual spaces.

- X is a G-space if there are a compact Hausdorff space K and $(s_i, t_i, \lambda_i) \in K \times K \times \mathbb{R}$, i in some index set I, such that X is isometric to

$$\{x \in C(K) \mid x(s_i) = \lambda_i \cdot x(t_i) \text{ for all } i \in I\}.$$

- X is a C_σ-space if there are a compact Hausdorff space K and an involutory homeomorphism $\sigma : K \to K$ (i.e. $\sigma^2 = id_K$) such that X is isometric to

$$\{x \in C(K) \mid x(t) = -x(\sigma(t)) \text{ for all } t \in K\}.$$

- X is a C_Σ-space if it is a C_σ-space for some fixed point free involution σ on some K.

Obviously, a C_σ-space is a G-space. Lindenstrauss [414, p. 79ff.] showed that G-spaces are L^1-preduals, but that the converse does not hold, thus disproving a conjecture of Grothendieck who had introduced the class of G-spaces – now termed after his initial – in [282]. (Previously, Stone [584, p. 465ff.] and Kakutani [364, p. 1005] had shown that the closed subalgebras (resp. sublattices) of a real $C(K)$-space are exactly those spaces isometric to G-spaces with $\lambda_i \in \{0, 1\}$ (resp. $\lambda_i \geq 0$) throughout.) A detailed analysis of these spaces can be found in [390] and [424], see also Lacey's monograph [386].

Below we shall present characterisations of G- and C_σ-spaces in terms of their M-structure properties. This will also lead to a (in our context) natural argument to show that the class of G-spaces is strictly smaller than the class of L^1-preduals.
We first show:

Proposition 5.2 *Let*

$$X = \{x \in C(K) \mid x(s_i) = \lambda_i x(t_i) \text{ for all } i \in I\}$$

be a G-space. Then $J \subset X$ is an M-ideal if and only if there is a (not necessarily uniquely determined) closed set $D \subset K$ such that

$$J = \{x \in X \mid x|_D = 0\}.$$

PROOF: The "only if" part is a consequence of Proposition I.1.18. To prove the "if" part we verify the 3-ball property (Theorem I.2.2(iv)). Given $x \in B_X$, $y_1, y_2, y_3 \in B_J$ we consider the function

$$y = \max\{y_1, y_2, y_3, -x\} + \min\{y_1, y_2, y_3, -x\} + x$$

(defined pointwise) which of course vanishes on D. Also, $y \in X$ by [414, p. 79], and the desired inequality $\|x + y_i - y\| \leq 1 + \varepsilon$ is elementary to verify pointwise (even with $\varepsilon = 0$), since for real numbers the implication

$$|a_i| \leq 1 \quad (\text{all } i) \quad \Longleftrightarrow \quad \left| a_i - (\max_j a_j + \min_j a_j) \right| \leq 1 \quad (\text{all } i)$$

holds. □

Corollary 5.3 *In a G-space the intersection of an arbitrary family of M-ideals is an M-ideal.*

There is a partial converse to this result.

Theorem 5.4 *Let X be a separable L^1-predual with the property that the intersection of an arbitrary family of M-ideals is an M-ideal. Then X is a G-space.*

PROOF: The proof relies on the following fact (cf. e.g. [215] or [601]). An L^1-predual space X is a G-space if and only if

$$\overline{\text{ex}}^{w^*} B_{X^*} \subset [0,1] \cdot \text{ex}\, B_{X^*}.$$

To provide a proof of Theorem 5.4 we consider $x_0^* \in \overline{\text{ex}}^{w^*} B_{X^*}$. To show that $x_0^* \in [0,1] \cdot \text{ex}\, B_{X^*}$ it will be enough to show that $\text{lin}\,\{x_0^*\}$ is an L-summand, equivalently that $\ker x_0^*$ is an M-ideal (recall Lemma I.1.5). Since X is separable, B_{X^*} is weak* metrizable so that there is a sequence of distinct points (x_n^*) in $\text{ex}\, B_{X^*}$ with $x_0^* = w^*\text{-}\lim x_n^*$. (We are assuming $x_0^* \notin \text{ex}\, B_{X^*}$.) Now fix $m \in \mathbb{N}$ and consider the subspace

$$U_m = \overline{\text{lin}}^{\|\,\|}\{x_0^*, x_{m+1}^*, x_{m+2}^*, \ldots\} \subset X^*.$$

We wish to show that U_m is a weak* closed L-summand.
Define a linear operator $T : X \to c$ by $Tx = (\langle x_{n+m}^*, x \rangle)_{n \in \mathbb{N}}$. Let $e_k^* : (s_n) \mapsto s_k$ be the kth evaluation functional on c and $e_0^* : (s_n) \mapsto \lim s_n$ be the limit functional. We then have

$$T^*(e_k^*) = x_{m+k}^* \qquad \text{and} \qquad T^*(e_0^*) = x_0^*.$$

The x_{m+k}^*, being linearly independent extreme points in some L^1-space, generate a subspace isometric to ℓ^1 as do the e_k^*. From this we conclude that $U_m = \text{ran}(T^*)$ is norm closed, and thus, by a classical result, it is weak* closed. Hence

$$U_m \;=\; \overline{\text{lin}}^{w^*}\{x_{m+1}^*, x_{m+2}^*, \ldots\} \;=\; \left(\bigcap_{k \in \mathbb{N}} \ker x_{m+k}^* \right)^{\perp}.$$

But $\ker x_{m+k}^*$ is an M-ideal, and the assumption of Theorem 5.4 now yields that so is $\bigcap_{k \in \mathbb{N}} \ker x_{m+k}^*$. Therefore, U_m is in fact a weak* closed L-summand.
It remains to observe that $\text{lin}\,\{x_0^*\} = \bigcap_{m \in \mathbb{N}} U_m$ and to apply Proposition I.1.11(a) to obtain that $\ker x_0^*$ is an M-ideal which happens only if x_0^* is a multiple of an extreme functional (Lemma I.1.5). □

We remark that an L^1-predual always has infinitely many M-ideals (cf. Proposition 5.1) so that nontrivial intersections of M-ideals exist. It is an open problem as of this writing if

one can dispense with the separability assumption in Theorem 5.4. However, it would be enough to require that $\operatorname{ex} B_{X^*}$ is sequentially dense in its weak* closure. As the example of the disk algebra shows (Example I.1.4(b)), one cannot weaken the assumption "X is an L^1-predual" to "$\ker p$ is an M-ideal for all $p \in \operatorname{ex} B_{X^*}$".
We next present an example of a separable L^1-predual which does not fulfill the conclusion of Corollary 5.3, hence cannot be a G-space by Theorem 5.4. (We admit that there are other ways of proving the existence of such spaces; the first example of this kind was constructed by Lindenstrauss [414, p. 78 and 81].)

Example 5.5 *Fix $t_0 \in [0,1]$ and let λ denote Lebesgue measure on $[0,1]$. Put*

$$X = \left\{ x \in C[0,1] \;\middle|\; x(t_0) = \int x \, d\lambda \right\}.$$

Then X is an L^1-predual, and for $t \neq t_0$

$$J_t := \{x \in X \mid x(t) = 0\}$$

is an M-ideal. However, if the sequence (t_n) converges to t_0 and $t_n \neq t_0$ for all n, the intersection $\bigcap_n J_{t_n}$ fails to be an M-ideal. Consequently, X is not isometric to a G-space.

PROOF: That X is an L^1-predual is a special case of results given in [46], but we wish to sketch a direct argument. Let μ denote the signed measure $\lambda - \delta_{t_0}$. We have to prove that $X^* = M[0,1]/\operatorname{lin}\{\mu\}$ is an L^1-space. Note

$$X^* = L^1(|\mu|)/\operatorname{lin}\{1\} \oplus_1 M_{\text{sing}}(|\mu|) \tag{$*$}$$

where $M_{\text{sing}}(|\mu|)$ is the space of $|\mu|$-singular measures which is an L^1-space. Thus it is left to prove that $L^1(|\mu|)/\operatorname{lin}\{1\}$ is an L^1-space, or equivalently [282] that its dual $U := \{f \in L^\infty(|\mu|) \mid \int f d\mu = 0\}$ is an L^∞-space. This will follow if there is a contractive projection P from $L^\infty(|\mu|)$ onto U. (Cf. e.g. [386, §§ 11 and 21] for these matters.) Now it is not hard to show that

$$Pf = f + \int f d\mu \cdot \chi_{\{t_0\}}$$

is such a projection.
It is seen from ($*$) that $\delta_t|_X \in \operatorname{ex} B_{X^*}$ for $t \neq t_0$, hence its kernel J_t is an M-ideal by Proposition 5.1. To refute that $J = \bigcap_n J_{t_n}$ is an M-ideal we refute the 3-ball (in fact 2-ball) property (Theorem I.2.2). We let $x = 1$, $\varepsilon > 0$ small enough ($\varepsilon < \frac{1}{10}$ will do) and will construct $y_1 \in B_J$ in such a way that *no* $y \in J$ satisfies

$$\|x \pm y_1 - y\| \leq 1 + \varepsilon. \tag{$**$}$$

The construction of such a y_1 is quite easy to understand but a bit cumbersome to describe in detail. Here is the idea: Choose an open neighbourhood U of t_0 of Lebesgue measure $\leq \varepsilon$. Also, choose open neighbourhoods U_n of those t_n which are not in U, whose total length does not exceed ε, either. The complement of these neighbourhoods consists of finitely many closed intervals. The function y_1 is defined in such a way that $|y_1(t)| = 1$ on a large part of these intervals (total length $\geq 1 - 3\varepsilon$ say), $\|y_1\| \leq 1$, $\int y_1 d\lambda = 0$, and

$y_1(t_n) = 0$ for all n so that $y_1 \in J$. Let y be any continuous function satisfying (**). Then

$$\begin{aligned} |y(t)-1| &\leq \varepsilon \qquad \text{on a set of measure } \geq 1-3\varepsilon, \\ |y(t)| &\leq 2+\varepsilon \qquad \text{otherwise.} \end{aligned}$$

Consequently

$$\int y\,d\lambda \geq 1-10\varepsilon > 0$$

for sufficiently small ε. But if $y \in J$ we have

$$\int y\,d\lambda = y(t_0) = 0$$

so that indeed no $y \in J$ can satisfy (**). □

The following result characterises G-spaces among all Banach spaces by a richness condition concerning the M-ideal structure. This condition will be given in terms of the structure topology on ex B_{X^*}, which we defined in I.3.11. To appreciate it the reader is referred to the remarks following I.3.11. There E_X was defined to be the quotient space ex $B_{X^*}/\sim$ with $p \sim q$ if and only if $p = q$ or $p = -q$.

Theorem 5.6 *For a Banach space X, the following conditions are equivalent:*
(i) *The structure topology on E_X is Hausdorff.*
(ii) *X is a G-space.*

We found it convenient to single out part of the argument for the implication (i) ⇒ (ii) in the following lemma.

Lemma 5.7 *If the structure topology on E_X is Hausdorff, then X is an L^1-predual.*

PROOF: By a result of Lima's [400, Theorem 5.8] it is enough to prove
(a) If $p \in X^*$, $\|p\| = 1$ and $P(p) \in \{0, p\}$ for all L-projections P on X^*, then $p \in \text{ex}\, B_{X^*}$.
(b) lin $\{p\}$ is an L-summand in X^* for all $p \in \text{ex}\, B_{X^*}$.

Note that (b) is equivalent with saying that singletons are closed in E_X, i.e. E_X is a T_1-space in its structure topology. Thus we are left with showing (a).

The Hausdorff property easily implies:

(*) If $F \subset E_X$ is a structurally closed set containing at least two points, then there are structurally closed sets F_1, F_2 different from F such that $F = F_1 \cup F_2$.

With the help of the Krein-Milman and Krein-Smulian theorems, (*) translates into the language of L-summands as follows:

(**) If $N \subset X^*$ is a weak* closed L-summand with $\dim(N) \geq 2$, then there are weak* closed L-summands N_1, N_2 different from N such that $N = N_1 + N_2$.

Now let p be a point as in (a) and let N be the intersection of all weak* closed L-summands containing p. Then N itself is a weak* closed L-summand (cf. I.1.11(a)). If

$p \notin \mathrm{ex}\, B_{X^*}$ then $\dim(N) \geq 2$. Assuming this to be the case we could split N according to (**). Either $p \in N_1$ or $p \in N_2$ contradicts the choice of N, thus, if P_i is the L-projection onto N_i, $P_1(p) = P_2(p) = 0$ by the very choice of p. Observe that $P = P_1 + P_2 - P_1P_2$ is the L-projection onto N so that $P(p) = 0$, contradicting $p \in N$. This completes the proof of Lemma 5.7. $\square$

We remark that Lemma 5.7 remains true if "E_X is Hausdorff" is replaced by "E_X fulfills (*)"; examine the above proof. (Property (*) is quickly seen to imply the T_1-separation axiom.) The property (*), which might be called "total reducibility" in accordance with [88, Chap. II.4], is not necessary for a Banach space X to be an L^1-predual, since the space of Example 5.5 fails it. [One can show, using the 3-ball property, that exactly the subspaces $\{x \in X \mid x|_D = 0\}$, where $D \subset [0,1]$ is a nonvoid closed set not containing t_0, are the nontrivial M-ideals in X. Consequently, E_X can be identified as a set with $[0,1]\backslash\{t_0\}$, the structurally closed sets corresponding to those D above and the trivial sets $\emptyset$ and $[0,1]\backslash\{t_0\}$. Thus $F = [0,1]\backslash\{t_0\}$ is a counterexample to (*).]

Proof of Theorem 5.6:

(i) $\Rightarrow$ (ii): We already know from Lemma 5.7 that X must be an L^1-predual. By the result quoted in the proof of 5.4, it remains to show that the weak* closure of $\mathrm{ex}\, B_{X^*}$ is contained in $[0,1] \cdot \mathrm{ex}\, B_{X^*}$. So, let (p_i) be a net in $\mathrm{ex}\, B_{X^*}$ converging in the weak* sense to some $p \in B_{X^*}$. We may assume $p \neq 0$. Let N be the smallest weak* closed L-summand in X^* containing p. We now claim that (p_i) converges structurally to all $q \in \mathrm{ex}\, B_N$: Should this be false there would exist a structurally closed set $F \subset \mathrm{ex}\, B_{X^*}$, some $q \in (\mathrm{ex}\, B_N)\backslash F$ and a subnet $(p_j) \subset F$ of (p_i). Now F has the form $\mathrm{ex}\, B_L$ for some weak* closed L-summand, and $p = w^*\text{-}\lim p_j$ implies $p \in L$. Hence $N \subset L$ and therefore $q \in L$ which contradicts $q \notin F$. (We have relied on Lemma I.1.5 several times.) By the claim and since E_X is structurally Hausdorff we conclude that N must be one-dimensional, i.e. $p/\|p\| \in \mathrm{ex}\, B_N \subset \mathrm{ex}\, B_{X^*}$.

(ii) $\Rightarrow$ (i): Let $X \subset C(K)$ be a G-space, and let $[p] \neq [q]$ be two equivalence classes in E_X. We wish to find two M-ideals J_1, J_2 in X such that

$$p \notin J_2^\perp \quad \text{and} \quad q \notin J_1^\perp.$$

Observe that $p \in \mathrm{ex}\, B_{X^*}$ has the form

$$p(x) = \lambda x(s_1)$$

where $\lambda = \pm 1$, and we may assume $\lambda = 1$ since $[p] = [-p]$. In the same vein,

$$q(x) = x(s_2)$$

for some $s_2 \in K$. Since p and q are linearly independent, we may find $x_1, x_2 \in X$ such that

$$\begin{aligned} p(x_1) &= q(x_1) &= 1 \\ p(x_2) &= -q(x_2) &= 1. \end{aligned}$$

For the closed sets

$$D_1 = \{s \in K \mid x_1(s)x_2(s) \geq 0\}$$
$$D_2 = \{s \in K \mid x_1(s)x_2(s) \leq 0\}$$

we consider the M-ideals (Proposition 5.2)

$$J_i = \{x \in X \mid x_{|D_i} = 0\}.$$

Let $x_\pm = x_1 \pm x_2 - (\max\{x_1, \pm x_2, 0\} + \min\{x_1, \pm x_2, 0\})$. Then $x_\pm$ is the pointwise median of $x_1, \pm x_2$ and 0, hence $x_\pm \in X$ by [414, p. 79]. It is left to observe

$$p(x_+) = x_+(s_1) \neq 0, \text{ yet } x_{+|D_2} = 0$$

and

$$q(x_-) = x_-(s_2) \neq 0, \text{ yet } x_{-|D_1} = 0$$

so that $p \notin J_2^\perp$, $q \notin J_1^\perp$, as requested. □

Next we shall discuss the centralizer of an L^1-predual. First we give examples of L^1-preduals having trivial centralizers. (The centralizer of a Banach space was introduced in I.3.7.)

Let X be the L^1-predual of Example 5.5. It is quickly checked that

$$\text{ex}\, B_{X^*} = \{\pm\delta_{t|X} \mid t \neq t_0\},$$

hence $T \in Z(X)$ (recall that we are considering real spaces here) if and only if there exists $h \in C[0,1]$ such that

$$(Tx)(t) = h(t)x(t) \qquad \forall t \in [0,1].$$

However, the only continuous functions with

$$hx \in X \qquad \text{for all } x \in X$$

are the constants, which is not hard to see. Consequently,

$$Z(X) = \mathbb{R} \cdot Id.$$

We pointed out after I.3.12 that X has a nontrivial M-ideal provided $Z(X)$ is nontrivial. The preceding example shows that there are separable spaces for which the converse is not true.

For the following example, which presents a nonseparable G-space yielding the same effect, one needs more elaborate arguments. First we give a general representation of the centralizer of a G-space. Let

$$X = \{x \in C(K) \mid x(s_i) = \lambda_i \cdot x(t_i) \quad \forall i \in I\}$$

be a G-space. It is proved in [215, Lemma 4] that

$$\operatorname{ex} B_{X^*} = \{\pm\delta_s|_X \mid \|\delta_s|_X\| = 1\}.$$

[Let us give a quick argument for this fact. The inclusion "$\subset$" is proved by the usual extension argument. For "$\supset$" we consider the M-ideal $\{x \in X \mid x(s) = 0\}$ (Proposition 5.2), obtain the decomposition $X^* = J^* \oplus_1 \operatorname{lin}\{\delta_s|_X\}$ and use Lemma I.1.5.]
Continuing our discussion, we may assume that $\{s \in K \mid \|\delta_s|_X\| = 1\}$ is dense in K. In fact, this is just a matter of representation, since X may be represented in an obvious way as a G-subspace of $C(\overline{\operatorname{ex}}^{w^*} B_{X^*})$. Using the same method as above, one may associate to each $T \in Z(X)$ a bounded continuous function h on $K\backslash L$, where L is the set of common zeros of all the $x \in X$, with

$$(Tx)(t) = h(t)x(t).$$

In addition, h must be constant on each equivalence class pertaining to the equivalence relation

$$s \sim t \iff \exists\lambda \neq 0 \quad \forall x \in X \quad x(s) = \lambda x(t)$$

on $K\backslash L$. Conversely such a function h defines an element in $Z(X)$ if multiplication by h leaves X invariant. (Cf. Example I.3.4(g) for a general view of this procedure.)

Proposition 5.8 *There exists a nonseparable G-space with a trivial centralizer.*

We remark that as a consequence of results by N. Roy [534], every separable G-space necessarily has a nontrivial centralizer, see the Notes and Remarks section.

PROOF: Let Y be any nonseparable Banach space, and let K denote its dual unit ball with the weak* topology. Define

$$X = \{x \in C(K) \mid x(\lambda y^*) = \lambda x(y^*) \quad \forall\lambda \in [-1,1],\ y^* \in K\}.$$

Obviously, X is a G-space, and Y can be identified with a subspace of X. In view of our above remark, we have to prove the following:

If $h \in C^b(K\backslash\{0\})$ is constant on each punctured ray $\{\lambda y^* \mid 0 < |\lambda| \leq 1\}$ (where $\|y^*\| = 1$), then h is constant.

Suppose not. First of all, let us extend h to $Y^*\backslash\{0\}$ so as to be constant on each punctured one-dimensional subspace. This extension – still to be called h – is continuous with respect to the bounded weak* topology, i.e. weak* continuous on bounded sets. Recall [179, p. 427] that a neighbourhood base of 0 for the bounded weak* topology is given by absolute polars of sequences in Y tending strongly to 0. Now if h were not constant there would be y_1^*, $y_2^* \neq 0$ and scalars $a_1, a_2 \in \mathbb{R}$ such that

$$h(y_1^*) < a_1 < a_2 < h(y_2^*).$$

By continuity $h|_{y_1^*+U} < a_1$, where

$$U = \{y^* \mid |\langle y^*, y_n\rangle| \leq 1 \quad \forall n \in \mathbb{N}\}$$

with some $y_n \to 0$. In particular,

$$h(y_1^* + y^*) < a_1$$

for all $y^* \in (\text{lin}\,\{y_n | n \in \mathbb{N}\})^\perp =: V_1^\perp$. Since V_1 is separable and Y is not, $V_1^\perp \neq \{0\}$. Hence, for all $y^* \in V_1^\perp \backslash \{0\}$, we obtain

$$h\Big(\frac{1}{n}y_1^* + y^*\Big) = h(y_1^* + ny^*) < a_1$$

so that

$$h(y^*) = \lim_{n\to\infty} h\Big(\frac{1}{n}y_1^* + y^*\Big) \leq a_1$$

by continuity. Analogously, for a certain separable subspace V_2 and all $y^* \in V_2^\perp \backslash \{0\}$

$$h(y^*) \geq a_2.$$

Since the separable subspace $V = \text{lin}\,(V_1 \cup V_2)$ is different from Y, there exists $y^* \in V^\perp \backslash \{0\} = (V_1^\perp \cap V_2^\perp) \backslash \{0\}$, and for any such y^* we have

$$a_1 \geq h(y^*) \geq a_2,$$

contradicting $a_1 < a_2$. Thus, Proposition 5.8 is proven. □

It is interesting to compare Proposition 5.8 with Theorem 5.6 in view of the Dauns-Hofmann type Theorem I.3.12. Namely, the structure space E_X of the G-space X of 5.8 is a Hausdorff topological space, yet every real-valued continuous function on E_X is constant. Next we are going to characterise those Banach spaces which yield "sufficiently many" structurally continuous functions on E_X, equivalently whose centralizer is "sufficiently large". Recall the notation

$$Z_X = \overline{\text{ex}}^{w^*} B_{X^*} \backslash \{0\}$$

from Section I.3. Further, we identify $Z(X)$ with an algebra of functions on Z_X via $T \leftrightarrow a_T$. In the next theorem p and q are said to be nonantipodal if neither $p = q$ nor $p = -q$. (Recall that we assume the scalars to be real in this section.)

Theorem 5.9 *For a Banach space X, the following assertions are equivalent:*
(i) *$Z(X)$ separates nonantipodal points of Z_X.*
(ii) *X is a C_σ-space.*

PROOF: (i) ⇒ (ii): This is a corollary to Theorem I.3.10.
By assumption (i), the set of equivalence classes $\mathcal{F}(X, C_0(Z_X))$ (cf. the notation of Section I.3) consists exactly of the pairs $\{p, -p\}$, $p \in Z_X$ (note $a_T(p) = a_T(-p)$ for $T \in Z(X)$). Now Theorem I.3.10 states

$$\begin{aligned} X &= \{f \in C_0(Z_X) \mid f|_{\{p,-p\}} \in X|_{\{p,-p\}} \quad \forall p \in Z_X\} \\ &= \{f \in C_0(Z_X) \mid f(p) = -f(-p) \quad \forall p \in Z_X\} \end{aligned}$$

so that X is a C_σ-subspace of $C(\overline{\mathrm{ex}}^{w^*} B_{X^*})$.
(ii) $\Rightarrow$ (i): By [386, p. 73] X is isometric to

$$\{f \in C_0(Z_X) \mid f(p) = -f(-p) \quad \forall p \in Z_X\}.$$

Also, $Z(X)$ acts on X by multiplication with functions in

$$\{h \in C^b(Z_X) \mid h(p) = h(-p) \quad \forall p \in Z_X\},$$

cf. the discussion preceding Proposition 5.8. To achieve the desired conclusion, choose, given nonantipodal $p_1, p_2 \in Z_X$, any $g \in C^b(Z_X)$ with

$$\begin{aligned} g(p_1) &= g(-p_1) = 0, \\ g(p_2) &= g(-p_2) = 1, \end{aligned}$$

and let $h(p) = g(p) + g(-p)$ for $p \in Z_X$. Then h "is" in $Z(X)$ and $h(p_1) \neq h(p_2)$. □

In the final part of this section, which does not depend on the results presented so far, we propose the use of the intersection property IP introduced in Definition 4.1 in order to distinguish between C_σ- and C_Σ-spaces.
It is a well-known fact that a Banach space is a C_σ-space if and only if it is isometric to a norm one complemented subspace of a $C(K)$-space [386, p. 74]. It follows that the class of C_σ-spaces is stable with respect to forming ℓ^∞-sums, since the class of $C(K)$-spaces is. (This could also be deduced directly from the definition.) For the same reason, the class of C_σ-spaces is stable with respect to forming ultraproducts, cf. [300]. For the case of C_Σ-spaces, things look different.
Recall the definition of an ultraproduct of a family of Banach spaces X_i ($i \in I$): Let $\mathcal{U}$ be a free ultrafilter on I, and let X be the $\ell^\infty(I)$-sum of the X_i. Then the quotient space X/N where

$$N = \{(x_i) \in X \mid \lim_{\mathcal{U}} \|x_i\| = 0\}$$

is called an ultraproduct of the X_i. Note that the norm in X/N is given by

$$\|(x_i) + N\| = \lim_{\mathcal{U}} \|x_i\|.$$

The reader is referred to [299] or [570] for a detailed account.

Proposition 5.10 *The class of C_Σ-spaces is not stable with respect to forming ℓ^∞-sums or ultraproducts.*

PROOF: Let X be the Banach space of Example 4.6. By its very construction it is an ℓ^∞-sum of C_Σ-spaces. But X fails the IP so that X cannot be a C_Σ-space by Proposition 4.2(d).
We now turn to the assertion concerning ultraproducts. For that matter, fix a free ultrafilter $\mathcal{U}$ on $\mathbb{N}$. Then, with the above notation, X/N is an ultraproduct of the spaces $X_n = C_\Sigma(S^n)$ ($n \in \mathbb{N}$). We claim that X/N fails the IP and hence cannot be a C_Σ-space (Proposition 4.2(d) again).

Fix $\varepsilon_0 < \frac{1}{3}$. We wish to find, given equivalence classes $x^1 + N, \dots, x^m + N \in \operatorname{int} B_{X/N}$, a class $y + N$ such that

$$\|(y \pm x^k) + N\| \le 1 \qquad (k = 1, \dots, m),$$

yet

$$\|y + N\| > \varepsilon_0,$$

thus refuting the defining condition for the IP.
In fact, we may assume that the representatives $x^k = (x_n^k)_{n\in\mathbb{N}}$ fulfill

$$\|x^k\| < 1 \qquad (k = 1, \dots, m),$$

i.e.

$$\|x_n^k\| < 1 \qquad (k = 1, \dots, m;\ n \in \mathbb{N}).$$

By (∗) in the proof of Proposition 4.2(h), there is, for each $n \ge m$, $y_n \in X_n$ such that

$$\|y_n \pm x_n^k\| \le 1 \qquad (\text{all } k),$$

but

$$\|y_n\| > \frac{1}{3}.$$

Let $y_n = 0$ for $n < m$ and $y = (y_n)_{n\in\mathbb{N}} \in X$. Then

$$\|(y \pm x^k) + N\| \le \sup_n \|y_n \pm x_n^k\| \le 1 \quad \text{for } k = 1, \dots, m$$

and

$$\|y + N\| = \lim_{\mathcal{U}} \|y_n\| \ge \frac{1}{3} > \varepsilon_0,$$

as desired. □

II.6 Notes and remarks

General remarks. The fundamental Proposition 1.1 provides a convenient tool for studying approximation theoretic properties by M-ideal methods. Various proofs of this proposition have appeared in the literature. Alfsen and Effros [11, Cor. I.5.6] derive it from their work on "dominated extensions" already alluded to in the Notes and Remarks to Chapter I. Independently it was obtained by Ando [19] as a consequence of a variant of our Lemma 2.5 which he proves without using any intersection properties (see also [319] for an account of this proof). We have drawn upon the simple proofs devised by Behrends [51, Prop. 6.5] and Yost [648]. Also Lau [388] gives a proof based on a variant of Lemma 2.5, which was later rediscovered in [241]. More specifically, he calls a subspace J of a Banach space X *U-proximinal* if there exists a positive function ϕ on $\mathbb{R}^+$ with $\phi(\varepsilon) \to 0$ as $\varepsilon \to 0$ such that

$$(B_X + J) \cap (1 + \varepsilon) B_X \subset B_X + \phi(\varepsilon) B_J \qquad \forall \varepsilon > 0$$

and proves that U-proximinal subspaces are in fact proximinal. He goes on to give examples of U-proximinal subspaces, among them M-ideals.
The fact that the set of best approximants in an M-ideal J algebraically spans J was first shown by Holmes, Scranton and Ward [321], our simple proof is due to Behrends [51, Prop. 6.6] and Lima [404]. The notion of a pseudoball is implicitly contained already in Lima's paper, too, but the explicit definition appears only in [67] where it is attributed to R. Evans. There it is also shown in addition to Theorem 1.6 that B is a pseudoball of radius 1 if and only if for every subspace $V \subset Y$ with $\dim Y/V < \infty$ there is some $y_0 + V \in Y/V$ such that $\operatorname{int} B(y_0 + V, 1) \subset q(B) \subset B(y_0 + V, 1)$ (where $q : Y \to Y/V$ denotes the quotient map). Other papers where the concept of a pseudoball is employed are [54], [268] and [480].
Also Proposition 1.8 comes from [321], the proof we have presented is a modification of the one in [648]. Apart from these authors, H. Fakhoury, too, has obtained Theorem 1.9 [217]. His argument is quite different and more in the spirit of the Alfsen-Effros paper. Corollary 1.7 and Proposition 1.10 seem to be new (with 1.10 essentially appearing in [292]) as does Proposition 1.11. This proposition answers a question raised in [321] where it is conjectured that an M-ideal is an M-summand as soon as its metric complement has an interior point. The concept of a grade (studied in [292], see also [67] and below) is the decisive tool to provide a complete solution.
The approximation theoretic properties of M-ideals have turned out to be of particular interest in the context of best approximation of bounded linear operators by compact operators. We refer to the Notes and Remarks to Chapter VI for a more detailed discussion.
As already mentioned in the text, Theorem 2.1 is due to Ando [20] and Choi and Effros [121]. For details of the development of this theorem see below. The classical Corollary 2.6 can be found in [174], for a nice geometric argument in the case that K is compact see [318, p. 103]. Corollary 2.7 comes from [487] and Corollary 2.8 from [445]. The strange proof of Sobczyk's theorem (Corollary 2.9) was suggested in [621]. Another argument for Proposition 2.10 is given in [652].
Proper M-ideals are studied in [67] and [294]. The fact that an M-ideal is proper if and only if the sets of best approximants are not balls (Proposition 3.2) was independently observed in [404] and [55]. Theorem 3.4 is due to Evans [206], with the present proof being essentially taken from [55]. Many proofs of Corollary 3.6 have appeared: see [206], [227], [400], [404], [477]. We can't help mentioning the simplest of these arguments, due to Payá [477]. Let $J^\perp \subset X^*$ be a weak* closed M-ideal so that there is an L-projection P from X^{**} onto $J^{\perp\perp}$. To show $Px \in J$ for $x \in X$, which is our aim, we calculate, given $y \in J$,

$$\begin{aligned} \|x - y\| &= \|Px - y\| + \|x - Px\| \\ &= \|Px - y\| + d(x, J^{\perp\perp}) \\ &= \|Px - y\| + d(x, J), \end{aligned}$$

where the natural identification $(X/J)^{**} \cong X^{**}/J^{\perp\perp}$ is used. Now take the infimum over all $y \in J$ to obtain

$$d(x, J) = d(Px, J) + d(x, J),$$

i.e. $d(Px, J) = 0$ and $Px \in J$. (One may note that this reasoning yields another proof of the fact, mentioned in the Notes and Remarks to Chapter I, that an F-ideal is an F-summand provided $(0,1) \in \operatorname{ex} B_F$.) We have taken the chance to add another proof of Corollary 3.6 which ultimately builds on the simple Proposition I.1.2. The rest of Section II.3 is taken from [67] and [292], whereas Section II.4 relies on [67] and [294] save for Theorem 4.9 which appears in [655]. Concerning the intersection property of Section II.4 we would like to mention the problem that up to now no dual space is known which fails this property; however, we strongly suspect such a space to exist.
As regards the material of Section II.5 let us note that the definition of a biface is due to Effros [190] who defines a structure topology on $\operatorname{ex} B_{X^*}$ by means of the weak* closed bifaces of B_{X^*}. By virtue of Proposition 5.1 this is the same as the structure topology of Section I.3 if X is a real L^1-predual space. (The use of bifaces in the general theory of L^1-preduals has proved useful e.g. in [215], [248], [387] or [430].) 5.2 - 5.7 are taken from Uttersrud's paper [608] with the exception of Theorem 5.4 which was first proved by N. Roy [536]. Simpler proofs of 5.4 are due to Lima, Olsen and Uttersrud [409] whose argument we have presented and to Rao [516]. Example 5.5 has already been used by Perdrizet [492, Section 6] for essentially the same purpose (up to the nomenclature), and a variant appears in Bunce's paper [100]. Uttersrud gives a detailed analysis of the M-structure of spaces of the form $\{x \in C(K) \mid x(t_0) = \int x \, d\mu\}$; the case $K = [0,1]$, $t_0 = \frac{1}{2}$, $\mu = (\delta_0 + \delta_1)/2$ corresponds to Bunce's example. Theorem 5.6 was proved under the additional assumption that X is an L^1-predual by Effros [190] in the separable case and in the nonseparable case by Taylor [601] and Fakhoury [215] using arguments different from those presented here. That this additional assumption is in fact superfluous is shown in Lemma 5.7. It is not clear how much has to be added to the T_1-separation property of E_X in order to force X to be an L^1-predual. We have already remarked that the "splitting property" $(*)$ appearing in the proof of Lemma 5.7 is sufficient, but not necessary. Contributions to this problem can be found in [537] and [608].
In his paper [139] Cunningham suggested a method of representing a Banach space as a "function module" or a Banach bundle, cf. the Notes and Remarks to Chapter I. He proposed to call a Banach space *square* if all the fibres in such a representation are zero- or one-dimensional. (Admittedly, this notation is not too suggestive.) To put it another way, a square Banach space X is a sup-normed space of bounded scalar-valued functions on some compact Hausdorff space Ω such that, for $x \in X$ and $f \in C(\Omega)$, the pointwise product $f \cdot x$ belongs to X and the numerical function $|x(\cdot)|$ is upper semicontinuous. In [140] and [141] Cunningham shows that every square Banach space is a G-space, but that there are nonseparable G-spaces which fail to be square. To this end he proves Proposition 5.8. [To show that the nonseparable G-spaces of Proposition 5.8 are not square one has to take into account that the point evaluations $x \mapsto \pm x(\omega)$ are exactly the extreme functionals on a square space X [143] so that all the multiplication operators $x \mapsto f \cdot x$ for $f \in C(\Omega)$ belong to the centralizer $Z(X)$. In particular, the centralizer of a square Banach space cannot be trivial.] In contrast has N. Roy shown in [534] (see also [535]) that *separable* G-spaces are in fact square. She also proves that a Banach space X is square if and only if $Z(X)$ separates E_X, i.e. nonantipodal points of $\operatorname{ex} B_{X^*}$. This result should be compared to Theorem 5.6 and to Theorem 5.9 which comes from [630]. Note that for C_σ-spaces the equation $Z_X = \operatorname{ex} B_{X^*}$ holds; the above results do, of course, not imply that every square Banach space is a C_σ. Another result along these lines worth

mentioning is that a Banach space X is a C_σ-space if and only if the structure topology on ex B_{X^*} coincides with the weak* topology. This is proved (modulo Lemma 5.7) in [215].
Finally we remark that Proposition 5.10 was first proved, using topological arguments, in [301]. Our approach follows [294].

THE $1\frac{1}{2}$-BALL PROPERTY. Following D. Yost [648] we say that a closed subspace J of a Banach space X has the *$1\frac{1}{2}$-ball property* if the conditions

$$\|x\| < r_1 + r_2 \quad \text{and} \quad B(x, r_2) \cap J \neq \emptyset$$

imply that

$$B(0, r_1) \cap B(x, r_2) \cap J \neq \emptyset.$$

This is equivalent to requiring the (strict) 2-ball property subject to the restriction that one of the centres lies in J. The $1\frac{1}{2}$-ball property is, however, far less restrictive than the 2-ball property. Unlike the latter the $1\frac{1}{2}$-ball property is self-dual, i.e. J has it if and only if $J^\perp$ has it. Among the examples one finds apart from (semi) M-ideals

- L-summands and semi L-summands (for the definition cf. Section I.4),
- closed subalgebras of $C_{\mathbb{R}}(K)$,
- $K(L^1(\mu), \ell^1)$ as a subspace of $L(L^1(\mu), \ell^1)$,
- $K(C(S), C(T))$ as a subspace of $L(C(S), C(T))$ if S is scattered and T is extremally disconnected.

Furthermore $K(c)$ has the $1\frac{1}{2}$-ball property in $L(c)$ if c is considered to be the space of convergent sequences of real numbers, whereas $K(c)$ fails it if we consider the complex sequence space c! Also, closed subalgebras of $C_{\mathbb{C}}(K)$ need not have the $1\frac{1}{2}$-ball property. (All these examples are contained in Yost's papers [648] and [654].)
Among the complex Banach spaces with the $1\frac{1}{2}$-ball property the examples of $K(\ell^1)$ in $L(\ell^1)$ and its relatives were until recently the only known specimens except for M-ideals and L-summands. Now it is known that every complex Banach space can arise as a subspace with the $1\frac{1}{2}$-ball property of some superspace in a nontrivial way, i.e. without being an M-ideal or L-summand [441].
The importance of this notion stems from its approximation theoretic implications. In fact, the proof of Proposition 1.1 applies as well to show that subspaces with the $1\frac{1}{2}$-ball property are proximinal. Moreover, in analogy with Proposition 1.8 one can show that the metric projection onto a subspace with the $1\frac{1}{2}$-ball property is Lipschitz continuous, with the Lipschitz constant 2. Thus Theorem 1.9, too, extends to this greater generality. These results yield an interesting consequence, namely:

A Lipschitz continuous metric projection need not have a Lipschitz continuous selection.

(A merely continuous selection always exists by Michael's theorem, cf. Theorem 1.9.) To see this let D denote the sup-normed space of all real valued functions on [0,1] which are

continuous at each irrational point, which are continuous from the right everywhere and which have limits from the left everywhere. Since D is isometric to a $C(K)$-space (e.g. by Gelfand's theorem) and $C = C_{\mathbb{R}}[0,1]$ is a closed subalgebra of D it has the $1\frac{1}{2}$-ball property, and the metric projection P_C is Lipschitz continuous on D. However, there cannot exist a Lipschitz selection of P_C because otherwise there would be a Lipschitz lifting for the quotient map from D onto D/C, contradicting a result by Aharoni and Lindenstrauss [4]. For this example see also [656], where a selfcontained proof is given. In [650] Yost gives a new proof of the duality theorem for spaces with the 3-ball property (resp. 2-ball property) and L-summands (resp. semi L-summands). More specifically he proves that Y has the 2-ball property in X if and only if Y has $1\frac{1}{2}$-ball property and is Hahn-Banach smooth in X, and Y is a semi L-summand if and only if Y has the $1\frac{1}{2}$-ball property and is a Chebyshev subspace. By a theorem due to Phelps [495] this proves that J has the 2-ball property if and only if $J^\perp$ is a semi L-summand, and with a bit of extra work Yost obtains that J has the 3-ball property if and only if $J^\perp$ is an L-summand, i.e. the equivalence (i) $\iff$ (ii) $\iff$ (iii) $\iff$ (v) of Theorem I.2.2.
Other papers dealing with the $1\frac{1}{2}$-ball property include [274], [472], [473] and [482].

LINEAR EXTENSION OPERATORS. The genesis of the important Theorem 2.1 lies in Pełczyński's papers [487] and [488]. There he proved the existence of a contractive linear extension operator from $X|_D$ to X under the following assumptions: X is a closed subspace of $C(K)$ where K is metrizable, $D \subset K$ is closed, $X|_D = C(D)$ and the pair $(X|_D, X)$ enjoys the bounded extension property. (In fact, his assumptions were formally stronger, but actually equivalent to those stated above, see [445].) His original proof was long and technical, and it was greatly simplified by Michael and Pełczyński in [445]. They were also able to drop the assumption that K be metrizable and $X|_D = C(D)$ in favour of the assumption that D be metrizable and $X|_D$ be a π_1-space; see also [515] for a proof. [Here a Banach space Z is called a π_1-space if and only if there is an increasing sequence of finite dimensional subspaces $E_1, E_2, \dots$ whose union is dense and each of which is the range of a contractive projection. This means that Z not only has the metric approximation property, but the metric approximation property is realized by an increasing sequence of (commuting) contractive finite rank projections; in particular such a space must be separable. It was shown in [446] that $C(D)$ is a π_1-space if D is a compact metric space. For a comparison of the various refinements of the BAP see the interesting recent paper [114] and its references.]
A considerable improvement in the development of linear extension theorems is due to Ryll-Nardzewski (see [491]) who replaced the π_1-property by the mere metric approximation property, still assuming separability of $X|_D$. The idea here is to reduce the problem to the previous case by switching to vector valued sequence spaces; this idea is used in the proof of Theorem 2.1, too. Note that the metric approximation property for $C(D)$ is far easier to check than the π_1-property and that the disk algebra yields an example of a space with the metric approximation property which fails to be a π_1-space [640].
In a similar vein, Lazar [389] proved the existence of a contractive linear extension operator from $A(F)$ to $A(K)$, where K denotes a Choquet simplex and F a closed metrizable face, as a consequence of his selection theorem, and Andersen [17] treated the same problem for a general compact convex set and a closed metrizable split face where he assumed

that $A(F)$ has the metric approximation property. This can be converted into a linear lifting theorem for quotients of C^*-algebras by closed two-sided ideals.
The first abstract theorem of this kind was proved by Ando [19] who supposed that J is an M-ideal in X and that X/J is a π_1-space to obtain a linear contractive lifting from X/J to X. Finally he showed in [20] that it is enough to assume that X/J is a separable space with the bounded approximation property in order to obtain a continuous linear lifting. All the results previously mentioned can be deduced from this theorem. It was also obtained by a different proof by Choi and Effros [121] (see also [120]), who point out the relevance of Theorem 2.1 for questions in the cohomology theory of C^*-algebras (see Section V.4). We have largely followed Ando's proof in the text.
Various authors have investigated which additional properties of a linear extension operator can be obtained, for instance positivity (see [17], [19], [389], [445]). Vesterstrøm has proved a rather general result in [614] using M-ideals in certain ordered Banach spaces. In the other direction we have already remarked that the assumptions in Theorem 2.1 are quite sharp; Proposition 2.3, which shows that the approximation assumption cannot be dispensed with, is related to an example in [149]. It follows from Theorem 1.9 that without metrizability there is always a nonlinear continuous normpreserving extension operator from $C(D)$ to $C(K)$; Benyamini [71] has shown that every number between 1 and ∞ may arise as the norm of a linear extension operator in this setting. It follows from the principle of local reflexivity that one can always obtain simultaneous extensions of finite dimensional subspaces of $C(D)$; see [216] for a more general result.
If one wants to apply Theorem 2.1 for function spaces $X \subset C(K)$, then a natural problem is to find conditions on D so that $X|_D = C(D)$. In this case D is called an interpolation set. Bishop [79] has shown that D is an interpolation set if $|\mu|(D) = 0$ for all $\mu \in X^\perp \subset M(K)$. The simplest proof of this proceeds as follows: Using the assumption of Bishop's theorem it is not hard to show that the map $\mu \mapsto (x \mapsto \int_D x\, d\mu)$ from $M(D)$ to X^* is an isometry. By duality this means that the restriction operator $x \mapsto x|_D$ is a quotient map, hence the result. But more is true: Since $J = J_D \cap X$ is an M-ideal in X (Cor. I.1.19) and thus proximinal, the restriction operator even maps the *closed* unit ball of X onto the closed unit ball of $C(D)$ so that we even get norm preserving extensions. This observation yields a proof of the Rudin-Carleson theorem mentioned in the proof of Corollary 2.7, since Bishop's condition is fulfilled by the F. and M. Riesz theorem. (The intimate connection between M-ideals and F. and M. Riesz type theorems will be discussed in Section IV.4; for elementary proofs of the Rudin-Carleson and the Riesz theorem we refer to [169] and its references.) Corresponding results under different orthogonality conditions (e.g. $\chi_D \mu = 0$ for all $\mu \in X^\perp$) are discussed e.g. in the papers [12], [94], [239], [249]; see Chapter 4 of the monograph [35] for a detailed account. For the topic of linear extension operators see also the memoir [489].
Finally we present an application of Theorem 2.1 to potential theory. Let $U \subset \mathbb{R}^n$ be open and bounded. We put

$$H(U) = \{f \in C(\overline{U}) \mid f \text{ is harmonic on } U\}.$$

The classical Dirichlet problem requires to find, given $\varphi \in C(\partial U)$, some $f \in H(U)$ such that $f|_{\partial U} = \varphi$. This is generally impossible. However, there is always a generalised solution, called the Perron-Wiener-Brelot solution. Those points $x_0 \in \partial U$ such that for all $\varphi \in C(\partial U)$ and corresponding Perron-Wiener-Brelot solutions f the relation $\lim_{x \to x_0} f(x) = \varphi(x_0)$ is valid are called regular boundary points. The set of all regular boundary points is denoted by $\partial_r U$, and it is classical that $\partial_r U = \partial U$ if the boundary of U is sufficiently smooth or if U is simply connected and $n = 2$. Concerning the weak solvability of the Dirichlet problem we now have:

PROPOSITION. *Let $U \subset \mathbb{R}^n$ be open and bounded, and let $E \subset \partial_r U$ be compact. Then there is a contractive linear operator $L : C(E) \to H(U)$ such that $(L\varphi)|_E = \varphi$ for all $\varphi \in C(E)$.*

This result will turn out to be almost obvious after we have reformulated it in terms of convexity theory. The space $H(U)$ is an order unit space and can hence be represented as a space of affine continuous functions $A(K)$. The crux of the matter is that here K is a Choquet simplex; see [81] or [194]. Moreover, it is known from [44] that $\partial_r U$ can be identified with the Choquet boundary of $H(U)$, that is $\operatorname{ex} K$. If we regard E as a compact subset of $\operatorname{ex} K$, then a corollary to Edwards' separation theorem states that every continuous function on E has a norm preserving extension to an affine continuous function on K [7, p. 91]. Let us denote $F = \overline{\operatorname{co}}\, E$ so that F is a split face since K is a simplex. Consequently, by Example I.1.4(c),

$$\{f \in A(K) \mid f|_E = 0\} = J_F \cap A(K) =: J$$

is an M-ideal in $A(K)$, and the quotient space $A(K)/J$ is isometric with $C(E)$ by the above. By Theorem 2.1 there is a linear contractive lifting L for the quotient map, and reidentifying $A(K)$ with $H(U)$ yields the desired solution operator from $C(E)$ to $H(U)$.

GRADES OF M-IDEALS. In Propositions 1.10 and 1.11 we attached two numbers $g^*(J,X)$ and $g_*(J,X)$ to an M-ideal J in X. According to Proposition 3.9 these numbers can be calculated as

$$\begin{aligned} g^*(J,X) &= \sup\{g(J,Y) \mid J \subset Y \subset X,\ \dim Y/J = 1\}, \\ g_*(J,X) &= \inf\{g(J,Y) \mid J \subset Y \subset X,\ \dim Y/J = 1\}. \end{aligned}$$

The interpretation is that J behaves like a very proper M-ideal in every direction if $g_*(J,X)$ is close to 1, while $g^*(J,X)$ close to 1 only yields the existence of one such direction. As was already announced in the text, for the familiar M-ideals J_D in $C(K)$ one obtains that $g^*(J_D, C(K)) = 0$ if D is clopen and $g^*(J_D, C(K)) = 1$ otherwise; and $g_*(J_D, C(K)) = 0$ if D contains an interior point and $g_*(J_D, C(K)) = 1$ otherwise. Also, it follows from Remark 3.8(d) that $g_*(Y, Y^{**}) = 1$ if Y is an M-ideal in its bidual.
One-codimensional M-ideals offer appealing geometric descriptions. If $Y = \ker(p)$ for some $p \in X^*$, $\|p\| = 1$, is an M-ideal in X and $L = \{x \in X \mid \|x\| = p(x) = 1\}$, then the unit ball of X is kind of cylindrically shaped with the set L being the lid of that cylinder, since one can show [292]

$$\operatorname{int} B_X \subset \operatorname{aco} L \subset B_X.$$

The idea of a cylindrical unit ball is perfect if J is even an M-summand, in which case L is the translate of a ball in J. The general case, however, produces a certain flaw in this idea because now L is merely the translate of a pseudoball in J. This flaw has already appeared in the example of Remark I.2.3(d) where we presented an M-ideal containing an extreme point of B_X, and it shows up again in the following characterisation of extreme M-ideals from [292]: J is an extreme one-codimensional M-ideal if and only if for each $x \in X$, $\|x\| > 1$, there is some $y \in \operatorname{co}(B_X \cup \{x\}) \cap J$ with $\|y\| > 1$.

INTERSECTION PROPERTIES AND COMPLEX L^1-PREDUALS. In his memoir [414] Lindenstrauss introduced, extending previous work by Hanner [287], the *n.k.* intersection property as follows: A Banach space has the *n.k.*IP if, given a family of n closed balls each subfamily of which consisting of k balls has a point in common, the intersection of these n balls is nonempty. Then he proved that a *real* Banach space is an L^1-predual if and only if it satisfies the 4.2.IP. Moreover he shows that neither the 4.3.IP nor the 3.2.IP is sufficient to ensure that the space under consideration is an L^1-predual (every L^1-space is known to fulfill the 3.2.IP) and that, for each $n \geq 4$, the 4.2.IP is equivalent to the n.2.IP. This resembles, at least formally, our Theorem I.2.2 where M-ideals are characterised by an intersection condition. Unlike the latter theorem Lindenstrauss' theorem is clearly false for *complex* Banach spaces; not even the one-dimensional complex space $\mathbb{C}$ satisfies the 3.2.IP.
In order to provide a substitute for the n.2.IP, Hustad [328] proposed the following weak intersection property: A Banach space X is an $E(n)$-space if, for each family of n closed balls $B(x_1, r_1), \ldots, B(x_n, r_n)$, the implication

$$\bigcap_{i=1}^{n} B(x^*(x_i), r_i) \neq \emptyset \quad \forall x^* \in B_{X^*} \quad \Longrightarrow \quad \bigcap_{i=1}^{n} B(x_i, r_i) \neq \emptyset$$

holds. It is readily seen that every Banach space is an $E(2)$-space (by virtue of the Hahn-Banach theorem) and that a real Banach space is an $E(n)$-space if and only if it satisfies the n.2.IP. Thus Lindenstrauss' result reads: A real Banach space X is an L^1-predual if and only if it is an $E(4)$-space. In [328] Hustad was able to prove that a complex Banach space is an L^1-predual if and only if it is an $E(7)$-space if and only if it is an $E(n)$-space for some $n \geq 7$.
The obvious question, which numbers smaller than 7 would suffice as well, was answered by Lima who at first proved that $n = 4$ is enough [399] and then obtained the final result that complex L^1-predual spaces are characterised by the $E(3)$-property [398, Appendix]. Simplifications of his proofs appear in [533] and [410]; let us remark that one variant of the final step of the proof that $E(3)$-spaces are L^1-preduals in the latter paper depends on a precise knowledge of the L-summands in the dual of an $E(3)$-space. Lima also shows in [399] that the 4.3.IP characterises complex L^1-preduals. Both these results suggest that in the complex case weaker intersection conditions than in the corresponding real situation suffice. With this philosophy at hand it was believed for many years that complex M-ideals might be characterised by the 2-ball property, but, as we have pointed out in the Notes and Remarks to Chapter I, this belief has only recently turned out to be faulty. – Papers related to this subject include [290], [327], [400], [401], [403], [405], [411], [412], [521].
The general theory of complex L^1-preduals advanced in the 70s after complex versions of the Choquet-Meyer representation theorem were proved, for which we refer to the

survey [496]. Substantial contributions are due to Effros [191] and Hirsberg and Lazar [313]. Using tools provided by these authors Olsen extended the Lazar-Lindenstrauss-Wulbert classification scheme ([390], [424]) of C_σ-spaces, G-spaces etc. to the complex setting which is not a straightforward task ([468], see also [456], [469] and [532]). Here we shall report on some results concerning L^1-preduals involving the L-structure of the dual space.

Apart from Lima's fundamental theorem [400, Th. 5.8] that a real or complex Banach space X is an L^1-predual if and only if

- if $p \in X^*$, $\|p\| = 1$ and $P(p) \in \{0, p\}$ for all L-projections P on X^*, then $p \in \operatorname{ex} B_{X^*}$,
- $\operatorname{lin}\{p\}$ is an L-summand in X^* for all $p \in \operatorname{ex} B_{X^*}$,

(we had occasion to use this in the proof of Lemma 5.7), the following results seem to be of interest. The authors of [203] show that a separable complex Banach space is an L^1-predual if and only if $\operatorname{lin}_{\mathbb{C}} \overline{\operatorname{aco}}^{w^*} K$ is an L-summand in X^* whenever $K \subset \operatorname{ex} B_{X^*}$ is weak* compact. By way of example they point out that the separability assumption cannot be dispensed with. In the general case they prove that a complex Banach space is an L^1-predual if and only if $\operatorname{lin}_{\mathbb{C}} F$ is an L-summand in X^* whenever F is a weak* closed face in B_{X^*}. (Both these results hold in the real case, too.) Rao [516] characterises complex L^1-preduals as follows: For all $x \in X$, $\|x\| = 1$, the set $\{p \in \operatorname{ex} B_{X^*} \mid |p(x)| = 1\}$ is structurally closed and $F = \{x^* \in B_{X^*} \mid x^*(x) = 1\}$ is a split face in $\overline{\operatorname{co}}^{w^*}(F \cup -iF)$. (In the real case the first condition is sufficient.) Moreover he points out that Proposition 5.2 extends to the complex case.

Our proofs of Proposition 5.2 and Theorem 5.6 rely essentially on the "max + min" operation which is restricted to the real case. It was believed to constitute remnants of a lattice structure in a real G-space [424, p. 337], and no complex analogue was contrived until recently when Uttersrud [609] succeeded in doing so, thus proving Theorem 5.6 for complex G-spaces as well. Uttersrud's idea is to view, given 3 real numbers r, s, t, $\frac{1}{2}(\max\{r,s,t\} + \min\{r,s,t\})$ as the centre of the smallest closed interval containing these three numbers. Accordingly he defines, given 3 complex numbers r, s, t, the complex number $c(r,s,t)$ as the centre of the smallest closed disk containing r, s and t. Using this notion he proves that a closed subspace X of $C_{\mathbb{C}}(K)$ is a G-space if and only if for $x, y \in X$ the centre function $c(x, y, 0)$ (defined pointwise) belongs to X if and only if the structure topology on E_X is Hausdorff.

CHAPTER III

Banach spaces which are M-ideals in their biduals

III.1 Examples and stability properties

In this section we begin our study of Banach spaces X which are M-ideals in their biduals. The reason for the rich theory of these spaces is mainly their nice stability behaviour (Theorem 1.6) and the *natural* L-decomposition of X^{***} (Proposition 1.2).

We write i_X for the isometric embedding of a Banach space X into its bidual X^{**}. However, in most of the following we will simply regard X as a subspace of X^{**}.

Definition 1.1 *Let X be a Banach space.*

(a) *X is called an* M-embedded space *if X is an M-ideal in X^{**}.*

(b) *X is called an* L-embedded space *if X is an L-summand in X^{**}.*

REMARKS: (a) These names are chosen to reflect the M-(L-)type embedding of X in X^{**}.

(b) Since by Theorem I.1.9 M-summands in dual spaces are w^*-closed, no nonreflexive Banach space X can be an M-summand in X^{**}, that is nonreflexive M-embedded spaces are proper M-ideals (cf. Definition II.3.1).

(c) Because of the following we can (should it be necessary) restrict ourselves to real spaces.

If X is a complex Banach space, then X is M- (L-)embedded if and only if $X_{\mathbb{R}}$, i.e. X considered as a real Banach space, is M- (L-)embedded.

[PROOF: The mapping $X^* \longrightarrow (X_{\mathbb{R}})^*$, $x^* \longmapsto \operatorname{Re} x^*$ is an $\mathbb{R}$-linear isometry, hence $(X^*)_{\mathbb{R}} \cong (X_{\mathbb{R}})^*$. Applying this also to X^{**} yields

$$(X^{**})_{\mathbb{R}} \cong ((X^*)_{\mathbb{R}})^* \cong (X_{\mathbb{R}})^{**}.$$

Since the isomorphism between $(X^{**})_{\mathbb{R}}$ and $(X_{\mathbb{R}})^{**}$ maps $X_{\mathbb{R}}$ onto $i_{X_{\mathbb{R}}}(X_{\mathbb{R}})$ we get the claim by Proposition I.1.23.]

Recall for the following that the natural projection from the third dual X^{***} of a Banach space X onto $i_{X^*}(X^*)$ is the mapping $\pi_{X^*} := i_{X^*} \circ i_X{}^*$.

Proposition 1.2 *For a Banach space X the following are equivalent:*
(i) *X is an M-ideal in X^{**}.*
(ii) *The natural projection π_{X^*} from X^{***} onto $i_{X^*}(X^*)$ is an L-projection.*

PROOF: (ii) $\Rightarrow$ (i): Since the kernel of π_{X^*} is $X^\perp$, we have $X^{***} = X^* \oplus_1 X^\perp$.
(i) $\Rightarrow$ (ii): By assumption $X^\perp$ is the kernel of an L-projection P in X^{***}. Since π_{X^*} is a contractive projection with kernel $X^\perp$ we get by Proposition I.1.2 that $P = \pi_{X^*}$. □

For easy reference we state:

Corollary 1.3 *If X is an M-embedded space, then X^* is an L-embedded space.*

The converse of this statement is not true (consider e.g. $X = c$; cf. also Proposition 2.7). In Proposition IV.1.9 we will characterise those L-embedded spaces which are duals of M-embedded spaces.

Banach spaces which are M-ideals in their biduals appear naturally among classical spaces. In the following we will give examples of sequence spaces, function spaces, operator spaces, and spaces of analytic functions which are M-embedded. The underlying idea in most of the examples is that there is a "$o(\cdot)$-$O(\cdot)$" relation between the elements of X and those of X^{**}. (o and O are to denote the usual Landau symbols.) In this case the proofs use the 3-ball property with the following idea: Suppose X is a space of functions on some set S which we would like to prove to be M-embedded. To find $y \in X$ for given $x^{**} \in B_{X^{**}}$, $x_i \in B_X$, $(i = 1, 2, 3)$ and $\varepsilon > 0$ such that $\|x^{**} + x_i - y\| \le 1 + \varepsilon$ choose a subset $K \subset S$ with $|x_i| < \varepsilon$ off K (the o-condition) and try to define $y = x^{**}$ on K and $y = 0$ elsewhere. This idea works perfectly well for $c_0(S)$ which will turn out to be the archetype of an M-embedded space, but it needs some work to adapt it to other examples. Also, it is sufficient to assume that the x_i lie in some dense subset of B_X, since we have an $\varepsilon > 0$ at our disposal.
Since not all of the following spaces are equally well-known let us recall some definitions first.

THE ORLICZ SPACES h_M AND ℓ_M, $H_M(I)$ AND $L_M(I)$. Let $M : \mathbb{R}^+ \to \mathbb{R}^+$ be a continuous convex function such that $M(0) = 0$. We shall also assume that M is nondegenerate, i.e. $M(t) > 0$ if $t > 0$. The Orlicz sequence space ℓ_M consists of all sequences (x_n) for which

$$\sum_{n=1}^{\infty} M\left(\frac{|x_n|}{\rho}\right) < \infty \quad \text{for some } \rho > 0,$$

h_M denotes the subspace of ℓ_M where

$$\sum_{n=1}^{\infty} M\left(\frac{|x_n|}{\rho}\right) < \infty \quad \text{for all } \rho > 0.$$

ℓ_M is a Banach space under the (Luxemburg) norm

$$\|x\|_M = \inf\left\{\rho > 0 \;\middle|\; \sum_{n=1}^{\infty} M\left(\frac{|x_n|}{\rho}\right) \le 1\right\}.$$

(Actually, the infimum is attained.) Under suitable conditions $(h_M)^{**}$ is canonically isometric with ℓ_M, namely if the conjugate function M^* satisfies the Δ_2-condition at zero, which is $\limsup_{t\to 0} M^*(2t)/M^*(t) < \infty$. The conjugate function is defined by $M^*(u) = \max\{tu - M(t) \mid 0 < t < \infty\}$. We refer to Chapter 4 of [422] for these notions, basic facts (to be used below without further comment) and nontrivial results on Orlicz sequence spaces.
In the case of function spaces the investigations can be reduced to the intervals $I = (0,1)$ resp. $I = (0,\infty)$ equipped with Lebesgue measure. For an Orlicz function M as above, the Orlicz spaces $L_M(I)$ and $H_M(I)$ are defined by

$$L_M(I) = \left\{f \;\middle|\; \int_I M\left(\frac{|f(s)|}{\rho}\right) ds < \infty \text{ for some } \rho > 0\right\},$$

$$H_M(I) = \left\{f \;\middle|\; \int_I M\left(\frac{|f(s)|}{\rho}\right) ds < \infty \text{ for all } \rho > 0\right\}.$$

The norm in L_M is of course

$$\|f\|_M = \inf\left\{\rho > 0 \;\middle|\; \int_I M\left(\frac{|f(s)|}{\rho}\right) ds \le 1\right\}.$$

The relevant Δ_2-conditions have to be checked at ∞ if $I = (0,1)$ and at both zero and ∞ if $I = (0,\infty)$. If M^* satisfies the Δ_2-condition, then $(H_M)^{**} = L_M$, and $H_M = L_M$ if (and only if) M satisfies the Δ_2-condition (cf. [423, p. 120]).

THE LORENTZ SPACES $d(w,1)$ AND $L^{p,1}$. Let $w = (w_n)$ be a decreasing sequence of real numbers such that $w_1 = 1$, $w_n \to 0$ and $\sum_{n=1}^{\infty} w_n = \infty$. The Lorentz sequence space $d(w,1)$ consists of all sequences (u_n) such that

$$\|(u_n)\|_{w,1} = \sup_{\pi} \sum_{n=1}^{\infty} |u_{\pi(n)}| w_n < \infty$$

where π ranges over all the permutations of N. In the case $w_n = n^{\frac{1}{p}-1}$ $(1 < p \le \infty)$ this space is known as $\ell^{p,1}$. Under the canonical duality the dual of $d(w,1)$ can be represented as the space of sequences (x_n) for which $\sum x_n u_n$ is absolutely convergent for all $(u_n) \in d(w,1)$. This implies $(x_n) \in c_0$. If (x_n^*)[1] denotes the nonincreasing

[1] We adopt here the symbol x_n^* and f^* for the decreasing rearrangement of a sequence or a function since it is commonly used in the literature. No confusion should arise with the notation x^* used in this book to denote an element of the dual space X^*.

rearrangement of the sequence of moduli $(|x_n|)$, the dual norm can be calculated as follows:

$$\|(x_n)\| = \sup_n \frac{\sum_{j=1}^n x_j^*}{\sum_{j=1}^n w_j}$$

It is also known that the subspace $d(w,1)_*$ of $d(w,1)^*$ consisting of all sequences for which

$$\lim_n \frac{\sum_{j=1}^n x_j^*}{\sum_{j=1}^n w_j} = 0$$

is a predual of $d(w,1)$, which we call the canonical predual. Moreover, $d(w,1)_*$ coincides with the closed linear span of the unit vectors e_n in $d(w,1)^*$. (All the results mentioned above can be found e.g. in [242].)
For Lorentz function spaces let I be one of the intervals $(0,1)$ or $(0,\infty)$ equipped with Lebesgue measure. Although the following could be done for the spaces $L(W,1)$ considered in [423, p. 120], we confine ourselves to $L^{p,1}$.
For a measurable function f on I let f^* be its nonincreasing rearrangement and f^{**} its maximal function

$$f^{**}(t) = \frac{1}{t}\int_0^t f^*(s)\,ds.$$

If

$$\|f\|_{p,1} = \frac{1}{p}\int_I s^{\frac{1}{p}} f^*(s)\,\frac{ds}{s}\,,$$

then $L^{p,1}(I) = \{f \mid \|f\|_{p,1} < \infty\}$, and $\|\cdot\|_{p,1}$ is a complete norm on $L^{p,1}$. If $1 < p < \infty$ and $1/p + 1/q = 1$, then the dual of $L^{p,1}$ coincides with the "weak L^q-space" $L^{q,\infty}$, and the dual norm is

$$\|f\|_{q,\infty} = \sup_t t^{\frac{1}{q}} f^{**}(t).$$

(See Lorentz' paper [426] where $L^{p,1}$ is denoted by $\Lambda(\frac{1}{p})$ and $L^{q,\infty}$ by $M(\frac{1}{p})$, or compare Chapter IV.4 of the monograph [70].) It is known that the closure of the set of bounded functions whose support has finite measure in $(L^{p,1})^* = L^{q,\infty}$ is a predual of $L^{p,1}$; this follows for instance from Theorem I.4.1 in [70]. We denote this closure by $(L^{p,1})_*$ and call it the canonical predual of $L^{p,1}$.

The Banach algebra H^∞ and the Hardy spaces H^p. If F is a bounded analytic function on the open unit disk $\mathbb{D}$, then a well-known result in complex variables asserts that

$$\lim_{r\to 1} F(rz) =: f(z)$$

exists for almost every $z \in \mathbb{T}$. The mapping which assigns to each such F its boundary function $f \in L^\infty(\mathbb{T})$ then identifies the space $H^\infty(\mathbb{D})$ isometrically with a closed subalgebra of $L^\infty(\mathbb{T})$ which is denoted by H^∞. (Often we shall not explicitly distinguish between $H^\infty(\mathbb{D})$ and H^∞.) Also, H^∞ is the weak* closed linear span of the functions $1, z, z^2, \dots$ in $L^\infty(\mathbb{T})$. Likewise the boundary function f is known to exist almost everywhere if F is an analytic function on $\mathbb{D}$ subject to the growth condition

$$\sup_{r<1}\left(\frac{1}{2\pi}\int_0^{2\pi} |F(re^{it})|^p dt\right)^{1/p} =: \|F\|_p < \infty,$$

and $\|F\|_p$ coincides with the norm of f in $L^p(\mathbb{T})$. These boundary functions make up a closed subspace of $L^p(\mathbb{T})$ denoted by H^p, which is called a Hardy space.
Under the duality $\langle f, g\rangle = \int f\overline{g}\, dt/2\pi$ the dual of $H_0^1 = \{f \in H^1 \mid f(0) = 0\}$ is isometric with the quotient space $L^\infty(\mathbb{T})/(H_0^1)^\perp = L^\infty(\mathbb{T})/H^\infty$. On the other hand, H_0^1 is isometric with the dual of the quotient space $C(\mathbb{T})/A$ by the classical F. and M. Riesz theorem. (Here A denotes the disk algebra.) Consequently $L^\infty(\mathbb{T})/H^\infty$ "is" the bidual of $C(\mathbb{T})/A$, and the canonical copy of $C(\mathbb{T})/A$ in its bidual is the quotient space $(H^\infty + C(\mathbb{T}))/H^\infty$. To see this one has to take into account that the distance of a continuous function to A is the same as its distance to H^∞. (We remark in passing that this distance formula and hence the identification of the canonical copy of $C(\mathbb{T})/A$ in $L^\infty(\mathbb{T})/H^\infty$ is the essence of Sarason's proof that $H^\infty + C(\mathbb{T})$ is closed in $L^\infty(\mathbb{T})$, cf. [170, p. 163] or [244, p. 376].) For a detailed discussion of the above facts and more information we refer to the monographs [170], [181] or [244]. We also point out that it is convenient in this setting to abbreviate $C(\mathbb{T})$ by C and $L^\infty(\mathbb{T})$ by L^∞.

THE SPACES OF ANALYTIC FUNCTIONS WITH WEIGHTED SUPREMUM NORM $A_0(\phi)$ AND $A_\infty(\phi)$. We denote by $\mathbb{D}$ the open unit disk and let $\phi : [0,1] \to \mathbb{R}$ be a positive continuous decreasing function such that $\phi(0) = 1$ and $\phi(1) = 0$. We define

$$A_\infty(\phi) = \{f : \mathbb{D} \to \mathbb{C} \mid f \text{ analytic}, \|f\|_\phi = \sup_{z\in\mathbb{D}} |f(z)| \cdot \phi(|z|) < \infty\},$$

$$A_0(\phi) = \{f \in A_\infty(\phi) \mid \lim_{|z|\to 1} |f(z)| \cdot \phi(|z|) = 0\}.$$

These are Banach spaces under the norm $\|\cdot\|_\phi$, and $A_\infty(\phi)$ is canonically isometric with $A_0(\phi)^{**}$ ([540] or [78]).

THE BLOCH SPACES B_0 AND B. These are the spaces

$$B = \{f : \mathbb{D} \to \mathbb{C} \mid f \text{ analytic}, f(0) = 0, \|f\| = \sup_{z\in\mathbb{D}} |f'(z)| \cdot (1 - |z|^2) < \infty\},$$

and

$$B_0 = \{f \in B \mid \lim_{|z|\to 1} |f'(z)| \cdot (1 - |z|^2) = 0\}$$

(the "little Bloch space"). That B is the bidual of B_0 and more information on Bloch functions and the Bloch spaces can be found in the surveys [37] and [131].

Examples 1.4 *The following spaces are M-embedded spaces:*

(a) c_0 *(more generally $c_0(\Gamma)$).*
(b) *The Orlicz sequence space h_M if M^* satisfies the Δ_2-condition at zero while M fails it.*
(c) *The canonical predual $d(w,1)_*$ of the Lorentz sequence space $d(w,1)$.*
(d) *The Orlicz function space $H_M(I)$ if M^* satisfies the appropriate Δ_2-condition while M fails it.*
(e) *The canonical predual $(L^{p,1})_*$ of the Lorentz function space $L^{p,1}$.*
(f) $K(H)$ *if H a Hilbert space.*

(g) *$K(\ell^p, \ell^q)$ if $1 < p \le q < \infty$.*
(h) *$C(\mathbb{T})/A$ where $\mathbb{T}$ denotes the unit circle and A the disk algebra.*
(i) *The weighted spaces of analytic functions $A_0(\phi)$.*
(j) *The little Bloch space B_0.*

PROOF: All arguments except that for (h) use the restricted 3-ball property (Theorem I.2.2) with the idea outlined on page 102.
(a) This is immediate; see also Example I.1.4(a).
(b) By what was said above the bidual of h_M is ℓ_M in this situation. (The assumption on M is made only to exclude the reflexive case.) Suppose $x^{**} \in B_{\ell_M}$, i.e. $\sum M(|x^{**}(n)|) \le 1$, $x_i \in B_{h_M}$ $(i = 1,2,3)$ and $\varepsilon > 0$ are given. There is no loss of generality in assuming $\|x_i\| < 1$ and $x_i(n) = 0$ for $n > n_0$, since such sequences are dense in the unit ball of h_M. Then $\sum_{n=1}^{n_0} M(|x_i(n)|) < 1$. Choose $m_0 > n_0$ such that

$$\sum_{n=1}^{n_0} M(|x_i(n)|) + \sum_{n=m_0}^{\infty} M(|x^{**}(n)|) \le 1.$$

Put $y(n) = x^{**}(n)$ for $n < m_0$ and $y(n) = 0$ otherwise. Then $y \in h_M$ and

$$\sum M(|x^{**}(n) + x_i(n) - y(n)|) = \sum_{n=1}^{n_0} M(|x_i(n)|) + \sum_{n=m_0}^{\infty} M(|x^{**}(n)|) \le 1,$$

hence $\|x^{**} + x_i - y\| \le 1$.
(c) We have already remarked that the bidual of $d(w,1)_*$ can canonically be identified with $d(w,1)^*$ and that $\operatorname{lin}(e_n)$ is dense in $d(w,1)_*$. Given $x^{**} = (a(n)) \in B_{d(w,1)^*}$, $x_i = (x_i(n)) \in B_{d(w,1)_*}$ with finite support, and $\varepsilon > 0$ we first show the existence of $y \in d(w,1)_*$ with $\|x^{**} + x_i - y\| \le 1 + \varepsilon$ under the additional assumption

$$\max\{j \mid x_i^*(j) \neq 0\} =: k_i = k \qquad \text{for all } i$$

and

$$\sum_{j=1}^{k} x_i^*(j) \le \sum_{j=1}^{k} a^*(j) \qquad \text{for all } i.$$

Pick N such that $x_i(n) = 0$ for $n > N$ and

$$|a(n)| \le \min\{\delta, a^*(k)\} \qquad \text{for } n > N,$$

where $\delta = \min_i x_i^*(k)$. Then define the sequence $y = (y(n))$ by $y(n) = a(n)$ if $n \le N$ and $y(n) = 0$ otherwise. If $z_i(n) = a(n) + x_i(n) - y(n)$, then

$$\begin{aligned} z_i^*(j) &= x_i^*(j) \qquad \text{for } j \le k, \\ z_i^*(j) &\le a^*(j) \qquad \text{for } j > k \end{aligned}$$

by construction. Hence

$$\frac{\sum_{j=1}^n z_i^*(j)}{\sum_{j=1}^n w(j)} \le 1 \qquad \text{if } n \le k,$$

since $\|x_i\| \le 1$, and for $n > k$ we have by the second part of our assumption

$$\frac{\sum_{j=1}^n z_i^*(j)}{\sum_{j=1}^n w(j)} \le \frac{\sum_{j=1}^n a^*(j)}{\sum_{j=1}^n w(j)} \le \|x^{**}\| \le 1,$$

hence $\|x^{**} + x_i - y\| \le 1$.
To dispose of the extra assumption we may clearly demand $x^{**} \notin d(w,1)_*$ (because otherwise one could trivially choose $y = x^{**}$). In this case we cannot have $x^{**} \in \ell^1$ either. Hence we find $l \ge k_i$ such that

$$\sum_{j=1}^{k_i} x_i^*(j) < \sum_{j=1}^{l} a^*(j). \tag{$*$}$$

We now modify x_i as follows: If $x_i(n) \ne 0$, then let $\xi_i(n) = x_i(n)$. At $l - k_i$ indices where $x_i(n) = 0$ let $\xi_i(n) = \eta$ (here $\eta > 0$ is a small number to be specified in a moment); otherwise let $\xi_i(n) = 0$. The number η should be chosen so small that $\|x_i - \xi_i\| \le \varepsilon$ and

$$\sum_{j=1}^{l} \xi_i^*(j) \le \sum_{j=1}^{l} a^*(j)$$

hold; this is possible by $(*)$. By the first part of the proof, some $y \in d(w,1)_*$ such that

$$\left\| x^{**} + \frac{\xi_i}{1+\varepsilon} - y \right\| \le 1$$

exists, hence $\|x^{**} + x_i - y\| \le 1 + 2\varepsilon$.
(d) The idea of the proof is the same as in part (e) below, while the details, which we leave to the reader, follow – mutatis mutandis – the proof of part (b).
(e) We have already pointed out that the bidual is the space

$$L^{q,\infty}(I) = \{f \mid \|f\|_{q,\infty} < \infty\}$$

where $1/p + 1/q = 1$. We first deal with the case $I = (0,1)$. Suppose $f \in L^{q,\infty}$, $\|f\|_{q,\infty} \le 1$, bounded functions g_i with $\|g_i\|_{q,\infty} \le 1$ and some $\varepsilon > 0$ are given. (The boundedness assumption may be imposed on the g_i since the bounded functions form a dense subspace of $(L^{p,1})_*$.) The idea is to define the function g by $\chi_{\{|f| \le r\}} \cdot f$ for sufficiently large r.
Since the g_i and hence the g_i^{**} are bounded, there exists some $t_1 > 0$ such that

$$t^{\frac{1}{q}} \cdot g_i^{**}(t) \le \varepsilon \ \text{ for } t \le t_1,$$

whereas

$$t^{\frac{1}{q}} \cdot g_i^{**}(t) \le 1 \ \text{ for } t > t_1$$

anyway. Next we choose $t_0 \le t_1$ such that

$$\int_0^{t_0} f^*(s)\, ds \le \varepsilon\, t_1^{\frac{1}{p}},$$

and finally let $r = f^*(t_0)$. Then $g = \chi_{\{|f| \le r\}} \cdot f$ is bounded and hence in $(L^{p,1}(0,1))_*$. By the subadditivity of the maximal operator $(\phi + \psi)^{**} \le \phi^{**} + \psi^{**}$ [70, Theorem II.3.4] we have

$$\|f + g_i - g\|_{q,\infty} \le \sup_t \left(t^{\frac{1}{q}} \cdot g_i^{**}(t) + t^{\frac{1}{q}} \cdot (f-g)^{**}(t) \right)$$

and for $t > t_1$ one computes

$$\begin{aligned} t^{\frac{1}{q}} \cdot (f-g)^{**}(t) &= t^{-\frac{1}{p}} \int_0^t (f-g)^*(s)\, ds \\ &\le t_1^{-\frac{1}{p}} \int_0^{t_0} (f-g)^*(s)\, ds \\ &\le \varepsilon \end{aligned}$$

by the choice of t_0. Since for $t \le t_1$

$$t^{\frac{1}{q}} \cdot (f-g)^{**}(t) \le \|f - g\|_{q,\infty} \le \|f\|_{q,\infty} \le 1,$$

these estimates, together with the ones for g_i^{**}, prove the inequality

$$\|f + g_i - g\|_{q,\infty} \le 1 + \varepsilon.$$

In the case of the infinite interval $(0,\infty)$ one has to combine the above approach (at 0) with the method of part (c) (at ∞) to accomplish the proof. We omit the details. (Note that $L^{p,1}(0,1)$ and $L^{p,1}(0,\infty)$ are not even isomorphic for $1 < p < \infty$ [111].)

(f) It is well known that the bidual of $K(H)$ is $L(H)$ (see for example [593, p. 64] or [159, p. 247]). Hence the result follows from the coincidence of M-ideals and closed two-sided ideals in C^*-algebras: Theorem V.4.4.

(g) In Example VI.4.1 we will prove that $K(\ell^p, \ell^q)$ in an M-ideal in $L(\ell^p, \ell^q)$. This, however, is our assertion since the latter space is the bidual of $K(\ell^p, \ell^q)$ [159, p. 247]. (The condition $p \le q$ in (g) is imposed only to assure nontrivial, meaning nonreflexive examples; see [159, p. 248].)

(h) As mentioned above we will abbreviate $C(\mathbb{T})$ by C and $L^\infty(\mathbb{T})$ by L^∞. To show that $X = C/A = (H^\infty + C)/H^\infty$ is an M-embedded space we will prove that $X^\perp$ is an L-summand in X^{***}. Let K denote the maximal ideal space of L^∞. In the following we identify, using the Gelfand transform $f \mapsto \widehat{f}$, H^∞ and C with proper closed subalgebras of $C(K)$ ($\cong L^\infty$). Then our problem is to find an L-projection from from $X^{***} \cong (H^\infty)^\perp$ ($\subset M(K)$) onto $X^\perp \cong (H^\infty + C)^\perp$.

To exhibit such a projection, the basic idea is this. Let us denote by m the normalized Lebesgue measure on $\mathbb{T}$. Then there is, by the Riesz representation theorem, a uniquely determined probability measure $\widehat{m} \in M(K)$ satisfying

$$\int_{\mathbb{T}} f\, dm = \int_K \widehat{f}\, d\widehat{m} \qquad \forall f \in L^\infty.$$

The Lebesgue decomposition $\mu = \mu_a + \mu_s$ with respect to $\widehat{m}$ gives rise to an L-projection $P : \mu \mapsto \mu_s$ acting on $M(K)$ (Example I.1.6(b)). We intend to show that $(H^\infty)^\perp$ is an invariant subspace of P and that P maps $(H^\infty)^\perp$ onto $(H^\infty + C)^\perp$ which clearly settles our problem. For this the use of an abstract F. and M. Riesz theorem will turn out to be instrumental.
We will now give the details of this argument. The F. and M. Riesz theorem we have in mind reads as follows [95, Th. 4.2.2, Cor. 4.2.3] (a more general version can be found in [240, Th. II.7.6, Cor. II.7.9]):

THEOREM (Abstract F. and M. Riesz theorem) *Let B be a function algebra on the compact space K and ϕ be a multiplicative linear functional which admits a unique representing measure σ on K. Writing the Lebesgue decomposition of a measure $\mu \in M(K)$ with respect to σ as $\mu = \mu_a + \mu_s$ one obtains:*

(a) *If $\mu \in B^\perp$, then $\mu_a, \mu_s \in B^\perp$.*
(b) *If $\mu \in B_\phi^\perp := (\ker\phi)^\perp$, then $\mu_s \in B^\perp$ and $\mu_a \in B_\phi^\perp$.*

We shall apply this theorem to the algebra H^∞ (strictly speaking to its image under the Gelfand transform) with the above K and the functional

$$\phi(f) = \int_{\mathbb{T}} f\, dm.$$

Since $\phi(f) = F(0)$ if f is the boundary function of the bounded analytic function F on $\mathbb{D}$, ϕ is clearly multiplicative. Moreover, $\widehat{m}$ represents ϕ by definition, and $\widehat{m}$ is unique with respect to this property as shown for instance in [95, Lemma 4.1.1], [240, p. 38] or [244, p. 201]. (The crucial property to be used there is that H^∞ is a so-called logmodular algebra, see [240, p. 37] or [244, p. 66].) Part (a) of the above theorem now assures that $P((H^\infty)^\perp) \subset (H^\infty)^\perp$.
It is left to show that $P((H^\infty)^\perp) = (H^\infty + C)^\perp$. Since $P((H^\infty)^\perp) = M_{\rm sing}(K) \cap (H^\infty)^\perp$ (compare Lemma I.1.15) this amounts to showing that

$$(H^\infty)^\perp \cap M_{\rm sing}(K) = (H^\infty)^\perp \cap C^\perp.$$

To see "$\subset$" let $\mu \in (H^\infty)^\perp$ be $\widehat{m}$-singular. It will be convenient to denote the function $z \mapsto z^k$ ($z \in \mathbb{T}$, $k \in \mathbb{Z}$) simply by z^k. In the notation of part (b) of the abstract F. and M. Riesz theorem the measure $\widehat{z^{-1}}\mu$ belongs to $(H^\infty)_\phi^\perp$ (since $\mu \in (H^\infty)^\perp$) so that by $\widehat{m}$-singularity $(\widehat{z^{-1}}\mu)_s = \widehat{z^{-1}}\mu \in (H^\infty)^\perp$. Continuing inductively we find that $\widehat{z^{-k}}\mu \in (H^\infty)^\perp$ for all $k \in \mathbb{N}$, hence in particular $\int \widehat{z^{-k}}\, d\mu = 0$. Since $\int \widehat{z^n}\, d\mu = 0$ for $n \in \mathbb{N}_0$ anyway, the density of lin $\{z^k \mid k \in \mathbb{Z}\}$ (= the trigonometric polynomials) in $C = C(\mathbb{T})$ shows the desired inclusion.
For the proof of "$\supset$" take $\mu \in (H^\infty)^\perp$ with $\int \widehat{g}\, d\mu = 0$ for all $g \in C$. We want to show that its absolutely continuous part μ_a vanishes. By P-invariance of $(H^\infty)^\perp$ we have $\mu_s \in (H^\infty)^\perp$, so the inclusion already proved shows that $\int \widehat{g}\, d\mu_s = 0$ for all $g \in C$. Then by assumption this holds also for $\mu_a = \mu - \mu_s$. Thus, writing $\mu_a = f\widehat{m}$ for some $f \in L^1(\widehat{m})$, we obtain

$$\int \widehat{g} f\, d\widehat{m} = 0 \qquad \text{for all } g \in C.$$

Since the Gelfand transform of L^∞ extends uniquely to an isometric isomorphism between $L^1(\mathbb{T}, m)$ and $L^1(K, \widehat{m})$ (see [244, p. 202]) this allows us to conclude $f = 0$.

(i) Let $f \in A_\infty(\phi)$, $g_i \in A_0(\phi)$ with norm ≤ 1, and $\varepsilon > 0$ be given. For $r < 1$ let $f_r(z) = f(rz)$. Then $f_r \in A_0(\phi)$, $\|f_r\|_\phi \leq \|f\|_\phi$ (since ϕ is decreasing) and $\lim_{r\to 1} f_r = f$ uniformly on compact subsets of $\mathbb{D}$.
First choose $\varrho_1 < 1$ such that

$$|g_i(z)| \cdot \phi(|z|) \leq \varepsilon \qquad \text{for } |z| \geq \varrho_1$$

and then $r_1 < 1$ with

$$\sup_{|z|\leq\varrho_1} |f(z) - f_{r_1}(z)| \leq \varepsilon.$$

Since $f_{r_1} \in A_0(\phi)$, we have for some $\varrho_2 > \varrho_1$

$$|f_{r_1}(z)| \cdot \phi(|z|) \leq \varepsilon \qquad \text{for } |z| \geq \varrho_2$$

so that

$$|f(z) - f_{r_1}(z)| \cdot \phi(|z|) \leq 1 + \varepsilon \qquad \text{for } |z| \geq \varrho_2.$$

In the next step we pick $r_2 \geq r_1$ such that

$$\sup_{|z|\leq\varrho_2} |f(z) - f_{r_2}(z)| \leq \varepsilon.$$

In this way we find increasing sequences (ϱ_k) and (r_k) such that $|f(z) - f_{r_k}(z)| \cdot \phi(|z|) \leq 1 + \varepsilon$ for all z except in the annulus $\varrho_k < |z| < \varrho_{k+1}$. Now fix an integer $m \geq 2/\varepsilon$ and let $h = \frac{1}{m}\sum_{k=1}^m f_{r_k}$. Then $h \in A_0(\phi)$, and we claim

$$|f(z) + g_i(z) - h(z)| \cdot \phi(|z|) \leq 1 + 3\varepsilon \qquad \text{for all } z \in \mathbb{D}.$$

If $|z| \leq \varrho_1$, this is clear since $\|g_i\|_\phi \leq 1$ and $|f(z) - f_{r_k}(z)| \leq \varepsilon$ for all k. Otherwise, we have $|g_i(z)| \cdot \phi(|z|) \leq \varepsilon$ and $|f(z) - f_{r_k}(z)| \cdot \phi(|z|) \leq 1 + \varepsilon$ for all but possibly one k. However, by what was noted above $\|f - f_{r_k}\|_\phi \leq 2$ holds anyway so that

$$|f(z) - h(z)| \cdot \phi(|z|) \leq \frac{m-1}{m}(1+\varepsilon) + \frac{1}{m} \cdot 2 \leq 1 + 2\varepsilon.$$

(j) This is a special case of (i) since B_0 is isometric (by $f \mapsto f'$) to $A_0(\phi)$ with the weight function $\phi(r) = 1 - r^2$. □

Remarks: (a) The induction argument used in the proof of the part (h) above is similar to the one employed in [240, Th. 7.10, p. 45] to deduce the classical F. and M. Riesz theorem from the general one.

(b) We shall later obtain several proofs of the fact that $C(\mathbb{T})/A$ is M-embedded as consequences of more general results; see Remark 1.8 and Example IV.4.11. Although the above proof is not the shortest possible, it is very natural if one adopts the idea (compare I.1.15 and I.1.16) that the desired L-projection should be the restriction of an L-projection of the superspace $M(K)$ leaving the subspace $(H^\infty)^\perp$ invariant. The

abstract F. and M. Riesz theorem is in fact a statement about invariant subspaces of L-projections so that its use above seems quite appropriate.
(c) We will give more examples of Banach spaces which are M-ideals in their biduals in the Notes and Remarks section.

Some consequences of general results on M-embedded spaces for concrete examples will be explicitly noted in the following sections. We start with the answer to a question of Sarason and Adamjan, Arov and Krein; cf. [38, p. 608].

Corollary 1.5 *$H^\infty + C$ is proximinal in L^∞.*

Although $H^\infty + C$ is not an M-ideal in L^∞ (because it contains the constant functions), so that Proposition II.1.1 is not applicable right away, we will derive this by M-ideal methods.

PROOF: It was noted above that the canonical copy of C/A in its bidual is $(H^\infty + C)/H^\infty$. So this space is, as an M-ideal, proximinal in L^∞/H^∞ by Proposition II.1.1. Combined with the following elementary observation

> *If X is a Banach space and E and F are subspaces with $E \subset F \subset X$ such that F/E is proximinal in X/E and E is proximinal in X, then F is proximinal in X.*

and the fact that H^∞ is proximinal in L^∞ (as a w^*-closed subspace), this yields the claim. ◻

The next theorem describes the hereditary properties of M-embedded spaces.

Theorem 1.6 *The class of M-embedded Banach spaces is stable by taking*
(a) *subspaces,*
(b) *quotients,*
(c) *c_0-sums.*

PROOF: (a) For a subspace Y of X we have to show by Proposition 1.2 that π_{Y^*} is an L-projection. Denote by $i : Y \longrightarrow X$ the inclusion mapping and recall that $\pi_{Y^*} i^{***} = i^{***}\pi_{X^*}$. Moreover, given $y^{***} \in Y^{***}$ one can find $x^{***} \in X^{***}$ with $i^{***}x^{***} = y^{***}$ and $\|x^{***}\| \le \|y^{***}\|$. For y^{***} and x^{***} as above we have (using that π_{X^*} is an L-projection)

$$\begin{aligned}
\|y^{***}\| &\le \|\pi_{Y^*} y^{***}\| + \|y^{***} - \pi_{Y^*} y^{***}\| \\
&= \|\pi_{Y^*} i^{***}x^{***}\| + \|i^{***}x^{***} - \pi_{Y^*} i^{***}x^{***}\| \\
&= \|i^{***}\pi_{X^*} x^{***}\| + \|i^{***}x^{***} - i^{***}\pi_{X^*} x^{***}\| \\
&\le \|i^{***}\| \left(\|\pi_{X^*} x^{***}\| + \|x^{***} - \pi_{X^*} x^{***}\|\right) \\
&= \|x^{***}\| \\
&\le \|y^{***}\|.
\end{aligned}$$

This shows that π_{Y^*} is an L-projection.

(b) Replace i in the above proof by the quotient map $q: X \longrightarrow X/Y$ and note that q^{***} is isometric.

(c) If X_i ($i \in I$) are M-embedded spaces and X is the c_0-sum $(\oplus \sum X_i)_{c_0(I)}$ then $X^{**} = (\oplus \sum X_i^{**})_{\ell^\infty(I)}$. Direct verification of the 3-ball property shows the claim. □

Remarks 1.7 (a) Although subspaces and quotients of M-embedded spaces are again M-embedded, this property is *not* a 3-space property, i.e.

Y a subspace of X, Y and X/Y M-embedded $\not\Rightarrow$ X M-embedded.

[Consider e.g. $X = c_0 \oplus_1 c_0$ and $Y = c_0 \times \{0\}$. Since X^{**} has a nontrivial L-summand, it can't contain a nontrivial M-ideal by Theorem I.1.8.]

(b) The property of being an M-embedded space is clearly not isomorphically invariant, however it is invariant under almost isometric isomorphism. Recall that the Banach-Mazur distance of two Banach spaces X and Y is defined as $d(X,Y) = \inf\{\|T\|\|T^{-1}\| \mid T: X \to Y \text{ isomorphism}\}$ and that X and Y are said to be almost isometric if $d(X,Y) = 1$.

If X is an M-embedded space and Y a Banach space with $d(X,Y) = 1$, then Y is M-embedded.

[This is straightforward using the 3-ball property.]

Remark 1.8 The stability property of M-embedded spaces combined with the theorems of Nehari and Hartman (see below) immediately gives a second proof of Example 1.4(h): Recall that an operator $T \in L(H^2)$ is called a *Hankel operator* on the Hardy space H^2 if there is a sequence $(a_n)_{n\in\mathbb{N}_0}$ such that $t_{ij} := \langle T\gamma_j, \gamma_i \rangle = a_{i+j}$, where $i,j \in \mathbb{N}_0$ and $\gamma_k(z) = z^k$, $k \in \mathbb{Z}$. [The harmonic analysis type notation γ_k for the natural basis vectors is convenient also here; cf. Section IV.4.] The condition on T is easily seen to be equivalent to $S^*T = TS$, where S denotes the unilateral shift on H^2, i.e. $S\gamma_i = \gamma_{i+1}$, and S^* is its Hilbert space adjoint.

To give the precise statements of the quoted theorems we need some more notation: For $f \in L^\infty(\mathbb{T})$ we write M_f for the corresponding multiplication operator $M_f g := fg$ on $L^2(\mathbb{T})$. Further, $J \in L(L^2(\mathbb{T}))$ is the complex conjugation $Jf = \overline{f}$, P denotes the orthogonal projection from $L^2(\mathbb{T})$ onto H^2 and $i: H^2 \longrightarrow L^2(\mathbb{T})$ the inclusion map. Finally, we let $H_0^\infty = \{zf \mid f \in H^\infty\} \cong \{f \in H^\infty(\mathbb{D}) \mid f(0) = 0\}$.

THEOREM (Nehari and Hartman)

(a) *The mapping*

$$\begin{array}{rccl} \mathfrak{H} : & L^\infty/H_0^\infty & \longrightarrow & L(H^2) \\ & f + H_0^\infty & \longmapsto & H_f := PJM_f i \end{array}$$

is an isometric isomorphism onto the space of Hankel operators on H^2.

(b)

$$d(H_f, K(H^2)) = d(f, H_0^\infty + C)$$

In particular: H_f is a compact operator iff $f \in H_0^\infty + C$.

By the above the restriction of $\mathfrak{H}$ to $C/A_0 \cong (H_0^\infty + C)/H_0^\infty$ is an isometry into $K(H^2)$, so C/A_0 is an M-embedded space by Theorem 1.6(a) and Example 1.4(f). Observing that the isometrical isomorphism $M_{\gamma_1} : C(\mathbb{T}) \longrightarrow C(\mathbb{T})$ maps A onto A_0, we obtain Example 1.4(h). (We refer to the survey article [506] and the books [475] and [507] for a detailed exposition of the theory of Hankel operators. We explicitly mention the remarkably simple geometric proof of the above theorem due to Parrot which is outlined in [506, p. 430–432].)

We will complete our discussion of the stability properties of M-embedded spaces by showing that this is in fact a separably determined property of a Banach space. To show this we need a proposition which is of independent interest.

Proposition 1.9 *For a Banach space X, the following assertions are equivalent:*

(i) *X is an M-ideal in its bidual.*

(ii) *For all $x \in B_X$, all sequences (x_n) in B_X, all weak* cluster points x^{**} of (x_n) and all $\varepsilon > 0$ there is some $u \in \operatorname{co}\{x_1, x_2, \dots\}$ such that*

$$\|x + x^{**} - u\| \le 1 + \varepsilon.$$

(iii) *For all $x \in B_X$, all sequences (x_n) in B_X and all $\varepsilon > 0$ there is some $n \in \mathbb{N}$ and there are $u \in \operatorname{co}\{x_1, \dots, x_n\}$, $t \in \operatorname{co}\{x_{n+1}, x_{n+2}, \dots\}$ such that*

$$\|x + t - u\| \le 1 + \varepsilon.$$

(iv) *For all $x \in B_X$ and all $x^{**} \in B_{X^{**}}$ there is a net (x_α) in B_X weak* converging to x^{**} such that*

$$\limsup \|x + x^{**} - x_\alpha\| \le 1.$$

PROOF: (i) $\Rightarrow$ (ii): If this were false, then the inequality $\|x + x^{**} - u\| > 1 + \varepsilon$ would hold for all $u \in A := \operatorname{co}\{x_1, x_2, \dots\}$. Consequently, A and the ball $B_{X^{**}}(x + x^{**}, 1 + \varepsilon/2)$ could strictly be separated by some $x^{***} \in X^{***}$, with $\|x^{***}\| = 1$ say. By assumption we have a decomposition

$$x^{***} = x^* + x_s^{***} \in X^* \oplus X^\perp, \quad \|x^{***}\| = \|x^*\| + \|x_s^{***}\|.$$

Hence

$$\begin{aligned} 1 + \frac{\varepsilon}{2} &\le \operatorname{Re} x^{***}(u - (x + x^{**})) \\ &\le |x^*(x)| + |x^*(x^{**} - u)| + |x_s^{***}(x^{**})| \\ &\le \max\{\|x\|, \|x^{**}\|\}(\|x^*\| + \|x_s^{***}\|) + |x^*(x^{**} - u)| \\ &\le 1 + |x^*(x^{**} - u)| \end{aligned}$$

for all $u \in A$; however $|x^*(x^{**} - x_n)| < \varepsilon/2$ for some n since x^{**} is a weak* cluster point of the sequence (x_n). This leads to a contradiction.

(ii) $\Rightarrow$ (iii): Let x^{**} be a weak* cluster point of the sequence (x_n). We apply (ii) to obtain $\|x + x^{**} - u\| \le 1 + \varepsilon/3$ for some $u \in \operatorname{co}\{x_1, x_2, \dots\}$, say $u \in \operatorname{co}\{x_1, \dots, x_n\}$.

Suppose that $\|x+t-u\| > 1+\varepsilon$ holds for each $t \in \operatorname{co}\{x_{n+1}, x_{n+2}, \ldots\} =: A$. Again this implies that A and $B_X(u-x, 1+\varepsilon/2)$ can strictly be separated. Thus, for some $x^* \in X^*$ with $\|x^*\| = 1$

$$1 + \varepsilon/2 \leq \operatorname{Re} x^*(t - (u - x)) \qquad \forall t \in A.$$

But since $x^{**} \in \overline{A}^{w*}$, this yields

$$\begin{aligned} 1 + \frac{\varepsilon}{2} &\leq \operatorname{Re} x^*(x + x^{**} - u) \\ &\leq \|x + x^{**} - u\| \\ &\leq 1 + \frac{\varepsilon}{3}, \end{aligned}$$

a contradiction.

(iii) $\Rightarrow$ (iv): Again we argue by contradiction. Suppose that for some $x \in B_X$, $x^{**} \in B_{X^{**}}$ there is no such net. Consequently there is, for some $\varepsilon > 0$, a convex weak* neighbourhood V of x^{**} such that

$$\|x + x^{**} - v\| > 1 + \varepsilon \qquad \forall v \in V. \tag{1}$$

Pick $x_1 \in V \cap B_X$ and put $B_1 = -x + x_1 + (1+\varepsilon)B_{X^{**}}$. This is a weak* compact set not containing x^{**} (by (1)), so there is a convex weak* neighbourhood $W_1 \subset V$ of x^{**} such that $W_1 \cap B_1 = \emptyset$. This means

$$\|x + w - x_1\| > 1 + \varepsilon \qquad \forall w \in W_1.$$

Next choose $x_2 \in W_1 \cap B_X$, put $B_2 = -x + \operatorname{co}\{x_1, x_2\} + (1+\varepsilon)B_{X^{**}}$ and find a convex weak* neighbourhood $W_2 \subset W_1$ of x^{**} satisfying $W_2 \cap B_2 = \emptyset$, i.e.

$$\|x + w - u\| > 1 + \varepsilon \qquad \forall w \in W_2,\ u \in \operatorname{co}\{x_1, x_2\}.$$

Continuing in this manner, we inductively define a sequence of points (x_n) in B_X and a sequence of convex weak* neighbourhoods $V \supset W_1 \supset W_2 \supset \ldots$ such that $x_{n+1} \in W_n$ and

$$\|x + w - u\| > 1 + \varepsilon \qquad \forall w \in W_n,\ u \in \operatorname{co}\{x_1, \ldots, x_n\}.$$

for all $n \in \mathbb{N}$. This is a contradiction to (iii), since $\operatorname{co}\{x_{n+1}, x_{n+2}, \ldots\} \subset W_n$.

(iv) $\Rightarrow$ (i): We shall verify that the canonical projection from X^{***} onto X^* is an L-projection. To this end decompose a given $x^{***} \in B_{X^{***}}$ into $x^{***} = x^* + x_s^{***} \in X^* \oplus X^\perp$. For an arbitrary $\varepsilon > 0$ pick $x \in B_X$, $x^{**} \in B_{X^{**}}$ such that $\operatorname{Re} x^*(x) \geq \|x^*\| - \varepsilon$ and $\operatorname{Re} x_s^{***}(x^{**}) \geq \|x_s^{***}\| - \varepsilon$. By (iv) there is a net (x_α) such that, for sufficiently large α,

$$|x^*(x^{**} - x_\alpha)| \leq \varepsilon \quad \text{and} \quad \|x + x^{**} - x_\alpha\| \leq 1 + \varepsilon.$$

This yields

$$\begin{aligned} \|x\| + \|x_s^{***}\| - 2\varepsilon &\leq \operatorname{Re}\left(x^*(x) + x_s^{***}(x^{**})\right) \\ &\leq \operatorname{Re}\langle x^* + x_s^{***}, x + x^{**} - x_\alpha\rangle + \varepsilon \\ &\leq \|x^{***}\| \cdot (1 + \varepsilon) + \varepsilon \end{aligned}$$

so that in fact $\|x^*\| + \|x_s^{***}\| = \|x^{***}\|$. □

We remark that condition (iv) has implicitly appeared in the discussion of our examples at the beginning of this section. From a conceptual point of view, condition (iii) is the most interesting one. First of all, it is formulated in terms of X alone; no a priori knowledge of the bidual space is needed. (However, applying (iii) in concrete instances seems to be technically unpleasant.) As part (iii) involves just sequences so that only separable subspaces have to be checked, we immediately get:

Corollary 1.10 *A Banach space is an M-ideal in its bidual if and only if every separable subspace is.*

In the remainder of this chapter we will present several properties of M-embedded spaces. These results can be thought of as giving necessary conditions on a Banach space to be M-embedded. Let us conclude this first section with the reference to Proposition VI.4.4 where we will show that X is necessarily M-embedded if $K(X)$ is an M-ideal in $L(X)$.

III.2 Isometric properties

The first group of results in the following section is devoted to the study of surjective isometries and contractive projections in M-embedded spaces and their duals. After this we collect what is known about representation and characterisation of these spaces. At the end various other isometric properties – such as unique preduals and renormings – are discussed.

Proposition 2.1 *If X is an M-ideal in its bidual, then $\pi_{X^{**}}$ is the only contractive projection from $X^{(4)}$ onto X^{**}.*

PROOF: We will show the following stronger statement:

$$\|x^{**}\| \le \|x^{(4)} + x^{**}\| \quad \forall x^{**} \in X^{**} \quad \text{implies} \quad x^{(4)} \in X^{*\perp}. \qquad (*)$$

This shows the claim, because, for a contractive projection P from $X^{(4)}$ onto X^{**} and $x^{(4)} \in \ker P$, $(*)$ yields $x^{(4)} \in X^{*\perp}$, therefore $\ker P \subset X^{*\perp}$, hence $P = \pi_{X^{**}}$. To prove $(*)$ observe first that by assumption we have $X^{***} = X^* \oplus_1 X^\perp$, hence $X^{(4)} = X^{*\perp} \oplus_\infty X^{\perp\perp}$. Since $X \subset X^{\perp\perp}$ we have in particular

$$\|x^{*\perp} + x\| = \max\{\|x^{*\perp}\|, \|x\|\} \quad \text{for} \quad x^{*\perp} \in X^{*\perp},\ x \in X.$$

Under the hypothesis of $(*)$ we now decompose $x^{(4)} = x^{*\perp} - y^{**} \in X^{*\perp} \oplus X^{**}$ and easily get

$$\|y^{**} + x^{**}\| \le \|x^{*\perp} + x^{**}\| \quad \forall x^{**} \in X^{**}. \qquad (\dagger)$$

Assuming $y^{**} \ne 0$ we find $n \in \mathbb{N}$ such that $n\|y^{**}\| > \|x^{*\perp}\|$. By Goldstine's theorem and w^*-lower semicontinuity of the norm, there is a net (y_α) in X such that $y_\alpha \longrightarrow ny^{**}$ with respect to $\sigma(X^{**}, X^*)$ and $\|x^{*\perp}\| \le \|y_\alpha\| \le n\|y^{**}\|$. But then by $(\dagger)$

$$\|y^{**} + y_\alpha\| \le \|x^{*\perp} + y_\alpha\| = \max\{\|x^{*\perp}\|, \|y_\alpha\|\} = \|y_\alpha\| \le n\|y^{**}\|,$$

hence (passing to the limit)

$$\|y^{**} + ny^{**}\| \leq n\|y^{**}\|,$$

a contradiction. □

We remark that Proposition 2.1 finds its proper place among Godefroy's general results on uniqueness of preduals (in particular it is a consequence of Theorem 3.1 below and [259, Th. II.1 and Ex. II.2.3.b]). However the above direct proof using the geometric assumption is also interesting.

Proposition 2.2 *Let X be a Banach space which is an M-ideal in its bidual. Then every surjective isometry I of X^{**} is the bitranspose of a surjective isometry of X.*

PROOF: Let such an $I : X^{**} \longrightarrow X^{**}$ be given. As easily seen $P := I^{**-1}\pi_{X^{**}}I^{**}$ is a contractive projection of $X^{(4)}$ onto X^{**}, hence $P = \pi_{X^{**}}$ by Proposition 2.1. But then $\pi_{X^{**}}I^{**} = I^{**}\pi_{X^{**}}$, which is known to imply the w^*-continuity of I. Consequently $I = J^*$ for some surjective isometry $J : X^* \longrightarrow X^*$. To show the claim it is enough to prove that $I(X) = X$, which by the Hahn-Banach theorem means $I^*(X^\perp) = X^\perp$. But $I^* = J^{**}$ is a surjective isometry of X^{***} which maps the L-summand X^* onto itself; therefore it also maps the complementary L-summand $X^\perp$ onto itself. □

In our next theorem we collect some results on the M-structure of M-embedded spaces. Recall from Definition I.3.7 that we denote by $Z(X)$ the centralizer of a Banach space X.

Theorem 2.3 *Suppose X is an M-ideal in X^{**}.*

(a) *Every M-ideal J in X is an M-summand.*

(b) *If X has no nontrivial M-summands, then every nontrivial M-ideal of X^{**} contains X.*

(c) *Every M-projection in X^{**} is the bitranspose of an M-projection in X (in particular, every L-projection in X^* is w^*-continuous).*

(d) $Z(X) \cong Z(X^{**})$.

PROOF: (a) Let P be the M-projection in X^{**} onto $J^{\perp\perp}$. Since M-ideals are left invariant under M-projections (Lemma I.3.5) $P_{|X}$ is an M-projection in X onto $J^{\perp\perp} \cap X = J$ (see Lemma I.1.15).

(b) Let J be a nontrivial M-ideal in X^{**}. Then $J \cap X$ is an M-ideal in X (see Propositions I.1.11 and I.1.17), hence an M-summand by part (a). The assumption gives that $J \cap X$ equals $\{0\}$ or X. We will show that the first case cannot arise. In fact, then J and X are complementary M-summands in $J \oplus X$ by Proposition I.1.11(c). If now $J \neq \{0\}$ there is an $x^{**} \in J \cap (X^{**} \setminus X)$, so $P_X(x^{**})$ is a closed ball in X with radius $d(x^{**}, X)$ – see the introduction to Section II.1. But

> *For every Banach space X and $x^{**} \in X^{**} \setminus X$ the set $P_X(x^{**})$ has no interior points relative to X.*

[PROOF: Assuming that this is not the case, we may suppose after a suitable translation that 0 is an interior point of $P_X(x^{**})$, i.e. there is a $\delta > 0$ such that $y \in X$ and $\|y\| < \delta$ imply $\|x^{**} - y\| = \|x^{**}\|$ $(= d(x^{**}, X))$. Take $x^* \in S_{X^*}$ with $\operatorname{Re} x^{**}(x^*) > \|x^{**}\| - \delta/4$ and $x \in S_X$ with $\operatorname{Re} x^*(x) > 1/2$. For $x_\delta := (\delta/2)x$ we have $\|x_\delta\| = \delta/2$ and $\operatorname{Re} x^*(x_\delta) > \delta/4$. So

$$\|x^{**} + x_\delta\| \geq |(x^{**} + x_\delta)(x^*)| \geq \operatorname{Re} x^{**}(x^*) + \operatorname{Re} x^*(x_\delta) > \|x^{**}\|$$

gives a contradiction with $y = -x_\delta$.]
This shows that $J = \{0\}$ if $J \cap X = \{0\}$, so that $J \cap X = X$ as claimed.

(c) If P is an M-projection in X^{**}, then $I := 2P - Id_{X^{**}}$ is a surjective isometry of X^{**}, hence $I = I_0{}^{**}$ for some surjective isometry I_0 of X by Proposition 2.2. It easily follows that P is the bitranspose of an M-projection in X.

(d) Recall from Theorem I.1.10 that $\mathbb{P}_M(E)$ denotes the set of M-projections of a Banach space E. By part (c) the isometric injection $\phi : L(X) \longrightarrow L(X^{**})$, $T \longmapsto T^{**}$ maps $\mathbb{P}_M(X)$ onto $\mathbb{P}_M(X^{**})$, i.e.

$$\mathbb{P}_M(X^{**}) = \phi(\mathbb{P}_M(X)) \subset \phi\left(\overline{\operatorname{lin}}\, \mathbb{P}_M(X)\right).$$

Hence by Theorem I.3.14(c)

$$Z(X^{**}) = \overline{\operatorname{lin}}\, \mathbb{P}_M(X^{**}) \subset \phi\left(\overline{\operatorname{lin}}\, \mathbb{P}_M(X)\right) \subset \phi(Z(X)).$$

But the converse inclusion $\phi(Z(X)) \subset Z(X^{**})$ is true in arbitrary Banach spaces X by Corollary I.3.15(a). $\square$

REMARKS: (a) Part (c) of the above theorem shows that in the situation of (b) X^{**} has no nontrivial M-summands. However, there may be proper M-ideals between X and X^{**}. Take for example $X = K(H)$, H a nonseparable Hilbert space. Then X has no nontrivial M-ideals (Corollary VI.3.7) but $J := \{T \in L(H) \mid \operatorname{ran} T \text{ separable}\}$ is a closed two-sided ideal, hence an M-ideal (Theorem V.4.4) in $L(H) = K(H)^{**}$ and $X \subsetneqq J \subsetneqq X^{**}$. In Corollary 2.12 we will see that also for separable M-embedded spaces X there may exist M-ideals between X and X^{**}. We remark that $L(\ell^p), 1 < p < \infty$, contains no other M-ideals than $K(\ell^p)$ as will be shown in Corollary V.6.6.

(b) The above-mentioned examples also serve to show that there are Banach spaces with a trivial centralizer containing nontrivial M-ideals. (For the centralizer of $L(H)$ cf. Theorem V.4.7 or Theorem VI.1.2.)

(c) We refer to Lemma 4.1 for another result on the continuity of contractive projections in the dual of an M-embedded space.

Lemma 2.4 *Let X be a Banach space and recall that π_{X^*} denotes the natural projection from X^{***} onto X^*, i.e. $\pi_{X^*} = i_{X^*} \circ i_X{}^*$. The following are equivalent:*

(i) *π_{X^*} is the only contractive projection P in X^{***} with $\ker P = X^\perp$.*
(ii) *The only operator $T \in L(X^{**})$ such that $\|T\| \leq 1$ and $T_{|X} = Id_X$ is $T = Id_{X^{**}}$.*
(iii) *For every surjective isometry U of X, the only $T \in L(X^{**})$ such that $\|T\| \leq 1$ and $T_{|X} = U$ is $T = U^{**}$.*

Banach spaces which are M-ideals in their biduals have these properties.

Because of (iii) a space X fulfilling one (hence all) of the above statements is said to have the *unique extension property.*

PROOF: The last statement follows from Proposition I.1.2 and Proposition 1.2.
(i) $\Rightarrow$ (ii): With T as in (ii) define $P := T^*\pi_{X^*}$. Since $Tx = x$ for $x \in X$ one gets $T^*x^* - x^* \in X^\perp$ for $x^* \in X^*$. It is then easy to show that P is a projection with $\ker P = X^\perp$; of course $\|P\| \le 1$. By assumption $x^* = \pi_{X^*}x^* = Px^* = T^*\pi_{X^*}x^* = T^*x^*$, which shows $Tx^{**} = x^{**}$.
(ii) $\Rightarrow$ (i): With P as in (i) define $T := \left(P_{|X}\right)^* i_{X^{**}}$. Since $\ker P = X^\perp$ one gets $Px^*|_X = x^*|_X$ and this easily yields $T|_X = Id_X$; of course $T \in L(X^{**})$ and $\|T\| \le 1$. The assumption then shows $Px^* = x^*$. Therefore P and π_{X^*} have the same ranges and kernels, so they coincide.
(iii) $\Rightarrow$ (ii): Obvious.
(ii) $\Rightarrow$ (iii): With U and T as in (iii) define $V := (U^{**})^{-1}T$. Then $\|V\| \le 1$ and $V|_X = Id_X$, hence $V = Id_{X^{**}}$ by assumption. □

We refer to [270] for more information on the unique extension property. It is custom made for proving

Proposition 2.5 *If X is a Banach space with the unique extension property and X has the MAP [MCAP], then X^* has the MAP [MCAP] with adjoint operators.*
In particular: If X is an M-embedded space with the MAP [MCAP], then X^* has the MAP [MCAP].

PROOF: In the proof we use the symbols s_{op} and w_{op} for the strong and the weak operator topology.
Let (K_α) be a net in the unit ball of $F(X)$ $[K(X)]$ which converges to Id_X in the strong operator topology, in particular $K_\alpha \xrightarrow{w_{op}} Id_X$. We claim

$$K_\alpha^* \xrightarrow{w_{op}} Id_{X^*}, \quad \text{i.e. } x^{**}(K_\alpha^* x^*) \longrightarrow x^{**}(x^*) \quad \text{for all } x^* \in X^*,\ x^{**} \in X^{**}.$$

We will show that every subnet (K_β^*) has a subnet (K_γ^*) such that $K_\gamma^* \xrightarrow{w_{op}} Id_{X^*}$. Indeed, since $(X^*\widehat{\otimes}_\pi X^{**})^* \cong L(X^{**})$, there is by w^*-compactness of $B_{L(X^{**})}$ some $T \in L(X^{**})$ with $\|T\| \le 1$ and a subnet such that $K_\gamma^{**} \xrightarrow{w^*} T$, in particular

$$(K_\gamma^{**}x^{**})(x^*) \longrightarrow (Tx^{**})(x^*) \quad \text{for } x^* \in X^*,\ x^{**} \in X^{**}. \qquad (*)$$

Using this for $x^{**} = x \in X$ we find by the w_{op}-convergence of (K_α) that $Tx = x$. By the unique extension property we have $T = Id_{X^{**}}$, hence $(*)$ shows our claim.
To finish the proof recall that the topologies w_{op} and s_{op} yield the same dual space (e.g. [179, Theorem VI.1.4]), so

$$Id_{X^*} \in \overline{\text{co}}^{w_{op}}(K_\alpha^*) = \overline{\text{co}}^{s_{op}}(K_\alpha^*) \subset \overline{B_{F(X^*)}}^{s_{op}} \left[\overline{B_{K(X^*)}}^{s_{op}}\right]. \qquad □$$

We remark that the MAP passes from X^* to X [159, Cor. 9, p. 244], but the corresponding statement for the MCAP does not hold, as was recently shown by P. Casazza.

Except Proposition 1.9 there is no internal characterisation known – neither isometric, nor isomorphic – of Banach spaces which are M-ideals in their biduals (in contrast to what has been shown in Theorem II.3.10 and Theorem II.4.9 for proper M-ideals). However, the following proposition presents a general decomposition procedure for M-embedded spaces. We are also able to characterise the M-embedded spaces in some classes of Banach spaces completely (see Theorem 3.11 for an isomorphic result in this vein).

Proposition 2.6 *Let X be a Banach space which is an M-ideal in its bidual. Then there exists a set I and Banach spaces X_i ($i \in I$) which are M-ideals in their biduals and contain no nontrivial M-ideals such that $X \cong (\oplus \sum X_i)_{c_0(I)}$.*

PROOF: For $p \in \operatorname{ex} B_{X^*}$ denote by N_p the intersection of all w^*-closed L-summands of X^* containing p. By Proposition I.1.11 N_p is the smallest w^*-closed L-summand containing p. Note that every L-summand of X^* is w^*-closed by Theorem 2.3(c). Then

(1) $\qquad N_p = N_q$ or $N_p \cap N_q = \{0\}$ for $p, q \in \operatorname{ex} B_{X^*}$.

(2) $\qquad M_p$, the (w^*-closed) L-summand complementary to N_p is a maximal proper L-summand in X^*.

[To see (1) use Lemma I.1.5 and the Krein-Milman theorem. Further, assuming $X^* = N_p \oplus_1 M_p = N \oplus_1 M$ with $M_p \subset M$ one gets $N_p \supset N$ by Theorem I.1.10. Part (1) quickly shows that $N = \{0\}$ or $N = N_p$.]
By (1) an equivalence relation is defined on $\operatorname{ex} B_{X^*}$ by $p \sim q$ if $N_p = N_q$. Put $I = \operatorname{ex} B_{X^*}/\sim$ and $X_i = (M_p)_\perp$ for $p \in i$. By the above $(X_i)_{i \in I}$ is a family of minimal nontrivial M-summands in X with pairwise trivial intersection. Thus we have an isometric embedding of $c_{00}(I, X_i)$ into X (where $c_{00}(I, X_i)$ denotes the space of "sequences" of finite support), hence of $c_0(I, X_i)$ into X. But $c_0(I, X_i) \cong \overline{\operatorname{lin}} \bigcup X_i$ is an M-ideal in X (Proposition I.1.11(a)), hence an M-summand by Theorem 2.3(a). The assumption of a proper inclusion of this embedding would give

$$\{0\} \subsetneqq \left(\overline{\operatorname{lin}} \bigcup X_i\right)^\perp = \left(\bigcup X_i\right)^\perp = \bigcap X_i^\perp = \bigcap M_p$$

so that $\bigcap M_p$ is a nontrivial w^*-closed L-summand which contains no extreme point of B_{X^*}: a contradiction. By Theorems 2.3(a) and 1.6(a), the X_i are as desired. □

Proposition 2.7 *For an L^1-predual space X the following assertions are equivalent:*
(i) *X is an M-ideal in X^{**}.*
(ii) *$X \cong c_0(I)$ for some set I.*

PROOF: (ii) $\Rightarrow$ (i): Example 1.4(a).
(i) $\Rightarrow$ (ii): Since X^* is isometric to $L^1(\mu)$ we have $X^{**} \cong C(K)$ for some compact space K, so X, being an M-ideal in X^{**}, is of the form $J_D \cong C_0(K \setminus D)$ for some closed subset D of K (Example I.1.4(a)). For every closed subset A of $K \setminus D$ the M-ideal J_A – considered in $C_0(K \setminus D)$ – is an M-summand by Theorem 2.3(a), hence A is open by Example I.1.4(a). So $I := K \setminus D$ is discrete. □

As an application we get the following Banach space result:

Corollary 2.8 *Let X be a Banach space. Then $d(X, c_0) = 1$ implies $X \cong c_0$.*

PROOF: The assumption implies $d(X^*, \ell^1) = 1$ and, as easily seen, this yields that X^* is an $\mathcal{L}^{1,1+\varepsilon}$-space for all $\varepsilon > 0$. Corollary 5 to Theorem 7.1 in [419] entails $X^* \cong L^1(\mu)$ for some measure space (S, Σ, μ). But X is an M-embedded space by Remark 1.7(b), so Proposition 2.7 implies $X \cong c_0(I)$ for some set I. Clearly, I is countable by separability. □

We remark that it is possible to find two equivalent norms $|\,.\,|_1$ and $|\,.\,|_2$ on c_0 such that

$$d((c_0, |\,.\,|_1), (c_0, |\,.\,|_2)) = 1 \quad \text{but} \quad (c_0, |\,.\,|_1) \not\cong (c_0, |\,.\,|_2)$$

(see [484, p. 230] or Proposition 2.13 below). The above corollary shows that this cannot be achieved with the natural norm $\|\,.\,\|_\infty$ on c_0.

We now turn to a "noncommutative" analogue of Proposition 2.7.

Proposition 2.9 *Let A be a C^*-algebra. Then the following assertions are equivalent:*

(i) *A is an M-ideal in A^{**}.*

(ii) *A is isometrically $*$-isomorphic to the c_0-sum of algebras of compact operators on some Hilbert spaces.*

Note that by Theorem V.1.10 and Theorem V.4.4 below, (i) admits the following reformulation in purely algebraic terms:

(i*) *A is a two-sided ideal in its enveloping von Neumann algebra.*

Another algebraic characterisation of M-embedded C^*-algebras, proved in [76, Theorem 5.5] (and in part below as well) is:

(iii) *Every maximal commutative C^*-subalgebra of A is generated by its minimal projections.*

PROOF: (ii) $\Rightarrow$ (i): This implication follows from Theorem 1.6(c) and Example 1.4(f).

(i) $\Rightarrow$ (ii): By Proposition 2.6 there is no loss of generality in assuming that A contains no nontrivial M-ideals; that is we may suppose, anticipating Theorem V.4.4, that A contains no nontrivial closed two-sided ideals. In this case we shall show that A is isometrically and algebraically $*$-isomorphic to the C^*-algebra $K(H)$ for some Hilbert space, thus settling our claim. (To show the isomorphism in (ii) to be algebraic in the general case, too, we again invoke Theorem V.4.4.)

In what follows we call a self-adjoint idempotent element p of a C^*-subalgebra $B \subset A$ minimal with respect to B if the only idempotent elements $q \in B$ satisfying $0 \le q \le p$ are $q = 0$ or $q = p$.

Let M be a maximal abelian C^*-subalgebra of A. Thus $M \cong C_0(\Omega)$ for some locally compact Hausdorff space Ω by the Gelfand-Naimark theorem. Then we assert:

$$\begin{aligned} M &= \overline{\text{lin}}\,\{p \in M \mid p \text{ is minimal with respect to } M\} =: M_1 \\ &= \overline{\text{lin}}\,\{p \in M \mid p \text{ is minimal with respect to } A\} =: M_2. \end{aligned}$$

Indeed, since M is an M-ideal in M^{**} (by assumption and Theorem 1.6(a)) we conclude from Proposition 2.7 that $M \cong c_0(I)$ for some set I. Hence we have $M = M_1$, and clearly $M_2 \subset M_1$ holds. On the other hand, if an idempotent element $p \in M$ is minimal with respect to M and $0 \le q \le p$ for some idempotent $q \in A$, then p and q commute, and by maximality of M we infer that $q \in M$. Hence p is minimal with respect to A as well, and the inclusion $M_1 \subset M_2$ follows.
As a result, A is the closed linear span of its minimal idempotents, since every self-adjoint element is contained in some M as above.
Now let $p \in A$, $p \neq 0$ be a minimal idempotent of A. Then we claim:

$$pAp = \mathbb{C}p \tag{1}$$

To see this, note that $B := pAp$ is a unital C^*-subalgebra with unit p, and p is clearly minimal with respect to B. Now B is an M-ideal in B^{**} (Theorem 1.6), but on the other hand B has the IP by Proposition II.4.2(b). Hence B is reflexive as a result of Theorem II.4.4, consequently finite dimensional (cf. [361, p. 288]). The minimality of the unit implies that B must be one-dimensional. [If not, B would contain an abelian C^*-subalgebra M with $p \in M$ and $d := \dim(M) \ge 2$. Since M can be identified with $\mathbb{C}^d$, the product being pointwise multiplication, and p with $(1, \dots, 1)$, p cannot be minimal with respect to M, let alone B, if $d \ge 2$.]
Let $H = Ap$. We introduce a scalar product on H as follows. If $xp, yp \in H$, then $py^*xp \in pAp$ is a scalar multiple of p (by (1)) which we denote by $\langle xp, yp \rangle$:

$$\langle xp, yp \rangle p = (yp)^*(xp) = py^*xp \tag{2}$$

Then $\langle xp, xp \rangle = \|\langle xp, xp \rangle p\| = \|(xp)^*(xp)\| = \|xp\|^2$ so that the norm induced by $\langle \cdot, \cdot \rangle$ coincides with the C^*-algebra norm inherited from A. Hence H is a Hilbert space. Finally we define

$$T : A \to L(H), \qquad (Ta)(xp) = axp.$$

Clearly, this is a $*$-homomorphism; and T is injective since otherwise $\ker(T)$ would be a nontrivial closed two-sided ideal of A.
To complete the proof we wish to show that $\mathrm{ran}(T) = K(H)$. First of all, if $u : zp \mapsto \langle zp, yp \rangle xp$ is an operator on H with rank 1, then an easy computation reveals $T(xpy^*) = u$; and it follows that $K(H) \subset \mathrm{ran}(T)$. For the converse inclusion it is enough to show that $Tq \in K(H)$ if q is a minimal idempotent of A. (Recall from the first part of the proof that these q generate A.) Now observe $qAq = \mathbb{C}q$ (by (1)) and consequently

$$Tq\,K(H)\,Tq \subset Tq\,T(A)\,Tq = \mathbb{C}\,Tq.$$

Thus either $Tq\,K(H)\,Tq = 0$ in which case $Tq = 0$ or $\mathbb{C}\,Tq = Tq\,K(H)\,Tq \subset K(H)$. Hence $Tq \in K(H)$ in either case, and the proof is finished. □

In the remainder of this section we will collect various other isometric properties of M-embedded spaces. To start with, we recall the following definition: A Banach space X is said to be a *strongly unique predual* (of X^*), if, for every Banach space Y, every isometric isomorphism from X^* onto Y^* is w^*-continuous, hence is the adjoint of an isometric isomorphism from Y onto X. It is well-known (and routine to verify) that

X has this property iff π_{X^*} is the only contractive projection from X^{***} onto X^* with w^*-closed kernel.
Clearly a strongly unique predual is a *unique predual*, which of course means that $X^* \cong Y^*$ implies $X \cong Y$. This latter property is not easy to handle; in particular no example is known of a space which is a unique predual, but not a strongly unique predual. We refer to Godefroy's recent survey [259] for detailed information.

Proposition 2.10 *Let X be a nonreflexive Banach space which is an M-ideal in its bidual. Then*
(a) *X^* is a strongly unique predual of X^{**},*
(b) *X is not a strongly unique predual of X^*.*

PROOF: (a) By what was noted above this is an immediate consequence of Proposition 2.1.
(b) By assumption and Proposition 1.2 we have

$$X^{***} = X^* \oplus_1 X^\perp.$$

Consider $X^\perp$ as the dual space of X^{**}/X and choose $y^{***} \in X^\perp$, $\|y^{***}\| = 1$, and a w^*- (i.e. $\sigma((X^{**}/X)^*, (X^{**}/X))$-) closed hyperplane H in $X^\perp$ such that

$$1 \leq \|y^{***} + h^{***}\| \quad \text{for all } h^{***} \in H.$$

[Take for example a norm-attaining $y^{***} \in (X^{**}/X)^*$, $\|y^{***}\| = 1$, and put $H = \ker i_{x^{**}}$, where $y^{***}(x^{**}) = 1 = \|x^{**}\|$.] Because the w^*-topology on $(X^{**}/X)^*$ coincides with the relative w^*-topology of X^{***} on $X^\perp$, the hyperplane H is $\sigma(X^{***}, X^{**})$-closed in X^{***}. Hence for $0 < s \leq 1$ and $p^* \in X^*$, $\|p^*\| = 1$,

$$Z := H \oplus \mathbb{K}(y^{***} + sp^*)$$

is w^*-closed in X^{***}. Let P be the projection onto X^* associated with the decomposition

$$X^{***} = X^* \oplus Z.$$

For $x^* + z^{***} \in X^* \oplus Z$ we have

$$x^* + z^{***} = x^* + [h^{***} + \lambda(y^{***} + sp^*)] = (x^* + \lambda sp^*) + (h^{***} + \lambda y^{***}) \in X^* \oplus_1 X^\perp.$$

Since $\|\lambda sp^*\| = |\lambda| s \leq |\lambda| \leq \|h^{***} + \lambda y^{***}\|$ for all $\lambda \in \mathbb{K}$, $h^{***} \in H$ we get

$$\begin{aligned} \|P(x^* + z^{***})\| &= \|x^*\| \\ &\leq \|x^* + \lambda sp^*\| + \|\lambda sp^*\| \\ &\leq \|x^* + \lambda sp^*\| + \|h^{***} + \lambda y^{***}\| \\ &= \|x^* + z^{***}\|. \end{aligned}$$

So P is a contractive projection onto X^* with w^*-closed kernel Z and is different from π_{X^*}, since $Z \neq X^\perp$. □

REMARKS: (a) It is easy to verify that for the projection P in the above proof we have $\|Id - P\| = 1 + s$. For the predual Y of X^* obtained by $Y^\perp = Z$ we thus get by some formal calculations $\|Id_{Y^{***}} - \pi_{Y^*}\| = 1 + s$. This shows in particular that there are Banach spaces Y for which $\|Id_{Y^{***}} - \pi_{Y^*}\|$ can be any number between 1 and 2. (The first example of this kind was given in [353] using a renorming of c_0.)

(b) If X is an M-embedded space then we have by Corollary 1.3 and Proposition 2.10(a) that the L-embedded space X^* is a strongly unique predual. However the following is open:

PROBLEM: *Is every L-embedded space X a strongly unique predual of X^*?*

The answer is yes for the "classical" Examples IV.1.1 (see [259, section V]), but in general this problem seems to be difficult. Note that a counterexample would solve the long-standing question whether a Banach space which contains no isomorphic copy of c_0 is a strongly unique predual. (By Theorem IV.2.2 L-embedded spaces are weakly sequentially complete; a fortiori, they do not contain copies of c_0.)

(c) The authors of [27] give an isometric representation of the dual of the little Bloch space B_0 as the minimal Möbius invariant space $\mathcal{M}$, and they ask in [26] whether B_0 is the only isometric predual of $\mathcal{M}$ and whether $\mathcal{M}$ is the only isometric predual of B. An explicit solution of the case of $\mathcal{M}$ is contained in [450], whereas the first question is left open there. Proposition 2.10 and Example 1.4(j) answer both questions.

Incidentally, M-ideal arguments can also be used to reprove Theorem 2 of [133] which asserts that the onto isometries of $B = B_0^{**}$ coincide with the second adjoints of the onto isometries of B_0: Proposition 2.2. We also obtain from Proposition II.4.2 and Theorem II.4.4 that the unit ball of B_0 cannot contain any strongly extreme points, a fact first pointed out in [132] by ad-hoc methods.

In the next proposition we describe a renorming technique for M- (and L-) embedded spaces. In order not to obscure the main idea we don't give the most general form.

Proposition 2.11 *Let X be an M-embedded space, Y a Banach space and $T : Y \longrightarrow X$ a weakly compact operator. Then*

$$|x^*|^* := \|x^*\| + \|T^*x^*\|$$

defines an equivalent dual norm on X^* for which $(X, |\,.\,|)$ is an M-embedded space.

PROOF: Since $x^* \longmapsto \|T^*x^*\|$ is w^*-lower semicontinuous, $|\,.\,|^*$ is an equivalent dual norm (see e.g. [155, p. 106]). To calculate $|\,.\,|^{***}$ on X^{***} without using $|\,.\,|^{**}$ we employ the following argument:

By definition the operator

$$\begin{array}{rcll} S : (X^*, |\,.\,|^*) & \longrightarrow & X^* \oplus_1 Y^* \\ x^* & \longmapsto & (x^*, T^*x^*) \end{array}$$

is isometric, hence $S^{**} : (X^{***}, |\,.\,|^{***}) \longrightarrow X^{***} \oplus_1 Y^{***}$ is isometric, too. As easily seen $S^{***}x^{***} = (x^{***}, T^{***}x^{***})$, so $|x^{***}|^{***} = \|x^{***}\| + \|T^{***}x^{***}\|$. By w-compactness and w^*-continuity of T^* we get $T^{***} = \pi_{Y^*}T^{***} = T^{***}\pi_{X^*}$, therefore $T^{***}x^{***} = T^*x^*$

where $x^{***} = x^* + x^\perp \in X^* \oplus_1 X^\perp$. Hence

$$\begin{aligned} |x^{***}|^{***} &= |x^* + x^\perp|^{***} \\ &= \|x^* + x^\perp\| + \|T^{***}(x^* + x^\perp)\| \\ &= \|x^*\| + \|x^\perp\| + \|T^* x^*\| \\ &= |x^*|^* + \|x^\perp\| \end{aligned}$$

$x^* = 0$ yields $|x^\perp|^{***} = \|x^\perp\|$ (so that the two norms agree on $X^\perp$), and we can continue

$$= |x^*|^{***} + |x^\perp|^{***}$$

Hence $X^\perp$ is, with respect to the new norm, still an L-summand in X^{***}. □

Applying the above proposition to the formal identity $I : \ell^2 \to c_0$, we find:

Corollary 2.12 *There is an equivalent norm $|\,.\,|$ on c_0 such that $(c_0, |\,.\,|)^*$ is strictly convex, hence $(c_0, |\,.\,|)$ is smooth, and $(c_0, |\,.\,|)$ is an M-ideal in its bidual. Moreover, there are uncountably many M-ideals in $(c_0, |\,.\,|)^{**}$.*

PROOF: The injectivity of $I^* : \ell^1 \to \ell^2$ yields the strict convexity of $|\,.\,|^*$, and the smoothness of $|\,.\,|$ follows [155, p. 23, p. 100]. Every ideal in ℓ^∞ containing c_0 yields an M-ideal in ℓ^∞/c_0 (Prop. I.1.17) and $(c_0, |\,.\,|)^{**}/(c_0, |\,.\,|)$ is isometric to ℓ^∞/c_0 by what was noted in the above proof. □

In Theorem 4.6(e) we will see that a similar renorming is possible for all M-embedded spaces. Also, in Remark IV.1.17 we will point out an example of a strictly convex M-embedded space with a smooth dual, namely $C(\mathbb{T})/A$.

In the last renorming result we show that the existence of nontrivial M-ideals or M-summands is not preserved by almost isometries.

Proposition 2.13 *There is an M-embedded space X isomorphic to c_0 without nontrivial M-ideals such that $d(X, X \oplus_\infty X) = 1$.*

PROOF: Let $E_1 := \mathbb{K}$ and define inductively $E_{n+1} := E_n \oplus_{2^n} E_n$. Put $X_n := c_0(E_n)$ and denote the norm of X_n by $\|\;\|_n$. For $x \in c_0$ we obtain $\|x\|_n \le \|x\|_{n+1} \le 2^{1/2^n} \|x\|_n$, hence

$$\|x\|_n \le 2^{1/2^{n-1}} 2^{1/2^{n-2}} \ldots 2^{1/2} \|x\|_1 \le 2^{\sum_{k=1}^\infty 1/2^k} \|x\|_1 = 2\|x\|_1.$$

Thus $\|x\| := \sup_n \|x\|_n$ defines a norm equivalent to $\|\;\|_1$, the usual sup-norm of c_0. Since

$$\|x\|_n \le \|x\| \le 2^{\sum_{k=n}^\infty 1/2^k} \|x\|_n = 2^{2^{-n+1}} \|x\|_n \qquad (*)$$

we get $d(X_n, X) \to 1$ where $X = (c_0, \|\,.\,\|)$.

Since X_n is an M-embedded space by Theorem 1.6, the uniform estimate $(*)$ and the 3-ball property yield that X is M-embedded, too. From $X \overset{1+\varepsilon}{\simeq} X_n \cong X_n \oplus_\infty X_n \overset{1+\varepsilon}{\simeq} X \oplus_\infty X$ we deduce $d(X, X \oplus_\infty X) = 1$. Finally every $x \in c_{00}$ with unit norm is an extreme point of B_X: if $\operatorname{supp}(x) \subset \{1, \ldots, 2^n\}$ we get $\|x\|_{n+1} = \|x\|_{n+2} = \ldots = 1$ and the strict convexity of E_{n+k} implies $x \in \operatorname{ex} B_X$. So $\overline{\operatorname{ex}}^{\|\,\|} B_X = S_X$, hence X can't have nontrivial M-summands and by Theorem 2.3 no nontrivial M-ideals either. □

We now give a characterisation of the Hahn-Banach smoothness of a Banach space X, considered as a subspace of X^{**}, and then go on to apply it to M-embedded spaces.

Lemma 2.14 *For a Banach space X and $x^* \in S_{X^*}$ the following are equivalent:*

(i) *x^* has a unique norm preserving extension to a functional on X^{**}.*

(ii) *The relative w- and w^*-topologies on B_{X^*} agree at x^*, meaning that the function $Id_{B_{X^*}} : (B_{X^*}, w^*) \to (B_{X^*}, w)$ is continuous at x^*.*

Proof: (i) ⇒ (ii): Let (x^*_α) be a net in B_{X^*} with $x^*_\alpha \xrightarrow{w^*} x^*$. By $\sigma(X^{***}, X^{**})$-compactness of $B_{X^{***}}$ we find for every subnet (x^*_β) of (x^*_α) an $x^{***} \in B_{X^{***}}$ and a subnet (x^*_γ) such that $x^*_\gamma \longrightarrow x^{***}$ with respect to $\sigma(X^{***}, X^{**})$. Since $x^*_\gamma \longrightarrow x^*$ with respect to $\sigma(X^*, X)$ we infer that x^{***} is a norm preserving extension of x^*, hence $x^{***} = x^*$. So the $\sigma(X^{***}, X^{**})$-convergence of (x^*_γ) to x^* is the desired $\sigma(X^*, X^{**})$-convergence of (x^*_γ) to x^*.

(ii) ⇒ (i): Let $x^{***} = x^* + x^\perp$ be a norm preserving extension of x^*. By Goldstine's theorem there is a net (x^*_α) in B_{X^*} such that $x^*_\alpha \longrightarrow x^{***}$ with respect to $\sigma(X^{***}, X^{**})$. In particular $x^*_\alpha \xrightarrow{w*} x^*$ on B_{X^*}, hence by assumption $x^*_\alpha \xrightarrow{w} x^*$. This means $x^*_\alpha \longrightarrow x^*$ with respect to $\sigma(X^{***}, X^{**})$ so that $x^\perp = 0$. □

Corollary 2.15 *If X is an M-ideal in its bidual, then the relative w- and w^*-topologies on B_{X^*} agree on S_{X^*}, meaning that the function $Id_{B_{X^*}} : (B_{X^*}, w^*) \to (B_{X^*}, w)$ is continuous at all $x^* \in S_{X^*}$.*

Proof: This is an immediate consequence of Proposition I.1.12 and Lemma 2.14. □

Remarks: (a) It is an amusing exercise to give a direct proof of Corollary 2.15 using the 3-ball property.

(b) If the assumption in Corollary 2.15 is strengthened to "$K(X)$ is an M-ideal in $L(X)$" (see Proposition VI.4.4), then the conclusion that the relative $\|\,.\,\|$- and w^*-topologies agree on S_{X^*} obtains (Proposition VI.4.6). However, for this stronger conclusion it does not suffice to assume only that X is an M-ideal in X^{**}: renorm ℓ^2 by $\ell^2 \simeq \mathbb{K} \oplus_\infty \ell^2$; then $\|(1, e_n)\| = 1 = \|(1,0)\|$, $(1, e_n) \xrightarrow{w} (1,0)$, but $(1, e_n) \stackrel{\|\,\|}{\not\longrightarrow} (1,0)$.

Corollary 2.16 *If X is an M-ideal in its bidual, then X^* contains no proper norming subspace.*

Proof: If $V \subset X^*$ is norming, which means that $\|x\| = \sup_{x^* \in B_V} |x^*(x)|$ for all $x \in X$, then B_V is w^*-dense in B_{X^*} by the Hahn-Banach theorem. So every point $x^* \in S_{X^*}$ where the relative w- and w^*-topologies on B_{X^*} agree belongs to V. Hence the statement follows from Corollary 2.15. □

In Proposition 3.9 we will prove a stronger statement.

III.3 Isomorphic properties

By what was said in Remark (b) following Definition 1.1, nonreflexive M-embedded spaces are proper M-ideals, so they contain isomorphic copies of c_0 by Theorem II.4.7. This, and Theorem 3.1 below (M-embedded spaces are Asplund spaces, equivalently duals of M-embedded space have the Radon-Nikodým property), were the first results of isomorphic nature in M-structure theory.
The main part of the following section is devoted to the properties (V) and (u) of Pełczyński, two properties which imply the containment of c_0 in nonreflexive M-embedded spaces. Although in our situation property (u) is the stronger one, we will also give an independent proof of property (V) since it uses a completely different – and in our opinion, a very natural – approach to the problem. We will conclude this section with some more results of isomorphic nature, among them a characterisation of M-embedded spaces among separable $\mathcal{L}^\infty$-spaces.

First we deal with the Radon-Nikodým property (RNP). This property of a Banach space is defined and studied in detail in the monographs [93] and [159]. The RNP, originally defined in measure theoretic terms, can be equivalently characterised in a number of ways. The case of a dual Banach space is especially pleasing, since it is a theorem that X^* has the RNP if and only if every continuous convex function from an open convex subset O of X into $\mathbb{R}$ is Fréchet differentiable on a dense G_δ-subset of O. Moreover, either of these two properties is equivalent to the requirement that separable subspaces of X have separable duals. Banach spaces fulfilling the above differentiability hypothesis are called Asplund spaces, so X is an Asplund space if and only if X^* has the RNP. References for these results are [93, p. 91, p. 132], [159, p. 82, p. 195, p. 213] and [497, p. 34, p. 75].
Depending on the consequences one wants to draw from Theorem 3.1 below, it is sometimes more advantageous to use the Asplund property, sometimes one is better off using the RNP of the dual in order to give effective proofs. In particular, the result quoted in Corollary 3.2 is more easily derived from the Asplund property of X, whereas we shall have occasion to employ the (measure theoretic) RNP of the dual in Chapter VI.
The following theorem will frequently be used in the following. Its proof relies ultimately on Corollary 2.15.
For a Banach space X we denote the least cardinal $\mathfrak{m}$ for which X has a dense subset of cardinality $\mathfrak{m}$, the so-called density character of X, by $\operatorname{dens} X$. Thus $X \neq \{0\}$ is separable if and only if $\operatorname{dens} X = \aleph_0$.

Theorem 3.1 *If X is an M-embedded space and Y a subspace of X, then* $\operatorname{dens} Y = \operatorname{dens} Y^*$. *In particular, separable subspaces of X have separable duals. Consequently M-embedded spaces are Asplund spaces, and duals of M-embedded spaces have the RNP.*

PROOF: By Theorem 1.6 we may assume that $Y = X$. For a dense subset $\{x_i \mid i \in I\}$ of X we choose $x_i^* \in S_{X^*}$ such that $x_i^*(x_i) = \|x_i\|$. Then $\overline{\operatorname{lin}}\,\{x_i^* \mid i \in I\}$ is norming, hence equal to X^* by Corollary 2.16. For the final statement recall the above remarks. (Actually, here we only use the comparatively easy implications from [159, p. 82] or [497, p. 34].) □

A point x^* of a closed bounded convex subset K of a dual space X^* is called w^*-strongly

exposed (written $x^* \in w^*\text{-sexp}(K)$) if there is an $x \in X$ which strongly exposes x^*, meaning $\operatorname{Re} y^*(x) < \operatorname{Re} x^*(x)$ for $y^* \in K \setminus \{x^*\}$ and the sets $\{y^* \in K \mid \operatorname{Re} y^*(x) > \operatorname{Re} x^*(x) - \varepsilon\}$ $(\varepsilon > 0)$ form a neighbourhood base of x^* in K for the relative norm topology (see [93, Definition 3.2.1(w^*)] or [497, Definition 5.8]).
The following consequence of the above is occasionally useful.

Corollary 3.2 *If X is an M-embedded Banach space, then*

$$B_{X^*} = \overline{\text{co}}^{\|\ \|}\, w^*\text{-sexp}\, B_{X^*}.$$

PROOF: Since X is an Asplund space by Theorem 3.1 one knows from [497, Theorem 5.12] (or [93, Theorem 4.4.1]) that

$$B_{X^*} = \overline{\text{co}}^{w^*}\, w^*\text{-sexp}\, B_{X^*}.$$

Now, Corollary 2.15 yields for $C := \text{co}\, w^*\text{-sexp}\, B_{X^*}$ the inclusion $\overline{C}^{w} \supset S_{X^*}$, thus also

$$\overline{C}^{\|\ \|} = \overline{C}^{w} \supset \overline{S_{X^*}}^{w} = B_{X^*}$$

for infinite dimensional X. (The finite dimensional case is trivial.) □

Because the next property we want to prove for M-embedded spaces is possibly not so well-known, we wish to not only give the definition but also to list its most important consequences for easy reference.

3.3 The properties (V) and (V^*)

These properties were introduced by A. Pełczyński in [486], where also most of the following can be found. For further information see e.g. [512], [549], [268], [269] and [157]. Recall that a series $\sum x_n$ in a Banach space X is called *weakly unconditionally Cauchy (wuC)*, if for every $x^* \in X^*$ we have $\sum |x^*(x_n)| < \infty$. Equivalently, $\sum x_n$ is a wuC-series if there is some $M \geq 0$ such that $\|\sum \alpha_n x_n\| \leq M \max |\alpha_n|$ for all finitely supported scalar sequences (α_n) [158, p. 44]. We refer to [158] and [573] for basic properties of wuC-series.

DEFINITION. (a) A Banach space X is said to have *property* (V) if every subset K of X^* satisfying

$$\lim_n \sup_{x^* \in K} |x^*(x_n)| = 0$$

for every wuC-series $\sum x_n$ in X is relatively weakly compact.
(b) A Banach space X is said to have *property* (V^*) if every subset K of X satisfying

$$\lim_n \sup_{x \in K} |x_n^*(x)| = 0$$

for every wuC-series $\sum x_n^*$ in X^* is relatively weakly compact.

Note that (by considering suitably defined operators into ℓ^1) the converse implications in the above definition are always true, e.g. for a relatively weakly compact subset K of X^* and $\sum x_n$ wuC in X one has $\lim_n \sup_{x^* \in K} |x^*(x_n)| = 0$.
These properties were introduced in order to understand why operators from $C(K)$-spaces to spaces not containing c_0 are always weakly compact, and they are closely related to unconditionally converging operators which we define next.

DEFINITION. Let X and Y be Banach spaces. An operator $T \in L(X,Y)$ is called *unconditionally converging* if for every wuC-series $\sum x_n$ in X the series $\sum Tx_n$ is unconditionally convergent in Y. We write $U(X,Y)$ for the set of all unconditionally converging operators from X to Y. This makes U a closed Banach operator ideal (see [498, p. 48]).

These operators are characterised in the following lemma.

Lemma 3.3.A *Let X and Y be Banach spaces.*

(a) *For $T \in L(X,Y)$ one has*

$T \notin U(X,Y) \iff$ *there is a subspace X_0 of X which is isomorphic to c_0 such that $T|_{X_0}$ is an isomorphism.*

(b) *For $T \in L(Y,X)$ one has*

$T^* \notin U(X^*,Y^*) \iff$ *there is a subspace Y_0 of Y which is isomorphic to ℓ^1 such that $T|_{Y_0}$ is an isomorphism and $T(Y_0)$ is complemented in X.*

For a proof see [512, p. 270 and p. 272]. Maybe some readers will find it misleading to call this result a lemma, after all it contains two beautiful theorems of Bessaga and Pełczyński in the special case $X = Y$ and $T = Id_X$, namely "A Banach space X contains no copy of c_0 iff every wuC-series $\sum x_n$ in X is unconditionally converging" and "X contains a complemented copy of ℓ^1 iff X^* contains c_0". In our context, however, it serves only as a preparation to the following characterisation:

Theorem 3.3.B *Let X be a Banach space.*

(a) *The following are equivalent:*

(i) *X has property (V).*

(ii) $U(X,\,.\,) \subset W(X,\,.\,)$

(iii) *For all Banach spaces Y and for all operators $T: X \longrightarrow Y$ which are not weakly compact there is a subspace X_0 of X isomorphic to c_0 such that $T|_{X_0}$ is an isomorphism.*

(b) *The following are equivalent:*

(i*) *X has property (V^*).*

(ii*) $U^{dual}(\,.\,,X) \subset W(\,.\,,X)$

(iii*) *For all Banach spaces Y and all operators $T: Y \longrightarrow X$ which are not weakly compact there is a subspace Y_0 of Y isomorphic to ℓ^1 such that $T|_{Y_0}$ is an isomorphism and $T(Y_0)$ is complemented in X.*

[The statement (ii*) reads: for every Banach space Y every operator $T \in L(Y, X)$ is weakly compact if T^* is unconditionally converging – see [498, p. 67] for the dual of an operator ideal.] The equivalence of the second and the third statements is immediate from Lemma 3.3.A, "(i) $\iff$ (ii)" is proved in [486], the corresponding proof of part (b) is similar. [Note that $W(E, F) \subset U(E, F)$ for all Banach spaces E and F; see e.g. [498, p. 51].]

Corollary 3.3.C

(a) *A nonreflexive Banach space with property (V) contains a subspace isomorphic to c_0.*

(b) *A nonreflexive Banach space with property (V^*) contains a complemented subspace isomorphic to ℓ^1.*

The remaining statements are from [486]:

Proposition 3.3.D *Let X be a Banach space.*

(a) *If X has property (V), then X^* has property (V^*).*

(b) *If X^* has property (V), then X has property (V^*).*

Proposition 3.3.E *Let X be a Banach space and Y a subspace of X.*

(a) *If X has property (V), then X/Y has property (V).*

(b) *If X has property (V^*), then Y has property (V^*).*

Proposition 3.3.F *A Banach space X with property (V^*) is weakly sequentially complete.*

Clearly reflexive Banach spaces have the properties (V) and (V^*). To give at least one nonreflexive example we mention that $C(K)$-spaces have property (V). This is essentially a consequence of Grothendieck's characterisation of relatively weakly compact subsets of $M(K)$ [486, Theorem 1].

Now back to M-ideals.

Theorem 3.4 *Every M-embedded Banach space X has property (V).*

PROOF: We use Theorem 3.3.B, part (a), and show that for every Banach space Y every nonweakly compact operator $T \in L(X, Y)$ is not unconditionally converging. To achieve this we will construct a wuC-series $\sum x_n$ in X such that (Tx_n) does not converge to zero. Since T is not weakly compact there is $x^{**} \in X^{**}$ with $T^{**}x^{**} \in Y^{**} \setminus Y$. We may assume without restriction $\|x^{**}\| = 1$. Put $\alpha := d(T^{**}x^{**}, Y) > 0$ and take $\varepsilon_n > 0$ with $\prod(1+\varepsilon_n) < \infty$. We will inductively construct $x_n \in X$ such that

$$\begin{aligned} \|\pm x_1 \pm x_2 \cdots \pm x_n\| &\leq \prod_{k=1}^{2n}(1+\varepsilon_k), \qquad n \in \mathbf{N} &(1)\\ \|Tx_n\| &> \frac{\alpha}{2} \qquad \text{for } n \geq 2 &(2)\end{aligned}$$

Since (1) says that $\sum x_n$ is weakly unconditionally Cauchy this will finish the proof. The case $n = 1$ is trivial. So assume $x_1, \ldots, x_n$ are given satisfying (1) and (2). Put $\eta := \prod_{k=1}^{2n}(1+\varepsilon_k)$. Then $\|\frac{1}{\eta}(\pm x_1 \pm x_2 \cdots \pm x_n)\| \le 1$, $\|x^{**}\| = 1$ and, since X is an M-ideal in X^{**}, we find (Theorem I.2.2) $z \in X$ such that

$$\left\| x^{**} + \frac{1}{\eta}(\pm x_1 \pm x_2 \cdots \pm x_n) - z \right\| \le 1 + \varepsilon_{2n+1},$$

i.e.

$$\|\eta\,(x^{**} - z) + (\pm x_1 \pm x_2 \cdots \pm x_n)\| \le \eta\,(1 + \varepsilon_{2n+1}).$$

Put

$$E := \operatorname{lin}(x^{**}, z, x_1, \ldots, x_n) \subset X^{**}.$$

Because $d(T^{**}x^{**}, Y) = \alpha$ we have $\alpha \le \|T^{**}x^{**} - Tz\| = \|T^{**}(x^{**} - z)\|$. So we find $y^* \in B_{Y^*}$ such that

$$|T^{**}(x^{**} - z)(y^*)| > \frac{\alpha}{2}.$$

Put $G := \operatorname{lin}\{Ty^*\} \subset X^*$. By the principle of local reflexivity there is an operator $A : E \longrightarrow X$ such that

- $Ax = x$ for $x \in E \cap X$,
- $\|A\| \le 1 + \varepsilon_{2n+2}$,
- $g^*(Ae^{**}) = e^{**}(g^*)$ for $e^{**} \in E$, $g^* \in G$.

Put $x_{n+1} := A(\eta\,(x^{**} - z))$. Then

$$\begin{aligned} \|x_{n+1} \pm x_1 \cdots \pm x_n\| &= \|A[\eta\,(x^{**} - z) + (\pm x_1 \cdots \pm x_n)]\| \\ &\le \|A\|\eta\,(1 + \varepsilon_{2n+1}) \\ &\le \prod_{k=1}^{2(n+1)} (1 + \varepsilon_k) \end{aligned}$$

and

$$\begin{aligned} \|Tx_{n+1}\| &\ge |y^*(Tx_{n+1})| \\ &= |(T^*y^*)(x_{n+1})| \\ &= \eta\,|(T^*y^*)(A(x^{**} - z))| \\ &= \eta\,|(x^{**} - z)(T^*y^*)| \\ &= \eta\,|T^{**}(x^{**} - z)(y^*)| \\ &> \eta\,\frac{\alpha}{2} > \frac{\alpha}{2}. \end{aligned}$$

Hence $x_1, \ldots, x_n, x_{n+1}$ satisfy (1) and (2). □

We shall prove in Theorem IV.2.7 the companion result that L-embedded spaces enjoy property (V^*).

As a first application of the above theorem let us give an example which answers the question of Pełczyński [486, p. 646] whether the converse of Proposition 3.3.D holds:

Example 3.5 *The space $X = (\oplus \sum \ell^\infty(n))_{\ell^1}$ has property (V^*), but X^* fails property (V).*

PROOF: X is the dual of $X_* = (\oplus \sum \ell^1(n))_{c_0}$, which is an M-embedded space by Theorem 1.6. So X has property (V^*) by the above Theorem 3.4 and Proposition 3.3.D. But $X^* = (\oplus \sum \ell^1(n))_{\ell^\infty}$ contains a 1-complemented subspace isometric to ℓ^1 (see the proof of Example IV.1.7(b)). Since by Proposition 3.3.E property (V) passes to quotients (in particular to complemented subspaces) and ℓ^1 fails property (V) by Corollary 3.3.C, we get the desired conclusion. □

The first example to answer Pełczyński's question about the converse of Proposition 3.3.D was given in [549]. Let us mention in passing that the Bourgain-Delbaen spaces [91] may serve as examples of spaces X failing property (V), but whose duals have property (V^*).

By Propositions 3.3.D and 3.3.F the dual of a Banach space with property (V) is weakly sequentially complete. The following result gives a slightly stronger statement, which may also be viewed as an abstract version of Phillips's lemma.

Proposition 3.6 *If a Banach space X has property (V), then the natural projection π_{X^*} from X^{***} onto X^* is w^*-w-sequentially continuous.*

PROOF: It is enough to show $\pi_{X^*}(x_n^{***}) = x_n^{***}|_X \xrightarrow{w} 0$ for a w^*-null sequence (x_n^{***}) in X^{***}. The sequence (x_n^{***}) induces an operator $T \in L(X^{**}, c_0)$ by means of $Tx^{**} = (x_n^{***}(x^{**}))$, and $(x_n^{***}|_X)$ corresponds to $T|_X$. If $(x_n^{***}|_X)$ were not weakly null, then $T|_X$ would not be weakly compact, hence, by Theorem 3.3.B, there would exist a subspace X_0 of X isomorphic to c_0 such that the restriction of $T|_X$ to X_0 is an isomorphism. We would therefore obtain a diagram

$$\begin{array}{ccc} X^{**} & \xrightarrow{\quad T \quad} & c_0 \\ \Big\uparrow & & \Big\downarrow P \\ X_0 & \xleftarrow[\;T|_{X_0}{}^{-1}\;]{} & T(X_0) \simeq c_0 \end{array}$$

where P is a projection existing by Sobczyk's theorem (Corollary II.2.9). $T|_{X_0}{}^{-1} \circ P \circ T$ is then a projection from the dual space X^{**} onto a subspace isomorphic to c_0. But it is quickly checked that a complemented subspace V of a dual space is complemented in V^{**}, a property not shared by c_0. This contradiction completes the proof. □

It seems appropriate to gather some consequences of what has been shown so far.

Corollary 3.7 *Let X be a nonreflexive Banach space which is an M-ideal in its bidual. Then:*

(a) *Every subspace of X has property (V). In particular: X contains a copy of c_0, X is not weakly sequentially complete, and X fails the Radon-Nikodým property.*

(b) *X^* has property (V^*). In particular: X^* is weakly sequentially complete and contains a complemented copy of ℓ^1.*

(c) *X is not complemented in X^{**}.*

(d) *X^{**}/X is not separable.*

(e) *Every subspace or quotient of X which is isomorphic to a dual space is reflexive.*

(f) *Every operator from X to a space not containing c_0 (in particular, every operator from X to X^*) is weakly compact.*

PROOF: (a) This follows from Theorem 1.6 and Theorem 3.4.

(b) Theorem 3.4, Proposition 3.3.D, Proposition 3.3.F and Corollary 3.3.C.

(c) Assuming X to be complemented in X^{**} we infer from (a) and [529, Theorem 1.3] that X contains a subspace isomorphic to ℓ^∞. Hence X contains ℓ^1 – a separable space with nonseparable dual – contradicting Theorem 3.1.

(d) By (a) c_0 embeds into X, hence ℓ^∞/c_0 embeds into X^{**}/X.

(e) A subspace or quotient Y of X is again M-ideal in its bidual (Theorem 1.6). If Y is nonreflexive then Y is not complemented in Y^{**} by part (c); in particular Y is not isomorphic to a dual space.

(f) Theorem 3.4 and Theorem 3.3.B. □

That nonreflexive M-embedded spaces contain (arbitrarily good) copies of c_0 can be deduced from Theorem II.4.7, too. Let us also mention that we shall present an improvement of part (b) in Theorems IV.2.2 and IV.2.7.

The following property of a Banach space was introduced by Pełczyński in [485].

DEFINITION. A Banach space X is said to have *property* (u) if for every weak Cauchy sequence (x_n) in X there is a wuC-series $\sum y_k$ in X such that $(x_n - \sum_{k=1}^n y_k)$ converges to zero weakly.

A reformulation of this is: X has property (u) if and only if every x^{**} in the w^*-sequential closure of X in X^{**} is also the w^*-limit of a wuC-series $\sum y_k$ in X; we write $x^{**} = \sum^* y_k$ in this case.

REMARKS: (a) It is clear that weakly sequentially complete Banach spaces have property (u). To see other examples we mention that order continuous Banach lattices have property (u); cf. e.g. [423, Prop. 1.c.2].

(b) In the Notes and Remarks section of Chapter IV we will explain how property (u) may be viewed as a statement about the continuity of elements of the bidual considered as functions on (B_{X^*}, w^*).

Theorem 3.8 *Every M-embedded space X has property (u).*

PROOF: For a sequence (x_n) in X which is w^*-convergent to some $x^{**} \in X^{**}$ consider the separable M-embedded space $Y = \overline{\text{lin}}\,(x_n)$ and note that the σ-topology from Remark I.1.13 is just the w^*-topology in our situation. Hence the assertion is a special case of Theorem I.2.10. □

While property (u) always passes to subspaces (see Lemma I.2.9) it is generally not inherited by quotients. It seems noteworthy that quotients of M-embedded spaces do enjoy property (u) by Theorems 1.6 and 3.8.
Let us briefly comment on the relation between the properties (u) and (V). Pełczyński has shown in [486, Prop. 2] that a Banach space with property (u) in which every bounded sequence contains a weak Cauchy subsequence has property (V). The additional assumption is clearly fulfilled for M-embedded spaces, since $B_{Y^{**}}$ is w^*-metrizable if Y is a separable M-embedded space as a result of Theorem 3.1. By the way, the celebrated ℓ^1-theorem of Rosenthal (see e.g. [422, Theorem 2.e.5]) states that every bounded sequence in X contains a weak Cauchy subsequence if and only if ℓ^1 does not embed into X. Therefore a restatement of Pełczyński's above proposition is:

> *A Banach space with property (u) which doesn't contain an isomorphic copy of ℓ^1 has property (V).*

We refer to the Notes and Remarks section for consequences of Theorem 3.8 on property (X) of Godefroy and Talagrand.

The next result is a quantitative version of Corollary 2.16. Recall from p. 67 that the characteristic $r(V, X^*)$ of a subspace V of a dual space X^* is given by

$$r(V, X^*) = \max\{r \geq 0 \mid rB_{X^*} \subset \overline{B_V}^{w^*}\} = \inf_{x \in S_X} \sup_{x^* \in B_V} |x^*(x)|.$$

Proposition 3.9 *If X is an M-embedded space, then for every proper subspace V of X^* one has $r(V, X^*) \leq 1/2$.*

PROOF: Since $r(V, X^*) \leq r(W, X^*)$ if $V \subset W$, it is enough to show the claim for hyperplanes. Let $V = \ker x_0^{**}$, $\|x_0^{**}\| = 1$. Take $x^* \in X^*$ such that $\|x^*\| = 1/2 + \varepsilon$ and $x_0^{**}(x^*) > 1/2$. Assuming $(1/2 + \varepsilon)B_{X^*} \subset \overline{B_V}^{w^*}$ we could find $v_\alpha \in B_V$ with $v_\alpha \xrightarrow{w^*} x^*$. For a $\sigma(X^{***}, X^{**})$-accumulation point $x^{***} \in B_{X^{***}}$ of (v_α) we obtain $x^{***}|_X = x^*$. Therefore $x^{***} = x^* + x^\perp$ with $x^\perp \in X^\perp$ and $\|x^\perp\| = \|x^{***}\| - \|x^*\| \leq 1/2 - \varepsilon$. Also $x^{***}(x_0^{**}) = 0$ since $x_0^{**}(v_\alpha) = 0$. But then

$$|x^\perp(x_0^{**})| = |(x^{***} - x^*)(x_0^{**})| = |x_0^{**}(x^*)| > 1/2 > \|x^\perp\|,$$

which contradicts $\|x_0^{**}\| = 1$ and hence refutes the inclusion $(1/2 + \varepsilon)B_{X^*} \subset \overline{B_V}^{w^*}$. □

The example $X = c_0$ and $V = \{(\lambda_n) \mid \sum \lambda_n = 0\}$ shows that $1/2$ is optimal in the above proposition.
For unexplained notation in the following corollary we refer to [422] and [175].

Corollary 3.10 *Every basic sequence* (e_n) *in an* M*-embedded space with basis constant* $B < 2$ *is shrinking.*

PROOF: Since $\overline{\text{lin}}\,\{e_1, e_2, \ldots\}$ is M-embedded we may assume that (e_n) is a basis. The coefficient functionals (e_n^*) are a basis of $V := \overline{\text{lin}}\,\{e_1^*, e_2^*, \ldots\}$ and $r(V, X^*) \geq 1/B > 1/2$ (see [175, Propositions 6.1 and 6.3]). Hence $V = X^*$ by Proposition 3.9 which means that (e_n) is shrinking. □

To elucidate the connection with the unique extension property from p. 118 and Proposition 2.5 let us remark that according to the above corollary $P_n^* x^* \longrightarrow x^*$ $(x^* \in X^*)$ for the projections P_n associated to a basis (e_n) if $\sup \|P_n\| < 2$. (Note that $P_n x \longrightarrow x$ for all $x \in X$ anyway.)

We conclude this section with an isomorphic characterisation of M-embedded spaces within the class of separable $\mathcal{L}^\infty$-spaces. For the definition and basic properties of $\mathcal{L}^p$-spaces we refer to [419], [420], and [421].

Theorem 3.11 *Let* X *be a separable* $\mathcal{L}^\infty$*-space which is an* M*-ideal in its bidual. Then* X *is isomorphic to* c_0.

PROOF: We will show that X is a complemented subspace of any separable superspace Y containing X. Zippin's famous characterisation of separably injective spaces [661] then gives the claim.
As every separable Banach space embeds into $C[0,1]$ it is sufficient to prove that every subspace of $C[0,1]$ isometric to X is the range of a continuous linear projection. Since X is an $\mathcal{L}^\infty$-space, its bidual is injective [420]. Note that $X^{**} \cong X^{\perp\perp}$ and X is an M-ideal in $X^{\perp\perp}$. Now $X^{\perp\perp}$ is injective, and it is therefore complemented in $C[0,1]^{**}$. Let P denote a projection from $C[0,1]^{**}$ onto $X^{\perp\perp}$; hence

$$C[0,1]^{**} \simeq X^{\perp\perp} \oplus \ker P.$$

We now renorm $C[0,1]^{**}$ (and thus its subspace $C[0,1]$) to the effect that

$$(C[0,1]^{**}, |\,.\,|) \cong X^{\perp\perp} \oplus_\infty \ker P.$$

Observe that the norms $\|\,.\,\|$ and $|\,.\,|$ coincide on $X^{\perp\perp}$. We conclude that $X^{\perp\perp}$ is an M-summand in $(C[0,1]^{**}, |\,.\,|)$, hence X (being an M-ideal in $X^{\perp\perp}$) is an M-ideal in $(C[0,1]^{**}, |\,.\,|)$, a fortiori X is an M-ideal in $(C[0,1], |\,.\,|)$.
We remark that $(C[0,1]/X)^{**}$ is isomorphic to $\ker P$, which has the BAP (it is a complemented subspace of the space $C[0,1]^{**}$ which has the MAP in its original norm). Therefore $C[0,1]/X$ is a separable space with the BAP, hence, by Theorem II.2.1 and Remark II.2.2(b), X is a complemented subspace of $C[0,1]$. □

We do not know if the above theorem holds also in the nonseparable case. Its proof essentially used results which are limited to separable spaces; however the decomposition of M-embedded spaces given in Theorem 4.5 gives the impression that it might be valid in general.

Lewis and Stegall have shown that X^* is isomorphic to ℓ^1 if X is an $\mathcal{L}^\infty$-space with a separable dual [394, p. 182]. So Theorem 3.11 is equivalent to: If X is an M-embedded space such that $X^* \simeq \ell^1$, then $X \simeq c_0$. Here is yet another way to look at Theorem 3.11 which should be compared with the isometric result in Proposition 2.7.

Corollary 3.12 *Let X be a separable M-embedded space such that X^* is isomorphic to a complemented subspace of $L^1(\mu)$ for some measure μ. Then X is isomorphic to c_0.*

This corollary follows from Theorem 3.11, since the condition on X^* is equivalent to requiring that X is an $\mathcal{L}^\infty$-space (see [421, Theorem II.5.7]).
Applying this to $X = C(\mathbb{T})/A$, one recovers that H^1 is not complemented in $L^1(\mathbb{T})$ [543, Example 5.19].

III.4 Projectional resolutions of the identity in M-embedded spaces

Our next aim is to establish a set of isomorphic properties (Theorem 4.6) for M-embedded spaces which show that in some weak sense these spaces behave like subspaces of $c_0(\Gamma)$ and thus support the intuition that M-embedded spaces in a way "vanish at infinity". These properties can all be deduced from the (isometric) result in Theorem 4.5: M-embedded spaces admit a shrinking projectional resolution of the identity. The proof of this fact uses a variation of the Lindenstrauss compactness argument and is prepared in the next four lemmata.
We remark here that for a proof of the existence of a "nice" projectional resolution of the identity some degree of smoothness of the space, i.e. some degree of continuity of the support mapping, is always needed (see the Notes and Remarks section). It was the achievement of Fabian and Godefroy in [213] to show that the assumption "X is an Asplund space" via the Jayne-Rogers selection theorem gives a $\|\,.\,\|$-Baire-1 selector of the support mapping and that this is enough to prove the existence of a projectional resolution of the identity in X^*. However, when dealing with M-embedded spaces, we have much more geometrical structure at our disposal, which allows us – using the ideas of Sims and Yost from [571] and [572] – to give the following "elementary" proof avoiding the selection theorem.
We will work with a special kind of projection on the dual of a Banach space. So let us recall:

- If M is a subspace of the Banach space X, then a linear operator $T : M^* \to X^*$ such that Tm^* is a norm-preserving extension of m^* for all $m^* \in M^*$ is called a *Hahn-Banach extension operator*.

- There exists a Hahn-Banach extension operator $T : M^* \to X^*$ if and only if $M^\perp$ is the kernel of a contractive linear projection Q in X^* (in which case $Q = Ti^*$, where $i : M \to X$ is the inclusion mapping).

In the next lemma the geometric structure comes into play.

Lemma 4.1 *If M is a subspace of an M-embedded space X and $T : M^* \longrightarrow X^*$ is a Hahn-Banach extension operator, then T is w^*-continuous. The projection $Q = Ti^*$ is the adjoint of a contractive projection P in X with range M.*

PROOF: (We use two results which will be proved in the next chapter.) By Proposition IV.1.5 the range of Q is an L-embedded space, hence Proposition IV.1.10 shows that $\operatorname{ran} Q$ is w^*-closed in X^*. A projection in a dual space with w^*-closed kernel and range is w^*- continuous, i.e. $Q = P^*$ for a projection P in X. Since $M^\perp = \ker Q$ we get $\operatorname{ran} P = M$ by standard duality. It is easily deduced from this that also T is w^*-continuous. □

Lemma 4.2 *Let X be a Banach space, B a finite dimensional subspace of X, k a positive integer, ε a positive real number, and G a finite subset of X^*. Then there is a finite dimensional subspace Z containing B such that for every subspace E of X containing B and satisfying $\dim E/B \le k$, we find an operator $T : E \longrightarrow Z$ such that $Tx = x$ for $x \in B$, $\|T\| \le 1 + \varepsilon$ and $|f(Tx) - f(x)| \le \varepsilon\|x\|$ for $x \in E$, $f \in G$.*

PROOF: If there were only finitely many subspaces $E_i = B \oplus N_i$ we would put $Z := B \oplus \bigcup N_i$. The idea is now to reduce our problem to this "case": Find $(N_i)_{i \le i_0}$ such that for all $E = B \oplus N$ as in the statement of the lemma there is an N_i with $\|b+n\| \sim \|b+n_i\|$. The details are easier than expected:
Let $G = \{f_1, \dots, f_m\}$ and let P be a projection on X with range B. Put $U := \ker P$. So $X = B \oplus U$. Choose M so large that

$$M > \frac{5k\|Id - P\|}{\varepsilon} \qquad \text{and} \qquad \frac{M+1}{M-1} < 1 + \varepsilon.$$

Let

$$\begin{array}{ll} (b_\rho)_{\rho \le r} & \text{be a finite } 1/M\text{-net for } \{b \in B \mid \|b\| \le M\}, \\ (\lambda^\sigma)_{\sigma \le s} & \text{be a finite } 1/M\text{-net for } S_{\ell^1(k)}. \end{array}$$

Define

$$\begin{array}{rcccl} \Phi & : & (B_U)^k & \longrightarrow & \mathbb{K}^{rs} \times \mathbb{K}^{mk} = \mathbb{K}^{rs+mk} \\ & & (u_1, \dots, u_k) & \longmapsto & \left(\left(\|b_\rho + \sum_{\kappa=1}^k \lambda_\kappa^\sigma u_\kappa\| \right)_{\substack{\rho \le r \\ \sigma \le s}}, \ (f_\mu(u_\kappa))_{\substack{\mu \le m \\ \kappa \le k}} \right) \end{array}$$

Since $\Phi({B_U}^k)$ is totally bounded we find $u_1, \dots, u_n$ in ${B_U}^k$ such that

$$(\Phi u^\nu)_{\nu \le n} \quad \text{is a finite } 1/M\text{-net for } \Phi({B_U}^k)$$

where we may take any norm on $\mathbb{K}^{rs+mk}$ for which the coefficient functionals have norm ≤ 1. Put

$$Z := B \oplus \operatorname{lin} \{u_\kappa^\nu \mid 1 \le \kappa \le k, \ 1 \le \nu \le n\}.$$

Now given $E \supset B$ with $\dim E/B = k$ there are $u_1, \dots, u_k \in U$ such that $E = B \oplus \operatorname{lin}(u_1, \dots, u_k)$. By Auerbach's lemma (see e.g. [422, Prop. 1.c.3]) $u = (u_1, \dots, u_k)$ may be chosen so that

$$\|u_\kappa\| = 1, \ 1 \le \kappa \le k \quad \text{and} \quad \left\| \sum_{\kappa=1}^k \lambda_\kappa u_\kappa \right\| \ge \frac{1}{k} \sum_{\kappa=1}^k |\lambda_\kappa| \quad \text{for all } (\lambda_\kappa) \in \mathbb{K}^k. \tag{1}$$

So there is an $\nu \le n$ such that

$$\|\Phi u - \Phi u^\nu\| < \frac{1}{M},$$

i.e. with $\lambda u := \sum_{\kappa=1}^{k} \lambda_\kappa u_\kappa$ for $\lambda = (\lambda_1, \ldots, \lambda_k) \in \mathbb{K}^k$ we have

$$\Big| \|b_\rho + \lambda^\sigma u\| - \|b_\rho + \lambda^\sigma u^\nu\| \Big| < \frac{1}{M} \quad \text{for all } \rho \le r,\ \sigma \le s, \tag{2}$$

$$\Big| f_\mu(u_\kappa) - f_\mu(u_\kappa^\nu) \Big| < \frac{1}{M} \quad \text{for all } \mu \le m,\ \kappa \le k. \tag{3}$$

Define

$$\begin{array}{rccc} T & : & E & \longrightarrow & Z \\ & & b + \lambda u & \longmapsto & b + \lambda u^\nu. \end{array}$$

Obviously T fixes B. To show $\|T\| \le 1 + \varepsilon$ it is sufficient to prove

$$\|b + \lambda u^\nu\| \le (1+\varepsilon)\|b + \lambda u\| \qquad \text{for} \quad \|\lambda\|_{\ell^1(k)} = 1.$$

Assume first $\|b\| \le M$. Then $\|b - b_\rho\| < 1/M$ for some ρ and $\|\lambda - \lambda^\sigma\| < 1/M$ for some σ. Consequently

$$\begin{aligned} \|b + \lambda u^\nu\| &\le \|b_\rho + \lambda^\sigma u^\nu\| + \frac{1}{M} + \frac{1}{M} \\ &\overset{(2)}{\le} \|b_\rho + \lambda^\sigma u\| + \frac{3}{M} \\ &\le \|b + \lambda u\| + \frac{5}{M}. \end{aligned}$$

Also

$$\|b + \lambda u\| \ge \frac{1}{\|Id - P\|} \Big\| \sum \lambda_\kappa u_\kappa \Big\| \overset{(1)}{\ge} \frac{1}{k\|Id - P\|} \sum |\lambda_\kappa| = \frac{1}{k\|Id - P\|} > \frac{5}{\varepsilon M},$$

i.e. $5/M \le \varepsilon \|b + \lambda u\|$.
For $\|b\| > M$ we have

$$\|b + \lambda u^\nu\| \le \|b\| + 1 \quad \text{and} \quad \|b + \lambda u\| \ge \|b\| - 1,$$

so

$$\frac{\|b + \lambda u^\nu\|}{\|b + \lambda u\|} \le \frac{\|b\| + 1}{\|b\| - 1} < \frac{M+1}{M-1} < 1 + \varepsilon.$$

Finally for any $x = b + \sum \lambda_\kappa u_\kappa \in E$

$$\begin{aligned} |f_\mu(x) - f_\mu(Tx)| &= \Big| f_\mu\Big(\sum_\kappa \lambda_\kappa (u_\kappa - u_\kappa^\nu) \Big) \Big| \\ &\le \sum |\lambda_\kappa| \, |f_\mu(u_\kappa - u_\kappa^\nu)| \end{aligned}$$

$$\begin{aligned}
&\overset{(3)}{\leq} \frac{1}{M}\sum|\lambda_\kappa| \\
&\overset{(1)}{\leq} \frac{k}{M}\left\|\sum \lambda_\kappa u_\kappa\right\| \\
&= \frac{k}{M}\|(Id-P)x\| \\
&\leq \frac{k\|Id-P\|}{M}\|x\| \\
&< \frac{\varepsilon}{5}\|x\|.
\end{aligned}$$

□

We will find a shrinking projectional resolution of the identity (for the definition see Theorem 4.5) by constructing an increasing family of subspaces M_α which admit Hahn-Banach extension operators. To achieve this we employ a transfinite induction argument whose first step is the next lemma. Let us remark that Lindenstrauss has proved the above Lemma 4.2 with $G = \emptyset$. Our modification, suggested by D. Yost, proves useful when we want to show the ranges of the adjoint projections in the dual to be increasing.

Lemma 4.3 *If X is a Banach space, L a separable subspace of X, and F a separable subspace of X^*, then X has a separable subspace M containing L which admits a Hahn-Banach extension operator $T: M^* \longrightarrow X^*$ satisfying $TM^* \supset F$.*

PROOF: Let (x_n) be a sequence dense in L and (f_n) a sequence dense in F. Starting with $M_1 = \{0\}$ we inductively define subspaces M_n as follows: Putting $B_n := \operatorname{lin}(M_n, x_n)$, $G_n := \{f_1, \dots, f_n\}$ we let M_{n+1} be the subspace Z given by Lemma 4.2 when $B = B_n$, $k = n$, $\varepsilon = \frac{1}{n}$, and $G = G_n$. We may assume $\dim M_{n+1}/B_n \geq n+1$. Clearly $M := \overline{\bigcup M_n}$ is separable and contains L.
For $n \in \mathbb{N}$ define

$$I_n := \{E \subset X \mid B_n \subset E, \dim E/B_n \leq n\}$$

and put

$$I := \bigcup I_n.$$

Since $E \in I_n$, $F \in I_m$ implies $E + F + B_{\dim E + \dim F} \in I_{\dim E + \dim F}$, we have that I is a directed set. The condition $\dim M_{n+1}/B_n \geq n+1$ implies $\dim B_{n+1}/B_n \geq n+1$, and this easily gives that for each $E \in I$ there is a unique $n \in \mathbb{N}$ such that $E \in I_n$. So by Lemma 4.2 there exists $T_E : E \longrightarrow M_{n+1} \subset M$ such that $\|T_E\| \leq 1 + \frac{1}{n}$, $T_E|_{B_n} = Id_{B_n}$ and $|f_i(T_E x) - f_i(x)| \leq \frac{1}{n}\|x\|$ for all $x \in E$ and $1 \leq i \leq n$.
Extend T_E (nonlinearly) to X by setting

$$\overline{T_E}(x) := \begin{cases} T_E x & \text{if } x \in E \\ 0 & \text{otherwise.} \end{cases}$$

Since $\|\overline{T_E}x\| \leq 2\|x\|$ and regarding $\overline{T_E}x \in M \subset M^{**}$ as an element of M^{**} we have

$$(\overline{T_E}x)_x \in \prod_{x \in X} B_{M^{**}}(0, 2\|x\|).$$

Hence by compactness $((\overline{T_E x})_x)_E$ has a convergent subnet $((\overline{T_{E'} x})_x)_{E'}$, i.e. for all $x \in X$ there is $m_x \in M^{**}$ such that $\overline{T_{E'} x} \longrightarrow m_x$ with respect to $\sigma(M^{**}, M^*)$. Define

$$\begin{array}{rcl} T : M^* & \longrightarrow & X^* \\ m^* & \longmapsto & (x \longmapsto m_x(m^*) = \lim \overline{T_{E'} x}(m^*)) \end{array}$$

Noting that "for all $x \in X$ there is $E \in I$ such that $x \in E$" and "for $E, F \in I$ with $E \subset F$ and $E \in I_n$, $F \in I_m$ one obtains $n \leq m$" it is routine to verify that T is the required Hahn-Banach extension operator. For example, the following establishes that Tm^* is additive:

$$\begin{array}{rcl} Tm^*(x) + Tm^*(y) & = & \lim\limits_{E'} \overline{T_{E'} x}(m^*) + \lim\limits_{E'} \overline{T_{E'} y}(m^*) \\ & = & \lim\limits_{E'} (\overline{T_{E'} x} + \overline{T_{E'} y})(m^*) \;=\; \lim\limits_{E' \supset \{x,y\}} (T_{E'} x + T_{E'} y)(m^*) \\ & = & \lim\limits_{E' \supset \{x,y\}} T_{E'}(x+y)(m^*) \;=\; \lim\limits_{E'} \overline{T_{E'}}(x+y)(m^*) \\ & = & Tm^*(x+y) \end{array}$$

Since clearly $TM^* \supset F$, we have completed the proof of Lemma 4.3. □

We recall that the density character of a Banach space X, dens X, is defined to be the least cardinal $\mathfrak{m}$ for which X has a dense subset of cardinality $\mathfrak{m}$. If α, β are ordinals we use $\alpha < \beta$ as $\alpha \in \beta$, hence $\beta = \{\alpha \mid 0 \leq \alpha < \beta\} = [0, \beta)$, and we write card α for the cardinal number of α. The first infinite ordinal is denoted ω. We assume some familiarity with transfinite induction arguments and ordinal and cardinal arithmetic (to the extent given e.g. in [326]).

Lemma 4.4 *Let X be a Banach space, L a subspace of X, F a subspace of X^* with* dens $F \leq$ dens L. *Then there is a subspace M of X containing L and a Hahn-Banach extension operator $T : M^* \longrightarrow X^*$ such that* dens M = dens L *and $TM^* \supset F$.*

PROOF: We use transfinite induction on dens L. The base case dens $L = \aleph_0$ was treated in the previous lemma. So assume $\mathfrak{m} \leq$ dens L and that the assertion holds for all cardinals $< \mathfrak{m}$. We will show that it holds for $\mathfrak{m}$, too.
Let μ be the first ordinal of cardinality $\mathfrak{m}$, and let $\{x_\alpha \mid \alpha < \mu\}$ and $\{f_\alpha \mid \alpha < \mu\}$ be dense in L and F respectively (we repeat some f_α at the end if dens $F <$ dens L).

> CLAIM: *For all α with $\omega \leq \alpha < \mu$ there are subspaces M_α of X and Hahn-Banach extension operators $T_\alpha : M_\alpha^* \longrightarrow X^*$ such that*
>
> - *$M_\beta \subset M_\alpha$ for $\omega \leq \beta \leq \alpha$,*
> - dens $M_\alpha \leq$ card α,
> - *$M_\alpha \supset \{x_\beta \mid \beta < \alpha\}$, $\quad T_\alpha M_\alpha^* \supset \{f_\beta \mid \beta < \alpha\}$.*

This is again proved by transfinite induction: $\alpha = \omega$ is Lemma 4.3. Assuming M_β, T_β are given as in the claim for all β with $\omega \leq \beta < \alpha$ we put

$$\begin{array}{rcl} L & := & \overline{\mathrm{lin}}\left(\{x_\beta \mid \beta < \alpha\} \cup \bigcup\limits_{\beta < \alpha} M_\beta\right) \\ F & := & \overline{\mathrm{lin}}\,\{f_\beta \mid \beta < \alpha\}. \end{array}$$

Since $\operatorname{dens} L \leq \operatorname{card}\alpha + \sum_{\beta<\alpha}\operatorname{card}\beta = \operatorname{card}\alpha < \mathfrak{m}$ and $\operatorname{dens} F \leq \operatorname{dens} L$ we find M_α and T_α as in the claim by the (outer) induction hypothesis.
To finish the proof of the first induction with the help of the claim put

$$M := \overline{\bigcup_{\alpha<\mu} M_\alpha}.$$

Then $M \supset L$ and $\operatorname{dens} M \leq \sum_{\alpha<\mu}\operatorname{card}\alpha = \operatorname{card}\mu = \mathfrak{m}$, thus $\operatorname{dens} M = \operatorname{dens} L$. Clearly $\|T_\alpha i_\alpha^*\| \leq 1$ where $i_\alpha : M_\alpha \longrightarrow M$ stands for the inclusion map. So by compactness of $B_{L(M^*,X^*)}$ in the weak*-operator topology, there is a subnet $(T_{\alpha'}i_{\alpha'}^*)$ of $(T_\alpha i_\alpha^*)_{\alpha<\mu}$ such that

$$(T_{\alpha'}i_{\alpha'}^* m^*)(x) \longrightarrow (Tm^*)(x) \qquad \forall m^* \in M^* \quad \forall x \in X.$$

Again, it is routine to verify that T is the desired Hahn-Banach extension operator. □

After the above preparation we can finally prove the main result of this section.

Theorem 4.5 *Every M-embedded space X admits a shrinking projectional resolution of the identity. That is, there are contractive projections P_α on X ($\omega \leq \alpha \leq \mu$, where μ denotes the first ordinal with cardinality* dens X) *such that*

(a) $P_\mu = Id_X$,

(b) $P_\alpha P_\beta = P_\beta P_\alpha = P_\beta$ *for* $\beta \leq \alpha$,

(c) $\operatorname{dens} P_\alpha X \leq \operatorname{card}\alpha$.

(d) $\overline{\bigcup_{\beta<\alpha} P_\beta X} = P_\alpha X$ *for limit ordinals* α,

(e) $\overline{\bigcup_{\beta<\alpha} P_\beta^* X^*} = P_\alpha^* X^*$ *for limit ordinals* α.

Moreover the projections P_α come from subspaces M_α which admit Hahn-Banach extension operators $T_\alpha : M_\alpha^* \longrightarrow X^*$ in the sense that $P_\alpha^* = T_\alpha i_\alpha^*$, where i_α denotes the inclusion mapping from M_α to X.

PROOF: We will prove by transfinite induction:

> CLAIM. *For all α with $\omega \leq \alpha \leq \mu$ there are subspaces M_α of X and Hahn-Banach extension operators $T_\alpha : M_\alpha^* \longrightarrow X^*$ such that*
>
> (a') $M_\mu = X$,
>
> (b') $M_\beta \subset M_\alpha$ *and* $T_\beta M_\beta^* \subset T_\alpha M_\alpha^*$ *for* $\beta \leq \alpha$,
>
> (c') $\operatorname{dens} M_\alpha \leq \operatorname{card}\alpha$,
>
> (d') $\overline{\bigcup_{\beta<\alpha} M_\beta} = M_\alpha$ *for limit ordinals* α,
>
> (e') $\overline{\bigcup_{\beta<\alpha} T_\beta M_\beta^*} = T_\alpha M_\alpha^*$ *for limit ordinals* α.

Assuming the claim, we know from Lemma 4.1 that $T_\alpha i_\alpha^*$ is the adjoint of a contractive projection P_α on X with range M_α. Note that $\operatorname{ran} P_\beta \subset \operatorname{ran} P_\alpha$ gives $P_\alpha P_\beta = P_\beta$ and that $\operatorname{ran} P_\beta^* = T_\beta M_\beta^* \subset T_\alpha M_\alpha^* = \operatorname{ran} P_\alpha^*$ yields $P_\alpha^* P_\beta^* = P_\beta^*$, hence $P_\beta P_\alpha = P_\beta$. Trivially the other primed properties imply the corresponding unprimed ones in the statement of the theorem.

To establish the claim choose a set $\{x_\beta \mid \beta < \mu\}$ which is dense in X. First of all we find a separable subspace M_ω containing $\{x_\beta \mid \beta < \omega\}$ and a Hahn-Banach extension operator T_ω by Lemma 4.3. So assume M_β and T_β are given as in the claim for all $\beta < \alpha$. If α is a *successor ordinal* let $L := \overline{\operatorname{lin}}\,(M_{\alpha-1} \cup \{x_\beta \mid \beta < \alpha\})$ and $F := T_{\alpha-1}M_{\alpha-1}^*$. By Theorem 3.1 $\operatorname{dens} M_{\alpha-1} = \operatorname{dens} M_{\alpha-1}^*$, so $\operatorname{dens} F = \operatorname{dens} M_{\alpha-1}^* = \operatorname{dens} M_{\alpha-1} \leq \operatorname{dens} L \leq \operatorname{card}(\alpha - 1) + \operatorname{card} \alpha = \operatorname{card} \alpha$, and Lemma 4.4 provides us with M_α and T_α satisfying (b′) and (c′). Note $x_\beta \in M_\alpha$ for $\beta < \alpha$.

If α is a *limit ordinal* let $M_\alpha := \overline{\bigcup_{\beta<\alpha} M_\beta}$. For $\alpha = \mu$ we get $M_\mu = X$ since, for successor ordinals β, $M_\beta \supset \{x_\gamma \mid \gamma < \beta\}$. Denoting by $j_\beta : M_\beta \longrightarrow M_\alpha$ the inclusion maps we find a Hahn-Banach extension operator T_α as a w^*-operator topology accumulation point of $(T_\beta j_\beta^*)$ in $B_{L(M_\alpha^*, X^*)}$. As in the proof of Lemma 4.4 it is easily verified that M_α and T_α have the properties (a′) – (d′). The nontrivial part is property (e′).

Writing $Q_\beta = T_\beta i_\beta^* (= P_\beta^*)$, it is sufficient for the proof of (e′) to show that

$$Q_\beta x^* \longrightarrow Q_\alpha x^* \quad \text{for all } x^* \in X^*. \tag{$*$}$$

Actually (e′) and ($*$) are equivalent since property (e) is easily seen to be the same as the s_{op}-continuity of the mapping $\alpha \longmapsto P_\alpha^*$ from $[\omega, \mu]$ to $L(X^*)$, where $[\omega, \mu]$ is equipped with the order topology. A similar remark applies to (d) so that in particular $P_\beta x \xrightarrow{w} P_\alpha x$ for all $x \in X$, i.e.

$$(Q_\beta x^*)(x) \longrightarrow (Q_\alpha x^*)(x) \quad \text{for all } x^* \in X^* \quad \text{and for all } x \in X. \tag{$**$}$$

So our task is to improve w^*-convergence to $\|\,.\,\|$-convergence. To accomplish ($*$) it is enough to prove

$$Q_\beta x^* = 0 \quad \text{for } x^* \in \ker Q_\alpha, \tag{1}$$

$$Q_\beta x^* \longrightarrow x^* \quad \text{for } x^* \in \operatorname{ran} Q_\alpha. \tag{2}$$

Part (1) follows from the inclusion $M_\beta \subset M_\alpha$, which immediately yields

$$\ker Q_\alpha = M_\alpha^\perp \subset M_\beta^\perp = \ker Q_\beta.$$

To see (2) note that $\operatorname{ran} Q_\alpha = \operatorname{ran} P_\alpha^* = (\ker P_\alpha)^\perp = (X/\ker P_\alpha)^*$ is the dual of an M-embedded space, hence by Corollary 3.2

$$B_{\operatorname{ran} Q_\alpha} = \overline{\operatorname{co}}^{\|\,\|}\, w^*\text{-sexp}\, B_{(X/\ker P_\alpha)^*}. \tag{$\dagger$}$$

Since $T_\beta M_\beta^* \subset T_\alpha M_\alpha^*$ we may regard Q_β as an operator on $(X/\ker P_\alpha)^*$, in particular $Q_\beta x^* \in B_{(X/\ker P_\alpha)^*}$ for $x^* \in B_{(X/\ker P_\alpha)^*}$. With these identifications the convergence in ($**$) means

$$Q_\beta x^* \longrightarrow x^* \quad \text{weak}^* \text{ in } (X/\ker P_\alpha)^*.$$

So
$$Q_\beta x^* \xrightarrow{\|\,\|} x^* \quad \text{if } x^* \in w^*\text{-sexp}\, B_{(X/\ker P_\alpha)^*}.$$
Because of (†) this shows (2).
Thus the claim, hence the theorem, is completely proved. □

We remark that the existence of a (shrinking) projectional resolution of the identity is an isometric property which depends on the norm. However the important consequences we will deduce from it appeal to the isomorphic structure.
Recall that a norm $\|\,.\,\|$ on X is said to be *locally uniformly rotund* (LUR), whenever $\|x\| = \|x_n\| = 1$ and $\lim \|x + x_n\| = 2$ imply $x_n \to x$.
If $(X, \|\,.\,\|)$ is smooth, then by definition for every $x \in S_X$ there is a unique $f_x \in S_{X^*}$ such that $f_x(x) = 1$. This support mapping $x \mapsto f_x$ from S_X to S_{X^*} is then $\|\,.\,\|$-w^*-continuous. If it is even $\|\,.\,\|$-w-continuous, X is said to be *very smooth.* The norm is called *Fréchet differentiable* in case $x \mapsto f_x$ is $\|\,.\,\|$-$\|\,.\,\|$-continuous (the latter is equivalent to the more common requirement that $\lim_{t\to 0}(\|x+ty\|-\|x\|)/t$ exists uniformly in $y \in S_X$ for every $x \in S_X$). For more details on these notions we refer to [155].
Further recall that a Banach space is called *weakly compactly generated* if it is the closed linear span of some weakly compact set. A good source of information on these spaces is Chapter Five in [155].

Theorem 4.6 *Let X be an M-embedded space. Then:*

(a) *X has a shrinking Markuševič basis, i.e. there are $(x_i)_{i\in I}$ in X, $(f_i)_{i\in I}$ in X^* such that* card $I =$ dens X *and*
- $f_i(x_j) = \delta_{ij}$,
- $\overline{\text{lin}}\,(x_i) = X$,
- $\overline{\text{lin}}\,(f_i) = X^*$.

(b) *X is weakly compactly generated.*

(c) *There are operators $T : X \longrightarrow c_0(I)$ and $S : X^* \longrightarrow c_0(I)$,* card $I =$ dens X, *such that T^{**} is injective and S is w^*-w-continuous and injective.*

(d) *X has an equivalent LUR-norm whose dual norm is also LUR. In particular this norm is Fréchet-differentiable und LUR.*

(e) *X has an equivalent very smooth norm, whose dual norm is strictly convex and under which X is still M-embedded.*

PROOF: (a) We will establish this by transfinite induction on $\mathfrak{m} = \text{dens}\, X$. For $\mathfrak{m} = \aleph_0$ the dual X^* is separable by Theorem 3.1; so an old result of Markuševič, see [422, Proposition 1.f.3 and the remark thereafter] settles the base case.
Assume the statement holds for all M-embedded spaces Y with dens $Y < \mathfrak{m}$. Choose a shrinking projectional resolution of the identity $(P_\alpha)_{\omega\le\alpha\le\mu}$ for X as in Theorem 4.5 and put

$$\begin{aligned} Y_\omega &:= P_\omega X, \\ Y_{\alpha+1} &:= (P_{\alpha+1} - P_\alpha)X \qquad \text{for } \omega \le \alpha < \mu. \end{aligned}$$

Then $\operatorname{dens}(P_{\alpha+1} - P_\alpha)X \le \operatorname{dens} P_{\alpha+1}X \le \operatorname{card}(\alpha+1) = \operatorname{card}\alpha < \operatorname{card}\mu = \mathfrak{m}$, so Y_ω and $Y_{\alpha+1}$ are as in the induction hypothesis. Thus there exist shrinking Markuševič bases

$$(x_i^\omega,\, f_i^\omega)_{i\in I_\omega} \text{ in } Y_\omega \quad \text{and} \quad (x_i^{\alpha+1},\, f_i^{\alpha+1})_{i\in I_{\alpha+1}} \text{ in } Y_{\alpha+1}$$

with $\operatorname{card} I_{\alpha+1} = \operatorname{dens} Y_{\alpha+1} \le \operatorname{card}\alpha$. Defining $\overline{f_i^\omega} := f_i^\omega \circ P_\omega$ and $\overline{f_i^{\alpha+1}} := f_i^{\alpha+1}(P_{\alpha+1} - P_\alpha)$ we claim that

$$\{(x_i^\beta,\, \overline{f_i^\beta}) \mid \beta = \omega \text{ or } \beta = \alpha+1 \text{ for } \omega \le \alpha < \mu,\ i \in I_\beta\}$$

is the desired shrinking Markuševič basis. That $\overline{f_i^\beta}(x_j^\gamma) = \delta_{(\beta,i),(\gamma,j)}$ is clear since $Y_\beta \cap Y_\gamma = \{0\}$ if $\beta \ne \gamma$. An easy transfinite induction argument using $P_\alpha x = \lim_{\beta<\alpha} P_\beta x$ (i.e. property (d) in Theorem 4.5) shows

$$P_\alpha X = \overline{\operatorname{lin}}\left(P_\omega X \cup \bigcup_{\omega\le\beta<\alpha} (P_{\beta+1} - P_\beta)X \right) \qquad \omega \le \alpha \le \mu.$$

Applying this for $\alpha = \mu$ one obtains

$$\overline{\operatorname{lin}}\,\{x_i^\beta \mid \beta = \omega \text{ or } \beta = \alpha+1 \text{ for } \omega \le \alpha < \mu,\ i \in I_\beta\} = X.$$

The natural isomorphism $Y_{\alpha+1}^* = ((P_{\alpha+1} - P_\alpha)X)^* \simeq (P_{\alpha+1}^* - P_\alpha^*)X^*$ (similarly for Y_ω^*) gives $\overline{\operatorname{lin}}\,\{\overline{f_i^{\alpha+1}} \mid i \in I_{\alpha+1}\} = (P_{\alpha+1}^* - P_\alpha^*)X^*$. As above – now using property (e) in Theorem 4.5 – one establishes

$$P_\alpha^* X^* = \overline{\operatorname{lin}}\left(P_\omega^* X^* \cup \bigcup_{\omega\le\beta<\alpha} (P_{\beta+1}^* - P_\beta^*)X^* \right) \qquad \omega \le \alpha \le \mu,$$

and the rest is clear.

(b) By suitably scaling we may assume that the shrinking Markuševič basis from (a) satisfies $\|x_i\| = 1$ for all $i \in I$. We claim that

$$K := \{0\} \cup \{x_i \mid i \in I\}$$

is weakly compact; then $X = \overline{\operatorname{lin}}\, K$ is weakly compactly generated.

By the Eberlein-Smulian theorem it is sufficient to show that K is weakly sequentially compact. To prove this it is enough to establish that for all sequences (x_{i_n}) in K with each x_{i_n} appearing only finitely many times

$$x_{i_n} \xrightarrow{w} 0\,.$$

Given $f \in X^*$ and $\varepsilon > 0$, we find $g \in \operatorname{lin}(f_i)$ such that $\|f - g\| < \varepsilon$. Then $g(x_i) \ne 0$ only for finitely many $i \in I$. So

$$\Big| |f(x_{i_n})| - |g(x_{i_n})| \Big| \le |f(x_{i_n}) - g(x_{i_n})| \le \|f - g\| < \varepsilon$$

shows the claim.

(c) Assuming $\|f_i\| = 1$ for the shrinking Markuševič basis from (a) we define T by

$$\begin{array}{rcl} T : X & \longrightarrow & c_0(I) \\ x & \longmapsto & \big(f_i(x)\big). \end{array}$$

As in the proof of part (b) one verifies that $\{i \in I \mid |f_i(x)| > \varepsilon\}$ is finite for all $x \in X$ and $\varepsilon > 0$, so T is well-defined. Then $T^{**} : X^{**} \longrightarrow \ell^\infty(I)$ is given by $T^{**}x^{**} = (x^{**}(f_i))$. So it is clear that T has the required properties.
Assuming now $\|x_i\| = 1$ the desired operator S is given by

$$\begin{array}{rcl} S : X^* & \longrightarrow & c_0(I) \\ f & \longmapsto & \big(f(x_i)\big). \end{array}$$

(d) An inspection of the proof of Theorem 2.1 of [272] shows the following

> THEOREM (Godefroy, Troyanski, Whitfield, Zizler): *Let Y be an LUR Banach space and X a Banach space with a locally convex topology τ such that B_X is τ-closed. If there is a bounded linear operator $L : Y \to X$ with a $\|\,.\,\|$-dense range such that $L(B_Y)$ is τ-compact, then X admits an equivalent LUR-norm which is τ-lower semicontinuous.*

First we apply this to X with $\tau = \sigma(X, X^*)$: Assuming $\|x_i\| = 1$ for the shrinking Markuševič basis $(x_i, f_i)_{i \in I}$ from part (a), we have that $L : \ell^1(I) \to X$, $(\lambda_i) \mapsto \sum \lambda_i x_i$ is continuous with dense range. The w^*-continuity of $f_i \circ L$ and $\overline{\text{lin}}\,(f_i) = X^*$ provide the w^*-w-continuity of L, hence the weak compactness of $L(B_{\ell^1(I)})$. The existence of an equivalent dual LUR-norm $|\,.\,|$ on $\ell^1(I)$ (e.g. $|(\lambda_i)| = (\|(\lambda_i)\|^2_{\ell^1(I)} + \|(\lambda_i)\|^2_{\ell^2(I)})^{1/2}$, cf. [272, p. 348]) then shows that X admits an equivalent LUR-norm.
If we apply the renorming theorem quoted above to X^* with $\tau = \sigma(X^*, X)$ and take for L the adjoint of the operator $T : X \to c_0(I)$ defined in the proof of part (c), we find that X^* admits an equivalent dual LUR-norm (cf. [272, Remark 2.4]). So the Asplund averaging technique (see [155, Chapter Four, § 3 and Chapter Five, § 9] or [214]) yields a norm with both features. For the second statement in (d) see [155, Chapter Two, §2].

(e) By part (c) there is a w^*-w-continuous and injective operator $S : X^* \to c_0(I)$. In particular S is weakly compact and w^*-w^*-continuous when considered as an operator into $\ell^\infty(I)$. Hence, S is the adjoint of a weakly compact operator from $\ell^1(I)$ into X and we are in the situation of Proposition 2.11. Using Day's strictly convex norm on $c_0(I)$ [155, p. 94ff.] we find, with the help of that proposition, an equivalent smooth norm $|\,.\,|$ on X with $|\,.\,|^*$ strictly convex and X still M-embedded. Such a norm is very smooth: every support mapping from $S_{(X,|\,|)}$ to $S_{(X^*,|\,|^*)}$ is $\|\,.\,\|$-w^*-continuous [155, p. 22], hence $\|\,.\,\|$-w-continuous by Corollary 2.15, i.e. $|\,.\,|$ is very smooth. □

REMARKS: (a) The first step of the proof of part (d), the existence of an equivalent LUR-norm on X, is of course an immediate consequence of Troyanski's theorem that weakly compactly generated spaces can be equivalently LUR-renormed (see [155, Chapter Five, § 5]). However, we gave the above argument in order to point out that the Godefroy-Troyanski-Whitfield-Zizler theorem also yields a new and simpler proof of Troyanski's renorming result. [This is indicated in the introduction of [272], but inadvertently not in the text. To obtain the renorming for weakly compactly generated spaces X recall that by the Amir-Lindenstrauss theorem (see [155, Theorem 2, p. 147] or [583]) there is a w^*-w-continuous and injective operator $T : X^* \to c_0(I)$. Then $T^* : \ell^1(I) \to X$ is w^*-w-continuous and has dense range. So an application of the Godefroy-Troyanski-Whitfield-Zizler theorem indeed yields the renorming.]

(b) Parts (a) and (b) of Theorem 4.6 give in particular a new proof that every subspace of $c_0(I)$ is weakly compactly generated and has a shrinking Markuševič basis. This is a result obtained in [346, p. 10] by different techniques.

For the large number of consequences which can be deduced for weakly compactly generated spaces we refer to [155, Chapter Five]. For convenience we state some implications which may occasionally be useful.

Recall that a Banach space is said to have the *separable complementation property* if every separable subspace is contained in a separable and complemented subspace.

Corollary 4.7 *Let X be an M-embedded space. Then:*

(a) *w^*-compact subsets of X^* are w^*-sequentially compact.*

(b) *w^*-sequentially continuous elements of X^{**} are w^*-continuous.*

(c) *X has the separable complementation property.*

(d) *If X is nonreflexive then it contains a complemented copy of c_0, in fact every copy of c_0 is complemented.*

PROOF: (a) The operator S from Theorem 4.6(c) maps these sets homeomorphically to weakly compact subsets of $c_0(I)$. So the assertion follows from the Eberlein-Smulian theorem.

(b) Corollary 4 on p. 148 in [155].

(c) Theorem 3 on p. 149 in [155]. However, this property of M-embedded spaces follows already directly from Lemmata 4.1 – 4.3; it is even enough to use Lemma 4.2 with $G = \emptyset$.

(d) Combine the separable complementation property with Corollary 3.7(a) and Sobczyk's theorem (Corollary II.2.9). □

III.5 Notes and remarks

GENERAL REMARKS. The date of birth of M-embedded spaces is hard to fix. Before the paper [293], where the first systematic study of this class was initiated, this type of space appeared in Lima's work [402] and, under the name *class* L_0, in the article [97] by Brown and Ito; there were also some unpublished results by R. Evans. In [402] Proposition 2.7 was proved, and in [97] Theorem 1.6(b), (c) and Proposition 2.10(a) were established. The most important predecessor of [293], however, is Lima's paper [404] where the connection with M-ideals of compact operators was shown and the RNP for duals of M-embedded spaces was explicitly proved; see below for more information on this result. The study of M-embedded spaces marked an important point in the development of M-ideal theory: the point of view of considering *sub*spaces of given Banach spaces, C^*-algebras, L^1-preduals and ordered spaces was abandoned in favour of investigating pure Banach space

properties. In particular the fact that nontrivial isomorphic conclusions can be deduced from the geometric structure of M-embedded spaces has found some interest.
Many of the examples of M-embedded spaces in Section III.1 appeared in [624]. For C/A, however, we have followed Luecking's proof in [427]; another proof can be found in [241]. The assertion about Orlicz spaces was proved much earlier by Ando [18], establishing the L-decomposition in the dual directly. Even older is the result on $K(H)$ which was shown by Dixmier [165] in 1950. Corollary 1.5 was first proved in [38], the M-ideal argument we have presented is due to Luecking [427]. We will say more about best approximation from arbitrary Douglas algebras below and refer to [558, p. 109ff.] for the history of this problem and its relation to questions in function theory. Concerning Orlicz spaces let us mention that our proof shows that h_M is an M-ideal in ℓ_M (resp. H_M in L_M) no matter if M or M^* satisfies the Δ_2-condition; but in this generality ℓ_M need not be the bidual of h_M. There is another norm which is of interest in the theory of Orlicz spaces. The Orlicz norm is defined to be the dual norm of the Luxemburg norm with respect to the pairing $\langle f,g\rangle = \int f(t)g(t)\,dt$; i.e.,

$$\|f\|_M^0 = \sup\{|\langle f,g\rangle| \mid g \in L_{M^*}, \|g\|_{M^*} \le 1\}$$

in the case of function spaces and similarly for sequence spaces. H. Hudzik has shown us a proof that one doesn't obtain M-ideals for this norm. In Section VI.6 we shall discuss another renorming of Orlicz sequence spaces which is of importance in the theory of M-ideals of compact operators.
The stability of M-embedded spaces (Theorem 1.6) was fully established in [293]. In the proof of this result the main difference between M- and L-embedded spaces appears clearly: in the first case the *natural* projection π_{X^*} is an L-projection, in the second case we just know that there is one. The results and techniques in [264] have inspired Proposition 1.9 which is taken from [407]. We remark that the isomorphic version of Corollary 1.10 is false, since by a result of Johnson and Lindenstrauss [356, Ex. 2] there is a $C(K)$-space which is not a subspace of a weakly compactly generated space (hence in particular not isomorphic to an M-embedded space by Theorem 4.6), yet all separable subspaces of $C(K)$ are isomorphic to subspaces of c_0.
Proposition 2.1 is implicitly contained in [97] – see also [256, Th. 1] for a more general result. Using Proposition 2.1 we obtain a more selfcontained proof of Proposition 2.2 than the one in [293]. However, to get a better understanding of the question of automatic w^*-continuity of surjective isometries in dual spaces we refer to Godefroy's work in [250], [256] and the survey chapter VII in [259]. Theorem 2.3 is from [293], the unique extension property, Lemma 2.4, and Proposition 2.5 first appeared in [270]. By Proposition 2.5 of this last paper an important class of spaces enjoying the unique extension property are those whose duals don't contain proper norming subspaces. The decomposition of M-embedded spaces in Proposition 2.6 was first proved in [292] using function module representations. The proof of Proposition 2.7 is also from this work – the result, however, first appeared in [402]. Much more than the statement in Corollary 2.8 is true: in [72] Benyamini showed that, for a first countable compact space K, $d(X, C_\sigma(K)) = 1$ implies $X \cong C_\sigma(K)$. Besides the conditions characterising M-embedded C^*-algebras which we proved in Proposition 2.9 there are several others in [76]. The construction of the predual in Proposition 2.10 comes from [257, Th. 27]. Presumably the first paper to explicitly

note that the natural projection π_{X^*} from X^{***} onto X^* is not bicontractive is [96]. The example $\|Id_{(\ell^1)^{***}} - \pi_{(\ell^1)^*}\| = 2$ was extended in [103] and [264] to "If X contains a subspace isomorphic to ℓ^1 then $\|Id_{X^{***}} - \lambda\pi_{X^*}\| = 1 + |\lambda|$". In this last paper the so-called Godun set $G(X) = \{\lambda \in \mathbb{K} \mid \|Id_{X^{***}} - \lambda\pi_{X^*}\| = 1\}$ is studied. For instance it is proved that a separable space is Asplund provided that $G(X) \cap (1,2] \neq \emptyset$. The renorming of M-embedded spaces in Proposition 2.11 is from [294], the example in Proposition 2.13 essentially appeared in [338] – however, the arguments there are not at all clear; therefore, we have followed the exposition in [624].
It took a while to realize the connection between unique Hahn-Banach extension, equality of topologies on S_{X^*}, continuity of the support mapping, and the RNP for X^*. In [588] Sullivan introduced Hahn-Banach smooth spaces as those which admit unique Hahn-Banach extension from X to X^{**}, realized that no dual space is Hahn-Banach smooth and proved that the equality of norm and weak* topologies on S_{X^*} implies this property. Shortly after, Smith and Sullivan [574] showed that weakly Hahn-Banach smooth spaces X (only norm attaining functionals x^* are required to have unique Hahn-Banach extensions) have duals with the RNP. Studying the continuity of the support mapping $D : S_X \to 2^{S_{X^*}}$ Giles, Gregory, and Sims [247] proved that weak Hahn-Banach smoothness means $w^* = w$ on the set of norm attaining functionals in S_{X^*}. They also gave two arguments (one of them attributed to Phelps) that this implies the RNP for X^*. Also, their Theorem 3.3 contains that duals of M-embedded spaces have the RNP in a rather explicit manner. In [252] Godefroy called Namioka points those $x^* \in S_{X^*}$ where the w^*- and w-topologies agree on S_{X^*}, showed Lemma 2.14 and established connections to minimal norming subspaces and unique preduals – see also [646]. All this was unnoticed by those working in M-ideal theory so that the RNP for duals of M-embedded spaces was reproved twice in [404] and [253]. The argument in the text using proper norming subspaces now seems to be the shortest.
Property (V) for M-embedded spaces was first shown in [269]. The proof we have presented is a modification due to Lima of the argument in [293] which shows that M-embedded spaces contain c_0. Besides the references for properties (V) and (V^*) already cited in 3.3, we mention [116] and [117] for a study of (V) in vector valued function spaces, [83] representative of the work of Bombal and [513] for a weakening of (V) and its relation to Grothendieck spaces – see also [156]. A good summary is the survey article by E. and P. Saab [550]. The most remarkable recent result in this area is Pfitzner's theorem [494] that all C^*-algebras have property (V). Finally we mention that property (V) has occasionally been called *strict Dieudonné property*, e.g. in [452].
In [271] and [251] Godefroy and Talagrand (see also [259, p. 158ff.] and [257, p. 244ff.]) say that a Banach space X has *property* (X) if any $x^{**} \in X^{**}$ such that $x^{**}(\sum^* x_k^*) = \sum x^{**}(x_k^*)$ for every wuC-series $\sum x_k^*$ in X^*, must be in X. They show that every Banach space with property (X) is strongly unique predual of its dual; since property (X) is an isomorphic invariant this remains true for every equivalent norm on the space. In [184] Edgar introduced an ordering of Banach spaces by

$$X \prec Y \quad \text{iff} \quad X = \bigcap_{T \in L(X,Y)} (T^{**})^{-1}(Y)$$

and showed that X has property (X) iff $X \prec \ell^1$ [184, Prop. 10]. He also proved that property (X) implies property (V^*) [184, Th. 13]. Since it was shown in [251], [265] and

[259] that X^* has property (X) whenever X is a separable space with property (u) not containing ℓ^1, we get by Theorem III.3.8 that duals of separable M-embedded spaces X have property (X). By [259, Th. VII.8] we obtain that in this situation an operator $T : X^{**} \to Y^*$ is w^*-continuous if T is w^*-w^*-Borel measurable. We finish this digression by remarking that property (X) is strictly stronger than property (V^*) as shown in [595]. Accepting the existence of special cardinal numbers one also obtains an example of this from [184, Prop. 12]: $\ell^1(\Gamma) \prec \ell^1$ iff $\operatorname{card}\Gamma$ is not a real-valued measurable cardinal. (This also shows that some of the examples of spaces with property (X) in [271] and [251] have to be read with additional set-theoretic assumptions.)
Example 3.5 was first noted in [268] and Proposition 3.6 is an unpublished result of Pfitzner. Godefroy and Li proved property (u) for M-embedded spaces in [265], but only with the work in [397] the general background for this, presented in Theorem I.2.10, became apparent to us. Meanwhile the ultimate clarification was provided by [264] introducing the concept of u-ideals – see below. We mention that by the work of Knaust and Odell in [381] the hereditary Dunford-Pettis property (and equivalently property (S)) is sufficient for property (u), also the recent paper [368] studying property (u) in vector valued function spaces is of interest here. Let us remark that property (u) for M-embedded spaces can also be derived from recent deep results by Rosenthal who showed that X has (u) provided Y^* is weakly sequentially complete for all subspaces Y of X [531]. Proposition 3.9 and Corollary 3.10 are from [270]. Theorem 3.11 was obtained independently by Godefroy and Li [266] and D. Werner [621].
The main result of Section III.4, Theorem 4.5, is due to Fabian and Godefroy [213], but our proof largely follows ideas of Sims' and Yost's papers [571] and [572]. M. Fabian has shown us another proof of this result based on techniques from his article [211]. We shall have more to say on projectional resolutions of the identity later in this section.

FURTHER EXAMPLES OF M-EMBEDDED SPACES. We first consider the Schreier space S. This is the completion of the space of all sequences which are eventually 0 under the norm

$$\|(x_n)\|_S = \sup_E \sum_{n\in E} |x_n|.$$

Here the supremum has to be taken over all "admissible" finite sets E, meaning all sets $E = \{n_1, \dots, n_k\}$ where $k \le n_1 < \dots < n_k$. This Banach space has its origin in Schreier's paper [562], and some information on S can be found in [45] and [115]. Schreier used (a variant of) the space S to show that $C[0,1]$ fails the weak Banach-Saks property. (Recall that a Banach space X is said to have the *weak Banach-Saks property* if every weakly convergent sequence in X admits a subsequence with norm convergent arithmetic means.) We now present a result from [624].

- *The Schreier space S is an M-ideal in its bidual.*

To show this note first that the natural basis (e_n) is 1-unconditional and shrinking, hence $S^{**} \cong \{(x_n) \mid \|(x_n)\| := \sup_N \|\sum_{n=1}^N x_n e_n\| < \infty\}$ (cf. [422, Prop. 1.b.2]). Also $x_n \to 0$ for $(x_n) \in S^{**}$. For $x \in B_{S^{**}}$, $j_i \in B_S$ (without restriction with finite support) and $\varepsilon > 0$ choose first n_0 such that $j_i(n) = 0$ for $n > n_0$, then $m_0 \ge n_0$ such that $|x(m)| < \varepsilon/(n_0-1)$

for $m \geq m_0$. Put $j(n) = x(n)$ for $n < m_0$ and $j(n) = 0$ otherwise. Then

$$\sum_E |x(n) + j_i(n) - j(n)|$$

can be estimated by $\sum_E |x(n)| \leq \|x\| \leq 1$ if $\min E > n_0$, and by $\sum_{E\cap\{1,\dots,n_0\}} |j_i(n)| + \sum_{E\cap\{n_0+1,\dots\}} |x(n)| \leq \|j_i\| + (n_0-1)\frac{\varepsilon}{n_0-1} \leq 1+\varepsilon$ if $\min E \leq n_0$. Hence $\|x+j_i-j\| \leq 1+\varepsilon$.
Actually, the space originally considered by Schreier in [562] can be represented as a sequence space by means of the norm $\|(x_n)\| = \sup_E \left|\sum_{n\in E} x_n\right|$. The above proof can clearly be modified so as to show that this space is also M-embedded.
By Corollary 3.7 every nonreflexive subspace of a quotient of S contains a copy of c_0. In the recent paper [457] E. Odell proves the stronger result that *every* infinite dimensional subspace of a quotient of S contains a copy of c_0. Thus in order to obtain the full strength of Odell's theorem one "only" has to show that no infinite dimensional reflexive specimen exists among the subspaces of quotients of S. Odell also defines a hierarchy of Schreier spaces $S_1, S_2, \dots$ as follows. Let $(S_1, \|\cdot\|_1) = (S, \|\cdot\|_S)$ and suppose that $(S_m, \|\cdot\|_m)$ is already constructed. Then S_{m+1} is the completion of the finitely supported sequences with respect to the norm

$$\|(x_n)\|_{m+1} = \sup \sum_{j=1}^{k} \|P_{E_j}(x)\|_m$$

where the E_j are finite subsets of N satisfying

$$k \leq \min E_1 \leq \max E_1 < \min E_2 \leq \max E_2 < \dots \leq \max E_{k-1} < \min E_k,$$

where $(P_E x)_n = x_n$ if $n \in E$ and $(P_E x)_n = 0$ otherwise, and the supremum is taken over all k and all admissible strings $E_1, \dots, E_k$ as above. It follows in much the same way as above that each S_m is M-embedded and hence each infinite dimensional nonreflexive subspace of a quotient of S_m contains a copy of c_0. This gives a partial answer to Problem 6 in [457].
In [568] the extreme point structure of the unit ball of S is investigated. The knowledge that S is M-embedded allows us to add the information that one cannot expect, by Proposition II.4.2 and Theorem II.4.4, B_S to possess strongly extreme points, let alone strongly exposed points.
Let us now consider the function space $L^\infty + L^p$, $1 < p < \infty$, consisting of all measurable functions on $\mathbb{R}$ which have a representation $f = f_1 + f_2$ with $f_1 \in L^\infty(\mathbb{R})$, $f_2 \in L^p(\mathbb{R})$. We equip this space with the norm

$$\|f\| = \inf\{\|f_1\|_{L^\infty} \vee \|f_2\|_{L^p} \mid f = f_1 + f_2,\ f_1 \in L^\infty,\ f_2 \in L^p\},$$

which is equivalent to the usual one [423, p. 119]. Let $(L^\infty + L^p)_f$ denote the closed subspace generated by the characteristic functions χ_E where $\lambda(E) < \infty$.

- *$(L^\infty + L^p)_f$ is an M-ideal in its bidual.*

This time we will directly verify the norm condition on the projection. It is well known that the dual of $(L^\infty + L^p)_f$ is isometrically isomorphic to the space $L^1 \cap L^q$, normed by

$$\|g\| = \|g\|_{L^1} + \|g\|_{L^q}$$

(of course $1/p + 1/q = 1$), and the bidual can be identified with $L^\infty + L^p$. Let us consider $X = L^1 \cap L^q$ as the diagonal in $L^1 \oplus_1 L^q$. Then the annihilator of X in $(L^1 \oplus_1 L^q)^* = L^\infty \oplus_\infty L^p$ is

$$X^\perp = \{(f, -f) \mid f \in L^\infty \cap L^p\},$$

and one obtains

$$\begin{aligned} X^{\perp\perp} &= \\ &= \left\{ (\ell, g_1, g_2) \in L^1_s \oplus_1 L^1 \oplus_1 L^q \;\middle|\; \langle \ell, f\rangle + \int (g_1 - g_2) f \, d\lambda = 0 \quad \forall f \in L^\infty \cap L^p \right\} \\ &= \left\{ (\ell, g_1, g_2) \in L^1_s \oplus_1 L^1 \oplus_1 L^q \;\middle|\; \langle \ell, \chi_E\rangle + \int_E (g_1 - g_2)\, d\lambda = 0 \ \text{ if } \ \lambda(E) < \infty \right\}. \end{aligned}$$

Here we write $(L^1)^{**} = L^1_s \oplus_1 L^1$ where L^1_s is the space of "singular" functionals, cf. Example IV.1.1(a). Hence one deduces

$$X^{\perp\perp} = X_s \oplus_1 X$$

with

$$X_s \cong \{\ell \in (L^1)^{**} \mid \langle \ell, 1_E\rangle = 0 \ \text{ if } \ \lambda(E) < \infty\} \cong (L^\infty + L^p)_f{}^\perp$$

which shows our claim. Clearly, L^p can be replaced by a reflexive Köthe function space in the above proposition. This examples comes from [621].
In our next example we deal with the Bergman space

$$L^1_a = \left\{ f : \mathbb{D} \to \mathbb{C} \text{ analytic} \;\middle|\; \int_{\mathbb{D}} |f(x+iy)| \, dx dy < \infty \right\}.$$

By definition, this is a (closed) subspace of $L^1(\mathbb{D})$, and we are going to show:

- *The Bergman space L^1_a is L-embedded, in fact, it has a predual which is M-embedded.*

To show this we use a result proved in the next chapter, viz. Theorem IV.3.10. Let us first observe that L^1_a is a dual space. We consider the topology τ of uniform convergence on compact subsets of $\mathbb{D}$. Now, the unit ball of L^1_a is a normal family. Indeed, the mean value property of analytic functions implies

$$|f(z)| \le \frac{1}{\pi r^2} \int_{B_r(z)} |f(x+iy)| \, dx dy \le \frac{1}{\pi r^2} \|f\|$$

for $|z| \le R < 1$ and $r < 1 - R$. Moreover, an appeal to Fatou's lemma reveals that $B_{L^1_a}$ is τ-closed in the space of all analytic functions on $\mathbb{D}$; for if $f_n \xrightarrow{\tau} f$, then

$$\int |f_n(\varrho e^{i\varphi})| \, d\varphi \to \int |f(\varrho e^{i\varphi})| \, d\varphi \qquad \forall \varrho < 1,$$

hence

$$\begin{aligned}\|f\| &= \int \varrho\, d\varrho \int |f(\varrho e^{i\varphi})|\, d\varphi \;\le\; \liminf \int \varrho\, d\varrho \int |f_n(\varrho e^{i\varphi})|\, d\varphi \\ &= \liminf \|f_n\| \le 1.\end{aligned}$$

Consequently, $B_{L^1_a}$ is τ-compact, and by a classical result [318, p. 211] those $\ell \in (L^1_a)^*$ whose restriction to $B_{L^1_a}$ is τ-continuous forms a predual of L^1_a. Observe next that the embedding of L^1_a into $L^0(\mathbb{D})$, the space of all measurable functions on $\mathbb{D}$ equipped with the topology of convergence in measure, is continuous for the τ- and L^0-topologies so that the two topologies coincide on the unit ball. Hence the above predual coincides with the space $(L^1_a)^\sharp$ from Theorem IV.3.10, $(L^1_a)^\sharp$ separates L^1_a (since $f \mapsto f(z)$ is in $(L^1_a)^\sharp$), and the unit ball of L^1_a is L^0-closed. Thus Theorem IV.3.10 proves our claim. – This example is due to E. Werner.

The dual space of L^1_a can be identified with the Bloch space B (see e.g. [37]); however there is only an isomorphism and not an isometry between $(L^1_a)^*$ and B. One can check that the above predual is a renorming of the little Bloch space B_0. (In Example 1.4(j) we proved that B_0 is M-embedded in its most natural norm, which is different.) The papers [36] and [37] contain more information on renormings of B_0 and B, their relationship to operators on Hilbert space and the Bergman space. As a matter of fact, B_0 is known to be isomorphic to c_0 [567]. Let us take the chance to point out how to obtain this result from the material of this chapter: Since the Bergman projection maps $C_0(\mathbb{D})$ onto B_0 [566], we see that B_0 is a separable M-embedded $\mathcal{L}^\infty$-space; now apply Theorem 3.11. In this connection Lusky's recent paper [431] is also relevant. Similar arguments work for the case of weighted spaces of analytic functions on certain subsets of $\mathbb{C}^n$ [623].

We finally wish to connect the examples of M-ideals of Lorentz and Orlicz sequence spaces given in Example 1.4(b) and (c) to the corresponding ideals of compact operators on a separable Hilbert space H. Our setting will be as follows. Let E be a Banach space with a 1-symmetric basis. Then $\mathcal{J}_E$ denotes the set of compact operators T on H whose sequences $(s_n(T))$ of singular numbers belong to E. This is a Banach ideal of operators under the norm

$$\|T\|_E = \|(s_n(T))\|_E$$

Various properties of E are known to pass to $\mathcal{J}_E$ (see for instance [24] and its references); for the general theory of "symmetrically normed ideals" we refer to [276] and [569]. In this vein one can prove:

- *Let E have a 1-symmetric basis, and suppose $E \neq c_0$. Then E is an M-ideal in its bidual if and only if $\mathcal{J}_E$ is.*

The proof is contained in [624].

DOUGLAS ALGEBRAS. Let H^∞ denote the algebra of bounded analytic functions on the open unit disk. As usual we shall identify H^∞ via radial limits isometrically with a closed subalgebra of $L^\infty = L^\infty(\mathbb{T})$. The algebra H^∞ has the following maximal property: If B is a weak* closed subalgebra of L^∞ strictly containing H^∞, then $B = L^\infty$. (Note that H^∞ itself is weak* closed.) In 1969 R. G. Douglas, in his work on Toeplitz operators, was led to investigate certain norm closed subalgebras between H^∞ and L^∞; more precisely he became interested in those subalgebras B with the additional property that B is the closed algebra generated by H^∞ and those $f \in B$ for which $f^{-1} \in H^\infty$; equivalently, B is generated by H^∞ and those conjugates of inner functions which are in B. He conjectured that every closed algebra between H^∞ and L^∞ necessarily has this property, which was shown to be true by Chang and Marshall in 1976. Hence a closed algebra between H^∞ and L^∞ is called a *Douglas algebra.* (For references for this fact and related results we refer to the surveys [556], [557], [558] and the monograph [244].)

The simplest example of a Douglas algebra is $H^\infty + C$, the linear span of H^∞ and the space C of continuous functions on $\mathbb{T}$. (As we have already remarked, Sarason was the first to observe that $H^\infty + C$ is closed in that he shows that the canonical mapping $f + A \mapsto f + H^\infty$ from C/A to L^∞/H^∞ is an isometry. Here A denotes the disk algebra. Another proof is due to Rudin [544].) It is also known that $H^\infty + C$ is the smallest Douglas algebra (apart from H^∞).

In 1979 Axler, Berg, Jewell and Shields [38] proved that $H^\infty + C$ is a proximinal subspace of L^∞. In fact, they showed that the quotient space $(H^\infty + C)/H^\infty$ is proximinal in L^∞/H^∞, and the above result follows easily from this and the fact that H^∞, being weak* closed, is proximinal. Then Luecking [427] obtained the same conclusion from his result that $(H^\infty + C)/H^\infty$ is an M-ideal in L^∞/H^∞; see Corollary 1.5.

Subsequently M-ideal methods have proved useful for obtaining best approximation results for Douglas algebras. Here we shall survey some of them. Let M be the maximal ideal space of L^∞ so that $L^\infty = C(\mathsf{M})$. If $B \subset L^\infty$ is a Douglas algebra and $S \subset \mathsf{M}$ we let

$$B_S = \{f \in L^\infty \mid f_{|S} \in B_{|S}\}.$$

It is known that B_S is closed if S is a p-set for B. (See Section V.4 for the definition of a p-set.) Younis [658] shows that H^∞_S/H^∞ is an M-ideal in L^∞/H^∞ and thus that H^∞_S is proximinal if S is a p-set for H^∞. In [659] and [428] Younis and Luecking consider algebras of the form $H^\infty + L^\infty_F$ where for $F \subset \mathbb{T}$

$$L^\infty_F = \{f \in L^\infty \mid f \text{ is continuous at each } t \in F\}.$$

It is known that $H^\infty + L^\infty_F$ is always a Douglas algebra, and for closed F it is shown in [659] that $(H^\infty + L^\infty_F)/H^\infty$ is an M-ideal in L^∞/H^∞. One can translate this result as follows (cf. [428]): If $S \subset \mathsf{M}$ is a p-set for $H^\infty + C$, then $(H^\infty + C)_S/H^\infty$ is an M-ideal in L^∞/H^∞; and Luecking and Younis ask if $(H^\infty + C)_S$ is the only Douglas algebra with the above property. This was answered in the negative by K. Izuchi [331]. The paper [332] contains the general result that, for a Douglas algebra B and a p-set S, B_S/B is an M-ideal in L^∞/B. All these results contribute to the longstanding open question whether every Douglas algebra is proximinal in L^∞. Let us mention that this problem was finally solved by Sundberg [589] who constructed a shrewd counterexample in 1984. Also, the paper [241] is of interest in this connection.

There is another type of problem where M-ideal methods have turned out to be of importance. Axler et al. prove in [38] that the unit ball of $L^\infty/(H^\infty + C)$ fails to possess extreme points, thus $L^\infty/(H^\infty + C)$ is not isometric to a dual Banach space, which stands in marked contrast to L^∞/H^∞ which can be identified with the dual of the Hardy space H^1_0. In [659] and [428] it is shown that for closed or open $F \subset \mathbb{T}$, the quotient space $L^\infty/(H^\infty + L^\infty_F)$ shares this property with $L^\infty/(H^\infty + C)$ which represents the special case $F = \mathbb{T}$. The proof relies essentially on the non-uniqueness of best approximants from M-ideals. Meanwhile it has been shown by Izuchi [329] that the above statement holds for every subset F of $\mathbb{T}$. For related papers see [330], [333], [334] and [660].

PROJECTIONAL RESOLUTIONS OF THE IDENTITY. In 1966 Corson and Lindenstrauss stated the – in Namioka's words prophetic – conjecture that every weakly compact subset of a Banach space is homeomorphic to a weakly compact subset of $c_0(\Gamma)$ for a suitable set Γ. It is easily seen that in order to prove this it is sufficient to show the following result, which was established by Amir and Lindenstrauss in 1968.

THEOREM. (Amir-Lindenstrauss)
For every weakly compactly generated Banach space X there are a set Γ and a continuous injective operator $T : X \to c_0(\Gamma)$.

In fact, most of the properties of weakly compactly generated spaces already follow from the existence of such an injection T, cf. [16, Section 1]. The first step towards the above theorem was performed by Lindenstrauss in [416] and [417] where he proved it for reflexive X. As a decisive tool he showed the existence of a projectional resolution of the identity in X, and in order to get this he used what was later called a Lindenstrauss compactness argument. We recall that a Banach space X is said to admit a projectional resolution of the identity (PRI) if there is a family $(P_\alpha)_{\omega\le\alpha\le\mu}$ of projections in X, where μ is the first ordinal with cardinality dens X, satisfying

(a) $\|P_\alpha\| = 1$ for $\omega \le \alpha \le \mu$,
(b) $P_\mu = Id_X$,
(c) $P_\alpha P_\beta = P_\beta P_\alpha = P_\beta$ for $\beta \le \alpha$,
(d) $\operatorname{dens} P_\alpha X \le \operatorname{card} \alpha$ for $\omega \le \alpha \le \mu$,
(e) $\overline{\bigcup_{\beta<\alpha} P_\beta X} = P_\alpha X$ for limit ordinals α.

The proof of the general Amir-Lindenstrauss theorem in [16] is much harder than in the reflexive case, but again the existence of a PRI is instrumental. A survey of weakly compactly generated spaces up to 1967 is [418]. It still furnishes a good introduction to this area, though most of the problems stated there have meanwhile been solved, see [155, Chapter Five].
Vašák [613] generalized the Amir-Lindenstrauss theorem by constructing a PRI for every weakly countably determined Banach space X. (X is called weakly countably determined if there is a sequence $(A_n)_{n\in\mathbb{N}}$ of w^*-compact subsets of X^{**} such that for all $x \in X$ there is a subset $N_x \subset \mathbb{N}$ with $x \in \bigcap_{n\in N_x} A_n \subset X$.) In 1979 Gul'ko made a major innovation by exposing the purely topological core of the problem and by providing a very simple proof of the Amir-Lindenstrauss theorem using the notion of conjugate pairs. A recommendable account of Gul'ko's ideas for the weakly compactly generated case is the article of Namioka and Wheeler [449]. Also Stegall gave a short and selfcontained proof in [583].
An easy consequence of the Mackey-Arens theorem is that if K is any weakly compact subset of a Banach space X, then the restriction map $R : X^* \to C(K)$ is w^*-w-continuous. From this, one can show that a Banach space X is weakly compactly generated iff there exists a w^*-w-continuous linear injection $T : X^* \to Y$ for some Banach space Y; and

that a $C(K)$-space is weakly compactly generated iff K is an Eberlein compact space, i.e. homeomorphic to a weakly compact subset of a Banach space. Thus an essentially equivalent, yet more topological formulation of the Amir-Lindenstrauss theorem is: if K is an Eberlein compact space, then there is an injective operator $T : C(K) \to c_0(\Gamma)$ for a suitable set Γ. In this form it was extended by Argyros, Mercourakis and Negrepontis to Corson compact spaces K (a compact space K is called Corson compact if K is homeomorphic to a subset of $\Sigma(\mathbb{R}^I) := \{x \in \mathbb{R}^I \mid \text{supp}(x) \text{ is countable}\}$ for some set I); see [451, Section 6]. Further extensions – which reintroduce functional analysis and PRIs – can be found in [610] and [612]. In [470] Orihuela and Valdivia define the notion of a projective generator, thus providing a unified approach not only to the above questions but also to the problem of constructing PRIs in dual spaces, which we discuss below. In fact, the arguments in [470] and [611], when specialized to weakly compactly generated Banach spaces, seem to provide the simplest known proof of the Amir-Lindenstrauss theorem.

We now wish to relate the existence of a PRI to other properties of a Banach space. In fact, the assumption of a PRI neither implies the existence of a continuous injection $T : X \to c_0(\Gamma)$ nor the separable complementation property nor an equivalent locally uniformly rotund renorming. Let us give an example. Let X be a Banach space with $\text{dens}\, X = \mathfrak{m}$ admitting a PRI $(P_\alpha)_{\omega \le \alpha \le \mu}$, and suppose Z is an arbitrary Banach space with $\text{dens}\, Z = \mathfrak{n} < \mathfrak{m}$. Then the operators $Q_\alpha = P_\alpha \oplus 0$ for $\alpha \le \nu$ and $Q_\alpha = P_\alpha \oplus Id_Z$ for $\nu < \alpha \le \mu$ (where ν denotes the smallest ordinal with cardinality $\mathfrak{n}$) define a PRI on $X \oplus_\infty Z$. Suitable choices of Z provide the desired examples. (This simple construction was shown to us by D. Yost.) On the other hand, in many classes of Banach spaces for which PRIs can be constructed, the proofs automatically yield those stronger conclusions. By the way, having an injection $T : X \to c_0(\Gamma)$ is not sufficient for the existence of an equivalent locally uniformly rotund norm on X, either, e.g. $X = \ell^\infty$ even fails to have an equivalent weakly locally uniformly rotund norm [155, p. 120]. However, certain other properties less restrictive than being weakly compactly generated are enough, as shown in [603], [663], and [273]. On the other hand, the existence of an injective operator $T : X \to c_0(\Gamma)$ is sufficient for an equivalent strictly convex norm on X, but not necessary as shown by Dashiell and Lindenstrauss, see [144]. Finally, although used for isomorphic and topological problems, the existence of a PRI is an isometric property: $X = C[0, \omega_1]$ has a PRI with respect to the supremum norm, but when equipped with the equivalent Fréchet differentiable norm constructed by Talagrand in [598] it does not; see [213, p. 149] for details.

In some dual Banach spaces X^* it is possible to construct a projectional resolution of the identity even if X^* is not weakly compactly generated. Tacon [591] did this under the assumption that X is very smooth, i.e. X is smooth and has a $\|\,.\,\|$-w-continuous support mapping $S_X \to S_{X^*}$, cf. p. 142. It is easy to see that $\text{dens}\, Y = \text{dens}\, Y^*$ for every subspace Y of a very smooth space X, hence very smooth spaces are Asplund spaces. In [571] and [572] Sims and Yost tried to extend Tacon's result to arbitrary Asplund spaces X; however the projections P_α in X^* which they constructed could not be arranged to satisfy $P_\alpha X^* = \overline{\bigcup_{\beta<\alpha} P_\beta X^*}$ for limit ordinals α. Yet their approach gives an elementary (not model theoretic) proof of the result of Heinrich and Mankiewicz that the dual of every nonseparable Banach space contains uncountably many nontrivial complemented subspaces (i.e., they are neither finite dimensional nor finite codimensional).

In this connection let us mention the very recent spectacular counterexamples of Gowers and Maurey who found (separable) Banach spaces without nontrivial complemented subspaces.
As we said in the introduction to Section III.4 the ideas of Sims and Yost do prove successful in the special situation of M-embedded spaces. Fabian extended Tacon's result in [211] and [212]; he proves the crucial step [211, Lemma 1] without compactness arguments but with sophisticated geometrical reasoning, and he singles out the following

PROPOSITION. *Let X be a Banach space with a $\|\,.\,\|$-w lower semi-continuous set-valued mapping $D : X \to 2^{X^*}$ satisfying*

(a) *$D(x)$ is countable for every $x \in X$,*
(b) *$\overline{\text{lin}}\,\{x^*|_V \mid x^* \in D(x),\, x \in V\} = V$ for every closed subspace V of X.*

Then X^* has a projectional resolution of the identity (which comes from Hahn-Banach extension operators).

Recall that a set-valued map Φ from a topological space A into the subsets 2^B of a topological space B is said to be *lower* (*upper*) *semi-continuous*, if the set $\{a \in A \mid \Phi(a) \cap M \neq \emptyset\}$ is open (closed) whenever M is an open (closed) subset of B. Using the above proposition, Fabian and Godefroy [213] proved the general result that the dual of every Asplund space admits a projectional resolution of the identity. In order to find a suitable mapping D they use the following selection theorem of Jayne and Rogers [343, Th. 8]:

THEOREM. *Let M be a metric space and X^* a dual Banach space with the Radon-Nikodým property. Let $F : M \to 2^{X^*}$ be an upper semi-continuous set-valued map (X^* with its w^*-topology) which takes only nonempty and w^*-closed values. Then, with respect to the norm topology on X^*, the set-valued map F has a Borel measurable selector f of the first Baire class.*

This can be applied to the support map $F : X \to 2^{X^*}$, $F(x) = \{x^* \in B_{X^*} \mid x^*(x) = \|x\|\}$. The $\|\,.\,\|$-$\|\,.\,\|$-continuous functions $f_n : X \to X^*$ converging pointwise to the selector f are used to build the set-valued map $D : X \to 2^{X^*}$ required in the proposition by $D(x) = \{f_n(x) \mid n \in \mathbb{N}\}$. Note that establishing (b) for this map D is far from being obvious. In closing, we remark that the projections P_α in X^* are in general not w^*-continuous (cf. [213, p. 149]) and that clearly not every dual space admits a PRI (e.g. $X^* = \ell^\infty$).
The question under which conditions the adjoints P_α^* of a PRI (P_α) in X form a PRI in X^*, i.e. (P_α) is a shrinking PRI, was considered by John and Zizler in [346, Lemma 3]. This is the case if X has a Fréchet differentiable norm. Theorem 1 of this article says that a weakly compactly generated Banach space X admits a shrinking projectional resolution of the identity iff it has an equivalent Fréchet differentiable norm. The renorming results of Talagrand in [598] show that this equivalence fails without extra assumptions on X.

CHAPTER IV

Banach spaces which are L-summands in their biduals

IV.1 Basic properties

In this section we will mainly be concerned with the stability properties of Banach spaces which are L-summands in their biduals. Furthermore the relation of these spaces with M-embedded spaces is investigated, various examples and counterexamples are given, and some isometric properties (proximinality, extreme points) are studied.

Recall from Definition III.1.1 that an L- (M-) embedded space means a Banach space which is L-summand (M-ideal) in its bidual.
The spaces A and H_0^1 of analytic functions appearing in the following list of examples of L-embedded spaces have been defined on page 104. Furthermore recall that a von Neumann algebra M is a unital selfadjoint subalgebra of $L(H)$ (where H is a Hilbert space) which is closed for the weak (equivalently: strong) operator topology. It is easy to check that the unit ball of M is then compact with respect to the weak operator topology, hence the subspace

$$M_* = \{f \in M^* \mid f|_{B_M} \text{ is continuous for the weak operator topology}\}$$

of the dual M^* is a predual of M; that is, M is isometrically isomorphic to $(M_*)^*$ [593, p. 70], and the inclusion map coincides with the canonical embedding of M_* into its bidual M^*. Moreover, M_* is the only predual of M [593, p. 135]. A well known result in C^*-algebra theory states that conversely every C^*-algebra which is isometric to a dual Banach space (such an object is sometimes called a W^*-algebra) is a von Neumann algebra [593, p. 133].

Example 1.1 *The following (classes of) spaces are L-summands in their biduals:*

(a) *$L^1(\mu)$-spaces,*
(b) *preduals of von Neumann algebras,*
(c) *duals of M-embedded spaces,*
(d) *the Hardy space H_0^1, the dual of the disk algebra A^* and L^1/H_0^1.*

PROOF: (a) We first present a Banach lattice argument and refer to [560] for unexplained notation on this; see also Example I.1.6(a) and the text thereafter. $L^1(\mu)$ is a projection band in its bidual, which is again an AL-space. So the defining norm-condition of AL-spaces shows the claim (cf. [560, Section V.8]).
The idea of orthogonality also emerges if one employs the following measure theoretic argument: Represent $(L^1(\mu))^*$ as $C(K)$, hence $(L^1(\mu))^{**}$ as $M(K)$. Then μ has a canonical extension to a measure $\widehat{\mu}$ on K (see the proof of Example III.1.4(h)). Therefore, the Lebesgue decomposition with respect to $\widehat{\mu}$ yields the desired L-projection from $M(K)$ onto the copy of $L^1(\mu)$, along the singular measures on K.

(b) The proof relies on some results to be presented in Chapter V which we shall accept for the time being. Let M be a von Neumann algebra and M_* its predual. We consider the annihilator of M_* in M^{**}:

$$(M_*)^\perp = \{F \in M^{**} \mid F_{|M_*} = 0\}.$$

It will be shown in Theorem V.1.10 that M^{**} is a C^*-algebra in its own right and that multiplication both from the right and the left is weak* continuous. The product in M^{**} is the (by virtue of Theorem V.1.10(a) unique) Arens product as defined in Definition V.1.1.
We wish to show that $(M_*)^\perp$ is a two-sided ideal in M^{**}. To this end it suffices, by weak* continuity of the multiplications, to check that $Fx \in (M_*)^\perp$ and $xF \in (M_*)^\perp$ provided $x \in M$ and $F \in (M_*)^\perp$. To show the former we must prove

$$(Fx)(f) = F(xf) = 0 \qquad \forall f \in M_*. \tag{$*$}$$

Now, by definition of the Arens product, $(xf)(y) = f(yx)$ for $y \in M$ from which we deduce $xf \in M_*$. Hence $(*)$ obtains. The proof that $xF \in (M_*)^\perp$ is similar.
Using Theorem V.4.4 we conclude that $(M_*)^\perp$ is a weak* closed M-ideal, and Corollary II.3.6(b) implies that M_* is an L-summand in M^*.

(c) Corollary III.1.3.

(d) The Hardy space H_0^1 is the dual of the M-embedded space $C(\mathbb{T})/A$; see Example III.1.4(h). (We shall give another proof in Example 3.6.)
As for A^*, note that $A^* \cong M(\mathbb{T})/A^\perp \cong M(\mathbb{T})/H_0^1$ by the F. and M. Riesz theorem, so the claim about A^* follows from Corollary 1.3 below and the fact that $M(\mathbb{T})$ and H_0^1 are L-embedded spaces. (Note that A^* has no M-embedded predual because it fails the RNP.)
The same reasoning applies to L^1/H_0^1. □

We mention in passing that $H^1 \cong H_0^1$ and $L^1/H^1 \cong L^1/H_0^1$ so that these spaces are L-embedded, too.

In example (a) above the complementary L-summand for $X = L^1(\mu)$ in X^{**} is the set of elements which are singular, i.e. lattice or measure theoretically orthogonal with respect to X. Adopting this viewpoint also in the general case we will write X_s for the complementary L-summand of a Banach space X which is L-summand in its bidual, i.e.

$$X^{**} = X \oplus_1 X_s.$$

It is not only a formal matter to imagine the elements in X_s as "singular" elements of X^{**}. Proposition 2.1 shows that $x^{**} \in X^{**}$ belongs to X_s if x^{**}, viewed as a function on (B_{X^*}, w^*), is "extremely discontinuous", and for $X = L^1(\mu)$, $X^{**} = ba(\mu)$ a finitely additive measure ν is in X_s if it is "extremely non-σ-additive", namely purely finitely additive (see the discussion at the beginning of Section IV.3.)

Unlike to what has been shown in Theorem III.1.5 for M-embedded spaces, L-embeddedness does not pass to subspaces or quotients. (This is clear for quotients as every Banach space is a quotient of an L^1-space and follows for subspaces e.g. from Corollary 1.15.) Let us have a closer look at subspaces of L-embedded spaces:

Theorem 1.2 *For a Banach space X which is an L-summand in its bidual (so that $X^{**} = X \oplus_1 X_s$) and a closed subspace Y of X the following are equivalent:*

(i) *Y is an L-summand in its bidual.*

(ii) $Y^{\perp\perp} = Y \oplus_1 (Y^{\perp\perp} \cap X_s)$

(iii) $P\overline{Y}^{w^*} = Y$, *where P is the L-projection from X^{**} onto X.*

(iii') $P\overline{B_Y}^{w^*} = B_Y$

PROOF: (i) $\Rightarrow$ (ii): Embedding Y^{**} as $Y^{\perp\perp}$ in X^{**} we have

$$Y^{\perp\perp} = Y \oplus_1 Z \qquad (*)$$

for some subspace Z of X^{**} since by assumption $Y^{**} = Y \oplus_1 Y_s$ (of course $Z \subset Y^{\perp\perp}$). We want to show

$$Z = Y^{\perp\perp} \cap X_s.$$

The inclusion "$\supset$" is obvious. To prove "$\subset$" take $z \in Z$ and decompose $z = x + x_s$ in $X^{**} = X \oplus_1 X_s$. Because of $(*)$ we have for all $y \in Y$

$$\|y + z\| = \|y\| + \|z\| = \|y\| + \|x\| + \|x_s\|$$

and because of $X^{**} = X \oplus_1 X_s$

$$\|y + z\| = \|(y + x) + x_s\| = \|y + x\| + \|x_s\|,$$

hence

$$\|y + x\| = \|y\| + \|x\| \qquad \text{for all } y \in Y.$$

But then also

$$\|y^{**} + x\| = \|y^{**}\| + \|x\| \qquad \text{for all } y^{**} \in Y^{\perp\perp}.$$

(This follows from considering the space $G = Y \oplus_1 \mathbb{K}x$ and noting $G^{**} = Y^{\perp\perp} \oplus_1 \mathbb{K}x$.) Since $z \in Y^{\perp\perp}$ we get

$$\|x_s\| = \|z - x\| = \|z\| + \|x\| = 2\,\|x\| + \|x_s\|.$$

So $x = 0$, hence $y \in X_s$.
(ii) $\Rightarrow$ (iii): Obvious by $Y^{\perp\perp} = \overline{Y}^{w*}$.
(iii) $\Rightarrow$ (i): Since (iii) is equivalent to $PY^{\perp\perp} = Y$ it follows from Lemma I.1.15 that $P_{|Y^{\perp\perp}}$ is an L-projection from $Y^{\perp\perp} = Y^{**}$ onto Y.
(iii) $\Rightarrow$ (iii′): Trivial.
(iii′) $\Rightarrow$ (iii): Since $\overline{B_Y}^{w*} = B_{Y^{\perp\perp}}$ we have $\overline{Y}^{w*} = \bigcup_{n\in\mathbb{N}} \overline{nB_Y}^{w*}$. [Note, however, that for a subspace Y of an arbitrary dual space E^* the inclusion $\overline{Y}^{\sigma(E^*,E)} = \bigcup_n \overline{nB_Y}^{\sigma(E^*,E)}$ is *false* in general.] □

Corollary 1.3 *Let X and Y be Banach spaces which are L-summands in their biduals, Y a subspace of X. Then X/Y is an L-summand in its bidual.*

PROOF: If P is the L-projection from X^{**} onto X we have by assumption and Theorem 1.2 $PY^{\perp\perp} \subset Y^{\perp\perp}$. Lemma I.1.15 shows now that

$$\begin{array}{rcll} P/Y^{\perp\perp} : X^{**}/Y^{\perp\perp} & \longrightarrow & X^{**}/Y^{\perp\perp} \\ x^{**} + Y^{\perp\perp} & \longmapsto & Px^{**} + Y^{\perp\perp} \end{array}$$

is an L-projection onto $(X + Y^{\perp\perp})/Y^{\perp\perp} \cong X/(X \cap Y^{\perp\perp}) = X/Y$. Clearly $X^{**}/Y^{\perp\perp} \cong (X/Y)^{**}$. □

Easy examples show:

- *X and X/Y L-embedded spaces $\not\Rightarrow$ Y L-embedded.*
- *Y and X/Y L-embedded spaces $\not\Rightarrow$ X L-embedded.*

Before we proceed with the discussion of stability properties of L-embedded spaces we give a kind of quantitative version of Theorem 1.2 which will be used in the next section. Note that for L-embedded spaces X and Y, with Y a subspace of X and corresponding L-projections P and Q from the the biduals onto the spaces, Theorem 1.2 gives $P_{|Y^{\perp\perp}} = Q$.

Lemma 1.4 *Let X be an L-embedded space with L-projection P from X^{**} onto X. Let the subspace $Y \subset X$ be an almost L-summand in its bidual in the sense that there is a number $0 < \varepsilon < 1/4$ such that $Y^{**} = Y \oplus Y_s$ and $\|y + y_s\| \geq (1-\varepsilon)(\|y\| + \|y_s\|)$ for all $y \in Y$, $y_s \in Y_s$. Then $\|P_{|Y^{\perp\perp}} - Q\| \leq 3\varepsilon^{1/2}$, where Y^{**} and $Y^{\perp\perp} = \overline{Y}^{w*} \subset X^{**}$ are identified and Q denotes the projection from $Y^{\perp\perp}$ onto Y.*

PROOF: By assumption there is a subspace $Z \subset X^{**}$ such that $Y^{**} \cong Y^{\perp\perp} = \overline{Y}^{w*} = Y \oplus Z$ with $\|y + z\| \geq (1-\varepsilon)(\|y\| + \|z\|)$. Because of $\|Py^{\perp\perp} - Q\,y^{\perp\perp}\| = \|P(y+z) - Q(y+z)\| = \|Pz\|$ and because of $(\varepsilon^{1/2} + 2\varepsilon)\|z\| \leq \frac{\varepsilon^{1/2}+2\varepsilon}{1-\varepsilon}\|y+z\| \leq 3\varepsilon^{1/2}\|y+z\|$ for any $y \in Y$, $z \in Z$ it is enough to show $\|Pz\| \leq (\varepsilon^{1/2} + 2\varepsilon)\|z\|$ for each $z \in Z$. Decompose

$z = x + x_s$ in $X^{**} = X \oplus_1 X_s$. Since we are done if $\|x\| = \|Pz\| \le \varepsilon^{1/2}\|z\|$, we assume $\|x\| > \varepsilon^{1/2}\|z\|$ from now on. We obtain

$$\begin{aligned} \|y + x\| &= \|(y+x) + x_s\| - \|x_s\| = \|y + z\| - \|x_s\| \\ &\ge (1-\varepsilon)(\|y\| + \|z\|) - \|x_s\| \\ &= (1-\varepsilon)(\|y\| + \|x\| + \|x_s\|) - \|x_s\| \\ &= (1-\varepsilon)(\|y\| + \|x\|) - \varepsilon\|x_s\| \\ &\ge (1-\varepsilon)(\|y\| + \|x\|) - \varepsilon\|z\| \\ &\ge (1-\varepsilon)(\|y\| + \|x\|) - \varepsilon^{1/2}\|x\| \\ &\ge (1-2\varepsilon^{1/2})(\|y\| + \|x\|) \end{aligned} \tag{1}$$

for all $y \in Y$, which extends to all $y^{\perp\perp} \in Y^{\perp\perp}$, as will be shown in a moment:

$$\|y^{\perp\perp} + x\| \ge (1 - 2\varepsilon^{1/2})(\|y^{\perp\perp}\| + \|x\|). \tag{2}$$

For the time being we take (2) for granted and have in particular for $z \in Y^{\perp\perp}$

$$\|x_s\| = \| - z + x\| \ge (1 - 2\varepsilon^{1/2})(\|z\| + \|x\|) \ge (1 - 2\varepsilon^{1/2})(\|z\| + \varepsilon^{1/2}\|z\|)$$

and finally

$$\|Pz\| = \|x\| = \|z\| - \|x_s\| \le \|z\| - (1-2\varepsilon^{1/2})(1+\varepsilon^{1/2})\|z\| = (\varepsilon^{1/2} + 2\varepsilon)\|z\|.$$

It remains to prove (2). We first observe that $x \notin Y$ because otherwise

$$0 = \| - x + x\| \ge (1 - 2\varepsilon^{1/2})(\| - x\| + \|x\|) > 0$$

by our standing assumption that $\|x\| > \varepsilon^{1/2}\|z\|$. Thus it makes sense to consider the direct sum $G = Y \oplus \mathbb{K}x \subset X$, and we let ι denote the identity from G onto $\widetilde{G} = Y \oplus_1 \mathbb{K}x$. Therefore $\widetilde{G}^{**} \cong Y^{\perp\perp} \oplus_1 \mathbb{K}x$ and $\|\iota\| \le (1 - 2\varepsilon^{1/2})^{-1}$ by (1). Inequality (2) follows now with $y^{\perp\perp} + x \in G^{\perp\perp}$ from

$$\|y^{\perp\perp}\| + \|x\| = \|\iota^{**}(y^{\perp\perp} + x)\| \le (1 - 2\varepsilon^{1/2})^{-1}\|y^{\perp\perp} + x\|. \qquad \square$$

We continue with two more stability properties of L-embedded spaces.

Proposition 1.5 *The class of L-embedded spaces is stable by taking*
(a) *1-complemented subspaces,*
(b) *ℓ^1-sums.*

PROOF: (a) Let X be an L-summand in its bidual, $P : X^{**} \to X^{**}$ the L-projection onto X, Y a 1-complemented subspace of X, $Q : X \to Y$ the contractive projection (considered as a mapping onto Y), and $i : Y \to X$ the inclusion mapping. Then $Q^{**}Pi^{**}$ is the L-projection from Y^{**} onto Y.
(b) Let $(X_i)_{i\in I}$ be a family of L-embedded spaces and $X := (\oplus \sum X_i)_{\ell^1(I)}$. Then $X^* = (\oplus \sum X_i^*)_{\ell^\infty(I)}$. Putting

$$Y := \Big(\oplus \sum X_i^{**}\Big)_{\ell^1(I)} \quad \text{and} \quad Z := \Big\{x^{**} \in X^{**} \mid x^{**}|_{(\oplus \sum X_i^*)_{c_0(I)}} = 0\Big\}$$

it is standard to show that

- $Y \cap Z = \{0\}$,
- $\|(x_i^{**}) + x^{**}\| = \|(x_i^{**})\| + \|x^{**}\| \quad$ for $(x_i^{**}) \in Y$, $x^{**} \in Z$,
- $Y + Z = X^{**}$.

So we have the decomposition $X^{**} = Y \oplus_1 Z$. Using the assumption it is now easy to see that X is an L-summand in Y, hence X is an L-summand in X^{**}. □

Proposition 1.6 *Let X be an L-embedded space, $Y_1, Y_2, Y_i (i \in I)$ subspaces of X which are also L-embedded spaces. Then:*

(a) *$\bigcap_{i \in I} Y_i$ is an L-embedded space.*

(b) *$B_{Y_1} + B_{Y_2}$ is closed.*

(c) *If $Y_1 + Y_2$ is closed, then $Y_1 + Y_2$ is an L-embedded space.*

PROOF: (a) follows from Theorem 1.2.

(b) By w^*-compactness and w^*-continuity of addition we have

$$\overline{B_{Y_1} + B_{Y_2}}^{w*} = \overline{B_{Y_1}}^{w*} + \overline{B_{Y_2}}^{w*},$$

hence by Theorem 1.2(iii′) (P is the L-projection from X^{**} onto X)

$$P(\overline{B_{Y_1} + B_{Y_2}}^{w*}) = P(\overline{B_{Y_1}}^{w*} + \overline{B_{Y_2}}^{w*}) = P\overline{B_{Y_1}}^{w*} + P\overline{B_{Y_2}}^{w*} = B_{Y_1} + B_{Y_2}.$$

This implies the closedness of $B_{Y_1} + B_{Y_2}$.

(c) If $Y_1 + Y_2$ is closed, a standard application of the open mapping theorem shows that there is $\alpha > 0$ such that

$$B_{Y_1+Y_2} \subset \alpha(B_{Y_1} + B_{Y_2}),$$

hence

$$P(\overline{B_{Y_1} + B_{Y_2}}^{w*}) \subset \alpha P(\overline{B_{Y_1}}^{w*} + \overline{B_{Y_2}}^{w*}) = \alpha(B_{Y_1} + B_{Y_2}) \subset Y_1 + Y_2.$$

An appeal to Theorem 1.2 concludes the proof. □

The next example shows that one can't expect more in Proposition 1.6(b). For part (b) recall that the bidual of an L^1-space [of the predual of a von Neumann algebra] is again an L^1-space [the predual of a von Neumann algebra].

Example 1.7

(a) *If X is an L-embedded space and Y_1, Y_2 are subspaces of X which are also L-embedded spaces, then $Y_1 + Y_2$ need not be closed and $\overline{Y_1 + Y_2}^{\|\,\|}$ need not be an L-embedded space.*

(b) *The bidual of an L-embedded space need not be an L-embedded space.*

PROOF: (a) By Proposition 1.10 below a subspace of $X := \ell^1 \cong \ell^1 \oplus_1 \ell^1$ is an L-embedded space if and only if it is w^*-closed. So take two w^*-closed subspaces Y_1 and Y_2 of X such that $\overline{Y_1 + Y_2}^{\|\,\|}$ is not w^*-closed. [To be specific, consider $Y_1 := \ell^1 \times \{0\}$ and $Y_2 = \text{graph}(T)$ with

$$\begin{array}{rccl} T : & \ell^1 & \longrightarrow & \ell^1 \\ & (x_n) & \longmapsto & (-\sum \frac{x_k}{k}, x_1, \frac{x_2}{2}, \frac{x_3}{3}, \ldots). \end{array}$$

Note: T is w^*-continuous and $\overline{Y_1 + Y_2}^{\|\,\|} = \{(x,y) \mid \sum y_k = 0\}$, which is w^*-dense.]

(b) The space $X = (\oplus \sum \ell^\infty(n))_{\ell^1}$ is an L-embedded space by Proposition 1.5(b) and $X^* = (\oplus \sum \ell^1(n))_{\ell^\infty}$. Now X^* contains a 1-complemented subspace isometric to ℓ^1 (see below), hence X^{**} contains an isometric copy of ℓ^∞ as a 1-complemented subspace. If X^{**} were an L-embedded space, then ℓ^∞ would be one by Proposition 1.5(a) as well – but ℓ^∞ has nontrivial M-structure.

For the convenience of the reader let us specify a contractive projection P from X^* onto an isometric copy of ℓ^1: Write the elements of X^* as $(x_{kn})_{k,n \in \mathbb{N}}$ with $x_{kn} = 0$ if $k > n$, i.e. $(x_{1n}, \ldots, x_{nn}, 0, \ldots) = (x_{1n}, \ldots, x_{nn}) \in \ell^1(n)$. Fix a free ultrafilter $\mathcal{U}$ on N and put for $k \in \mathbb{N}$

$$y_k := \lim_{n,\mathcal{U}} x_{kn} \quad \text{and} \quad y_{kn} = \begin{cases} 0 & \text{if } k > n \\ y_k & \text{if } k \leq n. \end{cases}$$

Then $P : (x_{kn}) \mapsto (y_{kn})$ is the desired projection. (One needs a free ultrafilter to get that P maps onto the subspace $\{(y_{kn}) \mid (y_k) \in \ell^1\}$.) □

Concerning the first statement in Example 1.7 we remark that in every infinite dimensional L-embedded space X, which is the dual of an M-embedded space Z, there are subspaces Y_1 and Y_2 such that $Y_1 + Y_2$ is not closed. Indeed, by a construction of Wilansky [636, p. 12] there are closed subspaces Z_1 and Z_2 in Z with $Z_1 + Z_2$ not closed. Let $Y_i = Z_i{}^\perp$ and recall the result of Reiter mentioned in Lemma I.1.14(a).

Using L-embedded spaces we will now partially redeem what was promised in Remark (c) following Theorem I.1.9.

Example 1.8 *There is a nonreflexive L-embedded space Y such that Y^* contains no nontrivial M-ideals. Hence (putting $X := Y^*$): There is a Banach space X without nontrivial M-ideals such that X^* has a nontrivial L-summand.*

PROOF: Take $Y = A_*$, the predual of an infinite dimensional von Neumann algebra A which is a finite factor (see [166], [593] for definitions and examples). Then $Y^* = A$ has no nontrivial closed two-sided ideals ([166, p. 257, Cor. 3] or [593, p. 349, Cor. 4.7]), hence no nontrivial M-ideals either, by Theorem V.4.4 below. □

The next two results concern L-embedded spaces which are duals of M-embedded spaces. We characterise these spaces in the class of Banach spaces which are L-summands in their biduals and describe their subspaces which are L-embedded.

Proposition 1.9 *Let X be a Banach space which is an L-summand in its bidual.*

(a) *There is (up to isometric isomorphism) at most one predual of X which is an M-ideal in its bidual.*

(b) *There is a predual of X which is an M-ideal in its bidual if and only if X_s is w^*-closed in X^{**}.*

More generally:

(b′) *If Z is a subspace of X which is an L-summand in its bidual, then there is a predual of Z which is an M-ideal in its bidual if and only if $Z^{\perp\perp} \cap X_s$ is w^*-closed in X^{**}.*

PROOF: (a) Let Y_1, Y_2 be M-embedded spaces with $Y_1^* \cong X \cong Y_2^*$ and $I: Y_1^* \longrightarrow Y_2^*$ an isometric isomorphism. We claim

$$\pi_{Y_1^*} = I^{**-1} \pi_{Y_2^*} I^{**} \tag{$*$}$$

where $\pi_{Y_i^*}$ denotes the natural (L-) projection from Y_i^{***} to Y_i^*. Assuming ($*$) we have $I^{**}\pi_{Y_1^*} = \pi_{Y_2^*} I^{**}$, but this is equivalent to the w^*-w^*-continuity of I. Hence $I = J^*$ for some isometric isomorphism $J: Y_2 \longrightarrow Y_1$.

To prove ($*$) observe that $\pi_{Y_1^*}$ is an L-projection and $I^{**-1}\pi_{Y_2^*} I^{**}$ is a bicontractive projection (even an L-projection) with the same range, so they coincide by Proposition I.1.2.

(b) The "only if" part is obvious. So assume that $X_s = Y^\perp$ for some subspace Y of X^*, i.e.

$$X^{**} = X \oplus_1 Y^\perp. \tag{$\dagger$}$$

Is is easy to show that

$$\begin{array}{rcl} I: X & \longrightarrow & Y^* \\ x & \longmapsto & i_x|_Y \end{array}$$

is an isometric isomorphism (note that because of $X \cap Y^\perp = \{0\}$ the subspace Y is total). The isometric isomorphism I^* maps $i_Y(Y)$ onto Y, which shows, since by ($\dagger$) Y is an M-ideal in X^*, that $i_Y(Y)$ is an M-ideal in Y^{**}.

(b′) By assumption and Theorem 1.2 we have $Z_s = Z^{\perp\perp} \cap X_s$. Noting that (via $i^{**}: Z^{**} \longrightarrow X^{**}$; $i: Z \longrightarrow X$ the inclusion map) $\sigma(Z^{**}, Z^*) = \sigma(X^{**}, X^*)|_{Z^{\perp\perp}}$ we conclude using the special case (b): Z has an M-embedded predual iff Z_s is $\sigma(Z^{**}, Z^*)$-closed iff $Z^{\perp\perp} \cap X_s$ is $\sigma(X^{**}, X^*)|_{Z^{\perp\perp}}$-closed iff $Z^{\perp\perp} \cap X_s$ is $\sigma(X^{**}, X^*)$-closed. □

Proposition 1.10 *For a Banach space X which is an M-ideal in its bidual and a subspace Y of X^* the following are equivalent:*

(i) *Y is an L-summand in its bidual.*
(ii) *Y is the dual of a space which is an M-ideal in its bidual.*
(iii) *Y is $\sigma(X^*, X)$-closed.*

PROOF: (i) ⇒ (iii): Since X is an M-embedded space, the L-projection P is the natural projection from X^{***} onto X^*, i.e., considered as a map to X^* it is i_X^*. By Theorem 1.2 we have $P\overline{B_Y}^{\sigma(X^{***}, X^{**})} = B_Y$. So by w^*-continuity of i_X^* and $\sigma(X^{***}, X^{**})$-compactness of $\overline{B_Y}^{\sigma(X^{***}, X^{**})}$ we infer that B_Y is $\sigma(X^*, X)$-compact, hence Y is $\sigma(X^*, X)$-closed by the Krein-Smulian theorem.

(iii) ⇒ (ii): If $Y = Z^\perp$ for some $Z \subset X$, we have that X/Z is an M-embedded space (Theorem III.1.6) and $(X/Z)^* \cong Z^\perp = Y$.

(ii) $\Rightarrow$ (i): Corollary III.1.3. □

In connection with Theorem 1.2 and Proposition 1.10 the following general fact seems worth remarking:

If X is a Banach space and P denotes the natural projection from X^{***} onto X^*, then for a subspace Y of X^* the following are equivalent:

(i) $PY^{\perp\perp} = Y$
(ii) $Y^{\perp\perp} = Y \oplus (Y^{\perp\perp} \cap X^{\perp})$
(iii) *Y is $\sigma(X^*, X)$-closed.*

[PROOF: The equivalence of (i) and (ii) is straightforward. For "(i) $\Rightarrow$ (iii)" argue as in the above proof. For "(iii) $\Rightarrow$ (i)" with $Y = Z^{\perp}$ for some $Z \subset X$ one may consider the quotient map $q : X \longrightarrow X/Z$ and observe that $\operatorname{ran} q^* = Z^{\perp}, \operatorname{ran} q^{***} = Z^{\perp\perp\perp} = Y^{\perp\perp}$ and $Pq^{***} = q^{***}Q$, where Q is the natural projection from $(X/Z)^{***}$ onto $(X/Z)^*$.]

We conclude this section with the study of some isometric properties (proximinality, extreme points) of L-embedded spaces. We prepare our discussion of best approximation by the following simple lemma.

Lemma 1.11 *Let X be a Banach space and $P : X^{**} \longrightarrow X^{**}$ a contractive projection. Then every subspace Y of X for which $P\overline{Y}^{w^*} \subset Y$ is proximinal in PX^{**}.*

PROOF: Recall the well known fact that every w^*-closed subspace of a dual space is proximinal. Then take a best approximation y^{**} in $\overline{Y}^{w^*}$ for $x \in PX^{**}$. We get

$$d(x, Y) \geq d(x, \overline{Y}^{w^*}) = \|x - y^{**}\| \geq \|P(x - y^{**})\| = \|x - Py^{**}\| \geq d(x, Y) . \qquad (*)$$

Hence Py^{**} is a best approximation for x in Y and $d(x, Y) = d(x, \overline{Y}^{w^*})$. □

Applying this to our present situation we get:

Proposition 1.12 *Let X be a Banach space which is an L-summand in its bidual and Y a subspace of X, which is also an L-summand in its bidual. Then Y is proximinal in X and $P_Y(x)$, the set of best approximations to x from Y, is weakly compact for all $x \in X$.*

PROOF: The first part is obvious from the above lemma and Theorem 1.2. For the second part note that the last (in)equality in $(*)$ yields

$$\|x - Py^{**}\| = \|x - y^{**}\| = \|x - Py^{**} - (Id - P)y^{**}\| = \|x - Py^{**}\| + \|y^{**} - Py^{**}\| .$$

Hence $y^{**} = Py^{**}$, and $P_{\overline{Y}^{w*}}(x) = P_Y(x)$. But then (with $d := d(x, Y)$)

$$P_Y(x) = Y \cap B_X(x, d) = \overline{Y}^{w*} \cap B_{X^{**}}(x, d) = P_{\overline{Y}^{w*}}(x)$$

shows that the w^*-closure of $P_Y(x)$ stays in X and this gives the weak compactness. □

To put the next results into their proper perspective note the following two facts:

- *For a proximinal subspace Y of a Banach space X there is always a homogeneous lifting $f : X/Y \longrightarrow X$ of the quotient map q satisfying $\|f(\dot{x})\| = \|\dot{x}\|$.*
(See Theorem II.1.9 and the text thereafter. An easy two-dimensional example shows that such a lifting need not be continuous.)

- *$L^1(\mu)$ contains a nontrivial finite-dimensional subspace with a continuous selection for the metric projection iff μ has atoms.* ([391, Th. 1.4])

Hence for an L-embedded space X with an L-embedded subspace Y one cannot expect a continuous lifting $f : X/Y \longrightarrow X$ in general.

Proposition 1.13 *Let X be an L-embedded Banach space and Y an L-embedded subspace of X. Then every lifting $f : X/Y \longrightarrow X$ of the quotient map q satisfying $\|f(\dot{x})\| = \|\dot{x}\|$ maps relatively $\|\,.\,\|$-compact sets onto relatively weakly compact sets. If Y is Chebyshev, f is even norm-weak continuous.*

PROOF: Let $K \subset X/Y$ be relatively $\|\,.\,\|$-compact and $f(\dot{x}_\alpha) \in f(K)$. Choose a $\|\,.\,\|$-convergent subnet $(\dot{x}_\beta)$ with $\dot{x}_\beta \longrightarrow \dot{x}$. Since $(f(\dot{x}_\beta))$ is bounded we find another subnet $(\dot{x}_\gamma)$ such that $f(\dot{x}_\gamma) \xrightarrow{w*} x^{**} \in X^{**}$. We wish to show $x^{**} \in X$. Now $q^{**}(f(\dot{x}_\gamma)) \xrightarrow{w*} q^{**}(x^{**})$ and $q^{**}(f(\dot{x}_\gamma)) = q(f(\dot{x}_\gamma)) = \dot{x}_\gamma$. From this we conclude

$$q^{**}(x^{**}) = \dot{x} = q(f(\dot{x})) = q^{**}(f(\dot{x})),$$

i.e. $x^{**} - f(\dot{x}) \in Y^{\perp\perp} = Y \oplus_1 Y_s$. Writing $x^{**} - f(\dot{x}) = y_a + y_s$ according to this decomposition, we get from the w^*-lower semicontinuity of the norm of X^{**}

$$\begin{aligned} \|\dot{x}\| &= \lim \|\dot{x}_\gamma\| = \lim \|f(\dot{x}_\gamma)\| \geq \|x^{**}\| \\ &= \|f(\dot{x}) + y_a\| + \|y_s\| \geq \|f(\dot{x}) + y_a\| \\ &\geq d(f(\dot{x}), Y) = \|q(f(\dot{x}))\| = \|\dot{x}\|. \end{aligned} \tag{1}$$

Hence $y_s = 0$, so $x^{**} \in X$ and the w^*-convergence of $f(\dot{x}_\gamma)$ to x^{**} is the desired w-convergence.
Concerning the second part of the statement note that $\|\dot{x}\| = \|f(\dot{x})\|$ means that 0 is a best approximation for $f(\dot{x})$, and the (in)equality (1) shows that $-y_a$ is another one. By uniqueness we get $y_a = 0$, and this implies $x^{**} = f(\dot{x})$. We leave it to the reader to adjust the beginning of the proof appropriately. □

It is obvious that we can't expect unique best approximation in general. However let us mention that by a result of Doob H^1_0 *is* Chebyshev in $L^1(\mathbb{T})$. Kahane has even characterised the translation invariant subspaces Y of $L^1(\mathbb{T})$ which admit unique best approximation: $Y = L^1_\Lambda(\mathbb{T})$ with Λ an infinite arithmetical progression with odd difference [363, p. 795]. In Section IV.4 we will see that they are all L-embedded (Proposition 4.4).
The weak compactness of $P_Y(x)$ has a nice consequence for the extreme points of the unit balls. For completeness we include the relevant result.

Proposition 1.14 *Let Y be a proximinal subspace of a Banach space X for which $P_Y(x)$ is weakly compact for all $x \in X$. Then*

$$\text{card}\,(\text{ex}\, B_{X/Y}) \leq \text{card}\,(\text{ex}\, B_X)\,.$$

PROOF: We first claim:

For every $\dot{x} \in \operatorname{ex} B_{X/Y}$ there is $y \in P_Y(x)$ such that $z = x - y \in \operatorname{ex} B_X$.

To see this observe that the weakly compact set $x - P_Y(x)$ has an extreme point $z = x - y$ by the Krein-Milman theorem. Now $x - P_Y(x) = B_X \cap (x + Y)$ and the latter set is a face of B_X (since $\dot{x} \in \operatorname{ex} B_{X/Y}$), i.e. $z = x - y \in \operatorname{ex} B_X$.
The mapping $\dot{x} \mapsto z$ from $\operatorname{ex} B_{X/Y}$ to $\operatorname{ex} B_X$, defined by the claim, is clearly injective, hence the proposition is proved. □

Combining Propositions 1.12 and 1.14 with the well known fact that $\operatorname{ex} B_{L^1(\mu)} = \emptyset$ if μ has no atoms, we get:

Corollary 1.15 *If μ has no atoms then no proper finite-codimensional subspace Y of $L^1(\mu)$ is an L-summand in its bidual. In particular, no such subspace is the range of a contractive projection in $L^1(\mu)$.*

We remark that a complete description of the ranges of contractive projections in $L^1(\mu)$-spaces is given by the Douglas-Ando theorem: these are the L^1-spaces over σ-subrings and measures absolutely continuous with respect to μ, cf. [386, p. 162]. In Section IV.3 we will characterise the L-embedded subspaces of $L^1(\mu)$.
The next result was first proved in [21, p. 37].

Corollary 1.16 $\operatorname{ex} B_{L^1/H^1_0} = \emptyset$

PROOF: Propositions 1.12 and 1.14, Example 1.1(d). □

Remark 1.17 (Smoothness and rotundity in L-embedded spaces)
(a) In Corollary III.2.12 we have proved using a renorming of M-embedded spaces:

There are nonreflexive strictly convex L-embedded spaces.

Concerning smoothness the following example is worth mentioning:

The Hardy space H^1_0 is smooth; hence there are nonreflexive smooth L-embedded spaces.

[PROOF: We consider H^1_0 as a real Banach space. Take an $f \in S_{H^1_0}$. Then $\mu\{z \in \mathbb{T} \mid f(z) = 0\} = 0$ (e.g. [317, p. 52]). Now also in $\mathbb{C}$-valued $L^1(\mu)$-spaces – considered as real Banach spaces – one has for $f \in S_{L^1(\mu)}$ the equivalence "f is a smooth point iff $\mu\{f = 0\} = 0$". So f is a smooth point even in $L^1(\mathbb{T})$, all the more in H^1_0.]

Note that this gives an easy and "natural" example of the following situation:

There is a Banach space X such that X is smooth and X^* is not strictly convex.

[In fact $H^{1\,*}_0 \cong L^\infty/H^\infty$ is not strictly convex, because it is the bidual of the M-embedded space $C(\mathbb{T})/A$; use Proposition I.1.7.] The first example of the above phenomenon was constructed by Klee in [380, Prop. 3.3]; another example is due to Troyanski [602]. We also note explicitly:

The space $C(\mathbb{T})/A$ is a strictly convex M-embedded space.

In fact, its dual H_0^1 is smooth.

(b) Passing to stronger smoothness properties we observe:

The norm of a nonreflexive L-embedded space is nowhere Fréchet-differentiable.

[PROOF: The following assertion is well known and easy to prove:

If the norm of a Banach space X is F-differentiable at x, then the norm of the bidual X^{**} is F-differentiable at i_x.

So assuming F-differentiability at x we would have in particular that i_x is a smooth point in X^{**} – but for nonreflexive L-embedded spaces this is clearly false.]

IV.2 Banach space properties of L-embedded spaces

We will start this section with two proofs of the weak sequential completeness of L-embedded spaces X. Both reveal interesting aspects: the first one the extreme discontinuity of the elements of X_s, the second one a weak form of w^*-closedness of arbitrary L-summands in dual spaces. We will then prove Pełczyński's property (V^*) for L-embedded spaces. Although this property is stronger than weak sequential completeness, its proof requires to have the latter in advance. We will conclude with a study of the Radon-Nikodým property for spaces which are L-summands in their biduals.

The following proposition provides the link between isometric and isomorphic properties and is the essential step for the first proof that Banach spaces X which are L-summands in their biduals are weakly sequentially complete. It characterises elements of X_s as maximally discontinuous, when considered as functions on (B_{X^*}, w^*). To measure the discontinuity of a function $f : K \to \mathbb{K}$ at a point $x \in K$ we will *not* work with the usual definition

$$\operatorname{osc}(f,x) := \inf_{U\in\mathfrak{U}(x)} \sup_{y,z\in U} |f(y) - f(z)|$$

where $\mathfrak{U}(x)$ denotes the neighbourhood system of x. Although this makes perfect sense for $\mathbb{K} = \mathbb{C}$, it causes – for complex scalars – some problems in the proof of the following proposition (see the Notes and Remarks). Instead we will use

$$MC(f,x) := \bigcap_{U\in\mathfrak{U}(x)} \overline{f(U)}.$$

For bounded functions this definition yields the usual properties of a "modulus of continuity":

- *f is continuous at x* $\iff$ $MC(f,x) = \{f(x)\}$
- *If f is continuous at x and $g : K \to \mathbb{K}$ is arbitrary, then*

$$MC(f+g,x) = f(x) + MC(g,x).$$

We omit the standard proofs of these statements. Note, however, that the above is false for unbounded functions and false for bounded functions if the closure of $f(U)$ is omitted.

Proposition 2.1 *Let X be a Banach space and consider $x^{**} \in X^{**}$ as a function on (B_{X^*}, w^*). The following statements are equivalent:*

(i) $\|x^{**} + x\| = \|x^{**}\| + \|x\|$ *for all* $x \in X$.

(ii) $MC(x^{**}, x^*) = B_{\mathbb{K}}(0, \|x^{**}\|)$ *for all* $x^* \in B_{X^*}$.

In particular: For a Banach space X which is an L-summand in its bidual all the $x^{**} \in X^{**} \setminus X$ are everywhere discontinuous on (B_{X^*}, w^*).

PROOF: (i) $\Rightarrow$ (ii): Since $x^{**}(U) \subset x^{**}(B_{X^*}) \subset B_{\mathbb{K}}(0, \|x^{**}\|)$ the inclusion "$\subset$" is clear. To prove "$\supset$" we have to show that for all $x^* \in B_{X^*}$ and for all w^*-neighbourhoods $U = \{y^* \in B_{X^*} \mid |x^*(x_k) - y^*(x_k)| < \eta,\ 1 \le k \le n\}$ of x^* the set $x^{**}(U)$ is dense in $B(0, \|x^{**}\|)$. So take $\theta \in B(0, \|x^{**}\|)$ and $\varepsilon > 0$. We are looking for a $y^* \in U$ such that $|x^{**}(y^*) - \theta| < \varepsilon$. Define a linear functional by

$$\begin{array}{rccc} \varphi : & X + \mathbb{K}x^{**} & \longrightarrow & \mathbb{K} \\ & x + \lambda x^{**} & \longmapsto & x^*(x) + \lambda\theta. \end{array}$$

Then by assumption

$$|\varphi(x + \lambda x^{**})| = |x^*(x) + \lambda\theta| \le \|x^*\|\|x\| + |\lambda||\theta| \le \|x\| + |\lambda|\|x^{**}\| = \|x + \lambda x^{**}\|,$$

hence $\|\varphi\| \le 1$. Let $\overline{\varphi} \in B_{X^{***}}$ be a Hahn-Banach extension of φ. By Goldstine's theorem we find $y^* \in B_{X^*}$ such that

$$|x^{**}(y^*) - \overline{\varphi}(x^{**})| < \varepsilon \tag{1}$$

$$|y^*(x_k) - \overline{\varphi}(x_k)| < \eta, \quad 1 \le k \le n. \tag{2}$$

Because $\overline{\varphi}(x^{**}) = \varphi(x^{**}) = \theta$ the inequality (1) implies $|x^{**}(y^*) - \theta| < \varepsilon$. By definition $\overline{\varphi}(x_k) = \varphi(x_k) = x^*(x_k)$, so inequality (2) gives $y^* \in U$.

(ii) $\Rightarrow$ (i): Take $x \in X$ and $\varepsilon > 0$. Choose $x^* \in B_{X^*}$ such that $x^*(x) = \|x\|$ and put $U := \{y^* \in B_{X^*} \mid |x^*(x) - y^*(x)| < \varepsilon/2\}$. For all $y^* \in U$ this yields $\operatorname{Re} y^*(x) \ge \|x\| - \varepsilon/2$. By assumption we have $x^{**}(U) = B_{\mathbb{K}}(0, \|x^{**}\|)$, so we find $y^{**} \in U$ such that $\operatorname{Re} x^{**}(y^*) > \|x^{**}\| - \varepsilon/2$. Combining these we get

$$\|x^{**} + x\| \ge |(x^{**} + x)(y^*)| \ge \operatorname{Re}(x^{**} + x)(y^*) \ge \|x^{**}\| + \|x\| - \varepsilon,$$

hence $\|x^{**} + x\| = \|x^{**}\| + \|x\|$.

Since $x^{**} = x + x_s$ is in $X^{**} \setminus X$ iff $x_s \ne 0$, the second statement follows from the first and the properties of the modulus of continuity MC. □

Theorem 2.2 *Every Banach space X which is an L-summand in its bidual is weakly sequentially complete.*

PROOF: For a weak Cauchy sequence (x_n) in X an element $x^{**} \in X$ is defined by $x^{**}(x^*) := \lim x^*(x_n)$, $x^* \in X^*$. This means $x_n \xrightarrow{w^*} x^{**}$, hence x^{**} is the pointwise limit of a sequence of continuous functions, i.e. of the first Baire class on (B_{X^*}, w^*). By Baire's theorem (see e.g. [158, p. 67]) x^{**} has a point of continuity on B_{X^*}, so by Proposition 2.1 x^{**} belongs to X and (x_n) converges to x^{**} weakly. □

By the above result and Example 1.1(c) we retrieve the Mooney-Havin theorem which asserts that L^1/H_0^1 is weakly sequentially complete. We refer to [490, Corollary 8.1, Theorem 7.1 and Proposition 6.1] for a "classical" proof.

Corollary 2.3 *Every nonreflexive subspace of a Banach space X which is an L-summand in X^{**} contains an isomorphic copy of ℓ^1.*

PROOF: Let Y be a nonreflexive subspace of X. By the Eberlein-Smulian theorem B_Y is not weakly sequentially compact, i.e. there is a sequence (y_n) in B_Y which doesn't have a weakly convergent subsequence. Theorem 2.2 shows that there are no weakly Cauchy subsequences either. Now Rosenthal's ℓ^1-theorem [422, Theorem 2.e.5] applies. □

For an extension, see Corollary 2.8.

Remark 2.4 Of course the conclusion of Corollary 2.3 holds in all weakly sequentially complete Banach spaces. We would like to mention, however, a local reflexivity argument of a more geometrical flavour to show that nonreflexive L-embedded spaces contain ℓ^1 isomorphically:
Take $x_1 \in S_X$ and $x_s \in S_{X_s}$. Then $\text{lin}\,(x_1, x_s) \cong \ell^1(2)$ and by local reflexivity we find $x_2 \in X$ such that $E_2 := \text{lin}\,(x_1, x_2) \overset{1+\varepsilon_1}{\simeq} \text{lin}\,(x_1, x_s)$. In the next step we start from $\text{lin}\,(x_1, x_2, x_s) \overset{1+\varepsilon_1}{\simeq} \ell^1(3)$ and find $x_3 \in X$ such that $E_3 := \text{lin}\,(x_1, x_2, x_3) \overset{1+\varepsilon_2}{\simeq} \text{lin}\,(x_1, x_2, x_s)$. The induction is now clear and a suitable choice of ε_n makes the sequence (x_n) equivalent to the unit vector basis of ℓ^1.

The second proof of the weak sequential completeness of L-embedded spaces will be an easy consequence of the following result, which we propose to call the "ace of ♢ lemma". It may be regarded as a weak substitute for the lacking w^*-closedness of L-summands in dual spaces (cf. Example I.1.6(b) and Theorem I.1.9). Its proof uses nothing but the interaction of the w^*-lower semicontinuity of the dual norm and the ace of ♢ shape of B_{X^*}, i.e. the L-decomposition. In particular it does not rely on a Baire argument.

Proposition 2.5 *Let U be an L-summand in a dual space X^* and (x_k^*) a weak Cauchy sequence in U. Then the w^*-limit x^* of (x_k^*) belongs to U, too.*

PROOF: Let V be the complementary L-summand for U, i.e. $X^* = U \oplus_1 V$. Note that by the Banach-Steinhaus theorem the w^*-limit x^* of (x_k^*) exists in X^* and $\|x^*\| \le \sup \|x_k^*\| < \infty$. Decompose $x^* = u^* + v^*$ in $X^* = U \oplus_1 V$. Passing to $(x_k^* - u^*)$ we may assume that (x_k^*) is a weak Cauchy sequence in U which is w^*-convergent to $v^* \in V$, and we want to show that $v^* = 0$.
Assuming $v^* \neq 0$ we will inductively construct a subsequence $(x_{k_n}^*)$ which is equivalent to the standard ℓ^1-basis:

$$m \sum_{i=1}^{n} |\alpha_i| \le \left\| \sum_{i=1}^{n} \alpha_i x_{k_i}^* \right\| \le M \sum_{i=1}^{n} |\alpha_i| \quad \forall\, n \in \mathbb{N} \;\; \forall\, \alpha_i \in \mathbb{K}$$

where $m = \|v^*\|/2$ and $M = \sup \|x_k^*\|$. This then gives a contradiction because such a sequence $(x_{k_n}^*)$ is not weakly Cauchy.
Choose a sequence of positive numbers ε_n such that $\sum_{n=1}^{\infty} \varepsilon_n < \|v^*\|/2$. By w^*-lower semicontinuity of the dual norm we have $\|v^*\| \leq \liminf \|x_k^*\| \leq M$. Thus there is an $x_{k_1}^*$ such that $\|x_{k_1}^*\| \geq \|v^*\| - \varepsilon_1$, which settles the beginning of the induction.
Assume we have found $x_{k_1}^*, \ldots, x_{k_n}^*$ satisfying

$$\left(\|v^*\| - \sum_{i=1}^{n} \varepsilon_i\right) \sum_{i=1}^{n} |\alpha_i| \leq \left\| \sum_{i=1}^{n} \alpha_i x_{k_i}^* \right\|.$$

Choose a finite (ε_{n+1}/M)-net $(\alpha^l)_{l \leq L}$ in the unit sphere of $\ell^1(n+1)$ such that $\alpha_{n+1}^l \neq 0$ for all $l \leq L$. The w^*-convergence (for $k \to \infty$) of $(\sum_{i=1}^{n} \alpha_i^l x_{k_i}^*) + \alpha_{n+1}^l x_k^*$ to $(\sum_{i=1}^{n} \alpha_i^l x_{k_i}^*) + \alpha_{n+1}^l v^*$ yields

$$\begin{aligned}
\liminf_k \left\| \left(\sum_{i=1}^{n} \alpha_i^l x_{k_i}^* \right) + \alpha_{n+1}^l x_k^* \right\| &\geq \left\| \left(\sum_{i=1}^{n} \alpha_i^l x_{k_i}^* \right) + \alpha_{n+1}^l v^* \right\| \\
&= \left\| \sum_{i=1}^{n} \alpha_i^l x_{k_i}^* \right\| + \|\alpha_{n+1}^l v^*\| \\
&\geq \left(\|v^*\| - \sum_{i=1}^{n} \varepsilon_i \right) \sum_{i=1}^{n} |\alpha_i^l| + |\alpha_{n+1}^l| \|v^*\| \\
&> \left(\|v^*\| - \sum_{i=1}^{n} \varepsilon_i \right) \sum_{i=1}^{n+1} |\alpha_i^l| \\
&= \|v^*\| - \sum_{i=1}^{n} \varepsilon_i.
\end{aligned}$$

Because only finitely many sequences are involved we find $x_{k_{n+1}}^*$ such that

$$\left\| \sum_{i=1}^{n+1} \alpha_i^l x_{k_i}^* \right\| \geq \|v^*\| - \sum_{i=1}^{n} \varepsilon_i \quad \forall\, l \leq L.$$

For arbitrary $\alpha \in S_{\ell^1(n+1)}$ choose α^l such that $\|\alpha - \alpha^l\|_1 \leq \varepsilon_{n+1}/M$. Then

$$\begin{aligned}
\left\| \sum_{i=1}^{n+1} \alpha_i x_{k_i}^* \right\| &= \left\| \sum_{i=1}^{n+1} \alpha_i^l x_{k_i}^* + (\alpha_i - \alpha_i^l) x_{k_i}^* \right\| \\
&\geq \left\| \sum_{i=1}^{n+1} \alpha_i^l x_{k_i}^* \right\| - \left\| \sum_{i=1}^{n+1} (\alpha_i - \alpha_i^l) x_{k_i}^* \right\| \\
&\geq \left(\|v^*\| - \sum_{i=1}^{n} \varepsilon_i \right) - \|\alpha - \alpha^l\|_1 M
\end{aligned}$$

$$\geq \quad \|v^*\| - \sum_{i=1}^{n+1} \varepsilon_i$$

$$= \left(\|v^*\| - \sum_{i=1}^{n+1} \varepsilon_i \right) \sum_{i=1}^{n+1} |\alpha_i|.$$

This clearly extends to all $\alpha \in \ell^1$ and thus finishes the induction and the proof. □

We have actually proved: If (x_k^*) is a w^*-Cauchy sequence in U with a non-zero w^*-limit in V, then (x_k^*) has a subsequence equivalent to the standard ℓ^1-basis. (The existence of the w^*-limit in X^* is again clear.) Note, however, that this does *not* show the w^*-sequential completeness of L-summands in dual spaces, since we need a *weak* Cauchy sequence to get a contradiction.
We will mention another proof of Proposition 2.5 in the Notes and Remarks section.

SECOND PROOF OF THEOREM 2.2: Let (x_n) be a weak Cauchy sequence in the L-embedded space X. Because then (x_n) is also a weak Cauchy sequence in X^{**}, Proposition 2.5 yields that the $\sigma(X^{**}, X^*)$-limit of (x_n) lies in X so that (x_n) is weakly convergent. □

We remark that the ace of diamonds lemma yields that an element of X_s cannot be of the first Baire class relative to (B_{X^*}, w^*). This should be compared to Proposition 2.1.

We now come to property (V^*) (see III.3.3) for L-embedded spaces. This was an open problem for quite a while and has only recently been proved in [493]. Property (V^*) was known to hold for all the concrete spaces in Example 1.1, yet the proofs also in these special cases are not at all easy. We again stress that we need the weak sequential completeness of L-embedded spaces, though formally a consequence of property (V^*) by virtue of Proposition III.3.3.F, as an instrumental ingredient in the proof of Theorem 2.7. We first provide a technical characterisation of property (V^*):

Lemma 2.6 *For a Banach space X the following assertions are equivalent:*

(i) *X has property (V^*), that is for each set $K \subset X$ which is not relatively w-compact there is a wuC-series $\sum x_i^*$ in X^* such that $\sup_{x\in K} |x_i^*(x)| \not\to 0$.*

(ii) *X is w-sequentially complete and for any (for some) number δ with $0 < \delta < 1$ the following holds: if (y_k) satisfies $(1-\delta)\sum|\alpha_k| \leq \|\sum \alpha_k y_k\| \leq \sum|\alpha_k|$ for all finite scalar sequences (α_k), then there are a subsequence (y_{k_n}), positive numbers $\rho = \rho(\delta,(y_k))$, $M = M(\delta,(y_k)) < \infty$ and for each $n \in \mathbb{N}$ there is a finite sequence $(y_i^{n*})_{i=1}^n$ in X^* such that*

$$|y_i^{n*}(y_{k_i})| \quad > \quad \rho \quad \forall\, i \leq n$$

$$\left\| \sum_{i=1}^n \alpha_i y_i^{n*} \right\| \quad \leq \quad M \max_{i\leq n} |\alpha_i| \quad \forall\, (\alpha_i) \subset \mathbb{K}.$$

PROOF: (i) $\Rightarrow$ (ii): Let δ be any number with $0 < \delta < 1$ and (y_k) an ℓ^1-basis in X satisfying $(1-\delta)\sum|\alpha_k| \le \|\sum \alpha_k y_k\| \le \sum|\alpha_k|$ for all (α_k). Since $K = \{y_k \mid k \in \mathbb{N}\}$ is not relatively w-compact, by (i) there are a wuC-series $\sum x_i^*$, a subsequence (y_{k_n}) and a number $\rho > 0$ such that $|x_n^*(y_{k_n})| > \rho$ for all $n \in \mathbb{N}$. Set $y_i^{n*} = x_i^*$ for all $i, n \in \mathbb{N}$. That property (V^*) implies w-sequential completeness was noted in Proposition III.3.3.F.
(ii) $\Rightarrow$ (i): Suppose (ii) holds for a fixed number δ with $0 < \delta < 1$. Let $K \subset X$ be not relatively w-compact. If K is not bounded, there are $x_n \in K$ and $x_n^* \in X^*$ such that $\|x_n^*\| = 1$ and $2^{-n}x_n^*(x_n) > 1$ for each $n \in \mathbb{N}$, and $\sum 2^{-n}x_n^*$ is trivially a wuC-series. Therefore we assume K to be bounded. Since X is weakly sequentially complete, by Rosenthal's ℓ^1-theorem K contains an ℓ^1-basis (x_n), i.e. $c\sum|\alpha_n| \le \|\sum \alpha_n x_n\| \le \sum|\alpha_n|$ for some $c > 0$. By James' distortion theorem [422, Prop. 2.e.3] (but for the present formulation the original reference [335] is slightly more appropriate) there are pairwise disjoint finite sets $A_k \subset \mathbb{N}$ and a sequence (λ_n) of scalars such that the sequence $y_k = \sum_{A_k} \lambda_n x_n$ satisfies

$$(1-\delta)\sum|\alpha_k| \le \left\|\sum \alpha_k y_k\right\| \le \sum|\alpha_k|, \quad \sum_{n\in A_k}|\lambda_n| < \frac{1}{c} \quad \forall k \in \mathbb{N}. \qquad (*)$$

Now observe that $(y_i^{n*})_n$ is a bounded sequence in X^*, for each i. Fix a free ultrafilter $\mathfrak{U}$ on $\mathbb{N}$ and let $x_i^* = w^*\text{-}\lim_{n\in\mathfrak{U}} y_i^{n*}$. It is clear that $x_i^*(y_{k_i}) \ge \rho$ for all i, and $\sum x_i^*$ is a wuC-series by virtue of

$$\begin{aligned}\left\|\sum_{i=1}^k \alpha_i x_i^*\right\| &= \left\|w^*\text{-}\lim_{n\in\mathfrak{U}} \sum_{i=1}^k \alpha_i y_i^{n*}\right\| \\ &\le \left\|\liminf \sum_{i=1}^k \alpha_i y_i^{n*}\right\| \\ &\le M \max|\alpha_i|.\end{aligned}$$

By $(*)$ there is x_{n_i} such that $|x_i^*(x_{n_i})| > \rho c$ for each $i \in \mathbb{N}$, because otherwise we would have

$$|x_i^*(y_{k_i})| \le \sum_{n\in A_{k_i}} |\lambda_n||x_i^*(x_n)| \overset{(*)}{<} \frac{1}{c}\rho c = \rho.$$

This proves (i). □

Theorem 2.7 *Every L-embedded Banach space X has property (V^*).*

PROOF: We will verify (ii) of Lemma 2.6 in order to prove property (V^*). We already know from Theorem 2.2 that X is w-sequentially complete.
Let ε, δ be numbers such that $0 < \varepsilon < 1/4$, $0 < \delta < \varepsilon^2/9^2$, and let (y_k) be an ℓ^1-basis in X as in (ii). We will find the subsequence (y_{k_n}) and the (y_i^{n*}) required in Lemma 2.6 by induction using the principle of local reflexivity. To make the construction work we need an $x_s \in X_s$ which is "near" to an accumulation point of the y_k. More precisely:

CLAIM: There is $x_s \in X_s$ such that

$$\|x_s\| \ge 1 - 4\delta^{1/2} \qquad (1)$$

and for all $\eta > 0$, all $x^* \in X^*$ and all $k_0 \in \mathbb{N}$ there is $k \geq k_0$ with

$$|x_s(x^*) - x^*(y_k)| \leq 3\delta^{1/2}\|x^*\| + \eta. \tag{2}$$

PROOF OF THE CLAIM: Denote the usual basis of ℓ^1 by (e_n^*) and denote by π_{ℓ^1} the canonical projection from $(\ell^1)^{**} = \ell^1 \oplus_1 c_0{}^{\perp}$ onto ℓ^1. The w^*-closure of the set $\{e_n^* \mid n \in \mathbb{N}\} \subset \ell^1$ in the bidual of ℓ^1 contains an accumulation point $\mu \in c_0{}^{\perp} = (\ell^1)_s$ of norm $\|\mu\| = 1$. (Actually, every such accumulation point has these properties.) We will map μ into X_s.
Put $Y = \overline{\text{lin}}\,\{y_k \mid k \in \mathbb{N}\}$. The canonical isomorphism $S : Y \to \ell^1$ satisfies $\|y^{**}\| \leq \|S^{**}y^{**}\| \leq \frac{1}{1-\delta}\|y^{**}\|$ for all $y^{**} \in Y^{**}$. In particular $1 - \delta \leq \|z_s\| \leq 1$ for $z_s = (S^{**})^{-1}(\mu)$. Consider $z_s \in X^{**}$ via the identification of Y^{**} and $Y^{\perp\perp} \subset X^{**}$. Denote by Q the canonical projection from $Y^{\perp\perp}$ onto Y (i.e. $Q = (S^{**})^{-1}\pi_{\ell^1}S^{**}$). Then $z_s \in Y_s = \ker Q$ follows from $\mu \in \ker \pi_{\ell^1}$, and z_s is a $\sigma(X^{**}, X^*)$-accumulation point of the set $\{y_k \mid k \in \mathbb{N}\}$ in Y_s. Finally put $x_s = (Id_{X^{**}} - P)(z_s) \in \ker P = X_s$, where P is as usual the L-projection from $X^{**} = X \oplus_1 X_s$ onto X.
For the decomposition $y^{\perp\perp} = y + y_s$ in $Y^{\perp\perp} = Y \oplus Y_s$ of any element $y^{\perp\perp} \in Y^{\perp\perp}$ we have

$$\begin{aligned} \|y + y_s\| &\geq (1-\delta)\|S^{**}y + S^{**}y_s\| \\ &= (1-\delta)(\|S^{**}y\| + \|S^{**}y_s\|) \\ &\geq (1-\delta)(\|y\| + \|y_s\|). \end{aligned}$$

Since $\delta < 1/4$ the assumptions of Lemma 1.4 are satisfied. We deduce from this result

$$\|x_s - z_s\| = \|Pz_s\| = \|Pz_s - Qz_s\| \leq 3\delta^{1/2}\|z_s\|. \tag{3}$$

Hence

$$\|x_s\| = \|z_s - Pz_s\| = \|z_s\| - \|Pz_s\| \geq (1 - 3\delta^{1/2})\|z_s\| \geq 1 - 4\delta^{1/2},$$

which is the first assertion of the claim. We use that z_s is an $\sigma(X^{**}, X^*)$-accumulation point of the y_k and (3) to find

$$\begin{aligned} |x_s(x^*) - x^*(y_k)| &\leq |x_s(x^*) - z_s(x^*)| + |z_s(x^*) - x^*(y_k)| \\ &\leq 3\delta^{1/2}\|x^*\| + \eta\,. \end{aligned}$$

This ends the proof of the claim.
Choose a sequence (ε_n) of positive numbers such that $\prod_{n\geq 1}(1-\varepsilon_n) \geq 1-\varepsilon$ and $\prod_{n\geq 1}(1+\varepsilon_n) \leq 1+\varepsilon$. We will construct by induction over $n = 1, 2, \ldots$ finite sequences $(y_i^{n*})_{i=1}^n \subset X^*$ and a subsequence (y_{k_n}) of (y_k) such that

$$|y_i^{n*}(y_{k_i})| > 1 - 9\delta^{1/2} \qquad \forall i \leq n \tag{4}$$

$$\begin{aligned} \Big(\prod_{i=1}^n (1-\varepsilon_i)\Big) \max_{i\leq n} |\alpha_i| &\leq \Big\|\sum_{i=1}^n \alpha_i y_i^{n*}\Big\| \\ &\leq \Big(\prod_{i=1}^n (1+\varepsilon_i)\Big) \max_{i\leq n} |\alpha_i| \qquad \forall (\alpha_i) \subset \mathbb{K}. \end{aligned} \tag{5}$$

(This means that we will fulfill (ii) of Lemma 2.6 with $\rho = 1 - 9\delta^{1/2}$ and $M \le 1 + \varepsilon$.) For $n = 1$ we set $k_1 = 1$ and choose y_1^{1*} such that $\|y_1^{1*}\| = 1$ and $y_1^{1*}(y_{k_1}) = \|y_{k_1}\| \ge 1 - \delta$; y_1^{1*} satisfies (5), too.
For the induction step $n \mapsto n+1$ we observe that $P^*|_{X^*}$ is an isometric isomorphism from X^* onto $X_s^\perp$, that $X^{***} = X^\perp \oplus_\infty X_s^\perp$ and that $(P^*x^*)|_X = x^*|_X$ for all $x^* \in X^*$. Choose $t \in \ker P^* \subset X^{***}$ such that

$$\|t\| = 1 \qquad \text{and} \qquad t(x_s) = \|x_s\|.$$

Put $E = \operatorname{lin}(\{P^*y_i^{n*} \mid i \le n\} \cup \{t\}) \subset X^{***}$ and $F = \operatorname{lin}(\{y_{k_i} \mid i \le n\} \cup \{x_s\}) \subset X^{**}$. An application of the principle of local reflexivity provides us with an operator $R: E \to X^*$ such that

$$(1 - \varepsilon_{n+1})\|e^{***}\| \le \|Re^{***}\| \le (1 + \varepsilon_{n+1})\|e^{***}\| \qquad \text{for all } e^{***} \in E$$

and

$$f^{**}(Re^{***}) = e^{***}(f^{**}) \qquad \text{for all } f^{**} \in F.$$

Let $y_i^{n+1*} = RP^*y_i^{n*}$ for $i \le n$ and let $y_{n+1}^{n+1*} = Rt$. Then the $(y_i^{n+1*})_{i=1}^{n+1}$ fulfill (5,n+1) by

$$\begin{aligned}
\left(\prod_{i=1}^{n+1}(1-\varepsilon_i)\right) \max_{i \le n+1} |\alpha_i| &\le (1-\varepsilon_{n+1}) \max\left(\left(\prod_{i=1}^{n}(1-\varepsilon_i)\right) \max_{i\le n}|\alpha_i|, |\alpha_{n+1}|\right) \\
&\le (1-\varepsilon_{n+1}) \max\left(\left\|\sum_{i=1}^{n} \alpha_i y_i^{n*}\right\|, \|\alpha_{n+1}t\|\right) \\
&= (1-\varepsilon_{n+1}) \max\left(\left\|\sum_{i=1}^{n} \alpha_i P^* y_i^{n*}\right\|, \|\alpha_{n+1}t\|\right) \\
&= (1-\varepsilon_{n+1}) \left\|\left(\sum_{i=1}^{n} \alpha_i P^* y_i^{n*}\right) + \alpha_{n+1}t\right\| \\
&\le \left\|\sum_{i=1}^{n+1} \alpha_i y_i^{n+1*}\right\| \\
&\le (1+\varepsilon_{n+1}) \left\|\left(\sum_{i=1}^{n} \alpha_i P^* y_i^{n*}\right) + \alpha_{n+1}t\right\| \\
&= (1+\varepsilon_{n+1}) \max\left(\left\|\sum_{i=1}^{n} \alpha_i P^* y_i^{n*}\right\|, \|\alpha_{n+1}t\|\right) \\
&= (1+\varepsilon_{n+1}) \max\left(\left\|\sum_{i=1}^{n} \alpha_i y_i^{n*}\right\|, \|\alpha_{n+1}t\|\right) \\
&\le (1+\varepsilon_{n+1}) \max\left(\left(\prod_{i=1}^{n}(1+\varepsilon_i)\right) \max_{i\le n}|\alpha_i|, |\alpha_{n+1}|\right) \\
&\le \left(\prod_{i=1}^{n+1}(1+\varepsilon_i)\right) \max_{i \le n+1} |\alpha_i|.
\end{aligned}$$

We use (2) with $\eta \leq 3\delta^{1/2}\varepsilon_{n+1}$ and $x^* = y_{n+1}^{n+1^*}$ in order to find $y_{k_{n+1}}$ such that

$$\left|x_s(y_{n+1}^{n+1^*}) - y_{n+1}^{n+1^*}(y_{k_{n+1}})\right| \leq 3\delta^{1/2}\|y_{n+1}^{n+1^*}\| + \eta \leq 3\delta^{1/2}(1+2\varepsilon_{n+1}) \leq 5\delta^{1/2};$$

here $\|R\| \leq 1+\varepsilon_{n+1}$ enters. Then (4,n+1) for $i = n+1$ follows from

$$\begin{aligned} |y_{n+1}^{n+1^*}(y_{k_{n+1}})| &\geq |x_s(y_{n+1}^{n+1^*})| - 5\delta^{1/2} \\ &\overset{\text{loc.refl.}}{=} |t(x_s)| - 5\delta^{1/2} \\ &\overset{(1)}{\geq} (1-4\delta^{1/2}) - 5\delta^{1/2} = 1 - 9\delta^{1/2}. \end{aligned}$$

Recall $P^*x^*(x) = x^*(x)$ to see that $y_i^{n+1^*}(y_{k_i}) = (P^*y_i^{n^*})(y_{k_i}) = y_i^{n^*}(y_{k_i})$. (4,n+1) for $i \leq n$ is consequence of this and (4,n). This completes the induction and the proof. □

We note the following consequence of the above theorem, Corollary III.3.3.C, and Proposition III.3.3.E:

Corollary 2.8 *Every nonreflexive subspace of an L-embedded space contains a complemented subspace isomorphic to ℓ^1.*

In the remainder of this section we will study the Radon-Nikodým property for L-embedded spaces (although this may seem not very promising at a first glance at the Examples 1.1). By Theorem III.3.1 L-embedded spaces X which are duals of M-embedded spaces have the RNP and for $L^1(\mu)$-spaces X this is even characteristic – use the well-known fact that *$L^1(\mu)$ has the RNP iff μ is purely atomic.* This carries over to the noncommutative case (cf. Example 1.1(b)).

Proposition 2.9 *Let X be the predual of a von Neumann algebra. Then X has the RNP if and only if X is the dual of a space which is an M-ideal in its bidual.*

PROOF: This is essentially a result of Chu who proved in Theorem 4 of [127] that if X has the RNP then X^* is a direct sum of type I factors, so

$$X^* \cong \Big(\oplus \sum L(H_i)\Big)_{\ell^\infty(I)} \qquad \text{and} \qquad X \cong \Big(\oplus \sum N(H_i)\Big)_{\ell^1(I)}.$$

The last-named space has the M-embedded space $(\oplus \sum K(H_i))_{c_0(I)}$ as a predual (see Example III.1.4(f) and Theorem III.1.6(c)). □

The following question is now tempting to ask.

QUESTION. *Is it true for an L-embedded space X that*

X has the RNP $\iff$ *X is the dual of an M-embedded space?*

We will see in Section IV.4 that the answer is **no**. [This can't be decided with the examples given so far, for instance the non-dual space L^1/H_0^1 (Corollary 1.16) fails the RNP.]

Remarks 2.10 (a) In spite of the negative answer to the above question there are some positive results. Note that by Proposition 1.9(b) the problem translates into

$$X \text{ } L\text{-embedded space, } X_s \text{ not } w^*\text{-closed} \quad \overset{?}{\Longrightarrow} \quad X \text{ fails the RNP}$$

Assuming even more than "X_s is w^*-dense" one gets:

If X is an L-summand in X^{**} and B_{X_s} is w^*-dense in $B_{X^{**}}$ then B_X has no strongly exposed point (in particular, X fails the RNP).

PROOF: If $x \in B_X$ is strongly exposed then $Id : (B_{X^{**}}, w^*) \to (B_{X^{**}}, \|\cdot\|)$ is continuous at i_x (see e.g. [250, proof of Theorem 10.2]). By w^*-density there is a net (x_α^{**}) in B_{X_s} such that $x_\alpha^{**} \xrightarrow{w^*} x$. Hence $x_\alpha^{**} \xrightarrow{\|\,\|} x$ and $x \in X_s$. See [159, p. 202] for the second statement of the proposition. □

Note that we only used the existence of a subspace $Y \subset X^{**}$ such that $X \cap Y = \{0\}$ and B_Y is w^*-dense in $B_{X^{**}}$.

(b) We mention a condition which guarantees the above w^*-density (of course the conclusion about the Radon-Nikodým property is trivial in this case).

If X is an L-embedded space with $\operatorname{ex} B_X = \emptyset$, then B_{X_s} is w^*-dense in $B_{X^{**}}$.

PROOF: We have by Lemma I.1.5 that $\operatorname{ex} B_{X^{**}} = \operatorname{ex} B_X \cup \operatorname{ex} B_{X_s}$, so by assumption and the Krein-Milman theorem

$$B_{X^{**}} = \overline{\operatorname{co}}^{w^*} \operatorname{ex} B_{X_s} = \overline{B_{X_s}}^{w^*}. \qquad \square$$

In this connection it is worth mentioning that for a nonreflexive L-embedded space X one has $\operatorname{ex} B_{X_s} \neq \emptyset$. [Otherwise, $\operatorname{ex} B_{X^{**}} = \operatorname{ex} B_X$ by Lemma I.1.5. Hence every $x^* \in X^*$ attains its norm on $\operatorname{ex} B_X$, so X is reflexive by James' theorem.]

There is another extreme point argument which has a consequence for the RNP.

Proposition 2.11 *If X is an L-embedded space with $\operatorname{ex} B_X = \emptyset$ (e.g. $X = L^1(\mu)$, μ without atoms) and $Y \subset X$ an L-embedded subspace, then X/Y fails the RNP.*

PROOF: By Propositions 1.12 and 1.14 we have $\operatorname{ex} B_{X/Y} = \emptyset$, so X/Y fails the RNP (see [159, p. 202]). □

The next result prepares a certain converse of the above proposition, but it is also of independent interest.

Proposition 2.12 *Let L be a Banach space such that L^* is injective, and X and Y L-embedded spaces with $Y \subset X$. Then every operator $T : L \to X/Y$ factors through the quotient map $q : X \to X/Y$:*

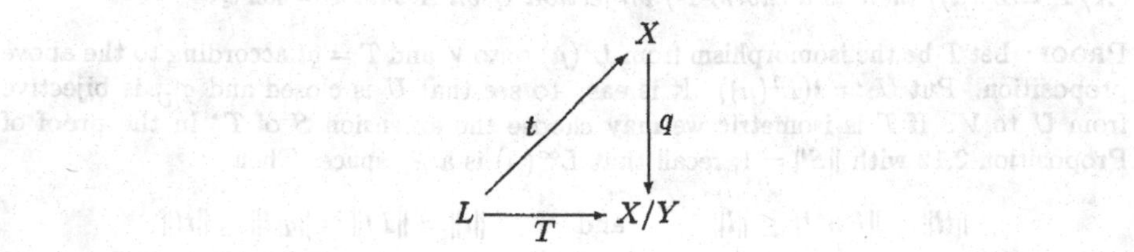

PROOF: Since q^* is a metric injection and L^* is injective there is an operator $S \in L(X^*, L^*)$ such that $Sq^* = T^*$:

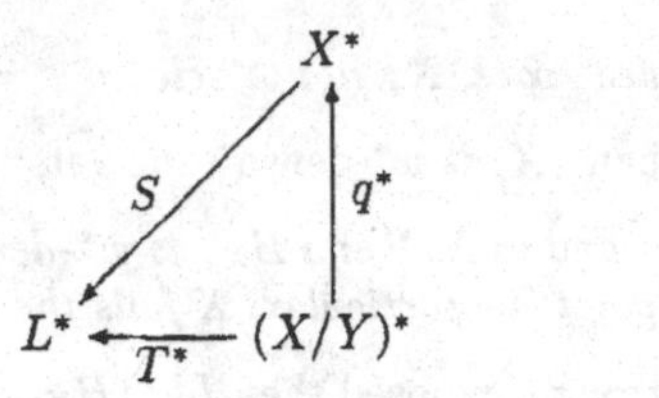

Put $t := PS^*|_L$ where P is the L-projection from X^{**} onto X:

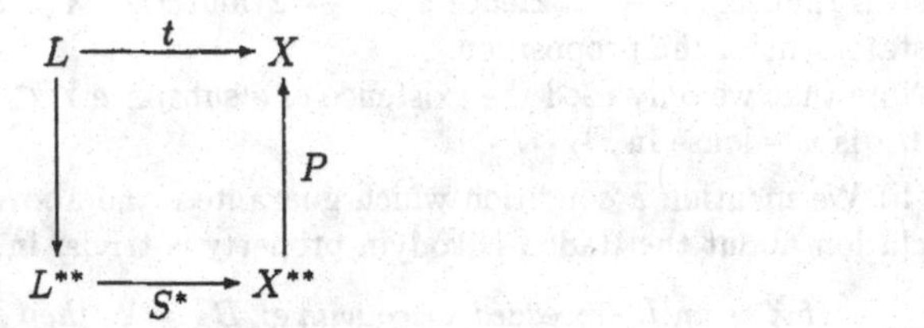

We will show that $T = qt$ is the wanted factorisation. Since

$$(S^* - t^{**})(L) = (S^* - t)(L) = (S^* - PS^*)(L) = (Id - P)S^*(L) \subset X_s$$

and

$$q^{**}(X_s) = (X_s + Y^{\perp\perp})/Y^{\perp\perp} = (X/Y)_s$$

(observe $q^{**}P = (P/Y^{\perp\perp})q^{**}$; Corollary 1.3) we get

$$q^{**}(S^* - t^{**})(L) \subset (X/Y)_s.$$

On the other hand $Sq^* = T^*$ gives $q^{**}S^* = T^{**}$, hence $q^{**}(S^* - t^{**}) = T^{**} - q^{**}t^{**}$. So

$$q^{**}(S^* - t^{**})(L) = (T^{**} - q^{**}t^{**})(L) = (T - qt)(L) \subset X/Y.$$

We obtain $(T - qt)(L) = \{0\}$, i.e. $T = qt$. □

The first part of the next result says that every $L^1(\mu)$-subspace of X/Y "comes from X".

Corollary 2.13 *If X and Y are L-embedded spaces with $Y \subset X$ and X/Y contains a subspace V isomorphic (isometric) to $L^1(\mu)$, then there is a subspace U of X such that $q(U) = V$ and $q|_U$ is an isomorphism (an isometric isomorphism). In case $X/Y \simeq L^1(\mu)$ ($X/Y \cong L^1(\mu)$) there is a (norm-1-) projection Q on X with $Y = \ker Q$.*

PROOF: Let T be the isomorphism from $L^1(\mu)$ onto V and $T = qt$ according to the above proposition. Put $U := t(L^1(\mu))$. It is easy to see that U is closed and $q|_U$ is bijective from U to V. If T is isometric we may choose the extension S of T^* in the proof of Proposition 2.12 with $\|S\| = 1$; recall that $L^\infty(\mu)$ is a $\mathcal{P}_1$-space. Then

$$\|tl\| = \|PS^*l\| \leq \|l\| \qquad \text{and} \qquad \|l\| = \|Tl\| = \|qtl\| \leq \|tl\|$$

show that t is isometric. From

$$\|qu\| = \|qtl\| = \|Tl\| = \|l\| = \|tl\| = \|u\|$$

we obtain that $q|_U$ is isometric.
If T is onto then $Q := tT^{-1}q$ is the desired projection. □

Finite-dimensional examples show that Y need not be the range of a norm-1-projection in X if $X/Y \cong L^1(\mu)$.

Corollary 2.14 *Let X and Y be L-embedded spaces with $Y \subset X$. If X has the RNP then X/Y has the RNP.*

PROOF: By the Lewis-Stegall theorem [159, p. 66] a Banach space Z has the RNP iff every operator from $L^1(\mu)$ to Z factors through ℓ^1. So the assumption and Proposition 2.12 give the following diagram:

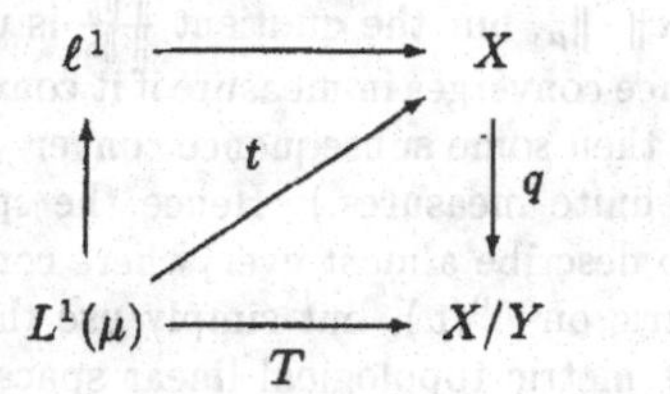

Now the proof is completed by another application of the Lewis-Stegall theorem. □

IV.3 Subspaces of $L^1(\mu)$ which are L-summands in their biduals

In this section we characterise the spaces appearing in its title and, among them, those which are duals of M-embedded spaces. The essential tool for doing this is the topology of convergence in measure on the space $L^1(\mu)$. Though not always easy to visualize and not very popular in (isometric) Banach space theory, it is this topology which allows the internal characterisation. We give various examples and applications.

Let us recall some facts about convergence in measure. Our setting in this section will always be a *finite* complete measure space (S, Σ, μ), that is a set S, a σ-algebra Σ of subsets of S, a finite nonnegative countably additive measure μ on Σ such that $B \in \Sigma$, $A \subset B$ and $\mu(B) = 0$ imply $A \in \Sigma$. (Completing the measure space has no effect on the Banach space $L^1(\mu)$ and avoids difficulties in the description of $L^1(\mu)^{**}$. See the Notes and Remarks for infinite measure spaces.) We denote by $L^0(\mu) := L^0(S, \Sigma, \mu)$ the space of $\mathbb{K}$-valued measurable functions on S with identification of equivalent (meaning

μ-almost everywhere equal) functions. A sequence (f_n) in $L^0(\mu)$ is said to *converge in measure* to $f \in L^0(\mu)$ ($f_n \xrightarrow{\mu} f$) if

$$\mu\{|f_n - f| \geq a\} \longrightarrow 0 \qquad \text{for all } a > 0.$$

Put for $f, g \in L^0(\mu)$

$$\|f\|_0 := \int \frac{|f|}{1+|f|}\, d\mu, \qquad d_0(f,g) := \|f-g\|_0$$

$$\|f\|_\mu := \inf\{a > 0 \mid \mu\{|f| \geq a\} \leq a\}, \qquad d_\mu(f,g) := \|f-g\|_\mu .$$

Then both d_0 and d_μ are invariant metrics on $L^0(\mu)$ for which $L^0(\mu)$ is complete and

$$f_n \xrightarrow{d_0} f \iff f_n \xrightarrow{d_\mu} f \iff f_n \xrightarrow{\mu} f, \quad f_n, f \in L^0(\mu)$$

(Note however that the metrics d_0 and d_μ are not Lipschitz equivalent, although there is $c > 0$ such that $\|\cdot\|_0 \leq c\|\cdot\|_\mu$, but the quotient $\frac{\|\cdot\|_\mu}{\|\cdot\|_0}$ is unbounded in general.) It is a standard fact that a sequence converges in measure if it converges almost everywhere, and if it converges in measure, then some subsequence converges almost everywhere. (Recall that we are dealing with finite measures.) Hence the space $L^0(\mu)$ is the topological vector space best suited to describe almost everywhere convergence. Mostly we will not work with an explicit metric on $L^0(\mu)$, but simply use that convergence in measure is convergence in a complete metric topological linear space. – All this can be found in text-books on measure theory.
Let us also recall that the bidual of $L^1(\mu)$ is isometrically isomorphic to

$$ba(\mu) := \{\nu \in ba(S,\Sigma) \mid \nu(E) = 0 \text{ if } E \in \Sigma \text{ and } \mu(E) = 0\},$$

see [179, Th. IV.8.16] or [647, Th. 2.2]. The embedded copy of $L^1(\mu)$ is the set of countably additive members of $ba(\mu)$, and $L^1(\mu)_s$ corresponds to the purely finitely additive measures [647, Th. 2.6].
In the following we collect some tools for proving the bidual characterisation of convex bounded subsets of $L^1(\mu)$ which are closed with respect to convergence in measure.

Lemma 3.1 *For every net $(f_\alpha)_{\alpha\in A}$ in $L^1(\mu)$ w^*-converging to $\nu \in L^1(\mu)^{**}$ there is a sequence $e_n \in \operatorname{co}\{f_\alpha \mid \alpha \in A\}$ such that $e_n \to P\nu$ μ – a.e. where P denotes the L-projection from $L^1(\mu)^{**}$ onto $L^1(\mu)$. Moreover, given $\alpha_n \in A$ we may assume $e_n \in \operatorname{co}\{f_\alpha \mid \alpha \geq \alpha_n\}$.*

PROOF: Let $f := P\nu$ and $\pi := \nu - f = (Id - P)\nu$. Then $\pi \in L^1(\mu)_s$, i.e. π is purely finitely additive. Put

$$F := \{g \in L^\infty(\mu) \mid |\pi|(|g|) = 0\}.$$

(Lattice-theoretically F is the absolute kernel of π.) Then F is a closed subspace of $\ker\pi$ and an order ideal in $L^\infty(\mu)$, i.e.

$$|h| \leq g,\ h \in L^\infty(\mu),\ g \in F \implies h \in F. \tag{$*$}$$

For $g \in F$ we have

$$\int f_\alpha g\, d\mu = g(f_\alpha) \longrightarrow \nu(g) = (f+\pi)(g) = f(g) = \int fg\, d\mu,$$

that is $f_\alpha \overset{\sigma(L^1,F)}{\longrightarrow} f$.
There is a $g_0 \in F$ such that $g_0 > 0$ μ-a.e. [By [647, Th. 1.19] there is for every $n \in \mathbb{N}$ an $A_n \in \Sigma$ such that $\mu(A_n) < \frac{1}{n}$ and $|\pi|(\complement A_n) = 0$. Assume without restriction that the A_n are decreasing and put $C_n := \complement A_n$. Then $g_0 := \chi_{C_1} + \sum_{n\geq 2} \frac{1}{n}\chi_{C_n \setminus C_{n-1}}$ has the desired properties.] Put $\lambda := g_0\mu$. The formal identity operator $T : L^1(\mu) \longrightarrow L^1(\lambda)$, $Tf = f$ is continuous, and looking at the involved dualities one finds $T^*h = hg_0$ for $h \in L^\infty(\lambda)$. Because of the ideal property $(*)$ of F one gets $\operatorname{ran} T^* \subset F$, hence $f_\alpha \longrightarrow f$ with respect to $\sigma(L^1(\lambda), L^\infty(\lambda))$. In particular

$$f \in \overline{\mathrm{co}}^{\sigma(L^1(\lambda),L^\infty(\lambda))}\{f_\alpha \mid \alpha \geq \alpha_n\} = \overline{\mathrm{co}}^{\|\;\|_{L^1(\lambda)}}\{f_\alpha \mid \alpha \geq \alpha_n\} \qquad \text{for all } n.$$

So we find $e_n \in \mathrm{co}\{f_\alpha \mid \alpha \geq \alpha_n\}$ such that $\|e_n - f\|_{L^1(\lambda)} \to 0$. Then we get for some subsequence (e_{n_k}) $e_{n_k} \to f$ λ-a.e., so also $e_{n_k} \to f$ μ-a.e. □

Corollary 3.2 *If $f, f_n \in L^1(\mu)$ and (f_n) converges to f almost everywhere, then for every w^*-accumulation point ν of (f_n) one obtains $P\nu = f$ where P denotes the L-projection from $L^1(\mu)^{**}$ onto $L^1(\mu)$.*

PROOF: There is a subnet, i.e. a directed set A and a monotone cofinal map $\alpha \mapsto n_\alpha$, such that (f_{n_α}) is w^*-convergent to ν. For every n choose α_n such that $n_{\alpha_n} \geq n$. By Lemma 3.1 there are $e_n \in \mathrm{co}\{f_{n_\alpha} \mid \alpha \geq \alpha_n\} \subset \mathrm{co}\{f_m \mid m \geq n\}$ with $e_n \overset{\text{a.e.}}{\longrightarrow} P\nu$. Hence $f = P\nu$. □

REMARK: There is a proof of Corollary 3.2 which avoids dealing with subnets and using the properties of purely finitely additive measures we employed in establishing Lemma 3.1. One may argue as follows:
Translation by f shows that it is sufficient to prove that

$$f_n \in L^1(\mu),\ f_n \overset{\text{a.e.}}{\longrightarrow} 0,\ \nu \text{ a } w^*\text{-accumulation point of } (f_n) \quad \Longrightarrow \quad \nu \in L^1(\mu)_s.$$

In this situation write $\nu = g + \pi \in L^1(\mu) \oplus_1 L^1(\mu)_s$. We claim $g = 0$. Let $\varepsilon > 0$ and put for $n \in \mathbb{N}$

$$A_n := \bigcap_{k\geq n} \{|f_k| \leq \varepsilon\}.$$

Then $A_n \subset A_{n+1}$ $(n \in \mathbb{N})$ and, because of the pointwise convergence, $S \setminus \bigcup A_n$ is a null set. Fix $n \in \mathbb{N}$ and denote by P_{A_n} the L-projection $h \longmapsto \chi_{A_n} h$ on $L^1(\mu)$. Since

$$\nu \in \{f_k \mid k \geq n\}^{-w^*}$$

and $P_{A_n}{}^{**}$ is w^*-continuous we get

$$P_{A_n}{}^{**}\nu \in \{P_{A_n} f_k \mid k \geq n\}^{-w^*}.$$

The set

$$K := \{h \in L^1(\mu) \mid |h| \leq \varepsilon\chi_{A_n}\}$$

is w-compact in $L^1(\mu)$, so w^*-compact in $L^1(\mu)^{**}$. Since K contains $P_{A_n}f_k$ $(k \geq n)$ we obtain

$$P_{A_n}{}^{**}\nu \in \overline{\{P_{A_n}f_k \mid k \geq n\}} \subset K \subset L^1(\mu).$$

Since $P_{A_n}{}^{**}P = PP_{A_n}{}^{**}$ by Theorem I.1.10 we get

$$P_{A_n}g = P_{A_n}{}^{**}P\nu = PP_{A_n}{}^{**}\nu = P_{A_n}{}^{**}\nu \in K,$$

which means $|g|_{A_n}| \leq \varepsilon$. Hence $|g| \leq \varepsilon$ a.e. and thus $g = 0$. □

We will be concerned with subsets of $L^1(\mu)$ which are closed in coarser topologies (e.g. in the topology of $L^0(\mu)$). The following proposition summarizes some easy observations in this connection.

Proposition 3.3 *For a bounded subset C of $L^1(\mu)$ the following are equivalent:*

(i) *C is closed in $L^1(\mu)$ with respect to convergence in measure.*
(ii) *C is closed in $L^0(\mu)$ with respect to convergence in measure.*
(iii) *C is closed in $L^1(\mu)$ with respect to convergence in $\|\,.\,\|_p$ for one (for all) $p \in (0,1)$.*

PROOF: (ii) ⇒ (i) is trivial.

(i) ⇒ (ii): Assume $f_n \in C$, $f \in L^0(\mu)$ and $f_n \xrightarrow{\mu} f$. Then there is a subsequence (f_{n_k}) of (f_n) which converges almost everywhere. Hence by Fatou's Lemma

$$\int |f|\,d\mu = \int \underline{\lim}\,|f_{n_k}|\,d\mu \leq \underline{\lim} \int |f_{n_k}|\,d\mu = \underline{\lim}\,\|f_{n_k}\|_1 < \infty$$

so that $f \in L^1(\mu)$. By assumption (i) we conclude now that $f \in C$.

(i) ⇒ (iii): This is a simple consequence of the Chebyshev-Markov inequality

$$\mu\{|f| \geq a\} \leq \frac{1}{a^p}\int |f|^p\,d\mu, \quad f \in L^p(\mu)$$

(which shows that the $\|\,.\,\|_p$-topology is finer than the topology of convergence in measure).

(iii) ⇒ (i): If for $M > 0$ we have $C \subset MB_{L^1(\mu)}$, $f_n \in C$, $f \in L^1(\mu)$ and $f_n \xrightarrow{\mu} f$ we get $f_n \xrightarrow{\|\,\|_p} f$ by (use Hölder's inequality with exponents $\frac{1}{1-p}$ and $\frac{1}{p}$)

$$\begin{aligned}
\|f_n - f\|_p^p &= \int |f_n - f|^p\,d\mu \\
&= \int_{\{|f_n - f| \leq a\}} |f_n - f|^p\,d\mu + \int_{\{|f_n - f| > a\}} 1 \cdot |f_n - f|^p\,d\mu \\
&\leq \mu(S)a^p + \left(\int_{\{|f_n - f| > a\}} 1\,d\mu\right)^{1-p}\left(\int_{\{|f_n - f| > a\}} |f_n - f|\,d\mu\right)^p \\
&\leq \mu(S)a^p + (2M)^p\mu\{|f_n - f| > a\}^{1-p}
\end{aligned}$$

and suitably choosing a. Use the hypothesis (iii) to conclude that $f \in C$. □

Because of the equivalence of (i) and (iii) in the above proposition we will henceforth simply write *μ-closed* for "closed in the (relative) topology of $L^0(\mu)$ (on $L^1(\mu)$)".
The next theorem, due to Buhvalov and Lozanovskii, is one of the main results from [99]; there a general lattice-theoretic approach in certain ideal function spaces is given, too (see also [369, Chapter X.5]).

Theorem 3.4 *For a convex bounded subset C of $L^1(\mu)$ the following are equivalent:*
(i) $P\overline{C}^{w^*} = C$, *where P is the L-projection from $L^1(\mu)^{**}$ onto $L^1(\mu)$.*
(ii) *C is closed in $L^1(\mu)$ with respect to convergence in measure.*

PROOF: (i) $\Rightarrow$ (ii): Take $f_n \in C$, $f \in L^1(\mu)$ with $f_n \xrightarrow{\mu} f$. We have to show that $f \in C$. Passing to a subsequence we may assume $f_n \xrightarrow{\text{a.e.}} f$. Since $\overline{C}^{w^*}$ is w^*-compact there is a w^*-accumulation point $\nu \in \overline{C}^{w^*}$ of (f_n). By Corollary 3.2 and the assumption (i) we have $f = P\nu \in P\overline{C}^{w^*} = C$.
(ii) $\Rightarrow$ (i): The inclusion $C \subset P\overline{C}^{w^*}$ is trivial. So we assume $\nu \in \overline{C}^{w^*}$ and we have to show that $f := P\nu$ belongs to C. Take a net (f_α) in C such that $f_\alpha \xrightarrow{w^*} \nu$. By Lemma 3.1 there is a sequence $e_n \in \operatorname{co}(f_\alpha) \subset C$ such that $e_n \xrightarrow{\text{a.e.}} f$. Since μ is finite this implies $e_n \xrightarrow{\mu} f$, so the μ-closedness of C gives $f \in C$. (Note that we didn't use the boundedness of C in this part of the proof.) □

Our aim for the first part of this section, the internal characterisation of subspaces of $L^1(\mu)$ which are L-embedded spaces, is now achieved.

Theorem 3.5 *For a subspace X of $L^1(\mu)$ the following are equivalent:*
(i) *X is an L-summand in its bidual.*
(ii) *B_X is μ-closed.*
(iii) *B_X is $\|\,.\,\|_p$-closed for one (for all) $p \in (0,1)$.*

PROOF: This follows from Proposition 3.3, Theorem 3.4 and Theorem 1.2 with $C = B_X$. (Note, of course, that $L^1(\mu)$ is an L-summand in its bidual.) □

We stress that statement (i) of the above theorem is independent of the embedding of the space X in an L^1-space, whereas the properties in (ii) and (iii) refer to the surrounding measure space and the particular embedding.
As a first application we will reprove the L-embeddedness of the Hardy space H_0^1 with the techniques of this section. Part (b) is included to show that the implication "(i) $\Rightarrow$ (ii)" in Theorem 3.4 doesn't hold for unbounded C; m denotes the normalized Lebesgue measure on $\mathbb{T}$.

Example 3.6
(a) *$B_{H_0^1}$ is m-closed (hence H_0^1 is an L-summand in its bidual).*
(b) *H_0^1 is m-dense in $L^1(\mathbb{T})$.*

PROOF: (a) We will prove a stronger statement which will be useful in the next section. To give the precise formulation we anticipate some notation from harmonic analysis (cf. Section IV.4): For $k \in \mathbb{Z}$ and $z \in \mathbb{T}$ put $\gamma_k(z) := z^k$. The k-th Fourier coefficient of $f \in L^1(\mathbb{T})$ is then

$$\widehat{f}(k) = \int_{\mathbb{T}} f\,\overline{\gamma_k}\,dm.$$

If we write for a subset Λ of $\mathbb{Z}$

$$L^1_\Lambda := \{f \in L^1(\mathbb{T}) \mid \widehat{f}(k) = 0 \quad \text{for } k \notin \Lambda\}$$

we have $H^1_0 = L^1_{\mathbb{N}}$.

CLAIM: *$B_{L^1_\Lambda}$ is $\|\,.\,\|_p$-closed $(0 < p < 1)$ in $L^1(\mathbb{T})$ for all subsets Λ of $\mathbb{N}$.*

For the proof it is appropriate to make a distinction between these spaces of classes of functions on $\mathbb{T}$ and the corresponding spaces of holomorphic functions on the open unit disk $\mathbb{D}$. We put

$$H^1(\mathbb{D}) := \left\{ F : \mathbb{D} \longrightarrow \mathbb{C} \;\middle|\; F \text{ holomorphic, } \sup_{r<1} \frac{1}{2\pi} \int_0^{2\pi} |F(re^{it})|\,dt < \infty \right\}$$

and recall that the Poisson transform $f \mapsto Pf$ yields a surjective isometry between H^1 and $H^1(\mathbb{D})$, the inverse being $F \mapsto F^*$, where $F^* \in L^1(\mathbb{T})$ denotes the boundary value function of F (see [545, Chapter 17]). Since $Pf(z) = \sum_{n=0}^\infty \widehat{f}(n) z^n$ for $z \in \mathbb{D}$ and $f \in H^1$, the space L^1_Λ corresponds to

$$H^1_\Lambda(\mathbb{D}) := \{F \in H^1(\mathbb{D}) \mid F^{(n)}(0) = 0 \quad \text{for } n \notin \Lambda\}.$$

Any $F \in H^1(\mathbb{D})$ is the Cauchy integral of its boundary values F^* (see [545, Th. 17.11]), so for $r < 1$

$$\sup_{|z| \le r} |F(z)| \le \|F^*\|_1 (1-r)^{-1}$$

and thus the unit ball of $H^1(\mathbb{D})$ is uniformly bounded on compact subsets of $\mathbb{D}$.
To prove the claim, let $(f_n) = (F_n^*)$ be a sequence in the unit ball of $L^1_\Lambda \cong H^1_\Lambda(\mathbb{D})$ which $\|\,.\,\|_p$-converges to $f \in L^1$. We have to show $f \in L^1_\Lambda$. By the above remark and Montel's theorem there is a subsequence (F_{n_k}) which converges uniformly on compact subsets of $\mathbb{D}$ (=: with respect to τ_K) to some holomorphic function G on $\mathbb{D}$. It is then easy to see that $G \in H^1(\mathbb{D})$ and $G^{(n)}(0) = 0$ for $n \notin \Lambda$, hence $G \in H^1_\Lambda(\mathbb{D})$. So proving $G^* = f$ will show the claim.
If we let $H_k := F_{n_k} - G$ we have

$$H_k \xrightarrow{\tau_K} 0 \quad \text{and} \quad H_k^* \xrightarrow{\|\,\|_p} f - G^*.$$

For all $\varepsilon > 0$ there is an $N \in \mathbb{N}$ such that

$$\int_0^{2\pi} |H_k^*(e^{it}) - H_l^*(e^{it})|^p\,dt < \varepsilon \quad \text{for all } k, l \ge N.$$

This implies

$$\int_0^{2\pi} |H_k(re^{it}) - H_l(re^{it})|^p \, dt < \varepsilon \quad \text{for all } k, l \geq N \text{ and } r < 1$$

since the left-hand side increases monotonically for $0 < r < 1$. Letting l tend to infinity this shows

$$\int_0^{2\pi} |H_k(re^{it})|^p \, dt \leq \varepsilon \quad \text{for all } k \geq N \text{ and } r < 1.$$

If now r tends to 1, we get

$$\int_0^{2\pi} |H_k^*(e^{it})|^p \, dt = \|H_k^*\|_p^p \leq \varepsilon \quad \text{for all } k \geq N.$$

This says $H_k^* \xrightarrow{\|\,\|_p} 0$, and we conclude $f = G^*$.

(b) Take functions $\varphi_n : \mathbb{T} \longrightarrow (0, \infty)$ such that

$$\varphi_n, \log \varphi_n \in L^1(\mathbb{T}), \quad \int \log \varphi_n \, dm = 0, \quad \text{and} \quad \varphi_n \xrightarrow{m} 0$$

(e.g. $\varphi_n = \frac{1}{n}\chi_{\mathbb{T}\setminus E_n} + a_n \chi_{E_n}$ with $m(E_n) = \frac{1}{n}$ and suitable a_n). Let Q_n be the outer function for φ_n, i.e.

$$Q_n(z) = \exp\left\{ \frac{1}{2\pi} \int_0^{2\pi} \frac{e^{it} + z}{e^{it} - z} \log \varphi_n(e^{it}) \, dt \right\}, \quad z \in \mathbb{D}$$

then (see [545, Th. 17.16]) $Q_n \in H^1(\mathbb{D})$, $|Q_n^*| = \varphi_n$, and

$$\log Q_n(0) = \frac{1}{2\pi} \int_0^{2\pi} \log \varphi_n(e^{it}) \, dt = 0.$$

Hence $Q_n^* \xrightarrow{m} 0$ and $Q_n(0) = 1$. Since H^1-functions are the Poisson integrals of their boundary values, we get

$$1 = Q_n(0) = \frac{1}{2\pi} \int_0^{2\pi} Q_n^*(e^{it}) \, dt = \widehat{Q_n^*}(0)$$

where $\widehat{Q_n^*}$ denotes the Fourier transform of Q_n^* – see Section IV.4. This yields

$$f_n := Q_n^* - \widehat{Q_n^*}(0)\, 1 \in H_0^1 \quad \text{and} \quad f_n \xrightarrow{m} -1.$$

Hence $H^1 \subset \overline{H_0^1}^{m}$. Since $f_n \gamma_{-1} \in H^1$ and $f_n \gamma_{-1} \xrightarrow{m} -\gamma_{-1}$, we find

$$\overline{\text{lin}}\, \{\gamma_k \mid k \geq -1\} \subset \overline{H^1}^{m} \subset \overline{H_0^1}^{m}.$$

Inductively one now proves easily that $\gamma_k \in \overline{H_0^1}^{m}$ for all $k \in \mathbb{Z}$, hence assertion (b). □

The above proof of Example 3.6(a) uses only elementary properties of H^1-functions. Allowing less trivial, yet standard, tools from H^p-theory, namely $H^p \cap L^1 \subset H^1$ (see

[244, Cor. II.4.3]) one gets its conclusion immediately. Also the stronger claim in the above proof can be obtained this way by observing that the inequality (see [181, p. 36])

$$|F(z)| \leq 2^{\frac{1}{p}} \|F^*\|_p (1-r)^{-\frac{1}{p}}$$

for $F \in H^p(\mathbb{D})$, $r < 1$ and $|z| = r$ shows that the topology τ_K is coarser than the $\|\cdot\|_p$-topology, hence $H^p_\Lambda(\mathbb{D})$ is $\|\cdot\|_p$-closed in $H^p(\mathbb{D})$.
Our next aim is to single out those X among the L-embedded subspaces of $L^1(\mu)$ which are even duals of M-embedded spaces. The extra condition on X is again given in terms of the topology of convergence in measure and is prepared by the following lemma and Proposition 3.9.

Lemma 3.7 *If $f_\alpha \in L^1(\mu)$, $\pi \in L^1(\mu)_s$, $\|f_\alpha\| \leq \|\pi\|$ and $f_\alpha \xrightarrow{w^*} \pi$, then $f_\alpha \xrightarrow{\mu} 0$.*

PROOF: For $a > 0$ we want to show that $\mu\{|f_\alpha| \geq a\} \to 0$. Let $\varepsilon > 0$. Since π is purely finitely additive there is, by [647, Th. 1.19], $E \in \Sigma$ such that $\mu(E) < \frac{\varepsilon}{2}$ and $|\pi|(E) = \|\pi\|$. Then also $\|\chi_E \pi\| = \|\pi\|$. Since $f_\alpha \xrightarrow{w*} \pi$ we have $\chi_E f_\alpha \xrightarrow{w*} \chi_E \pi$. But this, together with $\|\chi_E f_\alpha\| \leq \|f_\alpha\| \leq \|\pi\| = \|\chi_E \pi\|$ and the w^*-lower semicontinuity of the norm implies that $\|\chi_E f_\alpha\| \to \|\pi\|$. So there is α_0 such that $|\,\|\pi\| - \|\chi_E f_\alpha\|\,| < a\frac{\varepsilon}{2}$ for all $\alpha \geq \alpha_0$. We have for these α

$$\int_{\complement E} |f_\alpha|\, d\mu = \|f_\alpha\| - \int_E |f_\alpha|\, d\mu \leq \|\pi\| - \|\chi_E f_\alpha\| \leq a\frac{\varepsilon}{2},$$

so

$$\begin{aligned} \mu\{|f_\alpha| \geq a\} &= \mu(E \cap \{|f_\alpha| \geq a\}) + \mu(\complement E \cap \{|f_\alpha| \geq a\}) \\ &\leq \frac{\varepsilon}{2} + \frac{1}{a} \int_{\complement E \cap \{|f_\alpha| \geq a\}} |f_\alpha|\, d\mu \\ &\leq \varepsilon. \end{aligned}$$

□

Definition 3.8

(a) *If $S \subset L^0(\mu)$ and $f : S \to \mathbb{C}$ is a function, then f is called* μ-continuous *if it is continuous in the topology of convergence in measure.*

(b) *For a subspace X of $L^1(\mu)$ put*

$$X^\sharp := \{x^* \in X^* \mid x^*|_{B_X} \text{ is } \mu\text{-continuous}\}.$$

A standard $\frac{\varepsilon}{3}$-argument shows that $X^\sharp$ is a closed subspace of X^*. This space $X^\sharp$ is our candidate for an M-embedded predual of X. To get acquainted with the above definition we note:

REMARK.

(a) $L^1[0,1]^\sharp = \{0\}$

(b) *If X is nonreflexive, then $X^\sharp \neq X^*$.*

PROOF: (a) This follows from the following proposition and Remark 2.10(b), but a simple direct argument is possible, too.

(b) We show first:

CLAIM: A subspace X of $L^1(\mu)$ with $X^\sharp = X^*$ has the following property:

(*) Every bounded sequence (f_n) in X has a subsequence (g_n) which has weakly convergent arithmetic means $\left(\frac{1}{n}\sum_{k=1}^n g_k\right)$.

To prove this we need the following result of Komlós [382, Theorem 1a].

> THEOREM: *For every bounded sequence (f_n) in $L^1(\mu)$ there are an $f \in L^1(\mu)$ and a subsequence (g_n) of (f_n) such that the sequence of arithmetic means $\left(\frac{1}{n}\sum_{k=1}^n g_k\right)_n$ converges to f almost everywhere. Moreover, the conclusion remains true for every subsequence of (g_n).*

So, given a bounded sequence (f_n) in X we find by Komlós' theorem a subsequence (g_n) such that $h_n := \frac{1}{n}\sum_{k=1}^n g_k$ converges to some $f \in L^1(\mu)$ μ-a.e., hence in measure. Thus (h_n) is Cauchy in $L^0(\mu)$ so that $h_{n_{k+1}} - h_{n_k} \xrightarrow{\mu} 0$ for all subsequences (n_k). Now the assumption $X^\sharp = X^*$ gives $x^*(h_{n_{k+1}} - h_{n_k}) \longrightarrow 0$ for all $x^* \in X^*$. Thus (h_n) is weakly Cauchy, hence weakly convergent (by weak sequential completeness).

A well-known application of James' theorem (cf. [155, pp. 82–83]) shows that an arbitrary Banach space X which has the property (*) from the claim has to be reflexive. [Indeed, for $x^* \in S_{X^*}$ choose $x_n \in B_X$ such that $x^*(x_n) \to \|x^*\| = 1$. By (*) there is a subsequence (y_n) of (x_n) and $x \in X$ with

$$\frac{1}{n}\sum_{k=1}^n y_k \xrightarrow{w} x.$$

In particular

$$x^*\left(\frac{1}{n}\sum_{k=1}^n y_k\right) \longrightarrow x^*(x).$$

But

$$x^*\left(\frac{1}{n}\sum_{k=1}^n y_k\right) = \frac{1}{n}\sum_{k=1}^n x^*(y_k) \longrightarrow 1.$$

So x^* attains its norm at $x \in B_X$.] □

Because of the similarity of the above arguments with the Banach-Saks property it seems worth recalling that, in view of Szlenk's result that *$L^1(\mu)$ has the weak Banach-Saks property* (cf. [155, p. 85]), for subspaces X of $L^1(\mu)$ the Banach-Saks property is equivalent to reflexivity.

Actually, the converse to assertion (b) holds as well. This is a particular consequence of the following proposition. For more on Remark (b) see the Notes and Remarks section.

Proposition 3.9 *For a subspace X of $L^1(\mu)$ one has*

$$X^\sharp = (X^{\perp\perp} \cap L^1(\mu)_s)_\perp.$$

PROOF: "$\supset$": Assume there is $x^* \in (X^{\perp\perp} \cap L^1(\mu)_s)_\perp$ which is not continuous on B_X with respect to convergence in measure. Then there are $f_n, f \in B_X$ such that $f_n \xrightarrow{\mu} f$ and $x^*(f_n) \not\to x^*(f)$. We find a subsequence (g_n) of $(f_n - f)$ such that $g_n \xrightarrow{\text{a.e.}} 0$ and $x^*(g_n) \to a \neq 0$. Since (g_n) is a bounded sequence in X there is a w^*-accumulation point ν of (g_n) in $X^{\perp\perp}$. By Corollary 3.2 we also have $\nu \in L^1(\mu)_s$, so $\nu(x^*) = 0$ by assumption on x^*.
If now (g_α) is a subnet of (g_n) which converges to ν with respect to $\sigma(L^1(\mu)^{**}, L^1(\mu)^*)$, hence with respect to $\sigma(X^{**}, X^*)$, we have in particular $x^*(g_\alpha) \to \nu(x^*) = 0$. But since $(x^*(g_\alpha))$ is a subnet of $(x^*(g_n))$, we also have $x^*(g_\alpha) \to a \neq 0$. This contradiction shows the incorrectness of our assumption.
"$\subset$": Take $x^* \in X^\sharp$ and $\nu \in X^{\perp\perp} \cap L^1(\mu)_s$. We claim $\nu(x^*) = 0$. Since $\nu \in \|\nu\| B_{X^{\perp\perp}} = \|\nu\| B_{X^{**}}$, we find a net (f_α) in X such that $\|f_\alpha\| \le \|\nu\|$ and $f_\alpha \to \nu$ with respect to $\sigma(X^{**}, X^*)$, hence $f_\alpha \to \nu$ with respect to $\sigma(L^1(\mu)^{**}, L^1(\mu)^*)$. Since $\nu \in L^1(\mu)_s$ we infer from Lemma 3.7 that $f_\alpha \xrightarrow{\mu} 0$. So $x^* \in X^\sharp$ yields $x^*(f_\alpha) \to 0$. But the $\sigma(X^{**}, X^*)$-convergence of (f_α) to ν implies in particular $x^*(f_\alpha) \to \nu(x^*)$, hence the claim. □

The promised characterisation is the following.

Theorem 3.10 *For a subspace X of $L^1(\mu)$ the following are equivalent:*
(i) *X is isometrically isomorphic to the dual of a Banach space Y which is an M-ideal in Y^{**}.*
(ii) *B_X is μ-closed and $X^\sharp$ separates X.*
Moreover, if (i), (ii) *are satisfied Y is isometrically isomorphic to $X^\sharp$.*

PROOF: (i) $\Rightarrow$ (ii): By Corollary III.1.3 X is an L-summand in X^{**}, so B_X is μ-closed by Theorem 3.5 and $X^{\perp\perp} = X \oplus_1 (X^{\perp\perp} \cap L^1(\mu)_s)$ by Theorem 1.2. Neglecting for simplicity the isometric isomorphism, i.e. assuming $X = Y^*$, we get from Proposition III.1.2 that $X^{**} = Y^{***} = Y^* \oplus_1 Y^\perp$, hence $Y^\perp = X^{\perp\perp} \cap L^1(\mu)_s$. So by Proposition 3.9

$$Y = (Y^\perp)_\perp = (X^{\perp\perp} \cap L^1(\mu)_s)_\perp = X^\sharp.$$

Since Y separates Y^*, we get that $X^\sharp$ separates X, and the predual Y is $X^\sharp$.
(ii) $\Rightarrow$ (i): If B_X is μ-closed, X is an L-embedded space by Theorem 3.5 and

$$X^{\perp\perp} = X \oplus_1 (X^{\perp\perp} \cap L^1(\mu)_s) \qquad (*)$$

by Theorem 1.2. We know that $X^\sharp = (X^{\perp\perp} \cap L^1(\mu)_s)_\perp$ from Proposition 3.9, hence

$$(X^\sharp)^\perp = ((X^{\perp\perp} \cap L^1(\mu)_s)_\perp)^\perp = (X^{\perp\perp} \cap L^1(\mu)_s)^{-w^*}. \qquad (**)$$

But since $X^\sharp$ separates X we have that $(X^\sharp)^\perp \cap X = \{0\}$. This yields, together with $(*)$ and $(**)$, that $X^{\perp\perp} \cap L^1(\mu)_s$ is w^*-closed, hence X is the dual of an M-embedded space Y and $Y \cong (X^{\perp\perp} \cap L^1(\mu)_s)_\perp = X^\sharp$ by Proposition 1.9. □

IV.4 Relations with abstract harmonic analysis

In this section we will relate subsets Λ of the dual group Γ of a compact abelian group G with Banach space properties of certain spaces of Λ-spectral functions. In our context the spaces L^1_Λ (see below) appear as a natural class of nontrivial subspaces of $L^1(G)$. Most of the following material is based on Godefroy's important paper [258].

Let us fix our notation and recall some facts from harmonic analysis. The symbol G always denotes a compact abelian group with group operation written as multiplication and normalized Haar measure m. We write $\Gamma = \widehat{G}$ for the (discrete) dual group of G consisting of the continuous group homomorphisms $\gamma : G \to \mathbb{T}$, the so-called characters, where $\mathbb{T} = \{z \in \mathbb{C} \mid |z| = 1\}$ is the circle group. The spaces $L^1 := L^1(G) := L^1(G, \mathrm{Bor}(G), m)$ and $M := M(G)$ are commutative Banach algebras with respect to convolution. If we identify $f \in L^1$ with the m-continuous measure fm with density f, then $L^1(G)$ turns out to be an ideal in $M(G)$. [See [542, 1.1.7, 1.3.2 and 1.3.5] or [309]; these two books are our main references on harmonic analysis.] The Fourier (-Stieltjes) transform is the mapping

$$\begin{array}{rccl} \widehat{\ } : & M(G) & \longrightarrow & \ell^\infty(\Gamma) \\ & \mu & \longmapsto & \widehat{\mu} \qquad \text{with } \ \widehat{\mu}(\gamma) := \displaystyle\int_G \overline{\gamma(x)}\, d\mu(x). \end{array}$$

$\widehat{\ }$ is an injective, multiplicative ($\widehat{\mu * \nu} = \widehat{\mu}\widehat{\nu}$) and continuous ($\|\widehat{\mu}\| \le \|\mu\|$) operator [542, 1.3.3, 1.7.3]. If we identify Γ with the spectrum (= maximal ideal space) of $L^1(G)$ by $\gamma \longleftrightarrow \varphi_\gamma$ where

$$\begin{array}{rccl} \varphi_\gamma : & L^1(G) & \longrightarrow & \mathbb{C} \\ & f & \longmapsto & \varphi_\gamma(f) := \widehat{f}(\gamma) := \displaystyle\int f(x)\overline{\gamma(x)}\, dm(x) \end{array}$$

(see [542, 1.2.2]), then the restriction of $\widehat{\ }$ to $L^1(G)$ is the Gelfand transform of $L^1(G)$, [542, 1.2.3 and 1.2.4]. We denote by $T := T(G) := \mathrm{lin}\,\Gamma$ ($\subset C(G) \subset L^1(G)$) the space of so-called "trigonometric polynomials" on G.

Definition / Proposition 4.1 *For $\Lambda \subset \Gamma$ we write*

$$L^1_\Lambda := L^1_\Lambda(G) := \{f \in L^1(G) \mid \widehat{f}(\gamma) = 0 \text{ for all } \gamma \notin \Lambda\}$$

for the space of Λ-spectral functions in L^1. T_Λ, C_Λ and M_Λ are defined similarly.

(a) $T_\Lambda = \mathrm{lin}\,\Lambda$.
(b) *C_Λ, L^1_Λ and M_Λ are closed ideals in the corresponding convolution algebras.*
(c) *T_Λ is $\|\ \|_\infty$-dense in C_Λ and $\|\ \|_1$-dense in L^1_Λ.*
(d) $L^1_\Lambda = M_\Lambda \cap L^1$.
(e) $(C/C_{\Gamma\setminus\Lambda})^* \cong M_{\Lambda^{-1}}$.

PROOF: See [310, 35.7] for (a) – (c). (d) is clear from the definition and (e) follows from (c) by standard duality arguments. □

It is well-known in harmonic analysis that the spaces L^1_Λ are precisely the closed translation invariant subspaces and precisely the closed ideals of the convolution algebra $L^1(G)$ [310, Theorem 38.7].

Before we define the four types of subsets of Γ we will be concerned with, let us agree on some further notation: we write M_a $(= L^1)$ and M_s for the subspaces of m-absolutely continuous and m-singular measures in M, P_a and P_s $(= Id - P_a)$ denote the corresponding L-projections.

Definition 4.2 *A subset Λ of Γ will be called*

(a) nicely placed *if L^1_Λ is an L-summand in its bidual,*
(b) *a* Shapiro set *if all subsets Λ' of Λ are nicely placed,*
(c) Haar invariant *if $P_a M_\Lambda \subset M_\Lambda$,*
(d) *a* Riesz set *if $M_\Lambda = L^1_\Lambda$.*

G. Godefroy coined the first two terms in [258]. He used "sous-espace bien disposé" in [257] for subspaces X of $L^1(\mu)$ which are L-summands in their biduals. By Theorem 1.2 this means a nice, compatible placement of X^{**} in the bidual of $L^1(\mu)$. However we have avoided this expression in order to emphasize that being bien disposé is a property of the space X itself and not of the embedding of X in $L^1(\mu)$. (X is "nicely placed" in every space $L^1(\nu)$ which contains it – see the remark following Theorem 3.5.) Here we consider the fixed "embedding" $\Lambda \subset \Gamma$ and the use of the expression "nicely placed" seems appropriate by the above.
Let us recall the reason for the name "Riesz set": The F. and M. Riesz theorem says: If $\mu \in M(\mathbb{T})$ is such that $\widehat{\mu}(n) = 0$ for all $n < 0$, then μ is absolutely continuous with respect to the Haar measure m on $\mathbb{T}$. Using the above notations this translates into: $\mu \in M_{\mathbb{N}_0}(\mathbb{T})$ implies $\mu = fm$ for some $f \in L^1(\mathbb{T})$, i.e. $M_{\mathbb{N}_0}(\mathbb{T}) = L^1_{\mathbb{N}_0}(\mathbb{T})$. Riesz sets were studied systematically for the first time in [442].
Noting that Λ is a Riesz set if and only if $P_s M_\Lambda = \{0\}$, we see that Riesz sets are Haar invariant. To study the relation between these two families of subsets more closely we prove:

Lemma 4.3 *A subset Λ of Γ such that Λ' is Haar invariant for all $\Lambda' \subset \Lambda$ is a Riesz set.*

PROOF: Take $\mu \in M_\Lambda$. Since Λ is Haar invariant we have $\mu_s \in M_\Lambda$, i.e. $\widehat{\mu}_s(\gamma) = 0$ for all $\gamma \notin \Lambda$. Take now $\gamma \in \Lambda$ and consider $\Lambda' := \Lambda \setminus \{\gamma\}$ and $\nu := \mu - \widehat{\mu}(\gamma)\gamma$. We get $\nu \in M_{\Lambda'}$, hence by assumption $\nu_s \in M_{\Lambda'}$. But $\nu_s = \mu_s$, so $0 = \widehat{\nu}_s(\gamma) = \widehat{\mu}_s(\gamma)$. We have proved that $\widehat{\mu}_s = 0$, hence $\mu_s = 0$. □

In the next proposition we collect some easy examples and counterexamples of Riesz and nicely placed sets, as well as their basic stability properties.

Proposition 4.4

(a) *Finite sets Λ are Riesz and nicely placed.*
(b) *Cofinite sets $\Lambda \neq \Gamma$ are not Riesz and not nicely placed if G is infinite.*
(c) *If Λ is a Riesz resp. nicely placed subset, then Λ^{-1} and $\lambda\Lambda$ $(\lambda \in \Gamma)$ are Riesz resp. nicely placed.*
(d) *Subgroups Λ of Γ are nicely placed.*
(e) *$n\mathbb{Z}$ is not a Riesz subset of $\mathbb{Z}$ $(n \neq 0)$.*

PROOF: (a) is easy.
(b) For $\Lambda = \Gamma \setminus \{\gamma_1, \ldots, \gamma_n\}$ consider $\mu = \delta_e - (\gamma_1 + \ldots + \gamma_n)m$, $e \in G$ the neutral element, for the Riesz part. The space L^1_Λ is finite codimensional in L^1 in this case, so Corollary 1.15 shows that Λ is not nicely placed.
(c) We have $\widehat{\mu}(\gamma^{-1}) = \overline{\widehat{\overline{\mu}}(\gamma)}$, where $\overline{\mu}(E) := \overline{\mu(E)}$ (as a measure) or $\overline{\mu}(g) := \overline{\mu(\overline{g})}$ (as a functional). So $\mu \in M_{\Lambda^{-1}}$ implies $\overline{\mu} \in M_\Lambda$. Now Λ is Riesz, and this gives $\overline{\mu} \ll m$. Consequently $\mu \ll m$. For $f \in L^1$ one obtains $\widehat{f}(\lambda\gamma) = \widehat{f\overline{\lambda}}(\gamma)$, hence $f \in L^1_{\lambda\Lambda}$ iff $f\overline{\lambda} \in L^1_\Lambda$. So $M_{\overline{\lambda}} : L^1 \to L^1$, $f \mapsto \overline{\lambda} f$ is a surjective isometry which maps $L^1_{\lambda\Lambda}$ onto L^1_Λ. Thus $L^1_{\lambda\Lambda}$ is L-embedded since L^1_Λ is. The other two statements are proved in a similar way.
(d) Put $H := \Lambda_\perp := \{x \in G \mid \gamma(x) = 1 \text{ for all } \gamma \in \Lambda\}$. By the "bipolar theorem for groups" [542, 2.1.3] it follows that $L^1_\Lambda = \{f \in L^1 \mid f = f_x \text{ for all } x \in H\}$ where $f_x(y) := f(xy)$. The translations $T_x : L^1 \to L^1$, $f \mapsto f_x$ are isometric with respect to d_m (the metric of convergence in m-measure – see the beginning of Section IV.3), so $\{f \in L^1 \mid f = f_x\} = (Id - T_x)^{-1}(\{0\})$ is m-closed in L^1, hence L^1_Λ is m-closed. But then $B_{L^1_\Lambda}$ is m-closed.
(e) For $z \in \mathbb{T}$ and $k \in \mathbb{Z}$ we have $\widehat{\delta_z}(k) = z^{-k}$. Put $z := e^{-\frac{2\pi i}{n}}$ and $\mu := \frac{1}{n}(\delta_1 + \delta_z + \ldots + \delta_{z^{n-1}})$. Then $\mu \perp m$ and $\widehat{\mu} = \chi_{n\mathbb{Z}}$. □

Proposition 4.5 *For a subset Λ of Γ,*
(a) *if Λ is nicely placed then Λ is Haar invariant,*
(b) *if Λ is a Shapiro set then Λ is a Riesz set.*

PROOF: (a) We need the following result of Boclé ([82, Th. II], see also [565, Lemma 1.1]).

> LEMMA: *For a symmetric, open neighbourhood V of e in G put $u_V := \frac{1}{m(V)}\chi_V$. If $\nu \in M_s$ then $(u_V * \nu)_V$ converges in Haar measure to zero.*

[We wish to indicate the proof of this lemma in the classical case of the circle group $G = \mathbb{T}$ which we identify with the interval $]-\pi, \pi]$. If we let $F(t) = \int_0^t d\nu$, then a straightforward computation yields that

$$\left(\frac{1}{2\varepsilon}\chi_{[-\varepsilon,\varepsilon]} * \nu\right)(t) = \frac{1}{2\varepsilon}(F(t+\varepsilon) + F(t-\varepsilon))$$

which tends to 0 a.e. since by Lebesgue's differentiation theorem $F'(t)$ exists and equals 0 a.e. as ν is singular.]
If now $\Lambda \subset \Gamma$ is nicely placed, $\mu \in M_\Lambda$ and $\mu = f + \mu_s$, we have to show $f \in L^1_\Lambda$. Since (u_V) is an approximate unit in L^1 (see [542, 1.1.8]), we get $u_V * f \xrightarrow{\|\,\|_1} f$, hence $u_V * f \xrightarrow{m} f$. The lemma yields $u_V * \mu_s \xrightarrow{m} 0$, so

$$u_V * \mu = u_V * f + u_V * \mu_s \xrightarrow{m} f.$$

Because L^1 and M_Λ are ideals, $(u_V * \mu)$ is a (bounded) net in L^1_Λ. By assumption and Theorem 3.5 $B_{L^1_\Lambda}$ is m-closed, hence $f \in L^1_\Lambda$.

Part (b) follows from Lemma 4.3 and part (a). □

The following diagram contains not only the relations between the four types of sets established so far, but also their Banach space characterisations which we will prove next.

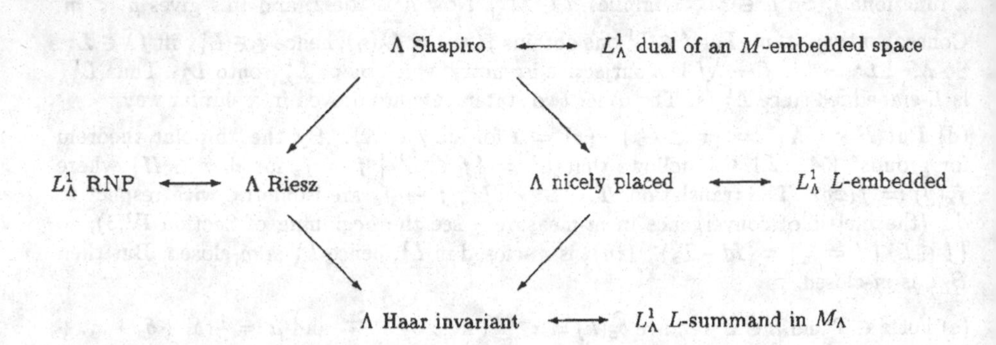

Proposition 4.6 *A subset Λ of Γ is Haar invariant if and only if L^1_Λ is an L-summand in M_Λ.*

PROOF: The "only if" part is trivial: Since $P_a M_\Lambda \subset M_\Lambda$, $P_a|_{M_\Lambda}$ is an L-projection in M_Λ with range L^1_Λ.
To prove the "if" part we consider the subspace M_Λ of the abstract L-space M. By Proposition I.1.21 M_Λ is invariant under the band projection P_B onto the band B generated by L^1_Λ in M. Since we may assume $\Lambda \neq \emptyset$ there is $\gamma \in L^1_\Lambda$ with $|\gamma(x)| = 1$ for all $x \in G$. So the closed (order) ideal generated by L^1_Λ in L^1 is L^1. But L^1 is a band in M, thus $B = L^1$. Of course $P_B = P_a$. □

There is no M- or L-structure characterisation of Riesz sets, however:

Theorem 4.7 *For a subset Λ of Γ the following are equivalent:*
- (i) *Λ is a Riesz set.*
- (ii) *L^1_Λ has the Radon-Nikodým property.*
- (iii) *$L^1_{\Lambda'}$ is a separable dual space for all countable subsets Λ' of Λ.*

PROOF: (i) ⇒ (iii): Every subset Λ' of Λ is Riesz, hence $L^1_{\Lambda'} = M_{\Lambda'} \cong (C/C_{\Gamma\setminus\Lambda'^{-1}})^*$ by Proposition 4.1(e). If Λ' is countable then $L^1_{\Lambda'}$ is separable.

(iii) ⇒ (ii): By [159, Th. III.3.2, p. 81] it is enough to show that every separable subspace X of L^1_Λ has the RNP. But every such subspace X is contained in some $L^1_{\Lambda'}$, Λ' countable. [PROOF: If (x_n) is dense in X, find $t_{n,k} \in T_\Lambda$ with $t_{n,k} \longrightarrow x_n$ $(k \to \infty)$ by Proposition 4.1(c). This yields countable subsets $\Lambda_n \subset \Lambda$ such that $x_n \in L^1_{\Lambda_n}$. Then $\Lambda' := \bigcup \Lambda_n$ will do.] So X has the RNP as a subspace of a separable dual space [159, Th. III.3.1, p. 79 and Th. III.3.2, p. 81].

(ii) $\Rightarrow$ (i): We have to show that every measure $\mu \in M_\Lambda$ is absolutely continuous with respect to the Haar measure m. By Proposition 4.1 the following mapping is well-defined for such a μ:

$$\begin{array}{rcccl} F & : & \mathrm{Bor}(G) & \longrightarrow & L^1_\Lambda \\ & & E & \longmapsto & F(E) := \chi_E * \mu. \end{array}$$

Now F is easily seen to be finitely additive and

$$\|F(E)\| = \|\chi_E * \mu\| \le \|\chi_E\| \|\mu\| = \|\mu\| m(E)$$

for $E \in \mathrm{Bor}(G)$ shows that F is m-continuous and of bounded variation. By the very definition of the Radon-Nikodým property there is a Bochner integrable function (!) $g : G \to L^1_\Lambda$ such that

$$F(E) = \int_E g\, dm \qquad \forall E \in \mathrm{Bor}(G).$$

We claim that there is a null set N such that

$$g(y)_y = g(x)_x \qquad \text{for } x, y \notin N. \tag{1}$$

(Recall that f_x denotes the translate of f by x, i.e. $f_x(y) = f(xy)$; $T_x : f \mapsto f_x$ is the corresponding operator on L^1.) Fix $z \in G$. Omitting the m for integration with respect to the Haar measure we obtain for $E \in \mathrm{Bor}(G)$

$$\begin{aligned} \int_E g(x)\, dx &= F(E) = \chi_E * \mu = (\chi_{zE} * \mu)_z = \Bigl(\int_{zE} g(x)\, dx\Bigr)_z \\ &= T_z\Bigl(\int \chi_{zE}(x) g(x)\, dx\Bigr) = \int T_z(\chi_{zE}(x) g(x))\, dx \\ &= \int \chi_{zE}(x) g(x)_z\, dx = \int \chi_E(z^{-1}x) g(x)_z\, dx \\ &= \int \chi_E(x) g(zx)_z\, dx = \int_E g(zx)_z\, dx. \end{aligned}$$

The uniqueness of the density function of a vector measure (see [159, Cor. II.2.5, p. 47]) now shows that $g(x) = g(zx)_z$ for $x \notin N_z$, a null set depending on z. Since $(x, z) \mapsto g(zx)_z$ is Bochner integrable, we deduce by a Fubini type argument the existence of a null set N such that $g(x) = g(zx)_z$ for almost all z if $x \notin N$. [In fact,

$$\int_E \int_F (g(x) - g(zx)_z)\, dz\, dx = \int_F \int_E (g(x) - g(zx)_z)\, dx\, dz = 0 \qquad \forall E, F \in \mathrm{Bor}(G)$$

so that off a null set N

$$\int_F (g(x) - g(zx)_z)\, dz = 0 \qquad \forall F \in \mathrm{Bor}(G),$$

whence the assertion.] This proves that for $x \notin N$

$$g(y)_y = g(x)_x \qquad \text{for almost every } y$$

and thus our claim.
So let us define

$$f := g(x_0)_{x_0} \in L^1(G)$$

where $x_0 \notin N$ is arbitrary. Hence $g(x)_x = f_{x^{-1}}$ for almost every x. We will now show $\mu = fm$ by proving

$$\chi_E * \mu = \chi_E * f \qquad \text{for all } E \in \mathrm{Bor}(G). \tag{2}$$

Indeed, if (2) holds then $\varphi * \mu = \varphi * f$ for all simple functions φ, hence – by continuity of multiplication (= convolution) in $M(G)$ – it holds for all $\varphi \in L^1(G)$; but then for all $\varphi \in C(G)$. Noting $C(G) * M(G) \subset C(G)$ we evaluate $\varphi * \mu$ and $\varphi * f$ at the neutral element and obtain

$$\int \varphi(y^{-1})\, d\mu(y) = \int \varphi(y^{-1})\, d(fm)(y),$$

hence $\mu = fm$, and μ is m-continuous.
In order to see (2) recall

$$\chi_E * \mu = F(E) = \int_E g(x)\, dx = \int_E f_{x^{-1}}\, dx$$

and

$$(\chi_E * f)(y) = \int \chi_E(x) f(x^{-1}y)\, dx = \int_E f(x^{-1}y)\, dx.$$

For every $\varphi \in L^\infty(G) = L^1(G)^*$ the following holds by Fubini's theorem:

$$\begin{aligned} \langle \varphi, \chi_E * f\rangle &= \int_G \varphi(y)(\chi_E * f)(y)\, dy = \int_G\int_E \varphi(y) f(x^{-1}y)\, dx\, dy \\ &= \int_E\int_G \varphi(y) f(x^{-1}y)\, dy\, dx = \int_E \langle \varphi, f_{x^{-1}}\rangle\, dx = \left\langle \varphi, \int_E f_{x^{-1}}\, dx \right\rangle. \end{aligned}$$

By the above this is what was claimed in (2). □

We need the following lemmata for the M-structure characterisation of Shapiro sets. Recall that φ_γ denotes the map which assigns to $f \in L^1$ the γ-th Fourier coefficient $\widehat{f}(\gamma)$. Also, recall from Definition 3.8 that $X^\sharp$ denotes the collection of those functionals on a subspace X of L^1 whose restrictions to the unit ball of X are m-continuous.

Lemma 4.8 *Let Λ be a nicely placed subset of Γ. Then Λ is a Shapiro set if and only if $\varphi_\gamma \in (L^1_\Lambda)^\sharp$ for all $\gamma \in \Lambda$.*

PROOF: For the "if" part we have to show by Theorem 3.5 that $B_{L^1_{\Lambda'}}$ is m-closed for all subsets Λ' of Λ. Now

$$B_{L^1_{\Lambda'}} = \{f \in B_{L^1_\Lambda} \mid \widehat{f}(\gamma) = 0 \text{ for all } \gamma \in \Lambda \setminus \Lambda'\} = \bigcap_{\gamma \in \Lambda\setminus\Lambda'} \varphi_\gamma|_{B_{L^1_\Lambda}}{}^{-1}\{0\}\,.$$

Hence $B_{L^1_{\Lambda'}}$ is m-closed in $B_{L^1_\Lambda}$, which is m-closed in L^1 since Λ is nicely placed. Therefore $B_{L^1_{\Lambda'}}$ is m-closed in L^1, too.

On the other hand, assuming that for some $\gamma \in \Lambda$ the restriction of φ_γ to the unit ball of L^1_Λ is not m-continuous, we find a bounded sequence (f_n) in L^1_Λ such that $f_n \xrightarrow{m} 0$, $a := \lim \widehat{f}_n(\gamma)$ exists, but $a \neq 0$. Consider $\Lambda' := \Lambda \setminus \{\gamma\}$ and $g_n := f_n - \widehat{f}_n(\gamma)\gamma$. Then (g_n) is a bounded sequence in $L^1_{\Lambda'}$ with $g_n \xrightarrow{m} -a\gamma$. Since by assumption Λ' is nicely placed, we get $-a\gamma \in L^1_{\Lambda'}$, a contradiction. □

The next lemma says "If there is an m-continuous functional in $(L^1_\Lambda)^*$ which doesn't vanish at $\gamma \in L^1_\Lambda$, then the natural functional not vanishing at γ, i.e. φ_γ, is m-continuous on the unit ball of L^1_Λ". This will be used to show that "If there is an M-embedded predual of L^1_Λ ($= M_\Lambda$), then it is the natural one, i.e. $C/C_{\Gamma\setminus\Lambda^{-1}}$".

Lemma 4.9 *If $\psi \in (L^1_\Lambda)^\sharp$, $\gamma \in \Lambda(\subset L^1_\Lambda)$, and $\psi(\gamma) \neq 0$, then $\varphi_\gamma \in (L^1_\Lambda)^\sharp$.*

PROOF: Without loss of generality we may assume $\psi(\gamma) = 1$. Define a new functional by

$$\begin{array}{cccc} \widetilde{\psi}: & L^1_\Lambda & \longrightarrow & \mathbb{C} \\ & f & \longmapsto & \int_G \psi(f_x)\,\overline{\gamma(x)}\,dx. \end{array}$$

Note that the space L^1_Λ is translation invariant and $x \mapsto f_x$ is continuous (see [542, Th. 1.1.5]); so the integral exists, and because of $\|f\| = \|f_x\|$ we have that $\widetilde{\psi}$ is bounded. We will show that $\widetilde{\psi}$ is m-continuous on the unit ball of L^1_Λ and $\widetilde{\psi} = \varphi_\gamma|_{L^1_\Lambda}$. The first statement follows since translation is isometric with respect to d_m, which was defined at the beginning of the previous section. Indeed, the supposed m-continuity of ψ implies: For every $\varepsilon > 0$ there is a $\delta > 0$ such that

$$|h|_m < \delta,\ h \in 2B_{L^1_\Lambda} \quad \Longrightarrow \quad |\psi(h)| < \varepsilon.$$

For $f, g \in B_{L^1_\Lambda}$ with $d_m(f,g) = |f-g|_m < \delta$, hence also $|f_x - g_x|_m < \delta$ for all $x \in G$, this yields $|\psi(f_x - g_x)| < \varepsilon$ ($x \in G$). This immediately gives the m-continuity of $\widetilde{\psi}$ at f. Observing $\lambda_x = \lambda(x)\lambda$ for $\lambda \in \Lambda$ we find

$$\widetilde{\psi}(\lambda) = \int \psi(\lambda_x)\,\overline{\gamma(x)}\,dx = \int \psi(\lambda(x)\lambda)\,\gamma^{-1}(x)\,dx = \psi(\lambda)\int (\lambda\gamma^{-1})(x)\,dx.$$

Now $\int \lambda\gamma^{-1}\,dm = 0$ for $\lambda \neq \gamma$ (see e.g. [309, Lemma 23.19]), hence $\widetilde{\psi}(\lambda) = \delta_{\lambda\gamma}$. This shows that $\widetilde{\psi}$ and φ_γ agree on lin Λ, so also on L^1_Λ (see Proposition 4.1). □

Theorem 4.10 *For a subset Λ of Γ the following are equivalent:*

(i) *Λ is a Shapiro set.*
(ii) *L^1_Λ is isometrically isomorphic to the dual of an M-embedded space.*
(ii') *M_Λ is isometrically isomorphic to the dual of an M-embedded space.*
(iii) *$C/C_{\Gamma\setminus\Lambda^{-1}}$ is an M-embedded space.*
(iv) *$B_{L^1_\Lambda}$ is m-closed and $\varphi_\gamma|_{B_{L^1_\Lambda}}$ is m-continuous for all $\gamma \in \Lambda$.*

In (ii) *or* (ii') *the M-embedded predual is isometrically isomorphic to $C/C_{\Gamma\setminus\Lambda^{-1}}$.*

PROOF: (ii) $\Leftrightarrow$ (ii'): If either L^1_Λ or M_Λ is the dual of an M-embedded space, it has the Radon-Nikodým property by Theorem III.3.1. Hence Λ is a Riesz set by Theorem 4.7, which means that $L^1_\Lambda = M_\Lambda$.
(i) $\Leftrightarrow$ (iv): This is Lemma 4.8.
(ii) $\Rightarrow$ (iv): By Theorem 3.10 the assumption is equivalent to

$$B_{L^1_\Lambda} \textit{ is } m\textit{-closed and } (L^1_\Lambda)^\sharp \textit{ separates } L^1_\Lambda.$$

In particular, for every $\gamma \in \Lambda$ there is a $\psi \in (L^1_\Lambda)^\sharp$ such that $\psi(\gamma) \neq 0$. By Lemma 4.9 this implies (iv). (Note that the M-embedded predual is isometrically isomorphic to $(L^1_\Lambda)^\sharp$ by Theorem 3.10.)
(iv) $\Rightarrow$ (ii): The second statement in (iv) says that $\varphi_\gamma \in (L^1_\Lambda)^\sharp$ for all $\gamma \in \Lambda$, consequently $(L^1_\Lambda)^\sharp$ separates L^1_Λ. Hence, by Theorem 3.10, L^1_Λ is isometrically isomorphic to the dual of the M-embedded space $(L^1_\Lambda)^\sharp$.
(iv) $\Rightarrow$ (iii): We have, as in "(ii) $\Leftrightarrow$ (ii')", that $L^1_\Lambda = M_\Lambda$. This shows that the "natural" predual $C/C_{\Gamma\setminus\Lambda^{-1}}$ of M_Λ is a predual of L^1_Λ. On the other hand, we know from "(iv) $\Rightarrow$ (ii)" that $(L^1_\Lambda)^\sharp$ is the M-embedded predual of L^1_Λ.
To prove $C/C_{\Gamma\setminus\Lambda^{-1}} \cong (L^1_\Lambda)^\sharp$ (and hence the statement (iii)) we use the general fact: *If Y_1 and Y_2 are two isometric preduals of X with the embedded copy of Y_1 in X^* contained in the one of Y_2, then the embedded copies coincide, hence $Y_1 \cong Y_2$.* Now for $\gamma \in \Lambda$ the equivalence class $\gamma^{-1} + C_{\Gamma\setminus\Lambda^{-1}}$ acts as an element of $(C/C_{\Gamma\setminus\Lambda^{-1}})^{**} = (L^1_\Lambda)^*$ as $f \mapsto \int f\gamma^{-1}\,dm = \varphi_\gamma(f)$ $(f \in L^1_\Lambda)$. Since by assumption φ_γ is m-continuous on the unit ball of L^1_Λ we find (as embedded copies)

$$\Lambda^{-1} + C_{\Gamma\setminus\Lambda^{-1}} \subset (L^1_\Lambda)^\sharp.$$

The linear span of the left hand side is dense in $C/C_{\Gamma\setminus\Lambda^{-1}}$ (cf. Proposition 4.1), so the above remark finishes the proof.
(iii) $\Rightarrow$ (ii'): Clear by Proposition 4.1(e). □

Having completed the Banach space description of the classes of subsets Λ of Γ introduced at the beginning of this section, one could now try to use techniques and methods from harmonic analysis to construct and further investigate nicely placed and Shapiro sets and thus special classes of L- and M-embedded spaces. However, we refrain from doing this here because it would lead us too far away; we refer to the Notes and Remarks section where some of the results in this connection are collected.
So far there doesn't seem to be much influence in the other direction, i.e. applications of Banach space properties of L- and M-embedded spaces to questions in harmonic analysis. However, we point out that the Shapiro sets which are defined by Banach space properties are, by means of convergence in measure, much easier to handle than the Riesz sets and that almost all known examples of Riesz sets are actually Shapiro sets – cf. [258, §3].
We conclude this section with two examples. The first one gives yet another proof of Example III.1.4(h), which is immediate after the above preparation. The second one applies the results proved in this section to the solution of the RNP-problem for L-embedded spaces mentioned on p. 176.

Example 4.11 N *is a Shapiro subset of* $\mathbb{Z}$, *hence* $C(\mathbb{T})/A$ *is an M-embedded space.*

PROOF: In the proof of Example 3.6(a) we actually showed that every subset Λ of N is nicely placed. □

Example 4.12 *For* $n \in \mathbb{N}_0$ *put* $D_n := \{l2^n \mid |l| \le 2^n\}$ *and* $\Lambda := \bigcup_{n\in\mathbb{N}_0} D_n$. *Then* Λ *is a nicely placed Riesz set which is not Shapiro. Hence* L^1_Λ *is an L-embedded Banach space with the Radon-Nikodým property, which is a dual space but not the dual of an M-embedded space.*

PROOF: The statements about L^1_Λ follow from the properties of Λ by Definition 4.2 and Theorems 4.7 and 4.10. Note that by Proposition 4.1(e) L^1_Λ is a dual space if Λ is Riesz. We prepare the actual proof by the following: Let for $m \in \mathbb{N}_0$

$$P_m := 2^m + 2^{m+1}\mathbb{Z}.$$

Observing that for $n \ne 0$

$$n \in P_m \iff 2^m \text{ divides } n \text{ and } 2^{m+1} \text{ doesn't divide } n$$

we see that (P_m) is a partition of $\mathbb{Z}\setminus\{0\}$. Note that $n \in D_k$ implies that 2^k divides n, hence 2^m divides n for $m \le k$, consequently $n \notin P_m$ for $m < k$. This gives for $m < k$

$$n \in P_m \implies n \notin D_k,$$

therefore

$$P_m \cap \Lambda \subset \bigcup_{k\le m} D_k,$$

so $P_m \cap \Lambda$ is finite for all $m \in \mathbb{N}_0$.
To prove that Λ is a Riesz set we have to show that $\mu = \mu_a + \mu_s \in M_\Lambda$ implies $\mu_s = 0$. The latter will be implied by $\widehat{\mu_s}(n) = 0$ for all $n \ne 0$, since the Fourier transform of a singular measure can't have finite support.
For $n \in \mathbb{Z}\setminus\{0\}$ we find $m \in \mathbb{N}_0$ such that $n \in P_m$. There is a discrete measure $\sigma \in M(\mathbb{T})$ with finite support such that $\widehat{\sigma} = \chi_{P_m}$ (cf. the proof of Proposition 4.4(e)). Since M_Λ and M_{P_m} are ideals, we get

$$\mu * \sigma \in M_{\Lambda\cap P_m}.$$

But $\Lambda \cap P_m$ is finite, consequently $\mu * \sigma$ is a trigonometric polynomial. In particular

$$(\mu * \sigma)_s = 0.$$

Now [442, Lemme 1]

$$(\mu * \sigma)_s = \mu_s * \sigma.$$

[Indeed, $(\eta * \delta_x)(E) = \int \eta(y^{-1}E)\, d\delta_x(y) = \eta(x^{-1}E)$ shows that $\eta \in M_s$ implies $\eta * \delta_x \in M_s$. Writing

$$\mu * \sigma = (\mu_a + \mu_s) * \sigma = \mu_a * \sigma + \mu_s * \sigma$$

we get $\mu_a * \sigma \in M_a = L^1$ (since L^1 is an ideal) and $\mu_s * \sigma \in M_s$. Hence the uniqueness of the Lebesgue decomposition gives the claim.]

Evaluating the Fourier transform of $(\mu * \sigma)_s$ at n we find

$$0 = \widehat{(\mu * \sigma)_s}(n) = \widehat{\mu_s * \sigma}(n) = \widehat{\mu_s}(n)\,\widehat{\sigma}(n) = \widehat{\mu_s}(n).$$

From this we deduce that $\mu_s = 0$, and Λ is a Riesz set.
To show that Λ is nicely placed, we have to prove by Theorem 3.5 that $B_{L^1_\Lambda}$ is m-closed. So let (f_k) be a sequence in the unit ball of L^1_Λ which converges in measure to some $g \in L^1$. We want to show that $\widehat{g}(n) = 0$ for $n \in \mathbb{Z} \setminus \Lambda$.
As above we find $m \in \mathbb{N}_0$ such that $n \in P_m$ and σ with $\widehat{\sigma} = \chi_{P_m}$. Since

$$(f * \delta_x)(y) = \int f(yz^{-1})\, d\delta_x(z) = f(yx^{-1}) = f_{x^{-1}}(y)$$

and σ is discrete with finite support, we conclude that

$$f_k * \sigma \longrightarrow g * \sigma$$

in measure. But $f_k * \sigma$ belongs to the finite-dimensional space $L^1_{\Lambda \cap P_m}$, so the convergence is also in norm. In particular

$$\lim_k \widehat{f_k * \sigma}(n) = \widehat{g * \sigma}(n) = \widehat{g}(n)$$

so that $\widehat{f_k * \sigma}(n) = \widehat{f_k}(n) = 0$ for all $k \in \mathbb{N}$ yields $\widehat{g}(n) = 0$.
Finally, to obtain that Λ is not a Shapiro set we will prove that φ_0 is not m-continuous on the unit ball of L^1_Λ (cf. Lemma 4.8).
It is easy to find $f_n \in L^1(\mathbb{T})$ such that

$$\|f_n\|_1 \le 2, \qquad \int f_n\, dm = \widehat{f_n}(0) = \varphi_0(f_n) = 0 \quad \text{and} \quad f_n \longrightarrow \chi_{\mathbb{T}} \;\; m\text{-a.e.}$$

Using the density of $T_{\mathbb{Z}\setminus\{0\}}$ in $L^1_{\mathbb{Z}\setminus\{0\}}$ (Proposition 4.1) we find an L^1-bounded sequence (t_n) in $T_{\mathbb{Z}\setminus\{0\}}$ with $t_n \xrightarrow{m} \chi_{\mathbb{T}}$. Each t_n has the form

$$t_n(z) = \sum_{\substack{l=-m_n \\ l \neq 0}}^{m_n} a_{l,n} z^l.$$

So

$$t_n(z^{2^{m_n}}) = \sum a_{l,n} z^{l2^{m_n}} =: \widetilde{t_n}(z)$$

and $\widetilde{t_n} \in T_{D_{m_n}} \subset T_\Lambda$. Since z and $z^{2^{m_n}}$ have the same distribution it follows easily that $m\{|\chi_{\mathbb{T}} - t_n| \ge a\} = m\{|\chi_{\mathbb{T}} - \widetilde{t_n}| \ge a\}$ and $\|t_n\| = \|\widetilde{t_n}\|$. So $(\widetilde{t_n})$ is a bounded sequence in $T_\Lambda \subset L^1_\Lambda$ with $\widetilde{t_n} \xrightarrow{m} \chi_{\mathbb{T}}$, but $\varphi_0(\widetilde{t_n}) = \int \widetilde{t_n}\, dm = 0$ and $\varphi_0(\chi_{\mathbb{T}}) = 1$. Therefore the restriction of φ_0 to the unit ball of L^1_Λ is not m-continuous. □

It is (at least implicitly) contained in the above proof that φ_n is m-continuous on the unit ball of L^1_Λ for all $n \in \Lambda \setminus \{0\}$ – so that all but one character are in $(L^1_\Lambda)^\sharp$. In this connection the following is open:

QUESTION. *Does there exist a nicely placed Riesz subset Λ of $\mathbb{Z}$ such that, for all $n \in \Lambda$, $\varphi_n|_{B_{L^1_\Lambda}}$ is not m-continuous?*

Let us indicate where the interest in this question comes from. We now know that for an L-embedded space X with X_s not w^*-closed it does not follow that X fails the RNP (cf. Remark 2.10). However it is unknown if w^*-density of X_s is sufficient for X to fail the RNP. It is not hard to show that a set Λ as in the above question would give a counterexample to that.
There is another offspring of the RNP-problem which remains open: Although the original question whether or not an L-embedded space with the RNP is the dual of an M-embedded space is settled in the negative by Example 4.12, the following is left unanswered.

QUESTION. *Is every L-embedded space with the RNP isometric to a dual space?*

Observe that for the L^1_Λ-spaces studied in this section the answer is *yes* by Theorem 4.7 – so harmonic analysis can't help here. This question is related to Talagrand's work on Banach lattices who proved in [594] that a separable Banach lattice with the RNP is a dual lattice. See [266, p. 260, Remark 6] for more comments on this.

IV.5 Notes and remarks

GENERAL REMARKS. Parallel to the study of spaces which are M-ideals in their biduals, some of the first results on L-embedded spaces were obtained in [292] and [293]. However the main contributions to this latter class were obtained by G. Godefroy in [257], [258] and [266], and by D. Li in [395] and [396]. The Example 1.1(b) dates back to [592, Th. 3], whereas the L-embeddedness of L^1/H^1_0 was first explicitly proved in [21, Th. 2]. The results on subspaces and quotients in Theorem 1.2 and Corollary 1.3 are from [395], where however the general invariance principle Lemma I.1.15 was not used. The special case where the bigger space is $L^1(\mu)$ was previously established in [257]. Lemma 1.4 is due to Pfitzner [493], the Propositions and Examples 1.5 to 1.10 are essentially from [292]. The fact that ℓ^1 is isometric to a 1-complemented subspace of $(\oplus\sum\ell^1(n))_{\ell^\infty}$ which was used in Example 1.7(b) was observed by W. B. Johnson [354, p. 303].
Let us have a closer look at the question whether the bidual of an L-embedded space is again L-embedded. Not even for "nice" examples, i.e. for L-embedded subspaces of L^1-spaces, is this true: The real Banach space $Y = (\oplus\sum\ell^1(n))_{c_0}$ is isometric to a subspace of c_0, so $Y^\perp$ is an L-embedded subspace of ℓ^1 by Proposition 1.10. However, $(Y^\perp)^{**} \cong Y^{\perp\perp\perp} \subset \ell^{1**}$ fails to be L-embedded, because otherwise $\ell^{1**}/Y^{\perp\perp\perp} \cong (\ell^1/Y^\perp)^{**} \cong (Y^*)^{**}$ would be. It is open for which finite dimensional spaces E_n the bidual of $(\oplus\sum E_n)_{\ell^1}$ is L-embedded and also whether the even duals of L^1/H^1_0 have this property. Note that by a result of Bourgain ([89] or [90, Cor. 5.4]) $(L^1/H^1_0)^{(2n)}$ is weakly sequentially complete.
A natural question which seems to have not yet been thoroughly investigated is the L-embeddedness of the Bochner space $L^1(X) := L^1([0,1], X)$. For some classes of L-embedded spaces this is known to be the case, e.g., if X is the predual of a von Neumann

algebra, then so is $L^1(X)$ (cf. [593, p. 263]), and if $X \subset L^1[0,1]$ is nicely placed, then so is $L^1(X) \subset L^1([0,1]^2)$. Less obvious is the case of reflexive X which are trivially L-embedded. In this case $L^1(X)$ is in fact L-embedded as can be deduced from results of Levin [393], who has presented a decomposition of $(L^\infty(Y))^*$ for arbitrary Banach spaces Y, as follows. Let M_1 denote the space of absolutely continuous Y^*-valued σ-additive vector measures and M_2 the space of absolutely continuous Y^*-valued purely finitely additive vector measures. Note that $M_1, M_2 \subset (L^\infty(Y))^*$ in a canonical fashion. Further, M_3 is to denote the annihilator of $\{\varphi \otimes y \mid \varphi \in L^\infty, y \in Y\}$. Then Levin's theorem asserts that

$$(L^\infty(Y))^* = M_1 \oplus_1 (M_2 \oplus M_3).$$

If Y is the dual of a reflexive space X, we obtain from the RNP of X and X^*

$$(L^1(X))^{**} = L^1(X) \oplus_1 (M_2 \oplus M_3),$$

which proves that $L^1(X)$ is L-embedded. As for the case of general L-embedded spaces X, we would like to mention Talagrand's result that $L^1(X)$ is weakly sequentially complete whenever X is [597]. Thus a positive answer to the problem of L-embeddedness of $L^1(X)$ would provide a proof of a special case of Talagrand's theorem. Going one step further, one might wonder about the projective tensor product of L-embedded spaces; recall that $L^1(X) = L^1 \widehat{\otimes}_\pi X$. Here even less is known. For example, it is not clear whether $L^p \widehat{\otimes}_\pi L^p$ is L-embedded for $1 < p < \infty, p \neq 2$, or can at least be so renormed. (The tensor product $\ell^p \widehat{\otimes}_\pi \ell^p$ is L-embedded since it is the dual space of $K(\ell^{p^*}, \ell^p)$, cf. Example III.1.4(g).) On the other hand, Pisier has shown in [500, Th. 4.1, Rem. 4.4] that $L^1/H^1 \widehat{\otimes}_\pi L^1/L^1_\Lambda$, for $\Lambda = \mathbb{N} \setminus \{3^n \mid n \geq 0\}$, contains a copy of c_0 and hence fails to be weakly sequentially complete. Thus this space cannot be renormed to be L-embedded; note that Λ is nicely placed because $\mathbb{N}$ is a Shapiro subset of $\mathbb{Z}$, hence L^1/L^1_Λ is L-embedded by Corollary 1.3. The remark following Example 1.7 was shown to us by T.S.S.R.K. Rao. It is unknown whether there is a commutative space as in Example 1.8, more precisely: does there exist a nonreflexive L-embedded subspace Y of an L^1-space such that Y^* has no nontrivial M-ideals? Lemma 1.11 to Proposition 1.13 are from [395] – we only have removed the unnecessary restriction in Proposition 7 of this article. The extreme point result in Proposition 1.14 is due to Deutsch [154, Th. 1]. The smoothness of H^1 (Remark 1.17) is a "folklore result". We haven't been able to locate its first appearance in the literature; the strict convexity of $C(\mathbb{T})/A$, however, appears e.g. in [333, Cor. 4], but the reasoning there is different.

In the Notes and Remarks to Chapter I we discussed several "semi" notions and observed that a Banach space which is a semi M-ideal in its bidual is necessarily M-embedded. It remained open for quite a while whether an analogous statement holds for L-embedded spaces. Let us say that a Banach space X is semi L-embedded if X is a semi L-summand in X^{**}, i.e., if there exists a (nonlinear) projection π from X^{**} onto X satisfying

$$\begin{aligned} \|x^{**}\| &= \|\pi(x^{**})\| + \|x^{**} - \pi(x^{**})\| \\ \pi(\lambda x^{**} + x) &= \lambda \pi(x^{**}) + x \end{aligned}$$

for $x \in X$, $x^{**} \in X^{**}$ and $\lambda \in \mathbb{K}$. Recently Payá and Rodríguez [480] succeeded in constructing semi L-embedded real Banach spaces which are not L-embedded. Their construction involves sets of constant width, which are defined as follows. A closed convex bounded subset S of a Banach space E is a set of constant width if $\mathrm{diam}(x^*(S)) = \mathrm{diam}(y^*(S))$ whenever $\|x^*\| = \|y^*\|$, $x^*, y^* \in E^*$. Let now E be M-embedded, $S \subset E^{**}$ be a weak* compact set of constant width and form the compact convex set $K = S^{\mathbb{N}}$. Then the space $A(K)$ of affine continuous functions is semi L-embedded, and if E^{**} enjoys the IP (Definition II.4.1) and S is not a ball, then $A(K)$ is not L-embedded. Let us mention that the proof of this result relies on several concepts studied in Chapter II, notably the IP and the notion of a pseudoball. Thus, the easiest way to produce a properly semi L-embedded space is to start with a triangle $S \subset \mathbb{R}^2$, or more generally a compact convex set $S \subset \mathbb{R}^n$ without a centre of symmetry, and to consider $A(S^{\mathbb{N}})$. By the way, in this case the resulting space $A(S^{\mathbb{N}})$ turns out to be a renorming of ℓ^1. For the time being it is not known whether a semi L-embedded space can be renormed to be L-embedded; note that semi L-embedded spaces are weakly sequentially complete and hence contain ℓ^1 unless they are reflexive [480]. In this connection a result due to Godefroy is worth mentioning, which says that every Banach space containing ℓ^1 can be renormed to be an L-summand in some subspace of its bidual [260]. Luckily there is no need to investigate more general norm decompositions (cf. the Notes and Remarks to Chapter I) in the case of the inclusion $X \subset X^{**}$, since results in [103] and [255] show that M- resp. semi L-embedded spaces are the only possibilities within a natural class of Banach spaces, namely where X is a so-called absolute subspace of X^{**}. The results of [480] are also surveyed in [526].

Godefroy has given several proofs of the weak sequential completeness of L-embedded spaces. The first one in [254] uses, among other things, admissible sets (a generalisation of weakly compact sets defined by a geometric condition in the bidual) and Baire's results on points of continuity of functions of the first Baire class. The second one in [257, Lemme 4] avoids all this and is completely elementary. The proof in [258, p. 308] uses upper envelopes and again properties of Baire-1 functions. The extreme discontinuity of elements of X_s (Proposition 2.1) was pointed out in [54], and this result was further simplified in [455, Lemma 6.1]. We used the set-valued modulus of continuity of this last article in order to obtain also the converse in Proposition 2.1. We don't state the technical problem one encounters if one uses the standard definition of oscillation in the complex case. However, we mention the following question which shows which kind of difficulty one faces: if X is a complex Banach space and $\mathrm{Re}\, x^{**}|_{B_{X^*}}$ is w^*-continuous at $x^* \in S_{X^*}$, is then also $x^{**}|_{B_{X^*}}$ w^*-continuous at x^*? The ace of diamonds lemma, Proposition 2.5, is an unpublished argument of Godefroy he suggested in a talk in Oldenburg. It has the following generalization [267, Th. II.1]: if (x_n) is a weak Cauchy sequence of fixed points of an isometric bijection $S : X^* \to X^*$, then the w^*-limit x^* of (x_n^*) is a fixed point too. The proof in [267] is very easy; however it doesn't yield the ℓ^1-subsequence one obtains with the argument in the text. A special case of [267, Th. II.1], namely that a space X admitting an isometric symmetry $S : X^{**} \to X^{**}$ with $X = \ker(Id - S)$ is weakly sequentially complete, had already been obtained in [254, Prop. 10]. Proposition 11 of this last paper (which gives the converse of the last-mentioned result for Banach lattices) and the connection with u-ideals (see below) are worth noticing. The simple local reflexivity argument, Remark 2.4, yielding ℓ^1-subspaces in nonreflexive L-embedded spaces, was first observed in [292]. Property (V^*) for L-embedded spaces, i.e. Lemma 2.6 and Theorem 2.7, is a recent achievement due to Pfitzner [493].

The RNP-problem for L-embedded spaces was studied in [395] and solved in [266]. Between the assumptions that X_s is not w^*-closed (which is not sufficient for X to fail the RNP) and that B_{X_s} is w^*-dense in $B_{X^{**}}$ (which suffices by Remark 2.10(a)) lies the condition that X_s has a positive characteristic in X^{**}. In [395, Th. 10] it is claimed that this last hypothesis already gives that X can't have the RNP. D. Li has asked us to mention that the proof contains a gap and it is open whether the result is true or not. The factorization property Proposition 2.12 appears in [396, Th. 1]. The special case of this last result for operators $T : L^1 \to L^1/H_0^1$ was proved in [172, Th. 2] by different methods – it is an important tool for the characterisation of the analytic Radon-Nikodým property. Besides the Corollaries 2.13 and 2.14 the article [396] contains many more applications of the factorization property.

We have already said in the text that the main result of the first part of Section IV.3, Theorem 3.4, is due to Buhvalov and Lozanovskii [99]. Lemma 3.1 and Corollary 3.2 have simply been extracted from their proof - see also [257, proof of Lemme 1]. The second proof of Corollary 3.2 is from [266, Lemma I.1]. Proposition 3.3 combines [257, Lemme 1] and [99, Lemma 1.1]. Note that this result fails for unbounded sets C, e.g. $L^1[0,1]$ is not μ-closed in $L^0[0,1]$. We remark also that for convex bounded subsets C of $L^1(\mu)$ a fourth equivalent condition in Proposition 3.3 is that C is closed in $L^1(\mu)$ with respect to convergence in $\|\,.\,\|_{1,w}$, where $\|f\|_{1,w} = \sup_{t>0} t \cdot \mu\{|f| > t\}$ denotes the weak-L^1 "norm". This was proved by Godefroy in [258, Lemma 1.7]. The characterisation of L-embedded subspaces of L^1-spaces in Theorem 3.5 is from [257, Lemme 1, Lemme 23]. In our development it is a simple consequence of Theorem 1.2; however this latter result wasn't available at the time of [257]. In [99, Th. 2.4′] it was already shown that subspaces of $L^1(\mu)$ with μ-closed unit balls are 1-complemented in their biduals.

We would now like to digress a bit and comment on the topology τ_μ of convergence in measure. We have said in the text that it is the vector space topology best suited to describe almost everywhere convergence. However, much deeper is the insight gained from the Buhvalov-Lozanovskii theorem: μ-closedness is a substitute for compactness. From the examples in [99] which substantiate this, we mention only the most striking one [99, Th. 1.3]: if (C_α) is a decreasing net of nonempty convex bounded sets in $L^1(\mu)$ which are μ-closed, then $\bigcap C_\alpha$ is nonempty. [For a proof simply use the w^*-compactness of $\overline{C_\alpha}^{w^*}$ in $L^1(\mu)^{**}$ and Theorem 3.4.] Note that the τ_λ-closed set $B_{L^1[0,1]}$ is not τ_λ-compact, since e.g. the sequence (r_n) of Rademacher functions has no τ_λ-convergent subsequence. On the other hand, the above weak form of compactness is strong enough to imply ([99, Th. 2.3]):

THEOREM. *A subspace Y of $L^1(\mu)$ is reflexive iff $\tau_\mu = \tau_{\|\,\|_1}$ on B_Y.*

[PROOF: If Y is reflexive and (f_n) is a sequence in B_Y τ_μ-convergent to $f \in B_Y$, we show that every subsequence (g_n) of (f_n) has a subsequence (h_n) which is $\|\,.\,\|_1$-convergent to f. In fact, by weak compactness (g_n) has a weakly convergent subsequence (h_n). Then $h_n \xrightarrow{w} f$ and [179, Th. IV.8.12] gives $h_n \xrightarrow{\|\,\|} f$. Conversely, let us establish Smulian's condition for reflexivity: each decreasing sequence (C_n) of nonempty, convex, closed, and bounded sets has nonempty intersection ([151, p. 58], nowadays one deduces this easily from James' theorem: for $y^* \in S_{Y^*}$ consider $C_n = \{y^* \geq 1 - 1/n\} \cap B_Y$). By assumption the C_n are μ-closed, so the conclusion obtains by what was observed above.]

We finish this intermission with some remarks on the case of infinite measures μ. First of all in this situation the "right" definition of convergence in measure is $f_n \xrightarrow{\mu} f$ iff (f_n) converges to f in measure on every set of finite measure. (This is suggested by the σ-finite case if one transforms μ to finite $\nu = h\mu$ with an integrable and strictly positive h.) But note that without further restrictions this convergence doesn't behave well, e.g. if there are atoms of infinite measure, the limit is not uniquely determined. The necessary assumptions and results can be found in [99]. We finally mention that in ℓ^1, τ_μ-convergence is pointwise convergence, so on bounded sets $\tau_\mu = \sigma(\ell^1, c_0)$ – another indication for the "compactness" of τ_μ-closed sets. In this way one finds examples of subspaces X of $L^1(\mu)$ such that B_X is μ-closed, but X is not: take a subspace X of ℓ^1 which is w^*-closed, but pointwise dense.

The less artificial Example 3.6 of this phenomenon is due to Godefroy, part (a) is [258, Ex. 3.1] and part (b) a personal communication from him. The paper [257] contains further examples of L-embedded subspaces of $L^1(\mu)$, for instance $H^1(U)$ for special subsets U of $\mathbb{C}^n$ [257, Th. 20] and certain abstract Hardy spaces $H^1(\mu)$ [257, Lemme 10]. Lemma 3.7 is from [257, Lemme 7]. A partial result pointing to a characterisation of duals of M-embedded spaces among L-embedded subspaces of $L^1(\mu)$ was [257, Th. 6]. The full version given in Proposition 3.9 and Theorem 3.10 was established in [266].

It was G. Godefroy who first applied concepts of M-structure theory to harmonic analysis in [257] and then extensively in [258]; see also the papers [266] and [267]. Definition 4.2, Lemma 4.3, Proposition 4.5, the equivalence of (i) and (iii) in Theorem 4.10, and essentially Lemma 4.8 come from [258], which contains many other results about which we shall say more in the next subsection. Theorem 4.7 is from [432] – if its proof is too complicated the blame is on us (we just wanted to be precise on the measure theoretical technicalities). Lemma 4.9, Theorem 4.10, and Example 4.12 were obtained in [266]. The set Λ in Example 4.12 is attributed to A. B. Aleksandrov in [258, p. 322]. For more information on Riesz sets one should consult the references listed in [258, p. 302] and the more recent articles [98], [171], [229] and [499].

Finally, we mention the note [77] where fixed points of nonlinear contractions on nicely placed subspaces of L^1 are investigated.

TECHNIQUES FOR CONSTRUCTING NICELY PLACED AND SHAPIRO SETS. Although Riesz, Shapiro, and nicely placed sets are not really "thin" sets, it is – in view of the fact that subsets of Riesz and Shapiro sets are again of this type – justified to consider them as in some sense small. So it is most interesting to find big examples and to investigate to what extent known examples can be enlarged. Note, however, the different behaviour of nicely placed sets: the set Λ from Example 4.12 is nicely placed, yet $\Lambda \setminus \{0\}$ is not. Let us begin by noting the stability under products:

THEOREM. *If Λ and Λ' are nicely placed, resp. Shapiro or Riesz subsets of countable discrete groups Γ and Γ', then $\Lambda \times \Lambda'$ is a nicely placed, resp. Shapiro or Riesz subset of $\Gamma \times \Gamma'$.*

For Riesz sets this was proved in [433, Th. 1.2], for the other cases see [258, Th. 2.7]. The theorem implies in particular that the quotient of $C(\mathbb{T}^2)$ by the bidisk algebra is M-embedded.

Next we will describe two techniques for constructing nicely placed and Shapiro sets. The first one uses ordered groups and the second one the localisation argument of Meyer from [442]. Recall that a subset Π of Γ with

$$\Pi\Pi \subset \Pi, \quad \Pi \cap \Pi^{-1} = \{e\}, \quad \text{and} \quad \Pi \cup \Pi^{-1} = \Gamma$$

induces a (total) ordering on Γ by $\gamma \geq \lambda$ iff $\gamma\lambda^{-1} \in \Pi$. Γ admits such an ordering if and only if $G = \widehat{\Gamma}$ is connected [542, 2.5.6(c) and 8.1.2(a)]. The main reference for ordered groups (always in the above sense) is Chapter 8 in [542]; but see also Chapter VII in [240]. Further recall that a subset Λ of Γ is said to be a $\Lambda(1)$*-set*, if there are $0 < q < 1$ and $C > 0$ such that

$$\|t\|_1 \leq C\|t\|_q \qquad \forall\, t \in T_\Lambda,$$

i.e. if the "norms" $\|\,.\,\|_1$ and $\|\,.\,\|_q$ are equivalent on T_Λ. It is known that Λ is a $\Lambda(1)$-set if and only if L^1_Λ is reflexive (see [40] together with [530], or [291] for an elementary proof). We refer to [541], [186], and [425] for more information on $\Lambda(1)$-sets. Building on similar results to those in [541] and [565] Godefroy proved in [258, Th. 2.1]:

THEOREM A. *A subset Λ of an ordered group Γ such that*

$$\Lambda \cap \{\gamma' \in \Gamma \mid \gamma' \leq \gamma\} \quad \textit{is a } \Lambda(1)\textit{-set for all } \gamma \in \Gamma$$

is a Shapiro set.

Note as a trivial application (finite sets are $\Lambda(1)$-sets) that this provides another proof of the fact that $\mathbb{N}$ is a Shapiro set, equivalently that $C(\mathbb{T})/A$ is an M-embedded space. We also remark (or recall) that proving that a subset Λ is a Shapiro set includes the proof of, or uses, an F. and M. Riesz theorem. In the demonstration of the above theorem G. Godefroy applies at this point the ideas of Helson and Lowdenslager to come to a function algebra setting where the abstract F. and M. Riesz theorem in the form of p. 109 can be used.

To exhibit at least a part of the argument leading to Theorem A let us prove that the positive cone Π of Γ is nicely placed. We use Lemma I.1.15 as in the first part of Example III.1.4(h) and show that $Q(L^1_\Pi)^{\perp\perp} \subset (L^1_\Pi)^{\perp\perp}$, where P is the L-projection from $L^1(G)^{**}$ onto $L^1(G)$. Write $L^1(G)^* = L^\infty(G) = C(K)$ with $K = M_{L^\infty(G)}$ and carry m to a measure $\widetilde{m}$ on K by means of $\int_G f\,dm = \int_K \widetilde{f}\,d\widetilde{m}$, $f \in L^\infty(G)$, where $f \mapsto \widetilde{f}$ of $L^\infty(G)$ denotes the Gelfand transform. After the identifications $L^1(G) = L^1(\widetilde{m})$ and $L^1(G)^{**} = M(K)$, the L-projection P is simply the Radon-Nikodým projection $P\mu = \mu_a$ where μ_a is the $\widetilde{m}$-absolutely continuous part of $\mu \in M(K)$. From the properties of Π it follows easily that $A := C_\Pi(G)$ is a function algebra in $C(G)$, that $\varphi : A \to \mathbb{C}$, $\varphi(f) = \int_G f\,dm$ is multiplicative, and that m is the unique representing measure for φ. By [95, Th. 4.1.2, p. 212] $H^\infty(m) := \overline{A}^{w*} \subset L^\infty(G)$ is a logmodular algebra in $C(K) = L^\infty(G)$. In particular $H^\infty(m)$ separates K, and multiplicative functionals on $H^\infty(m)$ have unique representing measures. We use this for the canonical extension ϕ of φ to $H^\infty(m)$ and its representing measure $\widetilde{m}$. Hence the abstract F. and M. Riesz theorem is applicable and yields $PH^\infty_0(m)^\perp \subset H^\infty_0(m)^\perp$, where $H^\infty_0(m) = \ker\phi = \overline{C_{\Pi\setminus\{e\}}}^{w*}$.

Since by definition and Proposition IV.4.1 $L^1_\Pi = \overline{A}^{\|\,\|_1}$ and $(L^1_\Pi)^\perp = H^\infty_0(m)$, we obtain the desired invariance $P(L^1_\Pi)^{\perp\perp} \subset (L^1_\Pi)^{\perp\perp}$.
We will now describe in more detail a second technique which actually achieves a local characterisation using the Bohr compactification of Γ. Let us recall the relevant facts: if we write G_d for $G = \widehat{\Gamma}$ equipped with the discrete topology, then $b\Gamma := \widehat{G_d}$ is a compact group which contains Γ as a dense subgroup [542, 1.8]. We will write τ for the topology induced on Γ by $b\Gamma$. A τ-neighbourhood of $\gamma_0 \in \Gamma$ is given by $U(\gamma_0, K, \varepsilon) = \{\gamma \in \Gamma \mid |\gamma(x) - \gamma_0(x)| < \varepsilon, \ \forall x \in K\}$ with $\varepsilon > 0$ and $K \subset G_d$ compact, i.e. finite. Hence τ is the topology of pointwise convergence on G. We write $\mathfrak{U}_\tau(\gamma_0)$ for the set of τ-neighbourhoods of γ_0. We will say a family $\mathcal{C}$ of subsets of Γ is *localizable* if for $\Lambda \subset \Gamma$

$$\Lambda \in \mathcal{C} \quad \Longleftrightarrow \quad \forall \gamma \in \Gamma \ \exists U \in \mathfrak{U}_\tau(\gamma) \ \ U \cap \Lambda \in \mathcal{C}$$

It is easy to check that if $\mathcal{C}$ is localizable then its hereditary family $\mathcal{C}_{\text{her}} := \{\Lambda \subset \Gamma \mid \Lambda' \in \mathcal{C}, \ \forall \Lambda' \subset \Lambda\}$ is localizable. Note that the families of Riesz and Shapiro sets are just the hereditary families of the Haar invariant and the nicely placed sets (see Lemma 4.3).

THEOREM B. ([442, Th. 1], [258, Th. 2.3]) *The families of Haar invariant, nicely placed, Riesz, and Shapiro sets are localizable. We even have, for* $\Lambda \subset \Gamma$, *that the condition*

$$\forall \gamma \notin \Lambda \ \exists U \in \mathfrak{U}_\tau(\gamma) \ \exists E \quad \gamma \notin E, U \cap \Lambda \subset E, E \text{ nicely placed} \qquad (*)$$

is sufficient for Λ *to be nicely placed.*

PROOF: The Riesz and the Shapiro parts follow from what was remarked above if we prove the assertion for the other two classes. In fact, for Haar invariant sets we will prove a stronger statement, namely: $\Lambda \subset \Gamma$ is Haar invariant if for all $\gamma \notin \Lambda$ there exists $U \in \mathfrak{U}_\tau(\gamma)$ with $U \cap \Lambda$ Haar invariant. We have to show that $\mu \in M_\Lambda$ implies $\mu_s \in M_\Lambda$, i.e. $\widehat{\mu_s}(\gamma) = 0$ for $\gamma \notin \Lambda$. Choose $U \in \mathfrak{U}_\tau(\gamma)$ as in the assumption and take $\widetilde{U} \in \mathfrak{U}_{b\Gamma}(\gamma)$ with $\widetilde{U} \cap \Gamma = U$. Using the regularity of the algebra $A(b\Gamma) = \{\widehat{f} \mid f \in L^1(G_d)\}$ we find some $f \in L^1(G_d)$ such that $\widehat{f}(\gamma) = 1$ and $\widehat{f}|_{b\Gamma \setminus \widetilde{U}} = 0$ [542, Th. 2.6.2]. Since $L^1(G_d) = \ell^1(G)$ we can view f as a discrete measure $\nu \in M(G)$ and we get

$$\widehat{\nu}(\gamma) = 1 \text{ and } \widehat{\nu}|_{\Gamma \setminus U} = 0.$$

Then $\mu * \nu \in M_{U \cap \Lambda}$ and

$$(\mu * \nu)_s = \mu_s * \nu.$$

[PROOF: Because $(\eta * \delta_x)(E) = \int \eta(y^{-1}E)\, d\delta_x(y) = \eta(x^{-1}E)$ we have that $\eta \in M_s$ implies that $\eta * \delta_x \in M_s$. Since M_s is a closed subspace of M and every discrete measure is a limit of measures of finite support we conclude from the above: $\eta \in M_s \Rightarrow \eta * \nu \in M_s$. Writing $\mu * \nu = (\mu_a + \mu_s) * \nu = \mu_a * \nu + \mu_s * \nu$, we get $\mu_a * \nu \in M_a = L^1$ and $\mu_s * \nu \in M_s$. The uniqueness of the Lebesgue decomposition now gives the claim.]
Using that $U \cap \Lambda$ is Haar invariant for $\mu * \nu$ we get $0 = \widehat{(\mu * \nu)_s}(\gamma) = \widehat{\mu_s * \nu}(\gamma) = \widehat{\mu_s}(\gamma)\widehat{\nu}(\gamma) = \widehat{\mu_s}(\gamma)$. This is what we wanted to show.
We prepare the proof of the case of nicely placed sets by showing the following

CLAIM: *For $f_n \in B_{L^1}$, $f \in L^1$, and a discrete measure ν the convergence $f_n \xrightarrow{m} f$ implies $f_n * \nu \xrightarrow{m} f * \nu$.*

[PROOF: It is clearly enough to show that, for a bounded sequence (f_n) in L^1, $f_n \xrightarrow{m} 0$ implies $f_n * \nu \xrightarrow{m} 0$. Assume $\|f_n\| \leq 1$ for all $n \in \mathbb{N}$. Observing $g * \delta_x = g_{x^{-1}}$, hence $\|g\|_m = \|g * \delta_x\|_m$, we get $f_n * \nu_0 \xrightarrow{m} 0$ for all measures ν_0 of finite support. Let $\varepsilon > 0$. Choose a finitely supported measure ν_0 with $\|\nu - \nu_0\| \leq \varepsilon^2$. Then

$$m\{|f_n * (\nu - \nu_0)| \geq \varepsilon\} \leq \frac{1}{\varepsilon}\|f_n * (\nu - \nu_0)\| \leq \frac{1}{\varepsilon}\|f_n\|\|\nu - \nu_0\| \leq \varepsilon$$

hence $\|f_n * (\nu - \nu_0)\|_m \leq \varepsilon$. Pick n_0 such that $\|f_n * \nu_0\|_m \leq \varepsilon$ for $n \geq n_0$. Then

$$\|f_n * \nu\|_m \leq \|f_n * \nu_0\|_m + \|f_n * (\nu - \nu_0)\|_m \quad \text{for} \quad n \geq n_0,$$

i.e. $f_n * \nu \xrightarrow{m} 0$.]

To prove the theorem for nicely placed sets assume $\Lambda \subset \Gamma$ is given as in assumption (*) of the theorem. We have to prove that $f_n \in B_{L^1_\Lambda}$, $f \in L^1$, $f_n \xrightarrow{m} f$ imply $f \in L^1_\Lambda$. For $\gamma \notin \Lambda$ choose U and E according to the assumption and a discrete measure ν as in proof of the first part. By the above claim we get $f_n * \nu \xrightarrow{m} f * \nu$. We have $f_n * \nu \in L^1_{U \cap \Lambda} \subset L^1_E$, and because E is nicely placed we deduce that $f * \nu \in L^1_E$. Finally $\gamma \notin E$ yields $0 = \widehat{f * \nu}(\gamma) = \hat{f}(\gamma)\hat{\nu}(\gamma) = \hat{f}(\gamma)$. □

Before we come to concrete applications let us note the following general consequence.

COROLLARY.

(a) *Let Λ_1 be a nicely placed subset of Γ and Λ_2 a τ-closed subset of Γ. Then $\Lambda_1 \cup \Lambda_2$ is nicely placed. In particular, every τ-closed subset is nicely placed.*

(b) *Let Λ_1 be a τ-closed Riesz subset of Γ and Λ_2 a Shapiro subset of Γ. Then every set Λ such that $\Lambda_1 \subset \Lambda \subset \Lambda_1 \cup \Lambda_2$ is Riesz and nicely placed.*

[PROOF: (a) By Theorem B it is enough to show that for $\gamma \notin \Lambda_1 \cup \Lambda_2$ there is $U \in \mathfrak{U}_\tau(\gamma)$ such that $U \cap (\Lambda_1 \cup \Lambda_2)$ is nicely placed. Clearly $U := (\Gamma \setminus \Lambda_2) \cup \Lambda_1$ has these properties.
(b) We can write $\Lambda = \Lambda_1 \cup \Lambda'$ for some $\Lambda' \subset \Lambda_2$, which by assumption is nicely placed. Hence Λ is nicely placed by part (a). Also, Λ' is a Riesz set by Proposition 4.5(b); [442, Th. 2] shows that Λ is Riesz.]

The following examples come from [258].

EXAMPLES.

(a) Let P be the set of prime numbers including 1. Then $\Lambda = \{-1\} \cup P$ is τ-closed and $-\mathbb{N} \cup P$ is Riesz and nicely placed.
(b) $Q = \{n^2 \mid n \in \mathbb{Z}\}$ is τ-closed and $-\mathbb{N} \cup Q$ is Riesz and nicely placed.
(c) If $A \subset \mathbb{Z}^2$ is such that (1) $\{n \in \mathbb{Z} \mid (n \times \mathbb{Z}) \cap A \neq \emptyset\}$ is bounded from below and (2) $(n \times \mathbb{Z}) \cap A$ is bounded from below or from above for all $n \in \mathbb{Z}$, then A is a Shapiro set.

(d) If B is the set of all $(n,m) \in \mathbb{Z}^2$ which are contained in a plane sector whose opening is less than π, then B is a Shapiro set.

We remark that the set P of prime numbers is not a $\Lambda(1)$-set [447], so part (a) can't be deduced from Theorem A. Bochner proved in 1944 that the sets B in part (d) are Riesz sets (see also [240, Th. VII.3.1] or [542, Th. 8.2.5]). That sets A as in (c) are Riesz sets was shown by Aleksandrov in [5].

PROOF: Observe first that any arithmetical progression $n+k\mathbb{Z}$ is τ-clopen. [For $z = e^{i\frac{2\pi}{k}}$ and $0 < \varepsilon < |z-1|$ we obtain that $U(0,\{z\},\varepsilon) = \{m \in \mathbb{Z} \mid |z^m - 1| < \varepsilon\} = k\mathbb{Z}$ is τ-open. An open subgroup of a topological group is also closed, and the map $m \mapsto m+n$ is a homeomorphism.] For $n \notin \Lambda$ and $|n|$ not prime we get $\Lambda \cap n\mathbb{Z} = \emptyset$. If $|n|$ is prime for $n \notin \Lambda$ then $n\mathbb{Z} \setminus \{-n\}$ is an open neighbourhood not meeting Λ. Hence $\mathbb{Z} \setminus \Lambda$ is τ-open. For the set Q of square numbers one has $(n+3n^2\mathbb{Z}) \cap Q = \emptyset$ if $n < 0$, and $(n+p^{2k+1}\mathbb{Z}) \cap Q = \emptyset$ if $n \geq 0$ and $n \notin Q$ (with p prime and $k \geq 0$ such that $n = p^{2k+1}n'$ and $p \not| n'$). We leave the elementary number theory to the reader.
As in the one-dimensional case one shows that $k\mathbb{Z} \times \mathbb{Z}$ is τ-closed. Thus $\{0\} \times \mathbb{Z}$ and $F \times \mathbb{Z}$ are τ-closed for $F \subset \mathbb{Z}$ finite. Since the assumptions (1) and (2) on the set A are hereditary, it suffices to prove that sets A as in (c) are nicely placed. We will show the weak localization property $(*)$ of Theorem B. Let $\gamma = (n,m) \notin A$ and assume that $(n \times \mathbb{Z}) \cap A$ is bounded below. Then $\{(n,l) \in A \mid l < m\}$ and $F = \{k \in \mathbb{Z} \mid k < n,\ (k \times \mathbb{Z}) \cap A \neq \emptyset\}$ are finite. So $(F \times \mathbb{Z}) \cup \{(n,l) \in A \mid l < m\}$ is τ-closed and its complement U is a τ-neighbourhood of (n,m). We have $U \cap A \subset E := \{(k,l) \mid (k,l) \geq (n,m+1)\} = (n,m+1) + \mathbb{Z}^2_+$ for the lexicographical order on $\mathbb{Z}^2$. By Proposition 4.4(c) and the fact, proved after Theorem A, that $\mathbb{Z}^2_+$ is nicely placed we deduce that E is nicely placed – this is what we wanted to show. Part (d) is most easily shown using Theorem A. We omit the details and refer to [258]. □

In [229] C. Finet generalized not only the results of Section IV.4, but also the techniques for constructing nicely placed and Shapiro sets described above, to the setting of compact groups and compact commutative hypergroups. The latter are compact spaces which have enough structure so that an abstract convolution on $M(K)$ can be defined. An extension to transformation groups was established in [230] and [231].

WEAK* CONTINUITY PROPERTIES OF FUNCTIONALS ON THE DUAL UNIT BALL.
In the first half of this subsection we consider the continuity of individual elements of X^{**} as functions on (B_{X^*}, w^*) in order to put Lemma III.2.14 and Proposition IV.2.1 in a general perspective. The second half relates properties of X with certain subsets of X^{**} defined in terms of w^*-continuity properties. This will give us an alternative approach to property (u) for M-embedded spaces, Theorem III.3.8.
An element $x^{**} \in X^{**}$ has a point of w^*-continuity iff x^{**} belongs to X. It is clearly a different story to ask whether $x^{**}|_K$ has a point of w^*-continuity for a subset $K \subset X^*$. In this case one has to intersect w^*-neighbourhoods in X^* with K, and geometric properties of K may enter to help establishing points of w^*-continuity relative to K. We exclusively consider $K = B_{X^*}$, and all topological concepts refer to the relative w^*-topology on K. By the Krein-Smulian theorem, $x^{**}|_{B_{X^*}}$ has a point of w^*-continuity in the open unit ball iff x^{**} belongs to X. Hence, apart from the trivial case, w^*-continuity can only

occur at points $x^* \in S_{X^*}$. A simple but important sufficient condition for this to happen is that the w^*- and the w-topology agree at x^* (which clearly is the case if and only if *all* $x^{**} \in X^{**}$ are w^*-continuous at x^*). A geometric characterisation of these $x^* \in S_{X^*}$ was given in Lemma III.2.14. That $w^* = w$ at x^* is certainly implied by the stronger condition that $w^* = \| \,.\, \|$ at x^*, which in turn is the case if $x^* \in w^*\text{-sexp}\, B_{X^*}$. Note, however, that $x^* = (0,1) \in \ell^1 \oplus_1 \mathbb{K} = C(\alpha\mathbb{N})^* = c^*$ is strongly exposed in the unit ball, yet all $x^{**} \in c^{**} \setminus c$ are w^*-discontinuous at x^*. In particular, there is no relation between extreme points of B_{X^*} and points of w^*-w-continuity. Nevertheless, by the above observation we have an important isomorphic condition which guarantees many points of w^*-continuity: if X is an Asplund space then $\overline{\text{co}}^{w^*} \mathcal{C}_{w^*,w} = B_{X^*}$, where $\mathcal{C}_{w^*,w}$ denotes the set of points of continuity of $id : (B_{X^*}, w^*) \to (B_{X^*}, w)$. In [252] Godefroy showed that also the assumption that X^{**} is smooth or that the norm of X is Fréchet differentiable on a dense set imply $\overline{\text{co}}^{w^*} \mathcal{C}_{w^*,w} = B_{X^*}$. By Corollary III.2.15 $\mathcal{C}_{w^*,w} = S_{X^*}$ holds for M-embedded spaces X; this can be understood as a maximal continuity relation. In the proof of Corollary III.2.16 we remarked that an $x^* \in \mathcal{C}_{w^*,w}$ belongs to every norming subspace V of X^*. This is easily seen to give that if $\overline{\text{co}}^{w^*} \mathcal{C}_{w^*,w} = B_{X^*}$, then the norm closed linear span of $\mathcal{C}_{w^*,w}$ is the smallest norming subspace of X^*. Hence the existence of a smallest norming subspace of X^*, called property (P) in [252], is necessary for the existence of "sufficiently many" points of w^*-continuity. For the relation of property (P) with strongly unique preduals we refer to [252] and [259, Section I]. A necessary geometric condition for the existence of many w^*-continuity points of a single $x^{**} \in X^{**}$ is provided by [254, Lemme 1.2] and [259, Lemma I.4]: if the points of w^*-continuity of x^{**} separate X then $\bigcap_{x \in X} B_{X^{**}}(x, \|x - x^{**}\|)$ contains at most one point. Of course points of w^*-continuity of x^{**} depend on the norm, yet the following renorming result is quite remarkable [260, Prop. II.1]:

For a separable Banach space X the following assertions hold:

(a) *X^* is separable iff there is an equivalent norm $|\,.\,|$ on X such that all $x^{**} \in X^{**}$ are w^*-continuous at all $x^* \in S_{(X^*, |\,.\,|)}$.*

(b) *X does not contain a copy of ℓ^1 iff for all $x^{**} \in X^{**}$ there is an equivalent norm $|\,.\,|$ on X such that x^{**} is w^*-continuous at all $x^* \in S_{(X^*, |\,.\,|)}$.*

Let us now turn to the points of w^*-discontinuity. Behrends proved in [54] that if $X = A(K)$, the space of affine continuous functions on a compact convex set K, or if X is the range of a contractive projection in X^{**}, then every $x^{**} \in X^{**} \setminus X$ has a point of w^*-discontinuity in S_{X^*}. He introduced the space $\text{cont}(X) = \{x^{**} \in X^{**} \mid x^{**}$ is w^*-continuous at every $x^* \in S_{X^*}\}$, which is a closed subspace of X^{**} containing X, and showed that the centers y_B of pseudoballs B in X (cf. Th. II.1.6) belong to $\text{cont}(X)$, hence spaces with $\text{cont}(X) = X$ don't contain proper pseudoballs and thus can't be proper M-ideals (Th. II.3.10). The equivalence, established in Proposition IV.2.1, between extreme discontinuity of $x^{**} \in X^{**}$ and x^{**} being ℓ^1-orthogonal to X, deserves special comments. Maurey showed in [436] that a separable real Banach space X contains a copy of ℓ^1 iff there is an $x^{**} \in X^{**} \setminus \{0\}$ satisfying $\|x^{**} + x\| = \|x^{**} - x\|$ for all $x \in X$. This was largely simplified and extended to the nonseparable case by Godefroy in the following form: X contains a copy of ℓ^1 iff there is an equivalent norm $|\,.\,|$ on X and an $x^{**} \in X^{**} \setminus \{0\}$ such

that $|x^{**} + x| = |x^{**}| + |x|$ for all $x \in X$, i.e., such that X becomes an L-summand in some subspace of X^{**}, [260, Th. II.4]. Incidentally, these results show that one has to use more than one ℓ^1-orthogonal direction in X^{**} to establish property (V^*) for L-embedded spaces. Note that the Maurey-Godefroy theorem provides an isomorphic characterisation of the existence of ℓ^1-orthogonal directions; and there is even a local one for separable X [262, Lemma 9.1]: there is an $x^{**} \in X^{**} \setminus \{0\}$ such that $\|x^{**} + x\| = \|x^{**}\| + \|x\|$ for all $x \in X$ iff for all finite dimensional subspaces E of X and all $\varepsilon > 0$ there is an $x_{E,\varepsilon} \in X$ with $\|x_{E,\varepsilon}\| = 1$ such that $\|x + x_{E,\varepsilon}\| \geq (1 - \varepsilon)(\|x\| + \|x_{E,\varepsilon}\|)$ for all $x \in E$. Norms satisfying this last condition were called octahedral in [260]. The proof of the quoted result in [262] is quite indirect and uses the so-called ball topology of X. All the above brings up the question: which $x^{**} \in X^{**}$ can be turned into ℓ^1-orthogonal directions by renorming? By Proposition IV.2.1 such an x^{**} is necessarily not of the first Baire class. In [260, p. 12] Godefroy remarked that if $x^{**} \in X^{**} \setminus X$ is such that there exists $Y \subset X$ with $Y \simeq \ell^1$ and $x^{**} \in Y^{\perp\perp}$, then there is an equivalent norm $|\,.\,|$ on X such that x^{**} is ℓ^1-orthogonal to X with respect to this new norm. However, not every ℓ^1-orthogonal element x^{**} is in the w^*-closure of a subspace Y isomorphic to ℓ^1, cf. [260, p. 12] and [627]. E. Werner obtained the following partial answer to the above question [627, Th. 2]: $x^{**} \in X^{**} \setminus \{0\}$ does not satisfy property (4) below iff for all $0 < \alpha < 1$ there is an equivalent norm $|\,.\,|$ on X such that $|x^{**} + x| \geq |x^{**}| + \alpha|x|$ for all $x \in X$.
In the following we want to relate some properties of a Banach space X with certain subsets $S \subset X^{**}$ defined by the continuity behaviour of their elements. The most important subset in this respect is the set functions of the first Baire class. To clarify this concept consider the following properties for a topological space K and $f : K \to \mathbb{R}$:

(1) f is the pointwise limit of a sequence of continuous functions on K.
(2) $f^{-1}(F)$ is a G_δ-subset of K for every closed set F.
(3) For all closed subsets $A \subset K$ the restriction $f_{|A}$ is continuous on a dense subset of A.
(4) For all closed subsets $A \subset K$ the restriction $f_{|A}$ is continuous at some point in A.

Then trivially (1) $\Rightarrow$ (2) and (3) $\Rightarrow$ (4), and by the results of Baire (3) $\Rightarrow$ (2) $\Rightarrow$ (1) if K is metrizable, and (2) $\Rightarrow$ (3) and (4) $\Rightarrow$ (3) if K is completely metrizable ([385, §31] and [295] – see also [92, Prop. 1E] and [429, p. 54f.]). Since we are mainly interested in $K = (B_{X^*}, w^*)$, let us record what is known for compact spaces: (4) $\Rightarrow$ (3) follows by an easy Baire category argument (cf. [444, Th. 3.1]), and (2) $\Rightarrow$ (3) is contained in [92, Lemma 1.C and Cor. 1.D]. That (3) $\not\Rightarrow$ (2) can be seen with $K = (B_{\ell^1(\Gamma)}, w^*)$, Γ uncountable and $f(x) = \sum x_\gamma$, where $x = (x_\gamma)$. (The RNP of $\ell^1(\Gamma)$ helps to show (4) for f.) The proof of (2) $\Rightarrow$ (1) in the metric case uses that closed sets are G_δ-sets, and for quite a while Baire's results were only extended to the setting of functions $f : X \to Y$ with metric X and suitable Y (cf. [288], [528]). Only recently Hansell proved the implication (2) $\Rightarrow$ (1) for normal K and real-valued f [289].
Today it seems to be a common convention – which we will follow – to call functions with property (1) functions of the first Baire class and those with property (2) functions of the first Borel class: [288], [289], [343], [528]. Note, however, that the important article [92] uses a different notation – also the class $B_r^{\S}(K)$ from [92] is in general not the set

of functions satisfying (3). The name *barely continuous* function introduced in [444] for those f fulfilling (4) was not accepted in the literature. For applications in Banach space theory the first Baire class functions seem to be most appropriate – see below. We define for a Banach space X

$$B_1(X) := \left\{ x^{**} \in X^{**} \;\middle|\; \begin{array}{l} \text{there is a sequence } (x_n) \text{ in } X \text{ such that} \\ x_n \xrightarrow{w*} x^{**} \end{array} \right\}$$

$$LWUC(X) := \left\{ x^{**} \in X^{**} \;\middle|\; \begin{array}{l} \text{there is a wuC-series } \sum x_n \text{ in } X \text{ such} \\ \text{that } x^{**} = \sum^* x_n \end{array} \right\}$$

and for a topological space K

$$\textit{Baire-1}(K) := \left\{ F : K \longrightarrow \mathbb{R} \;\middle|\; \begin{array}{l} \text{there is a sequence } (f_n) \text{ in } C(K) \text{ which} \\ \text{converges pointwise to } F \end{array} \right\}$$

$$DBSC(K) := \left\{ F : K \longrightarrow \mathbb{R} \;\middle|\; \begin{array}{l} \text{there is a sequence } (f_n) \text{ in } C(K) \text{ which} \\ \text{converges pointwise to } F \text{ and there is} \\ C > 0 \text{ such that, putting } f_0 = 0, \\ \displaystyle\sum_{n=0}^{\infty} |f_{n+1}(k) - f_n(k)| \le C \text{ for all } k \in K \end{array} \right\}$$

$LWUC(X)$ stands for "limits of weakly unconditionally Cauchy series" and $DBSC(K)$ for "differences of bounded semicontinuous functions". A justification has to be given for our definition of $DBSC$: of course one expects $DBSC$ to be $\{f_1 - f_2 \mid f_1, f_2$ semicontinuous and bounded on $K\}$ and this agrees indeed with the above if K is metrizable (see [87, Chap. IX, § 2, no. 7], [296, p. 3] or [375, p. 220]). We have introduced the above definitions in order to be able to formulate the following theorem succinctly.

THEOREM. *For a Banach space X the following holds:*

(a) $B_1(X) = X^{**} \cap \textit{Baire-1}(B_{X^*})$,
(b) $LWUC(X) = X^{**} \cap DBSC(B_{X^*})$.

[PROOF: Part (a) is well-known ([458, Lemma 1], [158, Basic Lemma, p. 235]) and part (b) is only a disguised form of Pełczyński's result that property (u) is inherited by subspaces ([485]; see also [423, Prop. 1.e.3] and Lemma I.2.9). To see the interesting inclusion "$\supset$" in (b) let (f_n) be a sequence in $C(B_{X^*})$ as in the definition of $DBSC(B_{X^*})$ which converges pointwise to $x^{**}|_{B_{X^*}}$. By part (a) there is a sequence (x_n) in X with $x_n \xrightarrow{w*} x^{**}$. Putting $g_n := f_n - f_{n-1}$, we obtain that the sequence $(\sum_{k=1}^n g_k)_n$ satisfies $\sum_{n=1}^\infty |g_n(x^*)| \le C$ for all $x^* \in B_{X^*}$ and converges pointwise to $x^{**}|_{B_{X^*}}$. So we are in the situation of Lemma I.2.9 with $E = C(B_{X^*})$, $F = X$, and $x_n = g_n$.]
Clearly

$$X \subset LWUC(X) \subset B_1(X) \subset X^{**}.$$

We are now in a position to formulate properties of X in terms of these spaces:

- $X = B_1(X) \iff X$ is weakly sequentially complete

- For separable spaces X: $B_1(X) = X^{**} \iff \ell^1 \not\hookrightarrow X$
- $X = LWUC(X) \iff c_0 \not\hookrightarrow X$
- $LWUC(X) = B_1(X) \iff X$ has property (u)

[The second statement is the Odell-Rosenthal theorem (see e.g. [422, Th. 2.e.7]), the third is a result of Bessaga and Pełczyński (see e.g. [158, Th. 8, p. 45]).] The last statement is only is trivial reformulation of property (u). To obtain a more serious characterisation we can use the equivalent description of *DBSC* for metric spaces mentioned above and recall that property (u) is separably determined.

PROPOSITION. *A Banach space X has property (u) if and only if for every separable subspace Y of X and every $y^{**} \in Y^{**}$ the following implication holds:*

If $y^{**}|_{B_{Y^*}}$ is Baire-1 then it is the difference of two bounded lower semicontinuous functions.

This together with Lemma I.2.8 and Lemma I.2.5 shows again that M-embedded spaces have property (u). It even gives $LWUC(X) = X^{**}$ for a separable M-embedded space X.
If one tries to pursue the idea of relating properties of a space X to subset of X^{**} further, it seems most promising to try to relate the existence of particular discontinuous functions in $B_1(X)$ to the existence of special subspaces of X. This was done by Haydon, Odell, and Rosenthal in [296]; also Rosenthal's recent paper [531] is relevant in this connection. A thorough study of subclasses of Baire-1 functions from a topological point of view can be found in [375]. We remark that it doesn't seem very appropriate to consider bigger sets than $B_1(X)$, e.g. $B_2(X) = \{x^{**} \in X^{**} \mid$ there is a sequence (x_n^{**}) in $B_1(X)$ such that $x_n^{**} \xrightarrow{w*} x^{**}\}$ is in general not norm closed in X^{**}, whereas $B_1(X)$ always is [437], [438]. Also $B_2(X) \neq X^{**} \cap Baire\text{-}2(X)$, cf. [176, p. 79]. However, there are important relations between properties of X and measure theoretic features of (the elements of) X^{**}, notably Haydon's nonseparable extension of the Odell-Rosenthal theorem using universally measurable functions. We refer to [182], [183], and [596] for these questions. In this regard [176] should also be mentioned, although this memoir focuses on the containment of ℓ^1.

u-IDEALS. Besides the more general norm decompositions extending M-ideals and L-summands which we discussed in the Notes and Remarks to Chapter I, there is a recent development which focuses only on the symmetry of the splitting $Y = X \oplus Z$. This concept, which leads to the notions of u-summands and u-ideals, was first introduced in [114] and later worked out in detail by Godefroy, Kalton, and Saphar in [264], see also [263]. Here the letter u stands for "unconditional", which turns out to be justified by the main results of [264]. In the following we report on those central results of the important paper [264] which are pertinent to M- and L-embedded spaces. Aspects relevant to M-ideals of compact operators and various approximation properties will be discussed in Notes and Remarks to Chapter VI.

A subspace X of a Banach space Y is called a *u-summand* if there is a subspace Z of Y such that $Y = X \oplus Z$ and

$$\|x+z\| = \|x-z\| \qquad \text{for all } x \in X,\ z \in Z .$$

The associated u-projection P is characterised by $\|Id - 2P\| = 1$. More precisely, X is a u-summand in Y iff X is the range of a u-projection iff X is the kernel of a u-projection. Since there is at most one u-projection onto a subspace X [264, Lemma 3.1], the complementary space Z is uniquely determined. Adopting another point of view we mention that u-projections P are in one-to-one correspondence with the isometric symmetries (involutions) S of the space Y, i.e. $S^2 = Id$, via $S = Id - 2P$. For example, this allows a complete description of the u-projections on a $C(K)$-space by the Banach-Stone theorem (cf. p. 341): An isometric symmetry S on $C(K)$ has the form $Sf = \theta(f \circ \varphi)$ with $\theta : K \to S_{\mathbb{K}}$ and $\varphi : K \to K$ continuous and satisfying $\theta^2 = 1_K$ and $\varphi^2 = Id_K$. Taking e.g. $K = [-1,1]$ and $\varphi(x) = -x$ one obtains the u-summands of even and odd continuous functions in $C([0,1])$.
The extension of the concept of a u-summand to that of a *u-ideal* proceeds in the obvious way: a subspace X of a Banach space Y is called a u-ideal if $X^\perp$ is a u-summand in Y^*. Since $X^\perp$ is then in particular the kernel of a contractive projection, there is a Hahn-Banach extension operator $L : X^* \to Y^*$ (see p. 135) which in turn induces a mapping $T : Y \to X^{**}$ with $Tx = x$ for $x \in X$. In this way the $\sigma(X^{**}, X^*)$-topology (in connection with the norm condition) is opened up as a tool for the study of u-ideals. If X is a u-ideal in Y, say $Y^* = V \oplus X^\perp$, the topological properties of V can be used to distinguish u-summands and "extreme" u-ideals. Namely, a u-ideal X in Y is a u-summand if and only if V is w^*-closed (compare the text preceding Proposition I.1.2), and a u-ideal X in Y is called *strict* if V is norming. Since this is equivalent to the condition $r(V, X^*) = 1$ (compare Definition II.3.7), strict u-ideals can be regarded as counterparts of the extreme M-ideals introduced in Definition II.3.7(b).
The main result on general u-ideals asserts that a u-ideal which is not a u-summand contains a subspace isomorphic to c_0 [264, Theorem 3.5]; this is an extension of Theorem II.4.7. Its proof uses a fundamental lemma from the theory of u-ideals which says in abridged form: if X is a u-ideal in Y, $T : Y \to X^{**}$ the operator defined above, and Ty is the $\sigma(X^{**}, X^*)$-limit of a sequence in X, then Ty is already the $\sigma(X^{**}, X^*)$-limit of a wuC-series $\sum x_k$ in X satisfying in addition $\|\sum_{k=1}^n \varepsilon_k x_k\| \le \|y\| + \varepsilon$ for all $n \in \mathbb{N}$ and $|\varepsilon_k| = 1$. A more elaborate version of this lemma together with the equivalence "X doesn't contain c_0 iff every wuC-series in X is unconditionally convergent" shows that if X is a u-ideal in Y and X contains no copy of c_0, then $T(Y) \subset X$, which easily gives the w^*-continuity of the u-projection in Y^*.
A much richer theory evolves in the special case when X is a u-ideal in X^{**}. Examples of this situation are M- and L-embedded spaces (the latter even being u-summands in X^{**}) and spaces for which the natural projection π_{X^*} in X^{***} with kernel $X^\perp$ is a u-projection (in which case X is a called a strict u-ideal). However the most remarkable class of examples are the order continuous Banach lattices. In this setting the extreme cases are easily characterised: an order continuous Banach lattice X is a u-summand in X^{**} iff X doesn't contain c_0, and X is a strict u-ideal in X^{**} iff X doesn't contain ℓ^1. It can be deduced from [254, Prop. 9] that X is weakly sequentially complete provided X is a u-summand in X^{**}.

By the above it only makes sense to compare M-embedded spaces with spaces which are strict u-ideals in their biduals. Clearly, they are easy to distinguish isometrically. The space c_0 with the norm $|x| = \|x\|_\infty + \|(x_n/2^n)\|_2$ is a strict u-ideal in its bidual which is not proximinal, and $X = c_0 \oplus_1 \mathbb{K}$ is one which lacks unique Hahn-Banach extension to X^{**}; thus they are not M-embedded. More interesting is that $\ell^2(c_0)$ is a space which is a strict u-ideal in its bidual and which can't be renormed to become an M-embedded space. The latter follows from

PROPOSITION [264, Prop. 4.4] *If a separable M-embedded space has a boundedly complete Schauder decomposition $\oplus \sum E_n$, then all but finitely many E_n are reflexive.*

A crucial distinction between M- and L-embedded spaces is that in the first case we deal with the natural projection π_{X^*}. In certain cases this carries over to u-ideals: if X is separable or doesn't contain ℓ^1, then X is a strict u-ideal in X^{**} iff π_{X^*} is a u-projection. In this case every subspace and every quotient of X is a strict u-ideal in its bidual, every isometric isomorphism of X^{**} is the bitranspose of one in X, X is Asplund and X is the only predual of X^* which is a strict u-ideal in its bidual ([264, Prop. 5.2, Prop. 2.8, Theorem 5.7], for the case of M-ideals compare Th. III.1.6, Prop. III.2.2, Th. III.3.1, Prop. IV.1.9). The main result on strict u-ideals is a characterisation in terms of a quantitative version of Pełczyński's property (u), introduced in Section III.3. Denote by $\varkappa_u(X)$ the least constant $C > 0$ satisfying: for all x^{**} in $B_1(X)$ (i.e., there is a sequence (x_n) in X with w^*-$\lim x_n = x^{**}$) there is a wuC-series $\sum y_k$ in X such that $x^{**} = \sum^* y_k$ and $\|\sum_{k=1}^n \theta_k y_k\| \le C\|x^{**}\|$ for all $n \in \mathbb{N}$ and all $|\theta_k| = 1$. [If complex scalars θ_k are allowed, we write $\varkappa_h(X)$.] Note that with this notation X has property (u) iff $\varkappa_u(X) < \infty$, and the proof of Theorem III.3.8 actually yields $\varkappa_u(X) = 1$ for M-embedded spaces. The following theorem provides in particular another approach to that result.

THEOREM [264, Theorem 5.4] *A Banach space X containing no copy of ℓ^1 is a strict u-ideal in its bidual iff $\varkappa_u(X) = 1$.*

There is a kind of quantitative version of this result which singles out the strict u-ideals among the u-ideals. It is shown in [264, Theorem 7.4] that for a Banach space X which is a u-ideal in its bidual and contains no copy of ℓ^1 the following conditions are equivalent: (1) X is a strict (u)-ideal in X^{**}, (2) $\varkappa_u(X) < 2$, (3) X^* contains no proper norming subspace. This can be used in the study of isomorphic (!) and isometric preduals. As an example we mention a special case of [264, Theorem 7.7]: If X is a separable M-embedded space enjoying the MAP and Y is a space which is a strict u-ideal in Y^{**} such that $d(X^*, Y^*) < 2$, then X is isomorphic to Y. We remark that the proof of this result relies on Theorem II.2.1.

For complex scalars there is a very natural subclass of u-projections and u-ideals. Let us note that a projection P on a complex Banach space Y is hermitian in the sense of p. 17 iff $\|Id_Y - (1+\lambda)P\| = 1$ whenever $|\lambda| = 1$. [This follows easily from the power series representation of e^{itP}.] This norm condition can be regarded as the natural complex analogue of the condition $\|Id - 2P\| = 1$ defining u-projections. Now, hermitian projections are termed *h-projections*, and *h-summands*, *h-ideals*, and *strict h-ideals* are defined in the obvious way. To see that M- and L-summands in complex spaces are h-summands we note that a decomposition $Y = X \oplus Z$ gives rise to h-summands if and only if

$$\|x + z\| = \|x + \lambda z\| \qquad \text{for all } x \in X,\ z \in Z,\ |\lambda| = 1.$$

The hermitian operators on $Y = C_{\mathbb{C}}(K)$ are multiplications by real valued functions [86, p. 91], so in this case h-summands and M-summands coincide. However, for noncommutative C^*-algebras $\mathfrak{A}$ they can differ since by a result of Sinclair [86, p. 95] a hermitian operator T on $\mathfrak{A}$ is of the form $Tx = hx + Dx$, where $h^* = h$ and D is a star derivation, i.e. $D(xy) = (Dx)y + xDy$ and $Dx^* = -(Dx)^*$. So taking e.g. $\mathfrak{A} = L(H)$, $D = 0$, and h a selfadjoint projection on H, one obtains nontrivial u-summands in $L(H)$; but $L(H)$ fails to have nontrivial M-summands as will be shown in Corollary VI.1.12.
Exploiting the theory of hermitian operators Godefroy, Kalton, and Saphar obtain even more satisfactory results for complex scalars than in the u-ideal setting. For example it can be shown that if X is an h-ideal in X^{**} then the induced mapping $T : X^{**} \to X^{**}$ satisfies $Tx^{**} = x^{**}$ not only for $x^{**} \in X$, but also for $x^{**} \in B_1(X) := \{x^{**} \in X^{**} \mid$ there is a sequence (x_n) in X such that $w^*\text{-}\lim x_n = x^{**}\}$. This gives

THEOREM [264, Theorem 6.4] *If X is an h-ideal in X^{**}, then $\varkappa_h(X) = 1$.*

Note that there is no strictness assumption in the above. The mapping T is not only the identity on $B_1(X)$, but a projection onto this space. Even more is true. A separable Banach space X is an h-ideal in X^{**} iff there is a hermitian projection T of X^{**} onto $B_1(X)$ such that for every $x^{**} \in S_{X^{**}}$ there is a net (x_d) in X with $w^*\text{-}\lim x_d = x^{**}$ and $\limsup \|x^{**} - (1+\lambda)x_d\| \le 1$ for $|\lambda| = 1$ [264, Theorem 6.5]. Without this extra norm condition the result is false; one can even show that there is a separable Banach space Y such that $\varkappa_h(Y) = 1$ and $B_1(Y)$ is complemented in Y^{**} by a hermitian projection, yet Y is not an h-ideal in Y^{**}; in fact one can take $Y = K(c_0 \oplus_1 \mathbb{C})$. The characterisation of strict h-ideals among h-ideals is particularly pleasing. It is shown in [264, Theorem 6.6] that the following conditions are equivalent for a Banach space X which is an h-ideal in X^{**}: (1) X is a strict h-ideal in X^{**}, (2) X^* is an h-ideal in X^{***}, (3) X is an Asplund space, (4) X contains no copy of ℓ^1.
We close this summary by mentioning some situations where h-ideals coincide with M- or L-embedded spaces. For example, if X is a separable (complex) M-embedded space, then a subspace Z of X^* is an h-ideal in Z^{**} iff Z is L-embedded (cf. Prop. IV.1.10). The same conclusion obtains for subspaces Z of $L^1[0,1]$ (see [264, Cor. 6.10, Cor. 6.13]). There is also a dual result stating that, for a separable (complex) Banach space X such that X^* is L-embedded, a quotient Y of X is M-embedded iff Y is a strict h-ideal in Y^{**}.

CHAPTER V
M-ideals in Banach algebras

V.1 Preliminary results

This chapter deals with M-ideals in Banach algebras. Quite surprisingly, their purely geometric definition forces M-ideals to get into contact with the underlying algebraic structure. To motivate the main idea behind the following, let us briefly come back to one of the very first examples in this book:
Take $\mathfrak{A} = C(K)$, the Banach algebra of continuous functions on the compact space K, and recall (Example I.1.4(a)) that here the M-ideals correspond precisely to the (closed) ideals, i.e. $J \subset C(K)$ is an M-ideal if and only if for some closed set $D \subset K$,

$$J = J_D := \{f \in C(K) \mid f|_D = 0\}.$$

Let us follow J_D's way through the higher duals of $C(K)$ just a little bit further: We have $C(K)^{**} = C(K'')$ for some suitable compact K'' and

$$C(K'') = J_D^{\perp\perp} \oplus_\infty J_2,$$

where the latter can only happen if for some clopen set $\widehat{D} \subset K''$

$$J_D^{\perp\perp} = J_{\widehat{D}} = \chi_{\widehat{D}} C(K'').$$

Hence, for each M-ideal J in $C(K)$ there is an idempotent element $p \in C(K'')$ such that

$$J^{\perp\perp} = p\, C(K)^{**}.$$

It is surely rather tempting to try to prove that – at least in some special cases – the algebraic and the geometric structure of a Banach algebra are linked in a similar way. And, as a matter of fact, there is such a connection: M-ideals are always subalgebras, in the case of the more "classical" Banach algebras, they are mostly generated by idempotents in the bidual or, at least, they are (algebraic) ideals.

Before we actually start to formulate and prove the central results of this chapter, we have to recall some definitions and a little more: To be able to pursue the above idea in the context of an arbitrary Banach algebra we need a product on the bidual $\mathfrak{A}^{**}$ of a Banach algebra. Fortunately, there is such a thing. What will make things a little bit more uncomfortable is the fact that one can define two products, both in quite a natural way, which are in general different from each other. For mnemonic reasons, we take the freedom of sometimes writing mappings on the right, when we now define these products.

Definition 1.1 *Let $\mathfrak{A}$ be a Banach algebra. Let $a, b \in \mathfrak{A}$, $f, g \in \mathfrak{A}^*$ and $F, G \in \mathfrak{A}^{**}$. We then put*

$$\begin{array}{rclrcl} fa(b) & := & f(ab) & (b)af & := & (ba)f \\ Ff(a) & := & F(fa) & (a)fF & := & (af)F \\ FG(f) & := & F(Gf) & (f)F.G & := & (fF)G \end{array}$$

The product FG is called the first, $F.G$ the second Arens multiplication.

When both products coincide, $\mathfrak{A}$ is called *Arens regular*. Examples of Arens regular Banach algebras are C^*-algebras as well as function algebras, whereas mere commutativity of $\mathfrak{A}$ generally does not suffice (see e.g. [134]). We also refer to the survey [177] for more information along these lines.

Let us agree upon some further notation:
Those operators on $\mathfrak{A}$ that multiply with a given element a from the left resp. right are denoted by L_a and R_a, respectively. When this distinction is not necessary, i.e. if a commutes with all elements in $\mathfrak{A}$, then we will also write M_a. To emphasize which Arens product in $\mathfrak{A}^{**}$ is under consideration we shall use the notation L_F^i resp. R_F^i to denote left or right multiplication with respect to the i-th Arens multiplication.
In the following theorem we collect some results in connection with the Banach algebra $\mathfrak{A}^{**}$. In the proofs of all of them nothing but Definition 1.1 is involved. Nevertheless, some of them are somewhat cumbersome to show.

Theorem 1.2 *Let $\mathfrak{A}$ be a Banach algebra.*

(a) *Whenever $\mathfrak{J}$ is a left (right) ideal in $\mathfrak{A}$, the same is true for $\mathfrak{J}^{\perp\perp}$ in $\mathfrak{A}^{**}$, independent of the Arens multiplication under consideration.*

(b) *If $H : \mathfrak{A} \to \mathfrak{B}$ is a homomorphism, then so is $H^{**} : \mathfrak{A}^{**} \to \mathfrak{B}^{**}$ if $\mathfrak{A}^{**}$ and $\mathfrak{B}^{**}$ are furnished with the same Arens product. Similarly, if $H : \mathfrak{A} \to \mathfrak{B}$ is an antihomomorphism (i.e. $H(ab) = H(b)H(a)$ for all $a, b \in \mathfrak{A}$), then H^{**} has the same property whenever $\mathfrak{A}^{**}$ and $\mathfrak{B}^{**}$ are provided with different Arens products.*

(c) *For every subalgebra $\mathfrak{B}$, both Arens products on $\mathfrak{B}^{**} = \mathfrak{B}^{\perp\perp}$ agree with the ones inherited from $\mathfrak{A}^{**}$. In particular, $\mathfrak{B}^{\perp\perp}$ is a subalgebra of $\mathfrak{A}^{**}$.*

(d) *For all $a \in \mathfrak{A}$ and $F \in \mathfrak{A}^{**}$ we have*

$$i_{\mathfrak{A}}(a)F = i_{\mathfrak{A}}(a).F = L_a^{**}(F)$$

and

$$Fi_{\mathfrak{A}}(a) = F.i_{\mathfrak{A}}(a) = R_a^{**}(F).$$

(e) *For fixed $F \in \mathfrak{A}^{**}$, the mappings R_F^1 and L_F^2 are weak* continuous.*

Note that according to (d), the mappings R_a^1, R_a^2, L_a^1 and L_a^2 are weak* continuous for each $a \in \mathfrak{A}$.
We will call a Banach algebra *unital* if it contains a two-sided unit which we usually denote by e. A *left λ-approximate unit* for $\mathfrak{A}$ is a net (p_α) such that $\|p_\alpha\| \leq \lambda$ and

$$\lim_\alpha p_\alpha a = a$$

for all $a \in \mathfrak{A}$. *Right approximate units* are defined in a similar manner. The connection between the concepts of unit and approximate unit is given by the next theorem.

Theorem 1.3 *Let $\mathfrak{A}$ be a Banach algebra.*

(a) *Suppose that $\mathfrak{A}$ has a bounded right approximate unit (q_α). Then each weak* cluster point of (q_α) is a right unit for $\mathfrak{A}^{**}$ with respect to the first Arens product. The analogous result is valid for left approximate units, when we substitute the first by the second Arens product.*
(b) *If $\mathfrak{A}^{**}$ has a right unit q with respect to either of the Arens multiplications then there exists a right approximate unit (q_α) in $\mathfrak{A}$ with*

$$\lim_\alpha \|q_\alpha\| = \|q\|.$$

Using Theorem 1.2(e) the proof of (a) is straightforward. It can also be found in [85], where a proof of (b) is given as well.
For reasons which will become clear later on, we would like to indicate an independent proof, which makes use of the following variant of the principle of local reflexivity. Compared to the more classical variants, it admits the additional degree of freedom (d).

Theorem 1.4 *Let X be a Banach space, $F \subset X^{**}$, $G \subset X^*$, $H \subset L(X)$ finite dimensional subspaces and put*

$$F_H := \operatorname{lin}\{h^{**}x^{**} \mid h \in H, x^{**} \in F\} + F.$$

Then for each $\varepsilon > 0$ there is an operator $T : F_H \to X$ with

(a) $T|_{F \cap X} = Id$
(b) *For all $g \in G$ and $f \in F$ we have $g(Tf) = f(g)$.*
(c) *For all $x \in F_H$,*
$$\|x\| \leq \|Tx\| \leq (1+\varepsilon)\|x\|.$$
(d) *For all $h \in H$,*
$$\|(hT - Th^{**})|_F\| < \varepsilon\|h\|.$$

PROOF: [58] □

Let us show how this result can be used to prove Theorem 1.3(b): To this end order the set

$$\mathsf{A} = \{(F,G,H,\varepsilon) \mid q \in F \subset \mathfrak{A}^{**}, G \subset \mathfrak{A}^*, H \subset \mathfrak{A}, \varepsilon > 0\}$$

where F, G and H are finite dimensional subspaces, by

$$(F,G,H,\varepsilon) \prec (\widetilde{F},\widetilde{G},\widetilde{H},\widetilde{\varepsilon}) \quad \Longleftrightarrow \quad F \subset \widetilde{F},\ G \subset \widetilde{G},\ H \subset \widetilde{H},\ \varepsilon > \widetilde{\varepsilon}$$

and identify a space $H \subset \mathfrak{A}$ with the operator space $\{L_h \mid h \in H\}$. We next choose for every $\alpha \in \mathsf{A}$ an operator T_α that satisfies the conditions (a) – (d) listed in Theorem 1.4 for the subspaces and the number ε determined by α. Let

$$q_\alpha := T_\alpha(q).$$

We have by condition (d) and since $L_x^{**} = L_{i_{\mathfrak{A}}(x)}$ for either of the Arens multiplications (Theorem 1.2(d)) that for every $a \in \mathfrak{A}$

$$\lim_\alpha \|aq_\alpha - T_\alpha(aq)\| = \lim_\alpha \|L_a T_\alpha(q) - T_\alpha L_a^{**}(q)\| = 0$$

By property (a) of Theorem 1.4 and the above,

$$\lim_\alpha \|aq_\alpha - a\| = \lim_\alpha \|aq_\alpha - T_\alpha(aq)\| = 0$$

for all $a \in \mathfrak{A}$, and accordingly, q_α is a right approximate unit. Finally, property (c) of Theorem 1.4 gives

$$\lim_\alpha \|q_\alpha\| = \|q\|.$$

Note that by the above, a right (left) unit with respect to the second (first) Arens product is a right (left) unit for the first (second) one as well. (□)

We will use Theorem 1.4 in much the same way in Section V.3.

Indispensable for our purposes is the theory of hermitian elements in an arbitrary Banach algebra as developed in [84] and [86]. Let us recall, for the sake of easy reference, some definitions and fundamental results.
The first concept we will use several times is the *state space* of a unital Banach algebra $\mathfrak{A}$. It is defined by

$$\mathsf{S}_{\mathfrak{A}} = \{\varphi \in \mathfrak{A}^* \mid \varphi(e) = 1 = \|\varphi\|\}.$$

We denote by

$$v(a, \mathfrak{A}) = \{\varphi(a) \mid \varphi \in \mathsf{S}_{\mathfrak{A}}\}$$

the *numerical range* of an element $a \in \mathfrak{A}$. An element a is called *hermitian* if and only if

$$v(a, \mathfrak{A}) \subset \mathbb{R}.$$

This is of course an interesting definition only when $\mathfrak{A}$ is a complex vector space. In general, we will omit the reference to the algebra in question and write $v(a)$ instead of $v(a, \mathfrak{A})$. Finally, we denote by $\mathbb{H}(\mathfrak{A})$ the (real) subspace of hermitian elements of $\mathfrak{A}$.

We close this introductory section with a small sample of the beautiful results which exist in connection with numerical ranges. They will turn out to be important cornerstones for our reasoning in the following section.

Theorem 1.5 *Let $\mathfrak{A}$ be a unital Banach algebra. Then, for each element $\psi \in \mathfrak{A}^*$, there are $\lambda_1, \ldots, \lambda_4 \geq 0$ and $\psi_1, \ldots, \psi_4 \in \mathsf{S}_{\mathfrak{A}}$ with $\lambda_1 + \lambda_2 + \lambda_3 + \lambda_4 \leq \sqrt{2}\mathrm{e}\|\psi\|$ such that*

$$\psi = \sum_{n=1}^{4} i^n \lambda_n \psi_n.$$

PROOF: [86, p. 100] □

Theorem 1.6 *Let $h \in B_{\mathfrak{A}}$ be hermitian and suppose that $\varphi \in S_{\mathfrak{A}}$ satisfies*

$$1 = \varphi(h) = \|h\|.$$

Then φ is multiplicative on the algebra generated by the unit e and h.

PROOF: [86, p. 77] □

Proposition 1.7 *An element h of a C^*-algebra is self-adjoint if and only if it is hermitian.*

PROOF: [84, p. 47] □

The following result, known as the Vidav-Palmer theorem, is an important means to distinguish C^*-algebras from general Banach algebras.

Theorem 1.8 *A unital Banach algebra $\mathfrak{A}$ is a C^*-algebra if and only if it is generated by its hermitian elements, that is if $\mathfrak{A} = \mathbb{H}(\mathfrak{A}) + i\mathbb{H}(\mathfrak{A})$. In this case the involution is given by $(h_1, h_2 \in \mathbb{H}(\mathfrak{A}))$*

$$(h_1 + ih_2)^* = h_1 - ih_2.$$

PROOF: [84, §6] □

As an application of the Vidav-Palmer theorem we would like to prove that the bidual of a C^*-algebra is a C^*-algebra. This fact will be used in Section V.4. We first need a lemma.

Lemma 1.9 *If $\mathfrak{A}$ is a unital Banach algebra, then for each $F \in \mathfrak{A}^{**}$ we have*

$$v(F, \mathfrak{A}^{**}) = \overline{\{F(\varphi) | \varphi \in S_{\mathfrak{A}}\}}.$$

PROOF: Theorem 12.2 in [84]. For a different approach see Theorem 3.4 and Corollary 5.8 of [23]. □

Theorem 1.10 *Let $\mathfrak{A}$ be a C^*-algebra.*

(a) *Then $\mathfrak{A}$ is Arens regular.*
(b) *The bidual algebra is a C^*-algebra as well.*
(c) *On $\mathfrak{A}^{**}$ multiplication from the left and the right and the involution are weak* continuous.*

PROOF: We put $f^*(x) := \overline{f(x^*)}$ and $F^*(f) := \overline{F(f^*)}$ for $f \in \mathfrak{A}^*$, $F \in \mathfrak{A}^{**}$. This defines an isometric linear involution in the sense of [85, p. 63] on $\mathfrak{A}^{**}$. Also, by definition of the Arens products we have

$$(FG)^* = G^*.F^* \qquad (*)$$

for $F, G \in \mathfrak{A}^{**}$.
Let us now suppose in addition that $\mathfrak{A}$, and hence $\mathfrak{A}^{**}$, is unital. We first claim:

$$F = F^* \quad \Longrightarrow \quad F \text{ hermitian}.$$

To show this, let $\varphi \in S_{\mathfrak{A}}$ so that $\varphi = \varphi^*$ (here Proposition 1.7 enters). It follows that

$$F(\varphi) = F^*(\varphi) = \overline{F(\varphi^*)} = \overline{F(\varphi)},$$

whence $F(\varphi) \in \mathbb{R}$. An appeal to Lemma 1.8 finishes the proof of our claim.
Now we can deduce that each $F \in \mathfrak{A}^{**}$ has a representation

$$F = \frac{F+F^*}{2} + i\frac{F-F^*}{2i} \in \mathbb{H}(\mathfrak{A}^{**}) + i\,\mathbb{H}(\mathfrak{A}^{**}).$$

Consequently, $\mathfrak{A}^{**}$ is a C^*-algebra for the above involution by the Vidav-Palmer theorem, no matter which Arens product on $\mathfrak{A}^{**}$ is chosen. But with this information at hand we derive from $(*)$ and the identity

$$(FG)^* = G^*F^*,$$

which is of course valid in any C^*-algebra, that in fact the two Arens products coincide. Thus we have shown both (a) and (b) for unital C^*-algebras.
In the nonunital case it is a simple matter to adjoin an identity to $\mathfrak{A}$ (cf. [85, p. 67]) so that $\mathfrak{A}$ embeds isometrically as a self-adjoint subalgebra into a unital C^*-algebra $\mathfrak{B}$ ($= \mathfrak{A} \oplus \mathbb{C}$ with a suitable norm). By what we already know, $\mathfrak{A}^{**}$ embeds isometrically as a self-adjoint subalgebra into the unital C^*-algebra $\mathfrak{B}^{**}$, hence (a) and (b) in the general case, too.
(c) follows now from Theorem 1.2 and the definition of the involution. □

Using representation theory for C^*-algebras, one may identify the bidual with an algebra of operators on some Hilbert space, the so-called universally enveloping von Neumann algebra. We refer to [593, p. 122] for this view of Theorem 1.10(b).

V.2 The general case of unital algebras

In this section we will follow the lane that we indicated at the beginning of the last section. As it should be expected, results won't be as strong in general as in the case of the algebra $C(K)$.
Let us retain for the moment the idea from the introduction to the previous section and suppose that a given M-ideal $\mathfrak{J}$ in a unital Banach algebra has the property that the associated M-projection P in the bidual operates by multiplication with a certain idempotent element p. Then, of course, we could gain information on p by looking at the element $P(e)$. In general, we would then have to decide whether multiplication with this element from an appropriate side and with the help of one of the Arens products gives rise to an M-projection. In this situation, Proposition I.1.2 will be helpful. We will hence start our investigation of M-ideals in Banach algebras with an examination of the properties of the mapping

$$\Delta_{\mathfrak{A}} : Z(\mathfrak{A}) \to \mathfrak{A}, \quad T \mapsto T(e).$$

(Here $Z(\mathfrak{A})$ denotes the centralizer of $\mathfrak{A}$ which was introduced in Section I.3.) We will occasionally omit the index and simply write Δ, when this is unlikely to cause any confusion.

Although there are Banach algebras with a trivial centralizer, yet containing a nontrivial M-ideal ($L(H)$ is an example of this, see Corollary VI.1.13 and Example I.1.4(d)), we may, in view of a general M-ideal structure theory of Banach algebras $\mathfrak{A}$, restrict our attention to studying the centralizer. This is due to the fact that M-ideals in $\mathfrak{A}$ correspond to certain M-projections in $Z(\mathfrak{A}^{**})$. In fact, a satisfactory characterisation of this correspondence is the subject of part (c) in the following result. (For part (a) recall the definition of $Z_{\mathbb{R}}(\mathfrak{A})$ from I.3.7.)

Theorem 2.1 *Let $\mathfrak{A}$ be a unital Banach algebra and denote by $\Delta_{\mathfrak{A}} : Z(\mathfrak{A}) \longrightarrow \mathfrak{A}$ the mapping $T \longmapsto T(e)$. Then the following assertions hold.*

(a) *For all $T \in Z(\mathfrak{A})$ the inclusion $v(\Delta(T)) \subset v(T)$ holds. It follows in particular that the range of $Z_{\mathbb{R}}(\mathfrak{A})$ under Δ is contained in the set of hermitian elements of $\mathfrak{A}$.*

(b) *$\Delta(P)$ is a projection for each M-projection $P \in Z(\mathfrak{A})$.*

(c) *An M-projection P on $\mathfrak{A}^{**}$ is associated with an M-ideal $\mathfrak{J}$ in $\mathfrak{A}$ (i.e. $P(\mathfrak{A}^{**}) = \mathfrak{J}^{\perp\perp}$) if and only if $\Delta_{\mathfrak{A}^{**}}(P)|_{\mathsf{S}_{\mathfrak{A}}}$ is lower semicontinuous.*

(d) *Δ is multiplicative.*

(e) *Δ is an isometry.*

(f) *Δ preserves spectral radii and spectra, relative to $Z(\mathfrak{A})$ and $\Delta(Z(\mathfrak{A}))$.*

For the proof we need the following lemma. To understand what is going on, recall the definition of a split face from Example I.1.4(c).

Lemma 2.2 *Suppose $\mathfrak{J}$ is an M-ideal in $\mathfrak{A}$, $\mathfrak{A}^* = \mathfrak{J}^{\perp} \oplus_1 \mathfrak{J}_1$. Then*

$$F_0 := \mathsf{S}_{\mathfrak{A}} \cap \mathfrak{J}^{\perp} \quad \textit{and} \quad F_1 := \mathsf{S}_{\mathfrak{A}} \cap \mathfrak{J}_1$$

are complementary split faces of $\mathsf{S}_{\mathfrak{A}}$. Furthermore,

$$\operatorname{lin} F_0 = \mathfrak{J}^{\perp} \quad \textit{and} \quad \operatorname{lin} F_1 = \mathfrak{J}_1.$$

PROOF: Let $\varphi \in \mathsf{S}_{\mathfrak{A}}$ and denote by Q the L-projection onto $\mathfrak{J}^{\perp}$. Suppose that both $\|Q\varphi\|$ and $\|(Id - Q)\varphi\|$ are strictly positive and write $\varphi = \|Q\varphi\|\varphi_0 + \|(Id - Q)\varphi\|\varphi_1$ where $\varphi_0 = \|Q\varphi\|^{-1}Q\varphi \in \mathfrak{J}^{\perp}$ and $\varphi_1 = \|(Id - Q)\varphi\|^{-1}(Id - Q)\varphi \in \mathfrak{J}_1$. Evaluation at e then gives

$$1 = \|Q\varphi\|\varphi_0(e) + \|(Id - Q)\varphi\|\varphi_1(e)$$

which is only possible when $1 = \varphi_0(e) = \varphi_1(e)$. Hence, φ_0, φ_1 are elements of $\mathsf{S}_{\mathfrak{A}}$ and thus

$$\mathsf{S}_{\mathfrak{A}} = \operatorname{co}(F_0 \cup F_1).$$

Uniqueness of the above representation is clear since $\mathfrak{A}^* = \mathfrak{J}^{\perp} \oplus_1 \mathfrak{J}_1$. It remains to show that F_0 and F_1 are faces. By symmetry, we may restrict ourselves to F_0. Suppose that $\psi \in F_0$ is a convex combination in $\mathsf{S}_{\mathfrak{A}}$, $\psi = \lambda\psi_1 + (1 - \lambda)\psi_2$. Then

$$\psi = \lambda Q\psi_1 + (1 - \lambda)Q\psi_2,$$

and we find $\|Q\psi_1\| = \|Q\psi_2\| = 1$. This gives $\|(Id - Q)\psi_1\| = \|(Id - Q)\psi_2\| = 0$ and consequently, $\psi_1, \psi_2 \in F_0$, i.e. F_0 is a face. The last part of the lemma follows directly from Theorem 1.5 and the fact that $S_{\mathfrak{A}} = \text{co}\,(F_0 \cup F_1)$. □

PROOF OF THEOREM 2.1:

(a) This follows from the fact that for each $\psi \in S_{\mathfrak{A}}$ the functional $T \mapsto \psi(T(e))$ belongs to $S_{L(\mathfrak{A})}$. The statement about $T \in Z_{\mathbf{R}}(\mathfrak{A})$ is a consequence of this observation and the fact that all operators in $Z_{\mathbf{R}}(\mathfrak{A})$ are hermitian (Lemma I.3.8).

(b) We retain the notation of Lemma 2.2 and put furthermore $z = P(e)$. By Lemma 2.2, z has norm one and is hermitian as a consequence of (a) (note that an M-projection belongs to $Z_{\mathbf{R}}(\mathfrak{A})$). Consequently, Theorem 1.6 applies, and each $\varphi \in F_1$ must be multiplicative on the algebra generated by e and z so that

$$\varphi(z^2) = \varphi(z)^2 = 1 = \varphi(z).$$

In a similar way, we find for $\psi \in F_0$

$$\psi\left((e - z)^2\right) = \psi(e - z)^2 = 1,$$

which implies $\psi(z^2) = \psi(z) = 0$. Now the above lemma applies and shows that z and z^2 coincide on $S_{\mathfrak{A}}$, a property that extends to all of $\mathfrak{A}^*$ by Theorem 1.5, and we are done.

(c) Observe first that by part (a)

$$z(S_{\mathfrak{A}}) = v(z) \subset v(P) = [0, 1].$$

Hence, the use of the expression "lower semicontinuous" makes sense. To prove the result, suppose that P is associated with some M-ideal $\mathfrak{J}$ in $\mathfrak{A}$, and let (s_α) be a net contained in the set $(r \geq 0)$

$$E_r := \{s \in S_{\mathfrak{A}} \mid z(s) \leq r\}$$

which we must prove to be closed. Suppose (s_α) has the weak* limit $s \in S_{\mathfrak{A}}$. We let F_0 and F_1 as in Lemma 2.2 and write

$$s_\alpha = (1 - t_\alpha) f_\alpha^{(0)} + t_\alpha f_\alpha^{(1)},$$

where $f_\alpha^{(i)} \in F_i$ and $t_\alpha \in [0, 1]$. Making the assumptions that

$$f_\alpha^{(0)} \xrightarrow{w^*} f^{(0)} \in F_0, \qquad s_\alpha \xrightarrow{w^*} s \in S_{\mathfrak{A}}, \qquad t_\alpha \longrightarrow t \in [0, 1]$$

and that, furthermore,

$$f_\alpha^{(1)} \xrightarrow{w^*} f^{(1)} = \tau\varphi_0 + (1 - \tau)\varphi_1,$$

where $\varphi_i \in F_i$ (note that everything is compact here), we obtain

$$s = (1 - t) f^{(0)} + t\,(\tau\varphi_0 + (1 - \tau)\varphi_1).$$

This gives

$$z(s) = t(1 - \tau) \leq t = \lim_\alpha t_\alpha = \lim_\alpha z(s_\alpha) \leq r,$$

and the first half of the proof is done. Now for the other one: Let P be an M-projection on $\mathfrak{A}^{**}$ such that $z := \Delta_{\mathfrak{A}^{**}}(P)$ is lower semicontinuous on S. By Theorem I.1.9 there is an L-projection P_* on $\mathfrak{A}^*$ such that $(P_*)^* = P$. We denote $F_0 = \mathrm{S}_{\mathfrak{A}} \cap \ker(P_*)$, and we have to show that $\ker(P_*)$ is weak* closed. To do so, we employ the Krein-Smulyan theorem and suppose that (ψ_α) is any bounded net in $\ker(P_*)$ with limit $\psi \in \mathfrak{A}^*$. Use Theorem 1.5 to write

$$\psi_\alpha = (Id - P_*)\psi_\alpha = \sum_{\nu=1}^{4} i^\nu \mu_\nu^{(\alpha)} \psi_\nu^{(\alpha)}, \tag{$*$}$$

with $\psi_\nu^{(\alpha)} \in \mathrm{S}_{\mathfrak{A}}$ and $0 \leq \mu_\nu^{(\alpha)} \leq C$ for some constant C; $\nu = 1, \dots, 4$. Note that we can choose the $\psi_\nu^{(\alpha)}$ in F_0. Now observe that the set $z^{-1}(0) \cap \mathrm{S}_{\mathfrak{A}} = \{s \in \mathrm{S}_{\mathfrak{A}} \mid z(s) \leq 0\}$ is weak* closed by lower semicontinuity. The next thing to note is that

$$z^{-1}(0) \cap \mathrm{S}_{\mathfrak{A}} = F_0,$$

which easily results from $z|_{\ker(P_*)} = 0$ and $z|_{\operatorname{ran}(P_*)} = 1$. Hence, F_0 is weak* compact. Consequently, we may again suppose convergence of the nets involved in $(*)$:

$$\mu_\nu^{(\alpha)} \longrightarrow \mu_\nu \in [0, C], \qquad \psi_\nu^{(\alpha)} \longrightarrow \psi_\nu \in F_0.$$

Hence,

$$\psi = \sum_{\nu=1}^{4} \mu_\nu \psi_\nu \in \operatorname{lin} F_0 \subset \ker(P_*),$$

and the proof of (c) is finished.

(d) Since by bitransposition $Z(\mathfrak{A})$ is a subalgebra of $Z(\mathfrak{A}^{**})$ (Corollary I.3.15) and since the map $\Delta_{\mathfrak{A}}$ is nothing else but $\Delta_{\mathfrak{A}^{**}}|_{Z(\mathfrak{A})}$, it is sufficient to prove the claim for dual Banach algebras. In this case, however, $Z(\mathfrak{A})$ is generated by elements of the form $\sum_{i=1}^{n} \alpha_i P_i$ where the P_i denote M-projections with $P_i P_j = P_j P_i = 0$ (see Theorem I.3.14). Put $\Delta(P_i) = z_i$. Now, it is clearly enough to show that $P_1 P_2 = P_2 P_1 = 0$ implies $z_1 z_2 = z_2 z_1 = 0$. But by assumption, $P_1 + P_2$ is an M-projection and so, according to (b), we have $(z_1 + z_2)^2 = z_1 + z_2$. This leads to $z_1 z_2 + z_2 z_1 = 0$ and

$$(-z_1 z_2)^2 = z_1 z_2 z_1 z_2 = -z_1 z_1 z_2 z_2 = -z_1 z_2.$$

In the same way it follows that $-z_2 z_1$ is idempotent. But $-z_1 z_2 = z_2 z_1$ and hence,

$$-z_1 z_2 = (-z_1 z_2)^2 = (z_2 z_1)^2 = -z_2 z_1.$$

This implies $z_1 z_2 = z_2 z_1 = 0$, as desired.

(e) Using the same reduction procedure as above, it is clearly enough to show that

$$\left\| \sum_{i=1}^{n} \alpha_i z_i \right\| = \max_{1 \leq i \leq n} |\alpha_i|,$$

where $z_i = \Delta(P_i)$ for some orthogonal projections $P_i \in Z(\mathfrak{A}^{**})$. But one inequality is clear since $\|\Delta\| = 1$ and the other follows, since for fixed z_j

$$\left\|\sum_{i=1}^{n} \alpha_i z_i\right\| \geq \left\|z_j \sum_{i=1}^{n} \alpha_i z_i\right\| = |\alpha_j|.$$

(f) With the help of the spectral radius formula $\rho(a) = \lim_n \sqrt[n]{\|a^n\|}$ this follows from (d) and (e). □

We now come to the most far-reaching result on M-ideals in arbitrary unital Banach algebras. The reader should compare part (b) below to what we said at the beginning of this section. We will use this part of the result in Section V.4.

Theorem 2.3 *Suppose that $\mathfrak{A}$ is a unital Banach algebra.*

(a) *Every M-ideal $\mathfrak{J}$ of $\mathfrak{A}$ is a subalgebra, but in general not an ideal.*

(b) *Let $\mathfrak{J}$ be an M-summand of a unital Banach algebra $\mathfrak{A}$, P the M-projection with range $\mathfrak{J}$, and put $z = P(e)$. Then $z\mathfrak{A}z \subset \mathfrak{J}$.*

PROOF: (a) Suppose first that $\mathfrak{J}_1$ and $\mathfrak{J}_2$ are complementary M-summands, that the M-projection P maps $\mathfrak{A}$ onto $\mathfrak{J}_1$, and put $z = P(e)$. Let $m_1, m_2 \in \mathfrak{J}_2$ have unit norm and suppose that

$$m_1 m_2 = j_1 + j_2$$

with $j_i \in \mathfrak{J}_i$ and $j_1 \neq 0$. For any $|\kappa| = 1$ we have

$$(z + m_1)(z + \kappa m_2) = z + m_1 z + \kappa z m_2 + \kappa j_1 + \kappa j_2.$$

By Lemma 2.2 we may select $\varphi \in S_{\mathfrak{A}} \cap \mathfrak{J}_2^{\perp}$ and fix κ_0, $|\kappa_0| = 1$, such that $\kappa_0 \varphi(j_1) > 0$. M-summands are invariant under hermitian operators (Corollary I.1.25), and, whenever $h \in \mathbb{H}(\mathfrak{A})$, the operators L_h and R_h are hermitian [84, p. 47]. Hence, $m_1 z + \kappa_0 z m_2 + \kappa_0 j_2 \in \mathfrak{J}_2$, since z is hermitian by Theorem 2.1(a), and we find

$$\varphi((z + m_1)(z + \kappa_0 m_2)) = \varphi(z) + \kappa_0 \varphi(j_1) > 1.$$

But this is impossible, because $\|(z + m_1)(z + \kappa_0 m_2)\| \leq 1$. Consequently, $m_1 m_2 \in \mathfrak{J}_2$ whenever $m_1, m_2 \in \mathfrak{J}_2$ which means that M-summands are subalgebras. Suppose now that $\mathfrak{J}$ is an M-ideal. By what we have just shown, $\mathfrak{J}^{\perp\perp}$ must be a subalgebra of $\mathfrak{A}^{**}$ (for any Arens product), a property that passes to $\mathfrak{J}$ due to the fact that $\mathfrak{A}$ is a subalgebra of $\mathfrak{A}^{**}$. The fact that $\mathfrak{J}$ need not be an ideal will be shown on page 231.

(b) Denote by $\mathfrak{J}_2$ the M-summand complementary to $\mathfrak{J}_1 := \mathfrak{J}$. By the above, $z\mathfrak{J}_i z \subset \mathfrak{J}_i$ for $i = 1, 2$, and the proof will be complete once it is shown that $z\mathfrak{J}_2 z = 0$. Suppose that there is $m \in \mathfrak{J}_2$ with $\|zmz\| = 1$. Since z is in $\mathfrak{J}_1$,

$$\|(\pm zmz + z)^2\| \leq \|\pm zmz + z\|^2 = 1.$$

This leads to $\|(zmz)^2 \pm 2zmz + z\| \leq 1$ and

$$\begin{aligned} 4 &= \left\|2zmz + \left(z + (zmz)^2\right) + 2zmz - \left(z + (zmz)^2\right)\right\| \\ &\leq \|(z + zmz)^2\| + \|(z - zmz)^2\| \leq 1 + 1, \end{aligned}$$

which is absurd. □

V.3 Inner M-ideals

In almost all the known cases, there is only one practicable way to obtain more information on the structure of an M-ideal in a given unital Banach algebra than the one delivered by Theorem 2.3: It consists in showing that the evaluation of the corresponding M-projection on the bidual at the unit element e yields an element which, to some extent, is responsible for the M-projection itself.
The concept which underlies this idea has been alluded to at the beginning of the previous section and will be made precise right now. We will study this type of M-ideal here in a rather abstract setting. The more concrete examples will be provided in future sections of the present and the following chapter.

Let $\mathfrak{A}$ be a Banach algebra and recall that $\mathrm{Mult}(\mathfrak{A})$ denotes the multiplier algebra of $\mathfrak{A}$ (Definition I.3.1). An element $T \in \mathrm{Mult}(\mathfrak{A})$ is called left (right) inner if there is an element $t \in \mathfrak{A}$ such that, for all $a \in \mathfrak{A}$,

$$T(a) = L_t(a) = ta$$

or, respectively,

$$T(a) = R_t(a) = at$$

for all $a \in \mathfrak{A}$. The subalgebra of $\mathfrak{A}$ consisting of all those elements of $\mathfrak{A}$ that give rise to a left (right) inner element in $\mathrm{Mult}(\mathfrak{A})$ is denoted by

$$\mathrm{Mult}^l_{inn}(\mathfrak{A}) \qquad (\mathrm{Mult}^r_{inn}(\mathfrak{A})).$$

Analogously we define

$$Z^l_{inn}(\mathfrak{A}) \qquad (Z^r_{inn}(\mathfrak{A})).$$

Note that $\mathrm{Mult}^l_{inn}(\mathfrak{A})$ and $\mathrm{Mult}^r_{inn}(\mathfrak{A})$ are subalgebras of $\mathfrak{A}$, whereas $\mathrm{Mult}(\mathfrak{A})$ is a subalgebra of $L(\mathfrak{A})$. The fact that there are two different reasonable multiplications on $\mathfrak{A}^{**}$ makes the corresponding definition for $\mathfrak{A}^{**}$ somewhat unpleasant: We denote by

$$\mathrm{Mult}^{l,i}_{inn}(\mathfrak{A}^{**}) \qquad (\mathrm{Mult}^{r,i}_{inn}(\mathfrak{A}^{**})) \qquad Z^{l,i}_{inn}(\mathfrak{A}^{**}) \qquad (Z^{r,i}_{inn}(\mathfrak{A}^{**}))$$

the respective subalgebra of $\mathfrak{A}^{**}$ furnished with the i-th Arens product.

Definition 3.1 *Let $\mathfrak{A}$ be a Banach algebra.*

(a) *An M-ideal $\mathfrak{J} \subset \mathfrak{A}$ is called* left (right) inner, *if the M-projection $P : \mathfrak{A}^{**} \to \mathfrak{J}^{\perp\perp}$ is right (left) inner with respect to the first (second) Arens multiplication on $\mathfrak{A}^{**}$.*

(b) *A subspace $\mathfrak{J}$ is called a* two-sided inner M-ideal *if and only if there is $p \in \mathfrak{A}^{**}$ such that L^2_p as well as R^1_p define M-projections $\mathfrak{A}^{**} \to \mathfrak{J}^{\perp\perp}$.*

(c) *We call an M-ideal* inner *if it is either left or right inner.*

We think that some remarks are in order:

(a) The necessity for using different multiplications in the definition of an inner M-ideal is due to the fact that M-projections in dual spaces are always weak* continuous (Theorem I.1.9), a property that is shared only by one of the respective Arens multiplications (Theorem 1.2). In fact, when e.g. also $a^{**} \mapsto a^{**}.p$ defines an M-projection on $\mathfrak{A}^{**}$, then by its weak* continuity, Theorem 1.2(d) and Theorem 1.2(e) we must have for all $a^{**} \in B_{\mathfrak{A}^{**}}$, being approximated in the weak* topology by (x_α) in $B_{\mathfrak{A}}$,

$$a^{**}.p = w^*\text{-}\lim_\alpha x_\alpha.p = w^*\text{-}\lim_\alpha x_\alpha p = a^{**}p.$$

Also, the reader might find it strange that we defined left inner M-ideals in terms of right inner M-projections. However, we decided to accept this asymmetry in favour of a symmetric characterisation of left inner M-ideals in terms of left ideals (Theorem 3.2).
(b) With the same reasoning as above, we find using Theorem I.3.14(c) that

$$Z_{inn}^{l,2}(\mathfrak{A}^{**}) \supset Z_{inn}^{l,1}(\mathfrak{A}^{**})$$

as well as

$$Z_{inn}^{r,1}(\mathfrak{A}^{**}) \supset Z_{inn}^{r,2}(\mathfrak{A}^{**}).$$

It is, for the time being, not clear whether a similar relation holds for the algebras $\mathrm{Mult}_{inn}^{l,i}(\mathfrak{A}^{**})$ and $\mathrm{Mult}_{inn}^{r,i}(\mathfrak{A}^{**})$.
Nevertheless, a glimpse at the proof of Lemma 3.4 below should convince the reader that $\mathrm{Mult}_{inn}^{l,2}(\mathfrak{A}^{**})$ and $\mathrm{Mult}_{inn}^{r,1}(\mathfrak{A}^{**})$ are the more pleasant objects to work with.
(c) Let $\mathfrak{J}$ be a two-sided inner M-ideal of $\mathfrak{A}$ and p be the element in $\mathfrak{A}^{**}$ that gives rise to the M-projections R_p^1 and L_p^2 onto $\mathfrak{J}^{\perp\perp}$. It follows by Proposition I.1.2 that

$$p.a^{**} = a^{**}p$$

for all $a^{**} \in \mathfrak{A}^{**}$ which is an "almost central" behaviour.
(d) Note further that, at least when $\mathfrak{A}^{**}$ is unital, every M-ideal $\mathfrak{J}$, which is a left as well as a right inner M-ideal is a two-sided inner M-ideal. This follows, since according to Proposition I.1.2, for every pair of M-projections L_p^2 and $R_{\widetilde{p}}^1$ with $\mathfrak{J}^{\perp\perp}$ as a common range we have

$$p = L_p^2(e) = R_{\widetilde{p}}^1(e) = \widetilde{p}.$$

(e) Let us mention some examples which show that (algebraic) ideals of $\mathfrak{A}$, which are M-ideals at the same time, are not necessarily inner:
For the first one we make a Banach algebra out of any Banach space X containing an M-ideal J by defining $xy = 0$ for any $x, y \in X$. A more serious example is this: Let J be an M-ideal of a Banach space X, the former of which does not enjoy the MAP. (To see that this is possible, fix e.g. a space Y without the MAP and note that $J := Y \oplus_\infty c_0$ is an M-ideal in the space $X := Y \oplus_\infty \ell^\infty$.) Then $A(X, J)$, the space of approximable operators from X to J, is a right ideal as well as, by Proposition VI.3.1, an M-ideal of $A(X) \cong X^* \widehat{\otimes}_\varepsilon X$. However, were $A(X, J)$ inner, there would be an approximate identity for this ideal by Theorem 3.2 below, which is impossible by the assumptions made on Y.

We will see in Sections V.4 and V.6 that this type of M-ideal is present in C^*-algebras, commutative Banach algebras and in the algebra $L(X)$ for a number of Banach spaces. In fact, in each of these cases, all M-ideals will turn out to be inner. We will prove in Proposition VI.4.10 that $K(X)$ is automatically an inner M-ideal once it is an M-ideal in $L(X)$.
The next theorem gives a characterisation of inner M-ideals which will be used in what follows.

Theorem 3.2 *Let $\mathfrak{A}$ be a unital Banach algebra and $\mathfrak{J} \subset \mathfrak{A}$ a closed subspace. Then the following are equivalent:*

(i) *$\mathfrak{J}$ is a left inner M-ideal.*

(ii) *$\mathfrak{J}$ is a left ideal as well as an M-ideal and contains a right 1-approximate unit for $\mathfrak{J}$.*

(iii) *$\mathfrak{J}$ is a left ideal and contains a right approximate unit (p_α) that satisfies*

$$\limsup_\alpha \|sp_\alpha + t(e - p_\alpha)\| \leq 1 \qquad \forall s, t \in B_{\mathfrak{A}}.$$

This equivalence remains true after exchanging "left" and "right" accordingly.
If, in addition, $\mathfrak{J}$ is a two-sided inner M-ideal, then the net (p_α) in (iii) can be chosen as a two-sided approximate unit and to satisfy

$$\limsup_\alpha \|sp_\alpha + t(e - p_\alpha)\| \leq 1 \qquad \forall s, t \in B_{\mathfrak{A}}$$

and

$$\limsup_\alpha \|p_\alpha s + (e - p_\alpha)t\| \leq 1 \qquad \forall s, t \in B_{\mathfrak{A}}$$

simultaneously.

Proof: (iii) $\Rightarrow$ (ii): The condition imposed on (p_α) implies $\limsup_\alpha \|p_\alpha\| \leq 1$ and so, by Theorem 1.3, we may suppose $\|p_\alpha\| \leq 1$ for all α. We have to check that $\mathfrak{J}$ is indeed an M-ideal.
We will apply Theorem I.2.2. With this goal in mind, suppose that $j_1, j_2, j_3 \in B_{\mathfrak{J}}$, $x \in B_X$ and $\varepsilon > 0$ are arbitrarily given. Select an index α such that $\|j_i p_\alpha - j_i\| < \frac{\varepsilon}{2}$ for $i = 1,2,3$ as well as $\|j_i p_\alpha + x(e - p_\alpha)\| < 1 + \frac{\varepsilon}{2}$ for $i = 1,2,3$ and put $j = xp_\alpha$. Then $j \in \mathfrak{J}$,

$$\|j_i + x - j\| \;\leq\; \|j_i p_\alpha - j_i\| + \|j_i p_\alpha + x(e - p_\alpha)\| \;<\; 1 + \varepsilon,$$

and $\mathfrak{J}$ must be an M-ideal.
(ii) $\Rightarrow$ (i): By Theorem 1.3, $\mathfrak{J}^{\perp\perp}$ contains a right unit with respect to the first Arens multiplication which we denote by p. Since $\mathfrak{J}^{\perp\perp}$ is a left ideal (Theorem 1.2(a)), the mapping R_p^1 is a projection onto this space. But M-projections are unique among norm one projections with the same range (Proposition I.1.2), and so $\mathfrak{J}$ is a left inner M-ideal.
(i) $\Rightarrow$ (iii): Suppose that $R_p^1 : \mathfrak{A}^{**} \to \mathfrak{J}^{\perp\perp}$ is an M-projection for some $p \in \mathfrak{A}^{**}$. Then $\mathfrak{J}^{\perp\perp}$ is a left ideal, and hence so is $\mathfrak{J}$. Let now $F \subset \mathfrak{A}^{**}$, $G \subset \mathfrak{A}^*$ and $H \subset \mathfrak{A}$ run through the respective sets of finite dimensional subspaces and order the set

$$B = \{(F, G, H, \varepsilon) \mid F \subset \mathfrak{A}^{**}, G \subset \mathfrak{A}^*, H \subset \mathfrak{A}, \varepsilon > 0\}$$

by

$$(F, G, H, \varepsilon) \prec (\widetilde{F}, \widetilde{G}, \widetilde{H}, \widetilde{\varepsilon}) \qquad \Longleftrightarrow \qquad F \subset \widetilde{F},\ G \subset \widetilde{G},\ H \subset \widetilde{H},\ \varepsilon > \widetilde{\varepsilon}.$$

Identify the space $H \subset \mathfrak{A}$ with the operator space $\{L_h \mid h \in H\}$ and choose for every $\beta \in B$ an operator T_β that satisfies the conditions (a) – (d) listed in Theorem 1.4 for the subspaces and the number ε determined by β. Let $p_\beta^0 := T_\beta(p)$. (Note that $T_\beta(a^{**})$

is eventually defined for all $a^{**} \in \mathfrak{A}^{**}$.) Then, as in the proof of Theorem 1.3, we may conclude that (p^0_β) is a right approximate unit. As was also shown there,

$$\lim_\beta \|ap^0_\beta - T_\beta(ap)\| = \lim_\beta \|L_a T_\beta(p) - T_\beta L^{**}_a(p)\| = 0$$

for all $a \in \mathfrak{A}$, and so, for $s, t \in B_{\mathfrak{A}}$, we have by property (c) of Theorem 1.4

$$\begin{aligned} \limsup_\beta \|sp^0_\beta + t(e - p^0_\beta)\| &= \limsup_\beta \|T_\beta(sp) + T_\beta(t) - T_\beta(tp)\| \\ &= \limsup_\beta \|T_\beta(sp + t(e-p))\| \\ &\leq 1. \end{aligned}$$

To finish the proof, we have to force the p^0_β into $\mathfrak{J}$. To this end, we take a net (q_β) from $B_{\mathfrak{J}}$ with $\sigma(\mathfrak{A}^{**}, \mathfrak{A}^*)$-$\lim_\beta q_\beta = p$. Note that we may use the same index set. Since by (b) of Theorem 1.4, w^*-$\lim_\beta p^0_\beta = p$, and since $p^0_\beta - q_\beta \in \mathfrak{A}$, we must have w-$\lim_\beta(p^0_\beta - q_\beta) = 0$. In passing to an appropriate $\|\,.\,\|$-convergent net of convex combinations we obtain

$$\|\,.\,\|\text{-}\lim_\alpha \sum_{i=1}^{N_\alpha} t_{i,\alpha}\left(p^0_{\beta_{i,\alpha}} - q_{\beta_{i,\alpha}}\right) = 0,$$

and

$$\lim_\alpha \max_{i \leq N_\alpha} \|xp^0_{\beta_{i,\alpha}} - x\| = 0.$$

We now put

$$p_\alpha = \sum_{i=1}^{N_\alpha} t_{i,\alpha} q_{\beta_{i,\alpha}}.$$

Hence we obtain

$$\begin{aligned} \lim_\alpha \|xp_\alpha - x\| &\leq \lim_\alpha \left(\left\| xp_\alpha - \sum_{i=1}^{N_\alpha} t_{i,\alpha} xp^0_{\beta_{i,\alpha}} \right\| + \left\| \sum_{i=1}^{N_\alpha} t_{i,\alpha} xp^0_{\beta_{i,\alpha}} - x \right\| \right) \\ &\leq \lim_\alpha \max_{i \leq N_\alpha} \|xp^0_{\beta_{i,\alpha}} - x\| \\ &= 0. \end{aligned}$$

In a similar fashion, for all $s, t \in B_{\mathfrak{A}}$,

$$\lim_\alpha \|sp_\alpha + t(e - p_\alpha)\| \leq 1,$$

and (iii) follows.

We are thus left with the case of a two-sided inner M-ideal $\mathfrak{J}$. Recall (Remark (d) after Definition 3.1) that there is $p \in \mathrm{Mult}^{l,2}_{inn}(\mathfrak{A}^{**}) \cap \mathrm{Mult}^{r,1}_{inn}(\mathfrak{A}^{**})$ with

$$L^2_p \mathfrak{A}^{**} = R^1_p \mathfrak{A}^{**} = \mathfrak{J}^{\perp\perp}.$$

To refine the properties of (p_α) in this situation, we use the index set

$$B^* = \{(F, G, H_L, H_R, \varepsilon) \mid p \in F \subset \mathfrak{A}^{**},\ G \subset \mathfrak{A}^*,\ H_L, H_R \subset \mathfrak{A},\ \varepsilon > 0\}$$

ordered in the now obvious way, and identify the spaces $H_L \subset \mathfrak{A}$ and $H_R \subset \mathfrak{A}$ with the operator spaces $\{L_h \mid h \in H_L\}$ and $\{R_h \mid h \in H_R\}$. Keeping these minor changes in mind, we may proceed as above. □

We continue our investigation of inner M-ideals with a proposition that deals with the behaviour of such M-ideals with respect to subalgebras.

As the case of the disk algebra $A(\mathbb{D})$, which we will now regard as a subalgebra of $C(\overline{\mathbb{D}})$, already shows, it is not to be expected that for subalgebras $\mathfrak{B}$ and inner M-ideals $\mathfrak{J}$ the space $\mathfrak{J} \cap \mathfrak{B}$ again is an M-ideal, much less an inner one. (To be more concrete, let $J = \{f \in A(\mathbb{D}) \mid f(0) = 0\}$. This cannot be an M-ideal in $A(\mathbb{D})$ by Example I.1.4(b), though it is of the above form.)

Proposition 3.3 *Suppose that $\mathfrak{A}$ is a unital Banach algebra and let $\mathfrak{B}$ be a subalgebra of $\mathfrak{A}$ with $e \in \mathfrak{B}$. If $p \in \mathfrak{B}^{\perp\perp} \cap \mathrm{Mult}^{r,1}_{inn}(\mathfrak{A}^{**})$ is idempotent, it gives rise to a right inner M-ideal $\mathfrak{J}$ in $\mathfrak{A}$ by virtue of*

$$\mathfrak{A}^{**}p = \mathfrak{J}^{\perp\perp}$$

if and only if

$$\mathfrak{B}^{**}p = \mathfrak{J}_0^{\perp\perp}$$

for some M-ideal $\mathfrak{J}_0$ in $\mathfrak{B}$. If one of these conditions is satisfied then $\mathfrak{J}_0 = \mathfrak{J} \cap \mathfrak{B}$, and $\mathfrak{J}$ is nontrivial if and only if $\mathfrak{J} \cap \mathfrak{B}$ is. In this case, one can find a left approximate unit (p_α) for $\mathfrak{J}$ in $B_{\mathfrak{J}\cap\mathfrak{B}}$ with the property that, for all $s, t \in B_{\mathfrak{A}}$,

$$\limsup_\alpha \|p_\alpha s + (e - p_\alpha)t\| \le 1.$$

A similar statement holds for left and two-sided inner M-ideals.

PROOF: Plainly, since $p \in \mathfrak{B}^{\perp\perp}$, we have for all $\psi \in \mathrm{S}_{\mathfrak{A}}$ and any $\rho \in \mathbb{R}$

$$p(\psi) \le \rho \quad \Longleftrightarrow \quad p(\psi|_{\mathfrak{B}}) \le \rho$$

which leads to

$$\{\psi \in \mathrm{S}_{\mathfrak{A}} \mid p(\psi) \le \rho\} = r^{-1}(\{\varphi \in \mathrm{S}_{\mathfrak{B}} \mid p(\varphi) \le \rho\})$$

where $r : \mathrm{S}_{\mathfrak{A}} \to \mathrm{S}_{\mathfrak{B}}$ denotes the restriction map. An examination of this equation quickly reveals that $\{\psi \in \mathrm{S}_{\mathfrak{A}} \mid p(\psi) \le \rho\}$ is weak* closed if and only if the set $\{\varphi \in \mathrm{S}_{\mathfrak{B}} \mid p(\varphi) \le \rho\}$ has the same property. This in turn implies that p as a function on $\mathrm{S}_{\mathfrak{A}}$ is lower semicontinuous if and only if p shows the same behaviour when considered as a function on $\mathrm{S}_{\mathfrak{B}}$. Hence, according to Theorem 2.1(c), $\mathfrak{A}^{**}p = \mathfrak{J}^{\perp\perp}$ for some M-ideal $\mathfrak{J}$ of $\mathfrak{A}$ is equivalent to $\mathfrak{B}^{\perp\perp}p = \mathfrak{J}_0^{\perp\perp}$ for some M-ideal $\mathfrak{J}_0$ of $\mathfrak{B}$. But then $\mathfrak{J}$ and $\mathfrak{J}_0$ are connected by

$$\mathfrak{J}_0 = \{b \in \mathfrak{B} \mid bp = b\} = \mathfrak{J} \cap \mathfrak{B}.$$

Next, since $e \in \mathfrak{B}$, we conclude that $\mathfrak{J} \ne \mathfrak{A}$ (for an ideal $\mathfrak{J}$ of $\mathfrak{A}$) is equivalent to $\mathfrak{J}\cap\mathfrak{B} \ne \mathfrak{B}$. Since $\mathfrak{J} = \{0\}$ if and only if $p = 0$ we have $\mathfrak{J} \cap \mathfrak{B} = \{0\}$ if and only if $\mathfrak{J} = \{0\}$. This proves that $\mathfrak{J}$ is not trivial if and only if $\mathfrak{J}_0$ is not either.

Let us finally construct the required net (p_α): To this aim, start with (p_α^0) in $B_{\mathfrak{J}}$ as delivered by Theorem 3.2, and select furthermore $q_\alpha \in B_{\mathfrak{B}\cap\mathfrak{J}}$ converging also to p in the $\sigma(\mathfrak{A}^{**}, \mathfrak{A}^*)$-topology. The claim then follows by an application of the same blocking technique as used in the proof of Theorem 3.2. □

As mentioned before, part of what follows will be devoted to the study of particular examples of inner M-ideals. We start this enterprise in this section with the description of a rather general situation in which these M-ideals appear quite naturally. We first need a lemma.

Lemma 3.4 *For every Banach algebra $\mathfrak{A}$,*

$$\mathrm{Mult}^l_{inn}(\mathfrak{A})^{\perp\perp} \subset \mathrm{Mult}^{l,2}_{inn}(\mathfrak{A}^{**})$$

This statement persists when "left" is changed to "right" and "2" to "1".

PROOF: We will apply Theorem I.3.6. To this end, choose $f^{**} \in \mathrm{Mult}^l_{inn}(\mathfrak{A})^{\perp\perp}$ and let $a^{**}, b^{**} \in \mathfrak{A}^{**}$ with

$$\|a^{**} + \lambda b^{**}\| \leq r$$

for all λ with $|\lambda| \leq \|f^{**}\|$. Use the principle of local reflexivity (Theorem 1.4) to find nets (a_α) and (b_α) with

$$\|a_\alpha + \lambda b_\alpha\| \leq r + \varepsilon_\alpha$$

for all $|\lambda| \leq \|f^{**}\|$, (ε_α) tending to zero, $w^*\text{-}\lim_\alpha a_\alpha = a^{**}$ and $w^*\text{-}\lim_\alpha b_\alpha = b^{**}$. If (f_β) denotes a net in $\mathrm{Mult}^l_{inn}(\mathfrak{A})$ with the property that $\|f_\beta\| \leq \|f^{**}\|$ and $w^*\text{-}\lim_\beta f_\beta = f^{**}$, then, by the weak* continuity of the mapping R_{b_α} (Theorem 1.2), we have that

$$\|a_\alpha + f^{**} b_\alpha\| \leq \limsup_\beta \|a_\alpha + f_\beta b_\alpha\| \leq r + \varepsilon_\alpha$$

for all α. By the weak* continuity of the map $L^2_{f^{**}}$ (Theorem 1.2),

$$\|a^{**} + f^{**}.b^{**}\| \leq \limsup_\alpha \|a_\alpha + f^{**}.b_\alpha\| \leq r,$$

and we are done. The proof of the "r,1"-case is similar. □

Proposition 3.5 *Let $\mathfrak{A}$ be a unital Banach algebra and $\mathfrak{B}$ be a subalgebra of $\mathrm{Mult}^l_{inn}(\mathfrak{A})$ such that $e \in \mathfrak{B}$. Suppose further that $\mathfrak{J}$ is an M-ideal in $\mathfrak{B}$.*

(a) *The space $\mathfrak{A}_{\mathfrak{J}} := \overline{\mathrm{lin}}\,\{ja_0 \mid j \in \mathfrak{J}, a_0 \in \mathfrak{A}\}$ is a right inner M-ideal, and there is a left approximate unit (p_α) in $B_{\mathfrak{J}}$ for $\mathfrak{A}_{\mathfrak{J}}$ with $\limsup_\alpha \|p_\alpha s + (e - p_\alpha)t\| \leq 1$ for all $s, t \in B_{\mathfrak{A}}$.*

(b) *We have $\mathfrak{A}_{\mathfrak{J}} = \{a \in \mathfrak{A} \mid \lim_\alpha p_\alpha a = a\}$ as well as $\mathfrak{J} = \mathfrak{B} \cap \mathfrak{A}_{\mathfrak{J}}$. Moreover, $\mathfrak{A}_{\mathfrak{J}}$ is trivial if and only if $\mathfrak{J}$ has this property.*

Also here we are in a position where the cases for "left" and, respectively, "right" can be treated in a completely analogous way.

PROOF: (a) Since $\mathfrak{B}$ is a function algebra, Theorem 4.1 below shows that

$$\mathfrak{J}^{\perp\perp} = p.\mathfrak{B}^{\perp\perp}$$

for some $p \in \mathfrak{B}^{\perp\perp}$. We have $p \in \mathrm{Mult}^l_{inn}(\mathfrak{A})^{\perp\perp}$ and so, by the above lemma, $p \in \mathrm{Mult}^{l,2}_{inn}(\mathfrak{A}^{**})$. Since $\mathfrak{B}$ is unital and $\mathfrak{J}^{\perp\perp} \subset \mathfrak{A}_{\mathfrak{J}}^{\perp\perp}$ we have $p \in \mathfrak{A}_{\mathfrak{J}}^{\perp\perp}$. Furthermore, $pja_0 = ja_0$ for all $j \in \mathfrak{J}$ and each $a_0 \in \mathfrak{A}$ and hence, $pa = a$ for all $a \in \mathfrak{A}_{\mathfrak{J}}$. By weak* continuity of the second Arens multiplication, we have for each $F \in \mathfrak{A}_{\mathfrak{J}}^{\perp\perp}$ with $F = w^*\text{-}\lim_\alpha f_\alpha$ for some net $f_\alpha \in \mathfrak{A}_{\mathfrak{J}}$

$$p.F = w^*\text{-}\lim_\alpha pf_\alpha = F.$$

This shows that p is a left unit for $\mathfrak{A}_{\mathfrak{J}}^{\perp\perp}$ with respect to the second Arens product. Since $\mathfrak{A}_{\mathfrak{J}}^{\perp\perp}$ is a right ideal, it follows that $\mathfrak{A}_{\mathfrak{J}}$ must be a right inner M-ideal. The net (p_α) is obtained just as in the proof of Theorem 3.2 with the aid of a net (q_α) in $B_{\mathfrak{J}}$ converging to p in the $\sigma(\mathfrak{A}^{**}, \mathfrak{A}^*)$-topology.

(b) The first equation is immediate from (a), and the second is clear by virtue of

$$\mathfrak{J} = \{b \in \mathfrak{B} \mid \lim_\alpha p_\alpha b = b\} = \mathfrak{A}_{\mathfrak{J}} \cap \mathfrak{B}.$$

Finally, $\mathfrak{J} = \mathfrak{B}$ if and only if $p = e$ which in turn is equivalent to $\mathfrak{A}_{\mathfrak{J}} = \mathfrak{A}$. Similarly, $\mathfrak{J} = \{0\}$ is the same as $p = 0$ and $\mathfrak{A}_{\mathfrak{J}} = \{0\}$. □

Corollary 3.6 *Let X be a Banach space and $\mathfrak{J}$ an M-ideal of a subalgebra $\mathfrak{B}$ of* $\mathrm{Mult}(X)$ *with $Id \in \mathfrak{B}$. Then*

$$X_{\mathfrak{J}} := \overline{\mathrm{lin}}\,\{j(x) \mid j \in \mathfrak{J},\ x \in X\}$$

is an M-ideal in X.

PROOF: Put $\mathfrak{A} = L(X)$ in the above proposition. Then, by Lemma VI.1.1 below, $\mathfrak{B} \subset \mathrm{Mult}^l_{inn}(\mathfrak{A})$, and, by the above, we find a net (P_α) in $B_{\mathfrak{J}}$ with

$$\limsup_\alpha \|P_\alpha T_1 + (Id - P_\alpha)T_2\| \le 1$$

for all $T_1, T_2 \in B_{L(X)}$. This easily implies that

$$\limsup_\alpha \|P_\alpha x_1 + (Id - P_\alpha)x_2\| \le 1$$

for all $x_1, x_2 \in B_X$. Moreover, for all $x \in X$, $(P_\alpha x)$ converges to x in norm, and since the range of each P_α is contained in $X_{\mathfrak{J}}$, one can proceed exactly as in the proof of Theorem 3.2 to settle our claim. □

It remains an open problem whether a nontrivial M-ideal in $\mathrm{Mult}(X)$ generates a nontrivial M-ideal in X in Corollary 3.6. What can be shown is that X^{**} always contains a nontrivial M-ideal as soon as $\mathrm{Mult}(X)$ is known to be nontrivial [639, Korollar 4.15].

Let us conclude this section with an example which shows that there are M-ideals in Banach algebras which are neither inner nor ideals: Let $\mathfrak{A}_1$ and $\mathfrak{A}_2$ be unital Banach algebras such that both

$$\mathrm{Mult}^l_{inn}(\mathfrak{A}_1) \setminus \mathrm{centre}(\mathfrak{A}_1)$$

and

$$\mathrm{Mult}^r_{inn}(\mathfrak{A}_2) \setminus \mathrm{centre}(\mathfrak{A}_2)$$

contain idempotent elements l and r, respectively. Put $\mathfrak{A} := \mathfrak{A}_1 \oplus_\infty \mathfrak{A}_2$. Under pointwise multiplication, $\mathfrak{A}$ becomes a Banach algebra. By uniqueness of M-projections with a prescribed range (Proposition I.1.2) and by the assumptions made on l and r,

$$L_{(l,0)} \neq R_{(l^*,0)} \qquad \text{and} \qquad R_{(0,r)} \neq L_{(0,r^*)}$$

for all idempotent elements l^* and r^* of $\mathfrak{A}_1$ resp. $\mathfrak{A}_2$. Then $(a,b) \longmapsto (la, br)$ defines an M-projection on $\mathfrak{A}$ which belongs neither to $\mathrm{Mult}^l_{inn}(\mathfrak{A})$ nor to $\mathrm{Mult}^r_{inn}(\mathfrak{A})$, and, consequently, the M-summand

$$\{(la, br) \mid a \in \mathfrak{A}_1, b \in \mathfrak{A}_2\}$$

is not inner, since it is not an ideal.
To be more specific, let $\mathfrak{A}_1 = L(\ell^\infty(2))$, $\mathfrak{A}_2 = L(\ell^1(2))$, and consider

$$l = \begin{pmatrix} 0 & 0 \\ 0 & 1 \end{pmatrix} = r.$$

Then $T \mapsto lT$ is an M-projection on $\mathfrak{A}_1$, and $T \mapsto Tr$ is an M-projection on $\mathfrak{A}_2$ so that l and r meet the above requirements. This choice even yields a semisimple Banach algebra so that there is no hope to extend the results on M-ideals in commutative or C^*-algebras to be proved in the next section to the class of semisimple Banach algebras.

V.4 Commutative and C^*-algebras

We investigate in this section the structure of the M-ideals in two rather common types of Banach algebras.
Whereas in the case of a C^*-algebra there is a neat correspondence between M-ideals and closed two-sided ideals, the situation for commutative Banach algebras is, due to the fact that commutativity generally doesn't have too deep an influence on the geometry of these spaces, less clear. The following result contains what can be said for this class of Banach algebras.

Theorem 4.1 *In a unital commutative Banach algebra $\mathfrak{A}$ all M-ideals of $\mathfrak{A}$ are inner. In particular, $\mathfrak{J} \subset \mathfrak{A}$ is an M-ideal if and only if it is a closed ideal and there is an approximate unit (p_α) in $B_\mathfrak{J}$ with*

$$\limsup_\alpha \|p_\alpha s + (e - p_\alpha)t\| \leq 1 \qquad \forall s,t \in B_\mathfrak{A}.$$

PROOF: We demonstrate that all M-ideals of $\mathfrak{A}$ are inner. The rest of the conclusion then follows from Theorem 3.2.
Now, by Theorem 2.3(b) we have $(e-z)\mathfrak{J}^{\perp\perp}(e-z) = 0$ – recall that z denotes the image of the unit under the associated M-projection in the bidual of $\mathfrak{A}$ – which implies that $zj = jz = j$ for all $j \in \mathfrak{J}$. Consequently, z is a unit for $\mathfrak{J}^{\perp\perp}$, whenever an Arens product is chosen that fits the direction from which z multiplies (remember that even for commutative Banach algebras, $\mathfrak{A}$ need not be Arens regular). Since $z \in \mathfrak{J}^{\perp\perp}$, we must

have that $L_z^{(2)} = R_z^{(1)}$ is a projection onto $\mathfrak{J}^{\perp\perp}$ whose norm does not exceed one. But there is only one such thing (Proposition I.1.2), and the result follows. □

The example of an ideal in the convolution algebra $L^1(G)$, where G denotes a locally compact abelian group, shows in light of Theorem I.1.8 that no substantially better result is to be expected here. However in the special case that $\mathfrak{A}$ is a function algebra, more can be done.

Theorem 4.2 *Let $\mathfrak{A}$ be a function algebra on a compact space K and $\mathfrak{J}$ a subspace of $\mathfrak{A}$. Then the following conditions are equivalent:*

(i) *$\mathfrak{J}$ is the annihilator of a p-set of $\mathfrak{A}$.*
(ii) *$\mathfrak{J}$ is an ideal of $\mathfrak{A}$ containing a bounded approximate unit.*
(iii) *$\mathfrak{J}$ is an M-ideal of $\mathfrak{A}$.*

We recall that a p-set of a function algebra $\mathfrak{A}$ is an intersection of a family of peak sets where this last expression means a set of the form $D = \{a = 1\} = a^{-1}(\{1\})$ for some $a \in \mathfrak{A}$ with $\|a\| = 1$. What we need to know about these sets is contained in the following result.

Lemma 4.3 *Let $\mathfrak{A}$ be a function algebra on a compact space K. Then a closed subset D of K is a p-set for $\mathfrak{A}$ if and only if for all $\varepsilon > 0$ and for every open set U containing D there exists $a \in \mathfrak{A}$ with*

$$\|e - a\| \leq 1 + \varepsilon, \qquad a|_D = 0 \qquad \text{and} \qquad \left|(e - a)|_{K\setminus U}\right| < \varepsilon.$$

PROOF: The sufficiency follows from Bishop's "$\frac{1}{4}$-$\frac{3}{4}$-condition" [586, p. 47], and the necessity is a consequence of [240, Lemma II.12.2]. □

PROOF OF THEOREM 4.2:

(i) ⇒ (ii): This conclusion follows easily from the above result.
(ii) ⇒ (iii): Since $\mathfrak{A}$ embeds isometrically as well as algebraically into a $C(K)$-space, it follows that $\mathfrak{A}^{**}$ furnished with one of the Arens products (in fact, in this case both products coincide) again is a function algebra.
For this reason, the unit, which $\mathfrak{J}^{\perp\perp}$ must contain by assumption, is necessarily a characteristic function. It follows that $\mathfrak{J}$ is an M-ideal.
(iii) ⇒ (i): Let $\mathfrak{J}$ be an M-ideal of $\mathfrak{A}$. The equation $\mathrm{Mult}^l_{inn}(C(K)) = C(K)$ permits us to apply Proposition 3.5, and there must be a closed set D in K with $J_D = C(K)_{\mathfrak{J}}$ and $\mathfrak{J} = J_D \cap \mathfrak{A}$. By the same result, we obtain an approximate unit (p_α) for the M-ideal J_D of $C(K)$ contained in $B_{\mathfrak{J}} = B_{J_D} \cap \mathfrak{A}$ such that $\|e - p_\alpha\| \to 1$. It is easy to check that (p_α) can be characterised by the condition

$$\forall \varepsilon > 0 \; \forall U \supset D, \; U \text{ open } \exists \alpha_0 \; \forall \alpha > \alpha_0 \quad \left|(e - p_\alpha)|_{K\setminus U}\right| < \varepsilon,$$

which is in light of Lemma 4.3 all we have to know in order to show that D is a p-set for $\mathfrak{A}$. □

Let us remark that the implication (i) ⇒ (iii) can also be deduced from Proposition I.1.20; see the remarks preceding that result. Also, we note that one even gets a 1-approximative unit in (ii).

Another close relative of the class of Banach algebras of type $C(K)$ is the class of C^*-algebras, which will be treated next.

Theorem 4.4 *Every M-ideal in a C^*-algebra $\mathfrak{A}$ is inner. There is, in addition, a one-to-one correspondence between the M-ideals and the closed two-sided ideals of $\mathfrak{A}$.*

PROOF: Let $\mathfrak{J}$ be an M-ideal of $\mathfrak{A}$. Since in this case an element is hermitian if and only if it is self-adjoint (Proposition 1.7), we find with the aid of Corollary I.3.15 that $h\mathfrak{J} \subset \mathfrak{J}$ and $\mathfrak{J}h \subset \mathfrak{J}$ for all $h \in \mathbb{H}(\mathfrak{A})$. Since $\mathfrak{A} = \mathbb{H}(\mathfrak{A}) + i\mathbb{H}(\mathfrak{A})$, the space $\mathfrak{J}$ must be a two-sided ideal. Since a two-sided closed ideal $\mathfrak{J}$ of a C^*-algebra is supported by a central idempotent z (which has to be of norm one) in $\mathfrak{A}^{**}$, i.e. $\mathfrak{J}^{\perp\perp} = z\mathfrak{A}^{**} = \mathfrak{A}^{**}z$ (see Proposition 4.5(b) below), $\mathfrak{J}$ is inner. For the converse, suppose that $\mathfrak{J}$ is a two-sided ideal, and again, denote by z the supporting idempotent. We have ($\mathfrak{A}^{**}$ always has a unit, see below)

$$\begin{aligned} \|a\|^2 &= \|(az + a(e-z))^*(az + a(e-z))\| \\ &= \|a^*az + a^*a(e-z)\| \\ &= \max\{\|a^*az\|, \|a^*a(e-z)\|\}, \end{aligned}$$

since the last equation takes place in the commutative C^*-algebra generated by z and a^*a. Because z is self-adjoint, central and idempotent, the last expression becomes

$$\max\{\|az\|^2, \|a(e-z)\|^2\},$$

and hence, multiplication with z yields an M-projection on $\mathfrak{A}^{**}$. It follows that $\mathfrak{J}^{\perp\perp}$ and $\mathfrak{A}^{**}(e-z)$ are complementary M-summands in $\mathfrak{A}^{**}$ and so, $\mathfrak{J}^{\perp}$ is an L-summand (see Theorem I.1.9) in $\mathfrak{A}^*$: $\mathfrak{J}$ must be an inner M-ideal. □

We still owe the following proposition. A C^*-algebra $\mathfrak{A}$ is called a W^*-algebra if, as a Banach space, it is a dual space. Equivalently (e.g. [593, p. 133]), a W^*-algebra is a unital C^*-subalgebra of $L(H)$ which is closed for the weak operator topology; usually the latter type of algebra is called a von Neumann algebra. It is known (e.g. [593, p. 135]) that a W^*-algebra has a uniquely determined predual so that we can unambiguously make reference to its weak* topology. In the next proposition we call a W^*-algebra a *D^*-algebra* if multiplication from the left and the right and the $*$-operation are weak* continuous. As a matter of fact, *every* W^*-algebra is a D^*-algebra [555, p. 18]. However, in order to keep the proof of Proposition 4.5 self-contained we did not want to rely on that result. Instead, what we will use here is the fact that, for a C^*-algebra $\mathfrak{A}$, the algebra $\mathfrak{A}^{**}$ is a D^*-algebra, which we proved in Theorem 1.10.

Proposition 4.5

(a) *A D^*-algebra has a unit. In particular, the bidual of any C^*-algebra is unital.*

(b) *For every weak* closed two-sided ideal $\mathfrak{J}$ of a D^*-algebra $\mathfrak{A}$ there is a central idempotent p in $\mathfrak{A}$ such that*

$$\mathfrak{J} = p\mathfrak{A}.$$

It follows in particular that for any two-sided ideal $\mathfrak{J}$ in an arbitrary C^*-algebra $\mathfrak{A}$ there is a central idempotent p in $\mathfrak{A}^{**}$ such that $\mathfrak{J}^{\perp\perp} = p\mathfrak{A}^{**}$.

PROOF: Since (b) is an easy consequence of (a) (let $p =$ the unit of $\mathfrak{J}$), we content ourselves with a proof of (a).
Fix a maximal abelian C^*-subalgebra $\mathfrak{M}$ of $\mathfrak{A}$, which is necessarily weak* closed by the D^*-condition. It follows that $\mathfrak{M}$ is a dual space, and, consequently, $B_{\mathfrak{M}}$ must contain an extreme point. Therefore, $\mathfrak{M} \cong C(K)$ for some compact (not only locally compact) K, so $\mathfrak{M}$ has a unit, e. Let $a \in \mathfrak{A}$ and put $h = (a - ae)^*(a - ae)$. Since e is a unit for $\mathfrak{M}$, we clearly have $h\mathfrak{M} = 0$, and likewise $\mathfrak{M}h = 0$. Hence, because $\mathfrak{M}$ is maximal abelian, $h \in \mathfrak{M}$. It follows that $h = 0$ and thus $ae = a$. Similarly $ea = a$, so we have found a unit for $\mathfrak{A}$. □

Theorem 4.4 above enables us to give an example which was promised in Section I.1, see Remark (c) following Theorem I.1.9.

Proposition 4.6 *There is a Banach space X without nontrivial M-ideals such that X^* contains a nontrivial L-summand. In fact, there is a Banach space X without nontrivial M-ideals such that X^* contains uncountably many nontrivial pairwise not isometrically isomorphic L-summands.*

PROOF: The first-mentioned example has already been produced in Example IV.1.8. For the harder example we rely on the deep results of R. T. Powers [509]. Let X be the UHF-algebra of type (2^n), that is the norm closure of all operators on ℓ^2 which have, for some $n \in \mathbb{N}$, a block diagonal matrix representation

$$\begin{pmatrix} A & 0 & 0 & \cdots \\ 0 & A & & \\ 0 & & \ddots & \\ \vdots & & & \end{pmatrix}$$

with some $2^n \times 2^n$-matrix A. It is known that X is a C^*-algebra without nontrivial two-sided ideals [362, Prop. 10.4.18], yet X^{**} has uncountably many weak* closed two-sided ideals. This is an easy corollary of the work of Powers who proved in [509, Th. 4.8] (see also Theorem 12.3.14 in [362]) that there is a family $\{\rho_\lambda |\ 0 \le \lambda \le \frac{1}{2}\}$ of states on X with corresponding representations $\pi_\lambda : X \to L(H_\lambda)$ such that the von Neumann algebra $M_\lambda :=$ the bicommutant of $\pi_\lambda(X)$ in $L(H_\lambda)$, i.e. $M_\lambda = \overline{\pi_\lambda(X)}^{w^*}$, is $*$-isomorphic to M_μ if and only if $\lambda = \mu$. Now π_λ has a unique weak* continuous extension to a $*$-homomorphism $\widehat{\pi}_\lambda$ from X^{**} onto M_λ (see e.g. [593, p. 121]) so that $\mathfrak{J}_\lambda := \ker(\widehat{\pi}_\lambda)$ is a weak* closed two-sided ideal in X^{**}. Powers's result entails $\mathfrak{J}_\lambda \neq \mathfrak{J}_\mu$ if $\lambda \neq \mu$ since otherwise ($\overset{*}{\cong}$ meaning "$*$-isomorphic to")

$$M_\lambda \overset{*}{\cong} X^{**}/\mathfrak{J}_\lambda = X^{**}/\mathfrak{J}_\mu \overset{*}{\cong} M_\mu.$$

Hence X^{**} contains uncountably many weak* closed M-ideals which have to be M-summands by Corollary II.3.6(b). Theorem I.1.9 now shows that there are uncountably many L-summands in X^*. No two of these L-summands are isometrically isomorphic, since otherwise two of the M_λ would be isometrically isomorphic and hence by a result due to Kadison [357] even $*$-isomorphic. $\square$

Our next aim is to show that also the centralizer admits a purely algebraic interpretation in the case of C^*-algebras. To this end recall that the centroid of an (arbitrary) algebra $\mathfrak{A}$ is defined as the set of all those linear mappings defined on $\mathfrak{A}$ which satisfy the relation

$$T(ab) = aT(b) = T(a)b$$

for all $a, b \in \mathfrak{A}$. Whenever $\mathfrak{A}$ is a C^*-algebra or more generally a Banach algebra where the multiplication separates points (meaning $\forall a \neq 0\ \exists b \neq 0 : \ ab \neq 0$), such a mapping is continuous, as a standard application of the closed graph theorem shows. Note further that the centroid can be identified with the centre of $\mathfrak{A}$ whenever $\mathfrak{A}$ has a unit.

Theorem 4.7 *Let $\mathfrak{A}$ be a C^*-algebra and $T \in L(\mathfrak{A})$. Then the following are equivalent:*
  (i) *T is in $Z(\mathfrak{A})$.*
  (ii) *T belongs to the centroid of $\mathfrak{A}$.*
  (iii) *There is an element a^{**} in the centre of $\mathfrak{A}^{**}$ such that $a^{**}\mathfrak{A} \subset \mathfrak{A}$ and $T = M_{a^{**}}$.*

PROOF: (i) $\Rightarrow$ (ii): Since the operators in $Z(\mathfrak{A})$ commute with L_h for any hermitian element h (Corollary I.3.15) we have $T(hb) = hT(b)$ for all $h \in \mathbb{H}(\mathfrak{A})$ and every $b \in \mathfrak{A}$. Since $\mathfrak{A} = \mathbb{H}(\mathfrak{A}) + i\mathbb{H}(\mathfrak{A})$ we must have $T(ab) = aT(b)$ for all $a \in \mathfrak{A}$. Similarly, $T(ab) = T(a)b$ for all $a, b \in \mathfrak{A}$, and this gives (ii).
(ii) $\Rightarrow$ (iii): Let T be an element in the centroid. Since $\mathfrak{A}$ is Arens regular (Theorem 1.10(a)), we have

$$T^{**}(FG) = T^{**}(F)G = FT^{**}(G)$$

for all $F, G \in \mathfrak{A}^{**}$, and because $\mathfrak{A}^{**}$ always has a unit (Proposition 4.5(a)), it follows

$$T^{**}(F) = T^{**}(e)F = FT^{**}(e).$$

Hence, putting $a^{**} = T^{**}(e)$, we clearly have $a^{**}\mathfrak{A} \subset \mathfrak{A}$, and so, a^{**} is what we were searching for.
(iii) $\Rightarrow$ (i): Since the centre of $\mathfrak{A}^{**}$ is weak* closed (this follows e.g. from the fact that $\mathfrak{A}$ is Arens regular (Theorem 1.10(a))), it is algebraically isomorphic to a dual $C(K)$-space. As a consequence, the elements of the form $\sum_{i=1}^{n} \lambda_i p_i$, where the λ_i are complex numbers and the p_i are idempotent, are dense. We are thus left with showing that each central idempotent $p \in \mathfrak{A}^{**}$ gives rise to an M-projection. But this has been done in the proof of Theorem 4.4. $\square$

We next wish to give an application to the homological theory of Banach algebras. To this end, we have to recall some definitions. Unfortunately, to keep things within reasonable bounds, we must refer the reader to the literature for almost all details.
A *cochain complex* C is a sequence of vector spaces C^n, $n \geq 0$, together with linear maps ∂^n, $n \geq -1$,

$$0 \xrightarrow{\partial^{-1}} C^0 \xrightarrow{\partial^0} C^1 \xrightarrow{\partial^1} C^2 \xrightarrow{\partial^2} \cdots$$

such that $\partial^n\partial^{n-1} = 0$ for all $n \in \mathbb{N}$. We put

$$\begin{aligned} Z^n(C) &= \ker \partial^n \\ B^n(C) &= \operatorname{ran} \partial^{n-1} \\ H^n(C) &= Z^n(C)/B^n(C) \end{aligned}$$

so that $H^n(C)$ measures failure of exactness of the cochain complex at stage n. In particular, $H^n(C) = 0$ if and only if C is exact at C^n, i.e. $\ker \partial^n = \operatorname{ran} \partial^{n-1}$. A *cochain homomorphism* ψ between cochain complexes C_1 and C_2 is a sequence $(\psi^n)_{n\geq 0}$ of linear mappings $\psi^n : C_1^n \to C_2^n$ such that $\partial_2^{n-1}\psi^{n-1} = \psi^n\partial_1^{n-1}$ for all $n \in \mathbb{N}$. A diagram of cochain complexes

$$\cdots \xrightarrow{\psi} C \xrightarrow{\varphi} \cdots$$

is said to be exact at C if $\operatorname{ran}\psi^n = \ker\varphi^n$ for all n. Let

$$0 \longrightarrow C_1 \xrightarrow{\psi} C_2 \xrightarrow{\varphi} C_3 \longrightarrow 0$$

be an exact sequence of cochain complexes. (Typically, C_1 is a complex of subspaces of the members of C_2, whereas C_3 is a complex of quotients of the C_2^n. Knowledge of $H(C_1)$ and $H(C_3)$ together with the procedure described below then facilitates the computation of $H(C_2)$.) Then there is always a *long exact cohomology sequence*

$$0 \longrightarrow H^0(C_1) \xrightarrow{H^0(\psi)} H^0(C_2) \xrightarrow{H^0(\varphi)} H^0(C_3) \xrightarrow{d^0} H^1(C_1) \xrightarrow{H^1(\psi)} \cdots$$

The maps $H^n(\psi)$ and $H^n(\varphi)$ are (well-) defined by

$$\begin{aligned} H^n(\psi)([z_1^n]) &= [\psi^n(z_1^n)] \qquad z_1^n \in Z^n(C_1), \\ H^n(\varphi)([z_2^n]) &= [\varphi^n(z_2^n)] \qquad z_2^n \in Z^n(C_2), \end{aligned}$$

where we used brackets to indicate cohomology classes. For the definition of d^n note that, by exactness, each φ^n is surjective, and this, together with a bit of chasing diagrams yields that

$$d^n([z_3^n]) = [(\psi^{n+1})^{-1}\partial^{n+1}(\varphi^n)^{-1}(z_3^n)] \qquad z_3^n \in Z^n(C_3)$$

is in fact well-defined. For more details, see [311, IV.2].

We now come to a more specific example. Let $\mathfrak{A}$ be a Banach algebra and denote by $\mathfrak{X}$ a two-sided Banach $\mathfrak{A}$-module. (By this we mean that $\mathfrak{A}$ operates uniformly continuously from the left and the right on the Banach space $\mathfrak{X}$ such that, for all $a, b \in \mathfrak{A}$ and any $x \in \mathfrak{X}$, the associative laws $(ab)x = a(bx)$ and $x(ab) = (xa)b$ hold.) We write $C^n(\mathfrak{A}, \mathfrak{X})$ for the space of n-linear maps from $\mathfrak{A}^n$ into $\mathfrak{X}$ whenever $n > 0$. We furthermore put $C^0(\mathfrak{A}, \mathfrak{X}) = \mathfrak{X}$. Define mappings $\delta^n : C^{n-1}(\mathfrak{A}, \mathfrak{X}) \longrightarrow C^n(\mathfrak{A}, \mathfrak{X})$ for every $n \in \mathbb{N}$ as follows: Whenever $T \in C^{n-1}(\mathfrak{A}, \mathfrak{X})$ then

$$\begin{aligned} \delta^n T(a_1, \ldots, a_n) &= a_1 T(a_2, \ldots, a_n) - T(a_1 a_2, \ldots, a_n) \\ &\quad + T(a_1, a_2 a_3, \ldots, a_n) - \ldots + (-1)^{n-1} T(a_1, \ldots, a_{n-1} a_n) \\ &\quad + (-1)^n T(a_1, \ldots, a_{n-1}) a_n. \end{aligned}$$

Note that $\delta^{n+1}\delta^n = 0$, and so $C(\mathfrak{A},\mathfrak{X}) := (C^n(\mathfrak{A},\mathfrak{X}))_{n\geq 0}$ becomes a cochain complex. Note also that each map δ^n is bounded. The resulting spaces

$$H^n(\mathfrak{A},\mathfrak{X}) := H^n(C(\mathfrak{A},\mathfrak{X}))$$

are usually called *n-th (Hochschild) cohomology groups.* For a nice explanation of the ideas underlying this construction, see for example [349] or [523]. Of course, in each of these groups some information on the $\mathfrak{A}$-bimodule $\mathfrak{X}$ and hence on $\mathfrak{A}$ itself is encoded. To illustrate this point by means of an example, we recall that, by a result of B. E. Johnson, a locally compact group G is amenable if and only if $H^1(L^1(G),\mathfrak{X}^*) = 0$ for every dual Banach $L^1(G)$-bimodule. (The latter means that the operation of $\mathfrak{A}$ on $\mathfrak{X}^*$ is in addition weak* continuous.) For more on this topic the reader should consult e.g. [348] or, for some more recent developments, [302] and [370].
Suppose that $\mathfrak{Y}$ is an $\mathfrak{A}$-submodule of $\mathfrak{X}$. Then there is an exact sequence

$$0 \longrightarrow C(\mathfrak{A},\mathfrak{Y}) \xrightarrow{\ i\ } C(\mathfrak{A},\mathfrak{X}) \xrightarrow{\ q\ } C(\mathfrak{A},\mathfrak{X}/\mathfrak{Y}) \longrightarrow 0,$$

where $i^n : C^n(\mathfrak{A},\mathfrak{Y}) \to C^n(\mathfrak{A},\mathfrak{X})$ simply enlarges the range of a multilinear mapping from $\mathfrak{A}$ to $\mathfrak{Y}$, and $q^n : C^n(\mathfrak{A},\mathfrak{X}) \to C^n(\mathfrak{A},\mathfrak{X}/\mathfrak{Y})$ compresses a map $T \in C^n(\mathfrak{A},\mathfrak{X})$ with the aid of the quotient map $\pi_{\mathfrak{Y}} : \mathfrak{X} \to \mathfrak{X}/\mathfrak{Y}$ to $\pi_{\mathfrak{Y}}T$. There are, in general, two points which deserve some additional attention in constructing connecting homomorphisms d^n as above: surjectivity of the mappings $H^n(\varphi)$ and continuity of the resulting homomorphisms. That these conditions can be satisfied is, in general, by no means clear. One situation in which these problems can be remedied is that in which M-ideals enter the scene.

Proposition 4.8 *Suppose that $\mathfrak{A}$ is a separable Banach algebra, that $\mathfrak{X}$ is a Banach $\mathfrak{A}$-bimodule and that $\mathfrak{Y}$ is a submodule which is an M-ideal in $\mathfrak{X}$ at the same time. Assume further that either*

(a) *$\mathfrak{A}$ has the λ-approximation property or*
(b) *$\mathfrak{Y}^*$ is an L^1-space.*

Then one has a long exact sequence

$$0 \longrightarrow H^0(\mathfrak{A},\mathfrak{Y}) \xrightarrow{H^0(i)} H^0(\mathfrak{A},\mathfrak{X}) \xrightarrow{H^0(q)} H^0(\mathfrak{A},\mathfrak{X}/\mathfrak{Y}) \xrightarrow{d^0} H^1(\mathfrak{A},\mathfrak{Y}) \xrightarrow{H^1(i)} \cdots$$

such that all the connecting homomorphisms $d^0, d^1, \dots$ are bounded.

PROOF: Let $T = \delta^n_{\mathfrak{X}/\mathfrak{Y}}T_0 \in Z^n(\mathfrak{A},\mathfrak{X}/\mathfrak{Y})$ be given. We shall tacitly make use of the identification $C^n(\mathfrak{A},\mathfrak{X}) = L(\mathfrak{A}\widehat{\otimes}_\pi \dots \widehat{\otimes}_\pi\mathfrak{A},\mathfrak{X})$. By hypothesis and the fact that separability as well as the bounded approximation property – with a different constant, though – pass to finite projective tensor products, we find, with the aid of Theorem II.2.1, a lifting $\widetilde{T} \in C^n(\mathfrak{A},\mathfrak{X})$ for T with $\|\widetilde{T}\| \leq \lambda^n\|T\|$ when (a) is satisfied and with $\|\widetilde{T}\| = \|T\|$ whenever (b) holds. It follows that

$$q^{n+1}\delta^{n+1}_{\mathfrak{X}}\widetilde{T} = \delta^{n+1}_{\mathfrak{X}/\mathfrak{Y}}q^n\widetilde{T} = \delta^{n+1}_{\mathfrak{X}/\mathfrak{Y}}\delta^n_{\mathfrak{X}/\mathfrak{Y}}T_0 = 0,$$

so $\delta_{\mathfrak{X}}^{n+1}\widetilde{T}$ takes its values in $\mathfrak{Y}$, and, formally, there is $T_1 \in C^{n+1}(\mathfrak{A}, \mathfrak{Y})$ with $i^{n+1}T_1 = \delta^{n+1}\widetilde{T}$. Put $d^n(T) = [T_1]$, the equivalence class of T_1, and convince yourself (if you didn't before) that d^n is well-defined. Since

$$\|d^n(T)\| = \|T_1\| = \|i^{n+1}T_1\| = \|\widetilde{T}\| \leq \lambda^n \|T\|$$

if (a) is valid and a similar estimate holds in case (b), d^n is bounded, and we are done. □

Using Theorem 4.4 and the above proposition one obtains:

Corollary 4.9 *Suppose that $\mathfrak{A}$ is a separable C^*-subalgebra of the C^*-algebra $\mathfrak{B}$ and that $\mathfrak{J}$ is a two-sided closed ideal of $\mathfrak{B}$. If either $\mathfrak{A}$ has the bounded approximation property or else, if $\mathfrak{J}$ is commutative, then one has a long exact sequence*

$$0 \longrightarrow H^0(\mathfrak{A}, \mathfrak{J}) \xrightarrow{H^0(i)} H^0(\mathfrak{A}, \mathfrak{B}) \xrightarrow{H^0(q)} H^0(\mathfrak{A}, \mathfrak{B}/\mathfrak{J}) \xrightarrow{d^0} H^1(\mathfrak{A}, \mathfrak{J}) \xrightarrow{H^1(i)} \cdots$$

with bounded connecting homomorphisms d^n.

We end our short digression to the homological theory of C^*-algebras with an illustration of how long exact sequences help one with calculations. (The proof below is, however, far from being self-contained.) Recall that a von Neumann algebra is called approximately finite dimensional if it is the (von Neumann) inductive limit of finite dimensional von Neumann algebras.

Corollary 4.10 *Suppose $\mathfrak{A} \subset L(H)$ is a separable C^*-algebra enjoying the bounded approximation property such that the closure of $\mathfrak{A}$ in the weak operator topology is an approximately finite dimensional von Neumann algebra. Then*

$$H^n(\mathfrak{A}, L(H)/K(H)) \cong H^{n+1}(\mathfrak{A}, K(H))$$

for all $n \in \mathbb{N}$.

PROOF: Write $\mathfrak{A}^-$ for the closure of $\mathfrak{A}$ in the weak operator topology. One of the main results in [351] (Theorem 6.1) shows that we have in the present situation

$$H^n(\mathfrak{A}, L(H)) = H^n(\mathfrak{A}^-, L(H)).$$

By a result of Kadison and Ringrose (see e.g. [523, Theorem 7.4]), $H^n(\mathfrak{A}^-, L(H)) = 0$. By Corollary 4.9 we then have an exact sequence

$$0 = H^n(\mathfrak{A}, L(H)) \to H^n(\mathfrak{A}, L(H)/K(H)) \to H^{n+1}(\mathfrak{A}, K(H)) \to H^{n+1}(\mathfrak{A}, L(H)) = 0.$$

But this clearly implies that $H^n(\mathfrak{A}, L(H)/K(H)) \cong H^{n+1}(\mathfrak{A}, K(H))$, which is what we had to show. □

It should be remarked that there is one important class of C^*-algebras which satisfies the hypothesis of the above result: It is known that a nuclear C^*-algebra $\mathfrak{A}$ always has the metric approximation property (see [123] or [376] for these matters) and that $\mathfrak{A}^{**}$ is approximately finite dimensional in this case ([122] and [201]). For more on this topic see the Notes and Remarks section.

V.5 Some further approximation theorems

The chief concern of this section is to present some results similar to Theorem 3.2. We preferred to collect them at this place, mainly because they all exhibit more or less the same idea: Whenever $P : \mathfrak{A}^{**} \to \mathfrak{J}^{\perp\perp}$ is the M-projection corresponding to an M-ideal $\mathfrak{J}$, then all of these results aim at finding a net (z_α) converging to $z = P(e)$ in the weak* sense subject to some constraints that have shown to be useful in special situations.
Our first result in this chain will find its application in the following section, when it will be shown that the M-ideals of $L(X)$ are inner whenever X is uniformly smooth or uniformly convex.

Proposition 5.1 *Suppose $\mathfrak{A}$ is a unital Banach algebra, and let $\mathfrak{J}$ be an M-ideal in $\mathfrak{A}$. Denote by P be the associated M-projection on $\mathfrak{A}^{**}$ and put $z = P(e)$. Then $\mathfrak{J}$ is left inner if and only if there is a net $(t_\alpha) \subset \mathfrak{J}$ approximating z in the weak* topology such that for all $a \in \mathfrak{A}$*

$$\lim_{\lambda\to 0} \limsup_{\alpha} \frac{\|t_\alpha + \lambda a(e - t_\alpha)\| - 1}{\lambda} = 0$$

and

$$\lim_{\lambda\to 0} \limsup_{\alpha} \frac{\|\lambda a t_\alpha + e - t_\alpha\| - 1}{\lambda} = 0.$$

An analogous result holds in the case of right inner M-ideals.

PROOF: That left inner M-ideals always admit nets of the above type is an immediate consequence of Theorem 3.2: If (t_α) is a net the existence of which was shown by this result, then

$$\lim_{\lambda\to 0} \limsup_{\alpha} \frac{\|t_\alpha + \lambda a(e - t_\alpha)\| - 1}{\lambda} \le \lim_{\lambda\to 0} \frac{\max\{\lambda\|a\|, 1\} - 1}{\lambda} = 0,$$

and one similarly obtains the other equation. For the converse, suppose $\mathfrak{A}^* = \mathfrak{J}^\perp \oplus_1 \mathfrak{J}^*$, and let us first prove that

$$\mathfrak{A}^{**}(e - z) \subset (\mathfrak{J}^*)^\perp.$$

By weak* continuity of the first Arens product, and since $(\mathfrak{J}^*)^\perp$ is weak* closed, we may reduce our efforts and show only that

$$\mathfrak{A}(e - z) \subset (\mathfrak{J}^*)^\perp.$$

This holds, in light of Lemma 2.2, once it has been demonstrated that $\varphi(a(e - z)) = 0$ for all $a \in \mathfrak{A}$ and $\varphi \in F_1$. (Here, F_1 stands for the set $S_{\mathfrak{A}} \cap \mathfrak{J}^*$.) Let us suppose to the contrary that for some $a \in \mathfrak{A}$ and an appropriate $\varphi \in F_1$

$$\varphi(a(e - z)) = \lambda_0,$$

with $0 < \lambda_0 < 1$. Put

$$a_n = z + \lambda_0^n a(e - z).$$

Then, by assumption on z,

$$\begin{aligned}\lambda_0 &= \frac{\varphi(a_n)-1}{\lambda_0^n} \\ &= \lim_\alpha \frac{\varphi(t_\alpha + \lambda_0^n a(e-t_\alpha))-1}{\lambda_0^n} \\ &\le \limsup_\alpha \frac{\|t_\alpha + \lambda_0^n a(e-t_\alpha)\|-1}{\lambda_0^n}\end{aligned}$$

and hence

$$\lambda_0 \le \lim_{n\to\infty} \limsup_\alpha \frac{\|t_\alpha + \lambda_0^n a(e-t_\alpha)\|-1}{\lambda_0^n} = 0,$$

which contradicts our assumption on λ_0. Repeating the same argument with the expression

$$\lambda_0 a t_\alpha + e - t_\alpha$$

will give $\mathfrak{A}^{**}z \subset \mathfrak{J}^{\perp\perp}$. But this permits us to infer that $R_z^{(1)}$ is actually onto, and the result once again follows by an application of Proposition I.1.2. □

We wish to present a second approximation theorem of this type (Theorem 5.4 below). This will be prepared by a result deserving independent interest (Theorem 5.3), which in turn requires the following lemma.

Lemma 5.2 *Let X be a Banach space and suppose $S \subset X^*$ is convex and weak* compact. If Ω is a compact convex subset of the complex plane, then*

$$\overline{\mathrm{co}}^{w^*}\,\Omega S = \overline{\mathrm{co}}^{\|\,\|}\,\Omega S.$$

PROOF: Fix $\delta > 0$ and pick $z_1, \dots, z_n \in \mathbb{C}$ with

$$\Omega \subset \Omega_\delta := \mathrm{co}\,\{z_1, \dots, z_n\} \subset (1+\delta)\Omega.$$

Since $\mathrm{co}\,\Omega_\delta S = \mathrm{co}\,(\bigcup_{j=1}^n z_j S)$ is weak* compact and thus norm closed, we have

$$\overline{\mathrm{co}}^{w^*}\,\Omega S \subset \mathrm{co}\,\Omega_\delta S \subset (1+\delta)\overline{\mathrm{co}}^{\|\,\|}\,\Omega S.$$

Letting $\delta \to 0$ gives $\overline{\mathrm{co}}^{w^*}\,\Omega S \subset \overline{\mathrm{co}}^{\|\,\|}\,\Omega S$, which is the less trivial inclusion. □

Theorem 5.3 *Let $\mathfrak{A}$ be a unital Banach algebra and suppose $F \in \mathfrak{A}^{**}$. Then, for some net (ε_α) of positive numbers tending to zero, there is a net $(f_\alpha) \subset \mathfrak{A}$ converging to F in the weak* topology with* $\limsup_\alpha \|f_\alpha\| \le \|F\|$ *and*

$$v(f_\alpha) \subset v(F) + B(0, \varepsilon_\alpha).$$

PROOF: Let M and M^* be subsets of a Banach space X and its dual, respectively. In the following, we denote the *real* polar of M by

$$M^\pi := \{x^* \in X^* \mid \mathrm{Re}\, x^*(x) \le 1\ \forall x \in M\}$$

and write M_π^* for the set

$$M_\pi^* := \{x \in X \mid \operatorname{Re} x^*(x) \leq 1 \ \forall x^* \in M^*\}.$$

In addition, $v_*(F)$ denotes the *lower* numerical range of $F \in \mathfrak{A}^{**}$, which is defined by

$$v_*(F) = \{F(\psi) \mid \psi \in S_{\mathfrak{A}}\}.$$

Suppose that Ω is a compact convex subset of the plane containing an interior point. Our claim is then – except for the additional norm condition – equivalent to showing that, for any such set,

$$\overline{\{a \in \mathfrak{A} \mid v(a) \subset \Omega\}}^{w^*} = \{F \in \mathfrak{A}^{**} \mid v(F) \subset \Omega\},$$

where the first set is identified with its canonical image in $\mathfrak{A}^{**}$. Since $v(\lambda e + a) = \lambda + v(a)$ for all $\lambda \in \mathbb{C}$ and every $a \in \mathfrak{A}$, we may suppose that $0 \in \operatorname{int} \Omega$ without spoiling the argument. Then $0 \in \{a \in \mathfrak{A} \mid v(a) \subset \Omega\}$ and hence, by the bipolar theorem, the above reduces to

$$\{a \in \mathfrak{A} \mid v(a) \subset \Omega\}^{\pi\pi} = \{F \in \mathfrak{A}^{**} \mid v(F) \subset \Omega\}.$$

Now, $\{a \in \mathfrak{A} \mid v(a) \subset \Omega\} = (\Omega^\pi S_{\mathfrak{A}})_\pi$ and similarly

$$(\Omega^\pi S_{\mathfrak{A}})^\pi = \{F \in \mathfrak{A}^{**} \mid v_*(F) \subset \Omega\} = \{F \in \mathfrak{A}^{**} \mid v(F) \subset \Omega\},$$

where the last equation holds by virtue of $\overline{v_*(F)} = v(F)$ (Lemma 1.9). By assumption, $0 \in \operatorname{int} \Omega$, so Ω^π is compact and convex, which together with Lemma 5.2 gives

$$\overline{\text{co}}^{w^*} \Omega^\pi S_{\mathfrak{A}} = \overline{\text{co}}^{\|\,\|} \Omega^\pi S_{\mathfrak{A}}$$

Putting all the pieces together, we obtain

$$\begin{aligned} \{a \in \mathfrak{A} \mid v(a) \subset \Omega\}^{\pi\pi} &= (\overline{\text{co}}^{w^*} \Omega^\pi S_{\mathfrak{A}})^\pi \\ &= (\Omega^\pi S_{\mathfrak{A}})^\pi \\ &= \{F \in \mathfrak{A}^{**} \mid v(F) \subset \Omega\}, \end{aligned}$$

which gives the claim. We now adjust the net thus obtained by a convex combinations argument so as to satisfy the condition $\limsup_\alpha \|f_\alpha\| \leq \|F\|$. To do so pick any net (g_α) in $\mathfrak{A}$ such that $\|g_\alpha\| \leq \|F\|$ and $g_\alpha \xrightarrow{w^*} F$. (One may employ the same index set.) Then $f_\alpha - g_\alpha \xrightarrow{w} 0$ so that there are $\overline{f}_\alpha \in \operatorname{co}\{f_\beta \mid \beta \geq \alpha\}$ and $\overline{g}_\alpha \in \operatorname{co}\{g_\beta \mid \beta \geq \alpha\}$ such that $\|\overline{f}_\alpha - \overline{g}_\alpha\| \to 0$. Hence $(\overline{f}_\alpha)$ still has the numerical range property, but also inherits the desired norm property from the net $(\overline{g}_\alpha)$. □

Since the $\overline{f}_\alpha$ are "convex blocks" of the f_α, we shall refer to the above convex combinations argument (which we shall use several times later and have first encountered in the proof of Theorem 3.2) also as the "blocking technique".

Theorem 5.4 *Let $\mathfrak{A}$ be a unital Banach algebra, and suppose that $\mathfrak{J}$ is an M-summand in $\mathfrak{A}^{**}$. Denote by P the corresponding M-projection. Then there is a net $(z_\alpha) \subset \mathfrak{A}$ with weak* limit $z = P(e)$ such that for a suitable net of positive numbers (ε_α) tending to zero*

$$v(z_\alpha) \subset [0, 1] + B(0, \varepsilon_\alpha)$$

and furthermore

$$\lim_\alpha \|\mu(e - z_\alpha) + \lambda z_\alpha\| = \max\{|\lambda|, |\mu|\}$$

for all complex numbers λ, μ. If $\mathfrak{J}$ is the bipolar of an M-ideal $\mathfrak{J}_0 \subset \mathfrak{A}$, then we may suppose that $z_\alpha \in \mathfrak{J}_0$ for all α.

PROOF: We first show how to produce a net for which the second relation holds. To do this, one has to observe that for an M-projection P the equation

$$\|\mu(Id - P) + \lambda P\| = \max\{|\mu|, |\lambda|\}$$

holds. Since Δ is isometric on $\mathfrak{A}^{**}$ (Theorem 2.1(e)), it follows that

$$\max\{|\mu|, |\lambda|\} = \|\mu(Id - P) + \lambda P\| = \|\mu(Id - P)(e) + \lambda P(e)\|.$$

Let (T_α) be a net of operators gained with the aid of the local reflexivity principle similarly to the proof of Theorem 3.2. Then it follows for $z_\alpha^0 := T_\alpha(P(e))$

$$\begin{aligned} \lim_\alpha \|\mu(e - z_\alpha^0) + \lambda z_\alpha^0\| &= \lim_\alpha \|\mu(e - T_\alpha(P(e))) + \lambda T_\alpha(P(e))\| \\ &= \lim_\alpha \|T_\alpha(\mu(e - P(e)) + \lambda P(e))\| \\ &= \max\{|\mu|, |\lambda|\}. \end{aligned}$$

A blocking technique similar to the one used in the proof of Theorem 3.2 allows us to pass to the required net (z_α) in $B_{\mathfrak{J}}$.
A net which fulfills the requirements concerning the numerical range is provided by Theorem 5.3 (note $v(P) = [0,1]$); and by again using our blocking technique we obtain a net sharing both properties simultaneously. □

V.6 Inner M-ideals in $L(X)$

In this last section of the present chapter we will investigate the M-ideal structure of the Banach algebra of operators $L(X)$. Spaces of bounded linear operators will also be the subject of the next chapter. Whereas in the final chapter we will put emphasis on Banach *space* methods and also consider spaces of operators which act between different Banach spaces, we will employ here Banach *algebra* techniques and apply the methods developed so far. Our knowledge on the M-ideal structure of $L(X)$ is rather incomplete. All that is known at this moment is contained in the following theorem.

Theorem 6.1 *Let J be an M-ideal of $L(X)$. Then*

(a) *J is left (right) inner if $\mathbb{K} = \mathbb{C}$ and X is uniformly convex (uniformly smooth),*
(b) *J is left (right) inner if $\mathbb{K} = \mathbb{C}$ and $X = L^1(\mu)$ or a predual of a function algebra (X is an L^1-predual or a function algebra on some compact Hausdorff space K),*
(c) *J is left (right) inner for arbitrary scalar fields if J is an M-summand and the centralizer $Z(X)$ (the Cunningham algebra $\operatorname{Cun}(X)$) is trivial,*
(d) *if $K(X)$, the space of compact operators on X, is an M-ideal in $L(X)$, then it is an inner M-ideal. Here, again, no restriction is imposed on the scalar field.*

The proof of part (d) will be presented in the next chapter (Proposition VI.4.10). Also, the proof of (c) will be postponed; (c) is a special case of Theorem VI.1.2. Finally, (a) and (b) can be attacked with tools that were provided in the foregoing sections. We mention in addition that one can read the following result between the lines of the proof of Theorem VI.2.3:

An M-ideal in $L(X)$ is right inner for the real space $X = C(K)$.

We now start preparing the proof of part (a). First, let us recall the definition of the modulus of convexity of a Banach space X: For $\varepsilon \in (0,2]$ this function is defined by

$$\delta_X(\varepsilon) := \inf\left\{1 - \frac{\|x+y\|}{2}\;\middle|\; x,y \in S_X, \|x-y\| = \varepsilon\right\}.$$

A Banach space is called *uniformly convex* if and only if $\delta_X(\varepsilon) > 0$ for all $\varepsilon \in (0,2]$. The proof of (a) in the following lemma can be found in [423, Section 1.e], whereas (c) is an immediate consequence of (a) and the definition involved. For part (b) we refer to [275, Lemma 5.1].

Lemma 6.2 *Let X be a Banach space.*

(a) *The modulus of convexity of X can equivalently be defined by*

$$\delta_X(\varepsilon) = \inf\left\{1 - \frac{\|x+y\|}{2}\;\middle|\; x,y \in B_X, \|x-y\| \ge \varepsilon\right\}.$$

Furthermore, the function $\varepsilon \mapsto \delta_X(\varepsilon)/\varepsilon$ is nondecreasing on $(0,2]$.

(b) *The function δ_X is continuous on $(0,2)$.*

(c) *If X is uniformly convex, then δ_X is strictly increasing as is its inverse function δ_X^{-1}, and*

$$\lim_{\varepsilon\to 0} \delta_X(\varepsilon) = \lim_{\varepsilon\to 0} \delta_X^{-1}(\varepsilon) = 0.$$

The modulus of smoothness is defined for $\tau > 0$ by

$$\rho_X(\tau) = \inf\left\{\frac{\|x+y\| + \|x-y\|}{2} - 1\;\middle|\; \|x\| = 1, \|y\| = \tau\right\},$$

and X is said to be *uniformly smooth* if and only if $\lim_{\tau\to 0} \rho_X(\tau)\tau^{-1} = 0$. We will make use of the following theorem. Again, the reader should consult [423] for proofs, examples and further literature.

Theorem 6.3

(a) *A Banach space X is uniformly convex if and only if X^* is uniformly smooth.*

(b) *Every uniformly convex (and thus also every uniformly smooth) Banach space is reflexive.*

Our first step towards the proof of Theorem 6.1(a) is the following lemma.

Lemma 6.4 *Let X be uniformly convex and suppose that there are operators $A, T \in L(X)$ with $\|A\| \le 1$, $\|T\| \le 1 + \varepsilon$, and $\|Id - \lambda T\| \le 1 + \varepsilon$ whenever $0 \le \lambda \le 2$. Then*

$$\|T + \lambda A(Id - T)\| \le 1 + \varepsilon + \lambda(1+\varepsilon)\delta_X^{-1}(\varepsilon + \lambda).$$

PROOF: Suppose that T is as above and fix $\lambda > 0$ as well as $y \in S_X$. If $\|Ty\| \leq 1-\lambda(1+\varepsilon)$ then, by the triangle inequality,

$$\|(T+\lambda A(Id-T))\,y\| \leq 1,$$

which permits us to assume that $\|Ty\| > 1-\lambda(1+\varepsilon)$. Set

$$u = \frac{Ty}{1+\varepsilon} \quad \text{and} \quad v = \frac{y-Ty}{1+\varepsilon}.$$

It follows by assumption that $\|u \pm v\| \leq 1$, and since $u = \frac{1}{2}[(u+v)+(u-v)]$, we have $\delta_X(\|2v\|) \leq 1 - \|u\|$, hence $\|u\| \leq 1 - \delta_X(\|2v\|)$. By assumption on y and definition of u, $[1-\lambda(1+\varepsilon)](1+\varepsilon)^{-1} \leq \|u\|$ and so

$$\frac{1-\lambda(1+\varepsilon)}{1+\varepsilon} \leq 1 - \delta_X(2\|v\|).$$

Regrouping terms and using the fact that δ_X^{-1} is increasing, we arrive at

$$\|v\| \leq 2\|v\| \leq \delta_X^{-1}\left(1 - \frac{1}{1+\varepsilon} + \lambda\right) \leq \delta_X^{-1}(\varepsilon + \lambda).$$

This gives

$$\|(T+\lambda A(Id-T))\,y\| \leq 1+\varepsilon+\lambda\|(Id-T)y\| \leq 1+\varepsilon+\lambda(1+\varepsilon)\delta_X^{-1}(\varepsilon+\lambda),$$

and we are done. □

We now come to the

PROOF OF THEOREM 6.1(a):
We first consider a uniformly convex Banach space X. Let $J \subset L(X)$ be an M-ideal and, as before, denote by $z \in J^{\perp\perp}$ the hermitian idempotent that corresponds to J. By Theorem 5.4, there is a net $(T_\alpha) \subset J$ of operators converging with respect to the weak* topology of $L(X)^{**}$ to z such that

$$\begin{aligned} \|Id - \lambda T_\alpha\| &= \|(Id - T_\alpha) + (1-\lambda)T_\alpha\| \\ &\leq (1+\varepsilon_\alpha)\max\{1, |1-\lambda|\} \\ &= 1+\varepsilon_\alpha \end{aligned}$$

for $0 \leq \lambda \leq 2$ and $\varepsilon_\alpha \to 0$. Using Lemma 6.2(b) and (c) and Lemma 6.4 we find for all $A \in B_{L(X)}$

$$\begin{aligned} \lim_{\lambda\to 0}\limsup_{\alpha} \frac{\|T_\alpha + \lambda A(Id-T_\alpha)\| - 1}{\lambda} &\leq \lim_{\lambda\to 0}\limsup_{\alpha} \frac{\varepsilon_\alpha + \lambda(1+\varepsilon_\alpha)\delta_X^{-1}(\varepsilon_\alpha+\lambda)}{\lambda} \\ &= \lim_{\lambda\to 0} \delta_X^{-1}(\lambda) \\ &= 0. \end{aligned}$$

Repeating the above argument with z replaced by $Id - z$ we get that for any $A \in B_{L(X)}$

$$\lim_{\lambda \to 0} \limsup_{\alpha} \frac{\|\lambda A T_\alpha + Id - T_\alpha\| - 1}{\lambda} = 0,$$

and the theorem follows by an appeal to Proposition 5.1.
The case of uniformly smooth spaces follows from this by duality (Theorem 6.3). □

Corollary 6.5 *Let X be a complex Banach space and J an M-ideal of $L(X)$. If X is uniformly convex then J is a left ideal. Is X uniformly smooth, then J must be a right ideal, and if X is both uniformly smooth and uniformly convex, then J is a two-sided ideal.*

PROOF: Use the fact that, by reflexivity of X, the mapping $T \mapsto T^*$ is an algebraic anti-isomorphism, and Theorem 3.2. □

Corollary 6.6 *Let $1 < p < \infty$. Then $K(\ell^p)$ is the only nontrivial M-ideal of $L(\ell^p)$.*

PROOF: That $K(\ell^p)$ is an M-ideal in $L(\ell^p)$ was mentioned already several times (Example I.1.4(d)) and will eventually be proved in Example VI.4.1. To prove uniqueness note that by a result from [277] (see also [308]), $K(\ell^p)$ is the only two-sided closed ideal in $L(\ell^p)$. Since ℓ^p is a well-known example of a uniformly smooth and uniformly convex Banach space we may now invoke Corollary 6.5. □

Whereas in the above the norm condition on the approximating net of Theorem 5.4 was used, the proof of Theorem 6.1(b) requires the restrictions put on the numerical range of the elements of the involved net.
Before we go into the details let us make some remarks on how the theory of numerical ranges for elements in Banach algebras transfers to the special case of the Banach algebra $L(X)$.

In the case of the algebra $L(X)$, the set $\mathrm{S}_{L(X)}$ as a whole is, of course, highly nonaccessible in general. However, concerning $v(T, L(X))$ one may restrict one's attention to the subset $\Pi(X)$ of $\mathrm{S}_{L(X)}$ which is defined by

$$\Pi(X) := \{(x^*, x) \in S_{X^*} \times S_X \mid x^*(x) = 1\}$$

and define the *spatial numerical range* of an operator T to be

$$V(T) := \{x^*(Tx) \mid (x^*, x) \in \Pi(X)\}.$$

In this way, one has access to the set $v(T, L(X))$ by virtue of

$$v(T, L(X)) = \overline{\mathrm{co}}\, V(T).$$

(See [84, Chapter 3] for further details.) In particular, the notion of a *hermitian operator* may also be defined by

$$T \text{ hermitian} \iff V(T) \subset \mathbb{R}.$$

Note that the above notion has already been used in Section I.1; with the previous remarks it will be possible to combine the results obtained in that section with those that were presented in the last one. More precisely, we will profit from the fact that sometimes operators with the property that $V(T)$ is distributed along a small strip around the real axis are perturbations of elements in $\mathrm{Mult}(X)$ or $\mathrm{Cun}(X)$:

Lemma 6.7 *Let $\mathfrak{A}$ be a function algebra on a compact space K and put $R(\varepsilon) = \{z \in \mathbb{C} \mid |\mathrm{Im}\, z| \le \varepsilon\}$. If $T \in L(\mathfrak{A})$ has the property that, for some $\varepsilon > 0$, $V(T) \subset R(\varepsilon)$, then there is an $a \in \mathfrak{A}$ such that $\|T - M_a\| \le 2\varepsilon$. Similarly, if $X = L^1(\mu)$ and an operator T on X has the above property, then there is an operator $S \in \mathrm{Cun}(X)$ such that $\|T - S\| \le 2\varepsilon$.*

For the proof we have to recall some details from the theory of function algebras. Those points $k \in K$ with $\delta_k \in \mathrm{ex}\, B_{\mathfrak{A}^*}$ are said to belong to the *Choquet boundary* of $\mathfrak{A}$, denoted by $\mathrm{ch}\,\mathfrak{A}$. It is known [586, Theorem 7.18] that k is in the Choquet boundary of $\mathfrak{A}$ if and only if $\{k\}$ is a p-set, and, consequently, for each such point, $\mathfrak{J}_k := \ker \delta_k$ is an M-ideal by Theorem 4.2.

PROOF: Put $a := T(e)$ and let $k \in \mathrm{ch}\,\mathfrak{A}$. Since $\mathfrak{J}_k$ is an M-ideal, and since for any $x \in \mathfrak{A}$ $\xi := \delta_k(x)e - x$ is in $\mathfrak{J}_k$, we have by Proposition I.1.24

$$\varepsilon\|\xi\| \ge d(T\xi, \mathfrak{J}_k) = |\delta_k(T\xi)| = |\delta_k(ax - Tx)|.$$

Since $\|\xi\| \le 2\|x\|$, this yields

$$\|ax - Tx\| = \sup_{k \in \mathrm{ch}\,\mathfrak{A}} |\delta_k(ax - Tx)| \le 2\varepsilon\|x\|.$$

Let us look at the case where $X = L^1(\mu)$: By the above, we can find an operator S which is contained in $\mathrm{Mult}(X^*) = Z(X^*)$ such that

$$\|T^* - S\| \le 2\varepsilon.$$

But S is a multiplication operator on $(L^1(\mu))^*$ (Example I.3.4(a)) and thus weak* continuous, hence the result. □

The way is now paved for the proof of Theorem 6.1(b). However, it turns out that more can be shown. We split this extended version of Theorem 6.1(b) into two pieces and begin with function algebras.

Theorem 6.8 *Let $\mathfrak{A}$ be a function algebra on some compact Hausdorff space K. Then every M-ideal in $L(\mathfrak{A})$ is right inner. Furthermore, the M-ideals of $L(\mathfrak{A})$ are precisely the subspaces of the form*

$$L(\mathfrak{A})_J = \overline{\mathrm{lin}}\,\{M_a T \mid T \in L(\mathfrak{A}), a \in J\} = \{T \in L(\mathfrak{A}) \mid \limsup_{k \to k_0} \|T^*\delta_k\| = 0\ \forall k_0 \in D\},$$

where J is an M-ideal of $\mathfrak{A}$ and D is the p-set corresponding to J according to Theorem 4.2.

PROOF: That all subspaces of the above form are in fact M-ideals was shown in Proposition 3.5. We are thus left with showing that all M-ideals look like this.
So, let $\mathfrak{J}$ be an M-ideal in $L(\mathfrak{A})$ and denote by P its defining M-projection $P: L(\mathfrak{A})^{**} \to \mathfrak{J}^{\perp\perp}$. Start with a net (T_α) in $\mathfrak{J}$ converging in the weak* topology to $P(e)$ and with the property that $V(T_\alpha) \subset R(\varepsilon_\alpha)$ for some net (ε_α) of positive numbers tending to zero; see Theorem 5.4. By Lemma 6.7, we may suppose without loss of generality that for some $a_\alpha \in \mathfrak{A}$ the operators T_α are of the form M_{a_α}. But then it follows that

$$P(e) = w^*\text{-}\lim_\alpha M_{a_\alpha} \in (\mathrm{Mult}^l_{inn} L(\mathfrak{A}))^{\perp\perp}$$

whence by Lemma 3.4,

$$P(e) \in \mathrm{Mult}^{l,2}_{inn} L(\mathfrak{A}^{**}).$$

It follows that $P(e)$ is an M-projection. By injectivity of the map Δ (Theorem 2.1), we must have $P = L^2_{P(e)}$, and so, $\mathfrak{J}$ is inner. Since $P(e) \in \mathfrak{A}^{\perp\perp}$, when $\mathfrak{A}$ is identified with a subalgebra of $\mathrm{Mult}^l_{inn} L(\mathfrak{A})$, we may apply Proposition 3.3 to see that $P(e)$ gives rise to an inner M-ideal J in $\mathfrak{A}$ as well. Next, observe that we are in a situation treated in Proposition 3.5. Consequently,

$$\mathfrak{J} = L(\mathfrak{A})_J = \overline{\mathrm{lin}}\, \{M_a T \mid T \in L(\mathfrak{A}), a \in J\},$$

since the inner projection pertaining to $L(\mathfrak{A})_J$ is also $P(e)$. It thus remains to verify that

$$L(\mathfrak{A})_J = \{T \in L(\mathfrak{A}) \mid \limsup_{k \to k_0} \|T^* \delta_k\| = 0 \ \forall k_0 \in D\}.$$

To this end, we observe that we have for an operator of the form $M_j T$ with $j \in J$ and for $k_0 \in D$,

$$\limsup_{k \to k_0} \|(M_j T)^* \delta_k\| = \limsup_{k \to k_0} \|j(k) T^* \delta_k\| \le \|T\| \limsup_{k \to k_0} |j(k)| = 0.$$

Now, recall that we saw in the proof of Theorem 4.2 that there is an approximate unit $(p_\alpha)_\alpha$ in J with

$$\forall \varepsilon > 0\ \forall U \supset D,\ U \text{ open } \exists \alpha_0\ \forall \alpha > \alpha_0\ \left|(e - p_\alpha)_{|K \setminus U}\right| < \varepsilon$$

which, of course, may be chosen such that M_{p_α} converges to $P(e)$ in the $\sigma(L(\mathfrak{A})^{**}, L(\mathfrak{A})^*)$-topology. Hence, by Theorem 1.2(d),

$$\mathfrak{J} = \{T \in L(\mathfrak{A}) \mid \lim_\alpha M_{p_\alpha} T = T\},$$

and so, each T in $L(\mathfrak{A})_J$ may be approximated uniformly by operators of the form $M_j T$, with $j \in J$. This yields

$$L(\mathfrak{A})_J \subset \{T \in L(\mathfrak{A}) \mid \limsup_{k \to k_0} \|T^* \delta_k\| = 0 \ \forall k_0 \in D\}.$$

On the other hand, for each $T \in L(\mathfrak{A})$ with $\limsup_{k\to k_0} \|T^*\delta_k\| = 0$ whenever $k_0 \in D$ we may, for all $\varepsilon > 0$, select an open neighbourhood U of D such that $\|T^*\delta_k\| < \varepsilon$ for all $k \in U$. Now, for sufficiently large α,

$$\left|(e - p_\alpha)_{|K\setminus U}\right| < \varepsilon$$

and consequently we have, regarding elements of $L(\mathfrak{A})$ as vector valued functions on K,

$$\|M_{p_\alpha}T - T\| = \max\{\|(M_{e-p_\alpha}T)_{|U}\|, \|(M_{e-p_\alpha}T)_{|K\setminus U}\|\} < \varepsilon,$$

whence $\lim_\alpha M_{p_\alpha}T = T$ and $T \in L(\mathfrak{A})_J$. □

We next treat M-ideals in the space $L(X)$ in the case where X is a predual of a function algebra.

Theorem 6.9 *Suppose that X is a predual of a function algebra $\mathfrak{A}$ on some compact space K.*

(a) *If $\mathfrak{J}$ is an M-ideal in $L(X)$ then it is left inner and, furthermore, there is a p-set $D \subset K$ for $\mathfrak{A}$ such that*

$$\mathfrak{J} = \mathfrak{J}^{(D)} := \{T \in L(X) \mid \limsup_{k\to k_0} \|T^{**}\delta_k\| = 0 \ \forall k_0 \in D\}.$$

(b) *In the special case that $X = L^1(\mu)$, for each closed subset D of K, where $C(K) = L^1(\mu)^*$, $\mathfrak{J}^{(D)}$ is an M-ideal.*

PROOF: Denote by Adj the antihomomorphism that maps an operator $T \in L(X)$ to $T^* \in L(\mathfrak{A})$. Since Adj** is an antihomomorphism from $L(X)^{**}$ equipped with the first Arens product to $L(\mathfrak{A})^{**}$ when the latter is furnished with the second Arens product (see Theorem 1.2), we may identify $L(X)^{**}$ with a subalgebra of $L(\mathfrak{A})^{**}$.
Let $P : L(X)^{**} \to \mathfrak{J}^{\perp\perp}$ be the M-projection given by $\mathfrak{J}$. As in the proof of Theorem 6.8, we start with a net (T_α) in $B_{L(X)}$ converging to $P(e)$ in the weak* topology and satisfying

$$V(T_\alpha) \subset R(\varepsilon_\alpha)$$

for some net (ε_α) of real numbers converging to zero. Clearly, (T_α) converges also in the $\sigma(L(\mathfrak{A})^{**}, L(\mathfrak{A})^*)$-topology to $P(e)$. Moreover, by Lemma 6.7, we may disturb T_α^* slightly in norm to obtain a net $(M_{a_\alpha}) \in L(\mathfrak{A})$ still converging to $P(e)$. Following the lines of Theorem 6.8's proof, we see that

$$P(e) \in \mathrm{Mult}_{inn}^{l,2}(L(\mathfrak{A})^{**})$$

which amounts to

$$P(e) \in \mathrm{Mult}_{inn}^{r,1}(L(X)^{**}).$$

Hence, $\mathfrak{J}$ must be left inner. Applying Proposition 3.3, we get an M-ideal $\mathfrak{J}_1$ in $L(\mathfrak{A})$ with

$$\mathfrak{J}_1 \cap L(X) = \mathfrak{J}.$$

An application of Theorem 6.8 now yields (a).
To prove (b), let $X = L^1(\mu)$ and choose a closed subset D of K. We must show that

$$\mathfrak{J}^{(D)} = \{T \in L(L^1(\mu)) \mid \limsup_{k \to k_0} \|T^{**}\delta_k\| = 0 \ \forall k_0 \in D\}$$

is an M-ideal. For this purpose, let $S_1, S_2, S_3 \in B_{\mathfrak{J}^{(D)}}$ and $T \in B_{L(L^1(\mu))}$ be given. Pick an open neighbourhood U of D such that $\|S_i^{**}\delta_t\| < \varepsilon$ for all $t \in U$ and a continuous function ψ with $0 \le \psi \le 1$, $\psi|_D = 0$ and $\psi|_{K \setminus U} = 1$. Putting $S = TM_\psi$ we have $S \in \mathfrak{J}^{(D)}$. Since, in addition,

$$\begin{aligned} \|S_i + T - S\| &= \max\{\|(S_i + (1-\psi)T)|_U\|, \|(S_i + (1-\psi)T)|_{K\setminus U}\|\} \\ &\le 1 + \varepsilon, \end{aligned}$$

the claim follows from Theorem I.2.2. □

It is unknown how the sets appearing in part (a) of the above theorem may be characterised, nor is it clear whether *all* p-sets D of $\mathfrak{A}$ arise in this way.
Let us come to our last theorem. Also here, the nature of the closed sets D involved is somewhat obscure.

Theorem 6.10 *If X is a (complex) L^1-predual then all M-ideals of $L(X)$ are right inner and possess the form*

$$\mathfrak{J} = L(X) \cap \mathfrak{J}^{(D)},$$

where D denotes a closed subset in the compact space K with $X^{**} = C(K)$ and $\mathfrak{J}^{(D)}$ is the M-ideal in $L(L^1(\mu))$ described in Theorem 6.9.

The proof of this theorem can be given in analogy to the first part of the proof of Theorem 6.9.

V.7 Notes and remarks

GENERAL REMARKS. Although the notion of an M-ideal was originally designed for real Banach spaces, it soon became clear that the case of complex scalars should be covered as well. This was first formulated in [312] for the explicit purpose of investigating M-ideals in function algebras. Some time afterwards the fundamental articles [575], [578] and [579] by R. Smith and J. D. Ward appeared, where a detailed analysis of M-ideals in Banach algebras and some applications are presented. In these papers the idea of associating the hermitian projection $P(e) \in \mathfrak{A}^{**}$ with an M-ideal $\mathfrak{J} \subset \mathfrak{A}$ was developed, which later led to the notion of an inner M-ideal in [631].
The background material on numerical ranges in Section V.1 can be found for example in the books [84], [85], and [86] by Bonsall and Duncan; for the Arens products we mention the useful survey [177]. Clearly, Theorem 1.10 is a standard result in the theory of C^*-algebras, but we decided to include a proof in order to be able to present a rather self-contained exposition of the M-structure of C^*-algebras. Actually, much more than the Arens regularity of a C^*-algebra is true. It is proved in [261] and [525] that *every*

bilinear operator from $X \times X$ to X is Arens regular provided X^* has property (V^*); note that this applies in particular to Banach spaces with L-embedded duals (Theorem IV.2.7) and thus to C^*-algebras (Example IV.1.1(b)).
As we said, the fact (Theorem 2.1) that, for an M-projection P associated with an M-ideal $\mathfrak{J} \subset \mathfrak{A}$, $\Delta_{\mathfrak{A}^{**}}(P)$ is a hermitian projection is a basic result from [578]; the sources of the remaining parts of Theorem 2.1 are [236] and [579] (part (c)), [631] (parts (d) and (f)) and [639] (part (e)). The rest of Section V.2 and the example at the end of Section V.3 are taken from [578]. Section V.3 is based on [631] where the phrase "inner M-ideal" is coined and this notion is investigated in detail. One aspect of the present chapter is to show that nearly everything that can be proved about M-ideals in Banach algebras has actually to do with inner M-ideals.
Theorem 4.1 is in essence a result of Smith [575] as is the equivalence (ii) $\iff$ (iii) of Theorem 4.2. The equivalence (i) $\iff$ (iii) of this theorem is due to Hirsberg [312] who found it as a special case of a much more general result characterising M-ideals in subspaces of $C(K)$ containing the constants in terms of so-called M-sets; see also [35, Chap. 4] for a presentation of this approach and [400, Th. 7.6] for another proof of the implication in question. Here we have tried to reconcile the two equivalences in a perspicuous manner; for the special case of the disk algebra cf. [130]. We shall discuss the noncommutative version of Theorem 4.1 in the next subsection.
As already mentioned in the Notes and Remarks to Chapter I, the coincidence of M-ideals and closed two-sided ideals in C^*-algebras can be traced back to work by Effros [187] and Prosser [510]. For the real Banach space of self-adjoint elements of a C^*-algebra it was noted explicitly by Alfsen and Effros [11, Prop. 6.18]. Their proof used only the by then well known correspondence between weak* closed ideals and central projections in von Neumann algebras, and so applied equally well in the complex case. Indeed, the complex case was attributed to them without further comment in [8, p. 237]. Another proof of the complex case (applicable to general Banach algebras) was later given by Smith and Ward [578]; yet another proof appeared in [590]. Our proof that M-ideals are ideals is taken from [478], for other arguments see [578], [580] or the following subsection. The proof of Proposition 4.6 was shown to us by R. Smith, and that of Theorem 4.7 by A. Rodríguez-Palacios; the latter theorem has its origin in [11, Cor. 6.17]. Proposition 4.8 and its corollaries are taken from [121].
The results in Section V.5 aim at approximating elements $a^{**} \in \mathfrak{A}^{**}$ by those of $\mathfrak{A}$ respecting norm and numerical range conditions. Clearly the simplest result of this type is Goldstine's theorem. A far more elaborate result is the principle of local reflexivity. The version we employ, Theorem 1.4, is proved in [58]; for a proof of the classical version see [152] and for various refinements [62]. Proposition 5.1 is a new result building on ideas from [119] and [579]. Theorem 5.3 is due to Smith [577], but our proof is taken from [435] where this result is proved in the abstract framework of numerical range spaces. In fact, its assertion holds with $\varepsilon_\alpha = 0$ if $v(F)$ has nonempty interior. A related result is discussed in [129]: If $\mathfrak{A}$ is a unital Banach algebra and $\mathfrak{J} \subset \mathfrak{A}$ is a two-sided ideal which is an M-ideal, then each $\xi \in \mathfrak{A}/\mathfrak{J}$ whose "essential" numerical range $v(\xi, \mathfrak{A}/\mathfrak{J})$ has nonempty interior has a representative $a \in \mathfrak{A}$ with $v(a) = v(\xi, \mathfrak{A}/\mathfrak{J})$ and $\|a\| = \|\xi\|$. The norm approximation in Theorem 5.4 is due to Smith and Ward [579] who offer an entirely different proof. Here are some applications proved in that paper. Suppose $\mathfrak{J}$ is a two-sided inner M-ideal in the unital Banach algebra $\mathfrak{A}$; for instance any M-ideal in a

commutative Banach algebra (Theorem 4.1), any closed two-sided ideal in a C^*-algebra (Theorem 4.4) or the ideal $K(\ell^p)$ in $L(\ell^p)$, $1 < p < \infty$ (Theorem 6.1 and Example VI.4.1). Then, for the natural quotient homomorphism $\pi : \mathfrak{A} \to \mathfrak{A}/\mathfrak{J}$, the formula

$$\|\pi(a^n)\| = \inf_{y\in\mathfrak{J}} \|(a+y)^n\| \qquad \forall a \in \mathfrak{A},\ n \in \mathbb{N}$$

holds, as does the spectral radius formula

$$r(\pi(a)) = \inf_{y\in\mathfrak{J}} r(a+y) \qquad \forall a \in \mathfrak{A}.$$

The same formula holds for an ideal in a commutative Banach algebra possessing a bounded approximative unit. In the context of C^*-algebras, these results are due to Pedersen [483].
Theorem 6.1(a) is the core of the proof from [119] that M-ideals in $L(X)$ are left ideals for uniformly convex X (Corollary 6.5). This result extends work in [579] on $L(\ell^p)$ (or $L(L^p(\mu))$) drawing on Clarkson's inequalities instead of Lemma 6.4. Corollary 6.6 comes from [579]; there is another argument in [235], completing a preliminary attack in [580]. It is still an open question whether $L(L^p[0,1])$ contains any nontrivial M-ideals; in [119] a uniformly convex Banach space X of the form $X = \ell^r(\ell^p(n_i))$ is constructed such that $K(X)$ is an M-ideal in $L(X)$, but $L(X)$ contains a closed ideal which is not an M-ideal. In fact, the closure of the ideal of operators on X which factor through a subspace of some $L^r(\mu)$-space works; though this is quite concrete an example, its verification remains involved. Theorem 6.1(b) is proved for $X = \ell^1$ in [576] and for $X = C(K)$ in [236]. This paper also covers the case of operators on $C_0(\Omega)$ for certain locally compact Ω. The remaining cases of Theorem 6.1(b), presented in detail in Theorems 6.8–6.10, are considered in [631] as is part (d). For part (c) see [629].

SUBALGEBRAS OF C^*-ALGEBRAS; NEST ALGEBRAS. Suppose $\mathfrak{A}$ is a unital (nonselfadjoint) subalgebra of a C^*-algebra and thus of $L(H)$. Such an algebra can be regarded as the noncommutative analogue of a function algebra. Adopting this point of view it seems natural to inquire whether a characterisation of the M-ideals of $\mathfrak{A}$ which parallels the corresponding result on function algebras (Theorem 4.2) can be obtained. This is indeed possible.

THEOREM. *Let $\mathfrak{A} \subset L(H)$ be a unital subalgebra. Then a closed subspace $J \subset \mathfrak{A}$ is an M-ideal if and only if it is a two-sided ideal containing a two-sided 1-approximative unit.*

This is proved in [196], and in [505] it is observed that the result extends easily to nonunital algebras. Let us sketch an argument, suggested by D. Yost, to show that the M-ideals of $\mathfrak{A}$ are actually ideals: We assume, without loss of generality, that J is an M-summand. Denote by P the M-projection from $\mathfrak{A}$ onto J, and let $Q = Id - P$ and $p = P(Id)$. Since P^* is an L-projection, one observes that $P^*f \geq 0$ and $Q^*f \geq 0$ whenever $f \in \mathfrak{A}^*$, $f \geq 0$ (i.e., $\|f\| = f(Id)$). Then one deduces that p must be a positive element of $L(H)$, and we know from Theorem 2.1 that p is a projection. Now the crucial step consists in proving that $ap \in J$ for all $a \in \mathfrak{A}$. To show this let $f \in \mathfrak{A}^*$, $f \geq 0$, and let $g \in L(H)^*$ be a positive norm-preserving extension of Q^*f. It follows that

$$\begin{aligned} |f(Q(ap))| &= |(Q^*f)(ap)| = |g(ap)| \\ &\leq |g(a^*a)|^{1/2} |g(p^2)|^{1/2} = 0, \end{aligned}$$

where we used the Cauchy-Schwarz inequality for C^*-algebras (note that a^* exists in $L(H)$, but not necessarily in $\mathfrak{A}$) and observed that

$$g(p^2) = g(p) = (Q^*f)(p) = f(Qp) = 0$$

since $Qp = 0$. An application of Theorem 1.5 yields that $f(Q(ap)) = 0$ for *all* $f \in \mathfrak{A}^*$ so that $Q(ap) = 0$ and $ap \in J$. Likewise, $pa \in J$, and J is a two-sided ideal.

A special class of subalgebras of $L(H)$ which has attracted some interest is the class of nest algebras. In the following, H denotes "a" separable infinite dimensional complex Hilbert space, and for convenience we put $\mathcal{K} = K(H)$ and $\mathcal{L} = L(H)$. A *nest* $\mathcal{N}$ is a strongly closed totally ordered set of orthogonal projections on H containing 0 and Id. The corresponding *nest algebra* $\mathcal{A} = \mathcal{A}(\mathcal{N})$ consists of all those operators on H that leave ran(P) invariant for each $P \in \mathcal{N}$. Examples include the nest of coordinate projections $x \mapsto \sum_{k=1}^n \langle x, e_k \rangle e_k$ with respect to a fixed orthonormal basis, in which case the corresponding nest algebra consists of those operators which have an upper triangular representation for that basis, and the Volterra nest, consisting of the projections $P_t(f) = \chi_{[0,t]} f$ on $L^2[0,1]$. Such algebras consist of operators which have, in some sense, a triangular representation, and are generally considered as noncommutative variants of the Hardy space H^∞. For information on the theory of nest algebras we recommend the survey [508] and the monograph [147].

A crucial fact about nest algebras is that the unit ball of $\mathcal{A} \cap \mathcal{K}$ is dense in that of $\mathcal{A}$ for the weak operator topology ([508, Th. 2.1] or [147, p. 36]). A variety of corollaries can be deduced from this result. First of all, one can show that $\mathcal{A} + \mathcal{K}$ is closed; operators in $\mathcal{A} + \mathcal{K}$, being compact perturbations of operators in $\mathcal{A}$, are called quasi-triangular. Clearly, this is the formal analogue of Sarason's theorem that $H^\infty + C(\mathbb{T})$ is a closed subalgebra of $L^\infty(\mathbb{T})$. Then, $\mathcal{A}$ is canonically isometric to the bidual of $\mathcal{A} \cap \mathcal{K}$, and thus, by Theorem III.1.6, $\mathcal{A} \cap \mathcal{K}$ is an M-ideal in $\mathcal{A}$. Note that the corresponding commutative result about $H^\infty \cap C$ ($= A$, the disk algebra) is false.

However, the result that C/A is isometric to $(H^\infty + C)/H^\infty$ which is an M-ideal in L^∞/H^∞ (Example III.1.4(h)) extends: $\mathcal{K}/(\mathcal{A} \cap \mathcal{K})$ is isometric to $(\mathcal{A} + \mathcal{K})/\mathcal{A}$ which is an M-ideal in $\mathcal{L}/\mathcal{A}$ [226]. The latter fact is most effectively proved using Proposition I.1.16, cf. [148]. One just has to observe that assumption (b$_1$) of that proposition, with $J = \mathcal{K}$, $X = \mathcal{L}$ and $Y = \mathcal{A}$, is fulfilled by the above-quoted density theorem. Since $\mathcal{L}/\mathcal{A}$ happens to be the bidual of $(\mathcal{A}+\mathcal{K})/\mathcal{A} \cong \mathcal{K}/(\mathcal{A} \cap \mathcal{K})$, the above result also follows from the stability of the class of M-embedded spaces with respect to quotients. It is noteworthy that the proof of this noncommutative M-ideal result is far easier than in the commutative case, since $\mathcal{K}$ is an M-ideal in $\mathcal{L}$; in the commutative case we had to deal with an L-projection in $(L^\infty)^*$ that did not derive from an M-ideal.

We just suggested using Proposition I.1.16 in the analysis of the quotient space $\mathcal{L}/\mathcal{A}$. The fact that one has (b$_1$) and thus (a) of this proposition might be considered as a version of the F. and M. Riesz theorem. For a deeper theorem of this type we refer to [210].

We wish to discuss one more aspect of the formal resemblance between C, H^∞, L^∞ and $\mathcal{K}$, $\mathcal{A}$, $\mathcal{L}$. In these strings of spaces the space L^1 is considered to be the counterpart of $N(H)$, the nuclear operators, and the Hardy space H^1 corresponds to $\mathcal{A}^1 := \mathcal{A} \cap N(H)$. The space H_0^1 teams up with the subspace $\mathcal{A}_0^1 \subset \mathcal{A}^1$ of "strictly" triangular operators, meaning

$$\mathcal{A}_0^1 = \{T \in \mathcal{A}^1 \mid (Id - P^-)TP = 0 \quad \forall P \in \mathcal{N}\}$$

where $P^- = \sup\{E \in \mathcal{N} \mid E < P\}$. Then one has indeed that $(\mathcal{K}/(\mathcal{A} \cap \mathcal{K}))^* \cong \mathcal{A}_0^1$ and $N(H)/\mathcal{A}_0^1 \cong (\mathcal{A} \cap \mathcal{K})^*$, see [508]. A lot of similarities between H_0^1 and $\mathcal{A}_0^1$ have been pointed out in [33], among them the noncommutative Havin theorem that $N(H)/\mathcal{A}_0^1$ is weakly sequentially complete; this holds true because $\mathcal{A} \cap \mathcal{K}$, being a subspace of $\mathcal{K}$, is M-embedded, see Corollary III.3.7(b).
For M-ideals and more general "analytic" subalgebras of $L(H)$ see [505].

JB-*ALGEBRAS AND *JB*-*TRIPLES. In an attempt to give an algebraic formalisation of the axioms of quantum mechanics Jordan, von Neumann and Wigner introduced the notion of a Jordan algebra in the thirties. Infinite-dimensional versions of their results were, however, first proved as late as 1978 when the landmark paper [13] by Alfsen, Shultz and Størmer appeared. This paper deals with a class of real algebras called JB-algebras; subsequently Kaplansky introduced their complex counterparts, the JB^*-algebras (see [645]). Here is the definition: A complex Banach space X, equipped with a bilinear map $(x,y) \mapsto x \circ y$ (the "Jordan product") and an isometric algebra involution $x \mapsto x^*$, is called a JB^*-algebra if

$$\begin{aligned} x \circ y &= y \circ x && (1)\\ (x^2 \circ y) \circ x &= x^2 \circ (y \circ x) && (2)\\ \|x \circ y\| &\le \|x\|\,\|y\| && (3)\\ \|\{xxx\}\| &= \|x\|^3 && (4) \end{aligned}$$

for all $x, y \in X$, where

$$\{xyz\} = x \circ (y^* \circ z) - y^* \circ (z \circ x) + z \circ (x \circ y^*). \qquad (5)$$

Note that the product is not supposed associative – (2) being a weak substitute – so that $(X, \circ)$ is not a Banach algebra proper. The main examples of JB^*-algebras are the norm-closed subspaces of C^*-algebras which are closed under the product

$$x \circ y = (xy + yx)/2.$$

(Here xy denotes the original product in the C^*-algebra.) But apart from these so-called special JB^*-algebras there are others, notably the properly complexified 27-dimensional algebra of self-adjoint 3×3-matrices over the Cayley numbers. This object is often denoted by C_6. Because any classification of JB^*-algebras has to take care of C_6, one cannot expect that the theory of irreducible representations of C^*-algebras extends literally to JB^*-algebras. All the same, many of the ideas figuring so prominently in C^*-algebra theory can be adapted to the nonassociative setting. In this regard factor representations, corresponding to the irreducible representations in C^*-algebra theory, ideal theory

and JBW^*-algebras (these are JB^*-algebras which are dual Banach spaces) have been studied; basic references are [185] and [645].
Payá, Pérez and Rodríguez [478] used M-ideals in a decisive manner to prove that every JB^*-algebra admits a faithful family of factor representations. For instance they prove that the weak* closed ideals in a JBW^*-algebra are in one-to-one correspondence with its M-summands, and that they are determined by central projections (cf. our Proposition 4.5). As a result, the closed ideals in a JB^*-algebra are in one-to-one correspondence with its M-ideals, and consequently the closed two-sided ideals in a C^*-algebra coincide with its closed Jordan ideals; this is a well-known result due to Civin and Yood. (We also mention the paper [285] for a geometric description of the ideals of a JB-algebra in terms of split faces.) To construct a factor representation the authors of [478] look at the largest M-ideal $M_p \subset \ker p$ for an extreme functional $p \in \operatorname{ex} B_{X^*}$ (a device introduced by Alfsen and Effros under the name primitive M-ideal) and prove that the M-summand orthogonal to $M_p^{\perp\perp}$ is a factor, i.e. has no weak* closed ideals, equivalently: no M-summands. They also classified the factors arising in their approach [479].
Another result relevant to M-structure theory in [478] is the L-embeddedness of preduals of JBW^*-algebras, which extends our Example IV.1.1(b). A complete list of those JB^*-algebras which are M-embedded, due to A. Rodríguez, can be found in [395, Th. 14], cf. Proposition III.2.9 for the C^*-algebra case. Actually, all the results given above were derived for noncommutative JB^*-algebras, here (1) is replaced by the requirement that

$$(x \circ y) \circ x = x \circ (y \circ x). \tag{1'}$$

Also, Rodríguez has proved [524] that the unital noncommutative JB^*-algebras are the largest class of (nonassociative) algebras for which the Vidav-Palmer theorem is valid, i.e. which are spanned by their hermitian elements.
A more general structure than a JB^*-algebra is that of a JB^*-triple which arises naturally in finite and infinite dimensional holomorphy. JB^*-triples were first used by Koecher and his school in order to give an alternative approach to the classification of symmetric domains in the finite dimensional case; their importance for infinite dimensions was realized by W. Kaup. Omitting all details – we refer to [25] and [615] for a more extended presentation and more on the relevant literature – we would like to briefly sketch the basic idea.
Denote by D a bounded symmetric domain in a (complex) Banach space X, which means that for all $x \in D$ there is an involutory and biholomorphic map φ which has x as an isolated fixed point. Extending a classical result of Cartan to this more general setting, Upmeier and Vigué independently proved that the group $\operatorname{Aut} D$ of all biholomorphic automorphisms naturally bears the structure of a real Banach Lie group, whose Lie algebra $\operatorname{aut} D$ can concretely be realized as the space of all *complete* holomorphic vector fields $F : D \to X$ by the mapping

$$\xi \in \operatorname{aut} D \longmapsto \left(x \in X \mapsto \lim_{t \to 0} \frac{\Phi_\xi(t)x - x}{t} \right)$$

where $\Phi_\xi(t)$ is the one-parameter subgroup of $\operatorname{Aut} D$ associated with $\xi \in \operatorname{aut} D$ via the exponential mapping. (We should add that a holomorphic vector field $F : D \to X$ is called complete if for each $u_0 \in D$ the initial value problem

$$\begin{aligned} y(0) &= u_0 \\ y'(t) &= F(y(t)) \end{aligned}$$

has a global solution $y : \mathbb{R} \to D$.) The point now is that there is no loss in generality to assume that D is actually balanced, and that for these domains all complete holomorphic vector fields are polynomials of degree at most 2. Moreover, if $F = p_0 + p_1 + p_2$ is the Taylor expansion of such a polynomial, then $p_0 + p_2$ and p_1 both belong to aut D, and $p_0 + p_2$ is uniquely determined by the constant term $p_0 \in X$. Also, the symmetry of D implies that Aut D acts transitively on D and that consequently *each* $p_0 \in X$ gives rise to a (unique) vector field $p_0 + p_2$. Hence, with each $y \in X$ one can associate a unique symmetric bilinear map $P_y : X \times X \to X$ leading to the triple product

$$\{x, y, z\} := P_y(x, z)$$

which is bilinear and symmetric in x and z and conjugate linear in y. It furthermore satisfies the Jordan triple identity

$$\{\{uvw\}yx\} + \{\{uvx\}yw\} - \{uv\{wyx\}\} = \{w\{vuy\}x\},$$

and it can be shown that the operators $x \,\square\, x \,:\, z \mapsto \{xxz\}$ are bounded, linear and hermitian. In addition, these operators have positive spectrum and satisfy the "C^*-condition"

$$\|x \,\square\, x\| = \|x\|^2,$$

which in turn is equivalent to the "JB^*-condition"

$$\|\{xxx\}\| = \|x\|^3.$$

Stated in a short and more notational way: A Banach space containing a bounded symmetric balanced domain can be given the structure of a *JB^*-triple* – the defining conditions for the triple product are listed above. Also the converse holds, and both statements together yield part of an important result of Kaup ([371], [372]) stating that, up to biholomorphic equivalence, bounded symmetric domains in complex Banach spaces are the open unit balls of JB^*-triples and that two such domains are equivalent iff the corresponding JB^*-triples are isometrically isomorphic (in which case they are isomorphic with respect to their JB^*-structure – this latter statement involves a result due to Kaup and Upmeier [374]).

There are prominent examples of JB^*-triples whose triple product does not refer explicitly to the analytic structure of the underlying space. Under the triple product (5) a JB^*-algebra becomes a JB^*-triple, in particular, $\{xyz\} = (xy^*z + zy^*x)/2$ introduces the structure of a JB^*-triple on a C^*-algebra. Further examples are the spaces $L(H, K)$ of bounded linear operators between (different) Hilbert spaces H and K, with the triple product $\{RST\} = (RS^*T + TS^*R)/2$. An important feature of JB^*-triples is the stability of this class under contractive projections ([237], [373]). This provides more examples, since in particular the range of a norm-one projection on a C^*-algebra, though generally not a C^*-algebra, is a JB^*-triple. (It is a C^*-algebra if the projection is completely positive.)

Similar to the case of a JB^*-algebra, the underlying geometrical structure often helps in understanding the rather puzzling algebraic behaviour of a given JB^*-triple. Of interest in this respect are results concerning the M-structure of JB^*-triples which again provides a coupling of algebraic and geometrical concepts. A subspace J of a JB^*-triple is called an ideal if $\{xyz\} \in J$ whenever x, y or z is in J. Horn [324] proves that the weak* closed ideals in a JBW^*-triple (a JB^*-triple which is a dual Banach space) coincide with its M-summands, and Barton and Timoney [43] deduce that the closed ideals of a JB^*-triple coincide with its M-ideals. These papers also contain proofs of the strong uniqueness of the predual of a JBW^*-triple and its L-embeddedness. Moreover, the bidual of a JB^*-triple X is a JBW^*-triple with separately weak* continuous triple product extending the one of X; it is this crucial result due to Dineen [161] that links the ideals of X to the weak* closed ideals of X^{**} and makes the approach to ideals via the bidual possible.

With these theorems in hand the theory of factor representations of [478] lends itself to generalisation. This is carried out in [43], too; and a Gelfand-Naimark theorem for JB^*-triples is achieved in [238] and [325]. For an M-ideal approach to the latter see [162, Example 38]. Dineen and Timoney characterise in [163] the centralizer of a JB^*-triple as its centroid, consisting of those bounded linear operators for which $T\{x,y,z\} = \{Tx,y,z\}$.

Several authors ([42], [101], [102], [128]) investigate the RNP for preduals of JBW^*-triples. Also here M-ideal methods turn out to be helpful; moreover one can read from the above papers that a JB^*-triple is M-embedded if and only if it is the c_0-sum of so-called elementary triples.

Finally, we mention the monographs [286] and [606] and the expository lectures [527] and [607] as general references on the subject. Also the recent paper [228] contains a survey of the general theory of JB^*-algebras.

Completely bounded maps and M-ideals. In recent years the study of "complete" notions has become a major subject in C^*-algebra theory. Here is the basic principle. Every (concrete) C^*-algebra $\mathfrak{A}$ of operators on a Hilbert space H gives rise to a sequence of matrix algebras as follows. Denote by M_n the space of complex $n \times n$-matrices and by $M_n(\mathfrak{A})$ the space of $n \times n$-matrices with entries from $\mathfrak{A}$. These spaces are endowed with C^*-algebra structures in a canonical (and unique) fashion, in that $M_n \cong L(\ell^2(n))$ and $M_n(\mathfrak{A})$ embeds into $M_n(L(H)) \cong L(H \oplus_2 \ldots \oplus_2 H)$. Now, given two C^*-algebras $\mathfrak{A}$ and $\mathfrak{B}$ and a bounded linear map $T : \mathfrak{A} \to \mathfrak{B}$, T gives rise to a string of bounded linear operators $T_n : M_n(\mathfrak{A}) \to M_n(\mathfrak{B})$, defined by $T_n((x_{ij})) = (Tx_{ij})$. The operator T is called completely bounded if $\sup_n \|T_n\| < \infty$, and the completely bounded norm $\|T\|_{cb}$ is defined to be this supremum. In general, given a property (P) such an operator might or might not have, T is said to have "complete (P)" whenever all the T_n fulfill (P); thus completely contractive, completely isometric, completely positive operators etc. are defined. For a detailed account and precise bibliographical references we refer to Paulsen's monograph [476].

In his address to the 1986 International Congress of Mathematicians [192] E. G. Effros suggested a research program entitled "Quantized Functional Analysis" to investigate matricially normed spaces and completely bounded maps from an abstract point of view.

Many steps in this program have been carried out so far (e.g. [80], [195], [197], [198]); here we would like to comment on M-ideal techniques employed in this setting. We first have to give some definitions.

Not only a C^*-algebra $\mathfrak{A}$ is equipped with a sequence of matrix norms, but also each of its closed subspaces. In addition, it is easily checked that for $X = (x_{ij}) \in M_n(\mathfrak{A})$ and $Y = (y_{ij}) \in M_m(\mathfrak{A})$ the block diagonal matrix $X \oplus Y := \operatorname{diag}(X, Y) \in M_{n+m}(\mathfrak{A})$ satisfies

$$\|X \oplus Y\|_{n+m} = \max\{\|X\|_n, \|Y\|_m\}.$$

This leads to the definition of an abstract operator space: An abstract operator space is a vector space V, together with a sequence of distinguished norms $\|\,.\,\|_n$ on each of the matrix spaces $M_n(V)$ such that

$$\|AXB\|_n \le \|A\|\,\|X\|_n\,\|B\| \quad \forall A, B \in M_n,\ X \in M_n(V),$$

$$\|X \oplus Y\|_{n+m} = \max\{\|X\|_n, \|Y\|_m\} \quad \forall X \in M_n(V),\ Y \in M_m(V).$$

Every normed space $(V, \|\,.\,\|)$ can be turned into an abstract operator space; but there are various ways to do this. One way is to equip $M_n(V)$, which can algebraically be identified with $M_n \otimes V$, with the injective tensor norm; another, different, method is to let

$$\|(v_{ij})\|_n^{\max} = \sup \|(\varphi(v_{ij}))\| \tag{1}$$

where the supremum ranges over all Hilbert spaces H and all contractive $\varphi : V \to L(H)$, the norm on the right hand side of (1) referring to the canonical C^*-norm of $M_n(L(H))$. The above considerations show that a subspace of a C^*-algebra is an abstract operator space. A fundamental result due to Ruan [538] states that, conversely, every abstract operator space is completely isometric to a subspace of $L(H)$, endowed with its C^*-matricial norms; see [200] for a simpler proof. This result can be regarded as a "quantized" version of the result from ordinary "commutative" functional analysis that every normed space is isometric to a subspace of some $C(K)$-space.

For a map T between abstract operator spaces $(V, \{\|\,.\,\|_n\})$ and $(W, \{\|\,.\,\|_n\})$ complete boundedness is defined as above; we denote by $CB(V, W)$ the space of completely bounded maps and by $\|\,.\,\|_{cb}$ the completely bounded norm. One of the most important results on completely bounded maps is the Arveson-Wittstock Hahn-Banach theorem [476, p. 100] which can be rephrased as follows: For an abstract operator space $(V, \{\|\,.\,\|_n\})$, a subspace $S \subset V$ and a completely bounded map $\varphi : S \to L(H)$ there is an extension $\psi : V \to L(H)$ such that $\|\psi\|_{cb} = \|\varphi\|_{cb}$. A simple proof of this theorem, building on factorisation techniques, was given by Pisier [503].

The dual of an abstract operator space $(V, \{\|\,.\,\|_n\})$ can be given the structure of an operator space as well. For this one identifies $M_n(V^*)$ with $CB(V, M_n)$ and gives it the corresponding completely bounded norm. We shall always tacitly assume that this dual structure is imposed on V^*. Then it turns out that the canonical embedding from V into V^{**} is completely isometric; for details see [80] or [199].

Effros and Ruan [198] define and study complete L- and M-projections on abstract operator spaces, the definition being of course that all the P_n are L- (resp. M-) projections on the $M_n(V)$. A complete L- (resp. M-) summand is the range of a complete L- (resp. M-) projection. A closed subspace $J \subset V$ is called a complete M-ideal if $J^\perp$ is a complete

L-summand in V^*. This can be shown to be equivalent with the requirement that all the $M_n(J)$ be M-ideals in $M_n(V)$. Effros and Ruan also give an example of an M-ideal which fails to be a complete M-ideal. In fact, if $\ell^\infty(3)$ is given an operator space structure by means of (1), then the M-summand $\ell^\infty(2)$ is not a complete M-ideal. (On the other hand, every M-ideal is complete for the operator space structure derived from $M_n \otimes_\varepsilon V$; this is a special case of Proposition VI.3.1 below. This is also the case in the C^*-algebra setting since the $M_n(J)$ are closed two-sided ideals if J is.)
The motivation to consider complete M-ideals lies in the first place in their use in the proof of lifting theorems, see Theorem II.2.1. The corresponding result for abstract operator spaces can, however, only be proved for those operator spaces for which a certain approximation condition is fulfilled. Since this condition corresponds to the equation $L(E, X^{**}) = L(E,X)^{**}$ for finite dimensional E in the Banach space setting and thus to the validity of the principle of local reflexivity, such operator spaces are termed locally reflexive; in the C^*-algebra case such a condition is investigated in [193] under the name property (C''). So, the lifting theorem of [198], which can be regarded as a general form of the ones in [120] and [193], states:

THEOREM. *If V and W are abstract operator spaces with V separable and W locally reflexive, if $J \subset W$ is a complete M-ideal, $q : W \to W/J$ the (complete) quotient map and $\varphi : V \to W/J$ a complete contraction, and finally if V enjoys the completely metric approximation property, then there exists a complete contraction $\psi : V \to W$ such that $q\psi = \varphi$:*

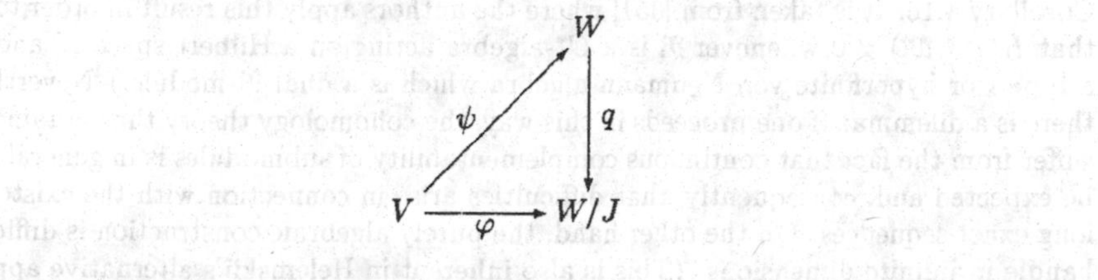

We also mention the paper [539] which studies uniqueness of preduals of abstract operator spaces from the M-structure point of view.

MORE ON COHOMOLOGY OF OPERATOR ALGEBRAS. The origins of homology theory are usually attributed to the paper "Analysis situs" by H. Poincaré [504]. Originally designed as a tool to distinguish between different topological spaces, this theory developed a purely algebraic branch on the basis of work by E. Noether, H. Hopf, L. Vietoris and W. Mayer. The diversity of algebraic (co)homology theories that emerged in the 40s and 50s found a unified and axiomatic foundation in the book of H. Cartan and Eilenberg [112]. For a somewhat more detailed exposition of this story we recommend Dieudonné's monograph [160].
The cohomology theory which is presented in Section V.4 was initiated in 1945 by Hochschild in [315]. (Hochschild himself attributes his construction to unpublished work of Eilenberg and MacLane, who had used similar constructions for groups.) One of his primary interests was an extension of Wedderburn's third structure theorem – an algebra $\mathfrak{A}$ splits as $\mathfrak{A} = \mathfrak{R} \oplus \mathfrak{A}_1$, where $\mathfrak{R}$ denotes the radical and $\mathfrak{A}_1$ is a *subalgebra* – in terms of

$H^2(\mathfrak{A},\mathfrak{R})$. He in fact proved that all singular extensions (i.e. extensions where $\mathfrak{R}^2 = 0$) of $\mathfrak{A}$ by $\mathfrak{R}$ are essentially parametrized by $H^2(\mathfrak{A},\mathfrak{R})$, and that a Wedderburn-type result holds for all extensions of $\mathfrak{A}$ iff $H^2(\mathfrak{A},\mathfrak{K}) = 0$ for every two-sided $\mathfrak{A}$-module $\mathfrak{K}$.
It is possible to relate low dimensional Hochschild groups to other known invariants. So $H^1(\mathfrak{A},\mathfrak{X})$ measures how many derivations from $\mathfrak{A}$ to $\mathfrak{X}$ exist which are not inner, and $H^2(\mathfrak{A},\mathfrak{X})$ is connected to extensions of $\mathfrak{A}$ by $\mathfrak{X}$ and to liftings of derivations. For $n = 3$, however, an appropriate interpretation is quite involved (see [316]) and a concrete meaning of $H^n(\mathfrak{A},\mathfrak{X})$ for $n > 3$ seems to be unknown. (It is, however, possible to "reduce dimensions" via the canonical isomorphism $C^{n+1}(\mathfrak{A},\mathfrak{X}) \cong C^n(\mathfrak{A}, C^1(\mathfrak{A},\mathfrak{X}))$ which yields $H^{n+1}(\mathfrak{A},\mathfrak{X}) \cong H^n(\mathfrak{A}, C^1(\mathfrak{A},\mathfrak{X}))$.) We should also mention that there is a corresponding sequence of homology groups $H_k(\mathfrak{A},\mathfrak{X})$, which are defined in terms of n-fold tensor-products $C_k(\mathfrak{A},\mathfrak{X}) = \mathfrak{X} \otimes_{\mathfrak{A}} \mathfrak{X} \ldots \otimes_{\mathfrak{A}} \mathfrak{X}$ in such a way that the adjoint of the boundary mapping $\delta_k : C_k(\mathfrak{A},\mathfrak{X}) \to C_{k-1}(\mathfrak{A},\mathfrak{X})$ equals δ^k as defined in Section V.4.
One of the first important results on Hochschild cohomology within the category of Banach algebras is due to Kadison [358] and, independently, Sakai [554], who showed that any derivation on a von Neumann algebra $\mathfrak{M}$ is inner or, equivalently, that $H^1(\mathfrak{M},\mathfrak{M}) = 0$. The whole theory was set to work within this frame mainly by Kadison, Ringrose and B. E. Johnson in a number of papers [347], [351], [359], [360]. (See also [348] or [523] for a survey of these results.) In the case of Banach algebras the idea to pass to multilinear operators which are *continuous* with respect to some natural topology suggests itself. (That it might be appropriate to use e.g. the weak* topology for dual modules is illustrated by the Johnson-Kadison-Ringrose theorem that was used in the proof of Corollary 4.10. It is taken from [351] where the authors apply this result in order to show that $H^k(\mathfrak{R},\mathfrak{M}) = 0$ whenever $\mathfrak{R}$ is a C^*-algebra acting on a Hilbert space H and $\mathfrak{M}$ is a type I or hyperfinite von Neumann algebra which is a dual $\mathfrak{R}$-module.) Nevertheless, there is a dilemma: If one proceeds in this way, the cohomology theory thus obtained will suffer from the fact that continuous complementability of submodules is in general not to be expected and, consequently, that difficulties arise in connection with the existence of long exact sequences. On the other hand, the purely algebraic construction is difficult to handle in infinite dimensions. (This is also inherent in Helemskii's alternative approach to Hochschild groups in terms of admissible projective resolutions – see Chapter III, especially Theorem 4.9 therein, of the monograph [302], which is the most exhaustive source on the topic we are aware of.)
Applications of continuous Hochschild cohomology cover – besides the already mentioned characterisation of amenability of topological groups – Wedderburn decompositions of Banach algebras with finite dimensional radicals [347] as well as deformations of Banach algebras [350], [514]. It is furthermore connected to cyclic cohomology, see below.
More recent progress on the calculation of cohomology groups for operator algebras was made by Christensen, Effros and Sinclair [124] by use of the completely bounded version of this theory. They show that $H^k_{cb}(\mathfrak{A},\mathfrak{M})$, the cohomology group based on completely bounded multilinear maps, vanishes whenever $\mathfrak{A}$ is a C^*-subalgebra of an injective von Neumann algebra $\mathfrak{M}$. The point here is that in some cases these groups coincide with the (norm-) continuous Hochschild groups. This permits one to show that in the above situation, with $\mathfrak{A} \subset \mathfrak{M} \subset L(H)$ and under the additional assumption that the weak closure of $\mathfrak{A}$ is either properly infinite or else isomorphic to a von Neumann tensor product $\mathfrak{R}_0 \widehat{\otimes} \mathfrak{N}$, where $\mathfrak{N}$ is the (Murray-von Neumann) hyperfinite factor of type II_1, it

is also true that $H^k(\mathfrak{A},\mathfrak{M}) = 0$.
In another direction, Christensen and Sinclair proved that $H^k(\mathfrak{A},\mathfrak{A}^*)$ vanishes for all k whenever the C^*-algebra $\mathfrak{A}$ is nuclear or has no bounded traces [125].
We would finally like to remark that the groups $H^k(\mathfrak{A},\mathfrak{A}^*)$ are of particular interest, since they are related to Connes' cyclic cohomology (cf. [136, pp. 100–102, II.4] and [113] for a general survey), as follows. A $(k+1)$-form ω with entries in $\mathfrak{A}$ is called *cyclic* if

$$\omega(a_1,\ldots,a_k,a_0) = (-1)^k\omega(a_0,a_1,\ldots,a_k),$$

and the space of cyclic $(k+1)$-forms is denoted by $C^k_\lambda(\mathfrak{A})$. By

$$\begin{aligned}\delta^k_\lambda\omega(a_0,a_1,\ldots,a_{k+1}) &= \sum_{\kappa=0}^{k}(-1)^\kappa\omega(a_0,\ldots,a_\kappa a_{\kappa+1},\ldots,a_{k+1}) + \\ &\quad + (-1)^{k+1}\omega(a_{k+1}a_0,a_1,\ldots,a_k)\end{aligned}$$

one defines linear mappings $C^k_\lambda(\mathfrak{A}) \to C^{k+1}_\lambda(\mathfrak{A})$ which satisfy $\delta^k_\lambda\delta^{k-1}_\lambda = 0$. The resulting cohomology groups are denoted by $H^k_\lambda(\mathfrak{A})$. Since there is an obvious one-to-one correspondence between $(k+1)$-forms on $\mathfrak{A}$ and elements of $C^k(\mathfrak{A},\mathfrak{A}^*)$, it can easily be shown that under this identification the $C^k_\lambda(\mathfrak{A})$ are a subcomplex of the Hochschild complex. Furthermore, there is a long exact sequence

$$\begin{aligned}0 \to H^0_\lambda(\mathfrak{A}) \to H^0(\mathfrak{A},\mathfrak{A}^*) \to H^{-1}_\lambda(\mathfrak{A}) \to H^1_\lambda(\mathfrak{A}) \to H^1(\mathfrak{A},\mathfrak{A}^*) \to H^0_\lambda(\mathfrak{A}) \to H^2_\lambda(\mathfrak{A}) \to \cdots \\ \cdots \to H^n_\lambda(\mathfrak{A}) \to H^n(\mathfrak{A},\mathfrak{A}^*) \to H^{n-1}_\lambda(\mathfrak{A}) \to H^{n+1}_\lambda(\mathfrak{A}) \to H^{n+1}(\mathfrak{A},\mathfrak{A}^*) \to \cdots\end{aligned}$$

connecting Hochschild groups with cyclic cohomology groups [136, p. 119]. The latter is very important for the calculation of $H^k_\lambda(\mathfrak{A})$ in general. It permits, for instance, using the Christensen-Sinclair result to prove that $H^k_\lambda(\mathfrak{A}) = 0$ for all k when $\mathfrak{A}$ has no bounded traces, and that $H^k_\lambda(\mathfrak{A}) = 0$ (k odd), respectively $H^k_\lambda(\mathfrak{A}) = H^0_\lambda(\mathfrak{A})$ for even k, whenever $\mathfrak{A}$ is nuclear.

CHAPTER VI

M-ideals in spaces of bounded operators

VI.1 The centralizer of $L(X,Y)$

It seems to be difficult to obtain a general description of the M-ideals of $L(X,Y)$: It is one of the aspects of Section VI.4 that there is for example no hope in trying to prove that each M-ideal in $L(X,Y)$ corresponds to an M-ideal in Y or X^* (as it will turn out for a special case in the following section). There is even no theorem known that produces in a one-to-one fashion an M-ideal in $L(X)$ for each M-ideal in X. (The foregoing section, however, gives a hint of what such a result probably could look like.)
In this section we are going to study the centralizer $Z(L(X,Y))$ and the multiplier algebra $\mathrm{Mult}(L(X,Y))$ of the space of bounded linear operators $L(X,Y)$. We first show that the centralizer $Z(Y)$ and the Cunningham algebra $\mathrm{Cun}(X)$ naturally give rise to operators in $Z(L(X,Y))$, and our main result, Theorem 1.2, essentially states the converse under the assumption that only one of these candidates is allowed to enter this game seriously, i.e. $\mathrm{Cun}(X)$ or $Z(Y)$ is trivial. This will then admit a handsome description of the M-ideals which are naturally connected with these algebras via Corollary V.3.6. The problem of characterising the product structure that arises when one treats the general case, in which both $\mathrm{Cun}(X)$ and $Z(Y)$ are supposed nontrivial, is illustrated at the end of Section VI.3, and for a little bit of more information on this the reader is referred to the Notes and Remarks section.

Let H be a closed subspace of $L(X,Y)$. We start this section with introducing some natural candidates for elements of $Z(H)$ and $\mathrm{Mult}(H)$.
In the sequel, we denote left multiplication on $L(X,Y)$ with an operator $T \in L(Y)$ by L_T and, in the same way, right multiplication with $T \in L(X)$ is denoted by R_T. Furthermore, the adjoint of an operator $T \in Z(X)$ will be denoted by $\overline{T}$. (Recall that this means that the eigenvalues of the two operators corresponding to a fixed eigenvector $p \in \mathrm{ex}\, B_{X^*}$ are conjugate to each other.) Also, recall from Theorem I.3.14 the equivalence $T \in \mathrm{Cun}(X)$

if and only if $T^* \in Z(X^*)$, and for $T \in \mathrm{Cun}(X)$ the symbol $\overline{T}$ stands for the operator whose adjoint (in the above sense) is $\overline{T^*}$.

Lemma 1.1

(a) *Let H be a closed subspace of $L(X,Y)$. If for an element T of $Z(Y)$ (or $\mathrm{Mult}(Y)$) the conditions $L_T H \subset H$ and $L_{\overline{T}} H \subset H$ ($L_T H \subset H$) are fulfilled, then L_T belongs to $Z(H)$ (or $\mathrm{Mult}(H)$, respectively).*

(b) *Similarly, if T^* is in $Z(X^*)$ (or $\mathrm{Mult}(X^*)$) and if $R_T H \subset H$ and $R_{\overline{T}} H \subset H$ ($R_T H \subset H$) then R_T belongs to $Z(H)$ (to $\mathrm{Mult}(H)$).*

(c) *In particular, the maps $S \mapsto P_\infty S$ and $S \mapsto SP_1$ are M-projections on $L(X,Y)$ and suitable subspaces H as above, if P_∞ is an M-projection on Y and P_1 is an L-projection on X.*

PROOF: The various invariance assumptions make sure that L_T resp. R_T maps H into H. Let now $T \in Z(Y)$. Since $Z(Y) = \mathrm{lin}\, Z_{0,1}(Y)$, we may suppose that even $T \in Z_{0,1}(Y)$. (This set of operators is defined in I.3.7.) Proposition I.3.9 yields

$$\|TUx + (Id - T)Vx\| \leq 1 \qquad \forall U, V \in B_H \ \forall x \in B_X,$$

and so,

$$\|TU + (Id - T)V\| \leq 1 \quad \forall U, V \in B_H,$$

which, again by Proposition I.3.9, implies $L_T \in Z(H)$.
If $T^* \in Z_{0,1}(X^*)$ then by the above

$$\|UT + V(Id - T)\| = \|T^*U^* + (Id - T^*)V^*\| \leq 1 \qquad \forall U, V \in B_H,$$

and hence $R_T \in Z(H)$.
The proof in the case that $T \in \mathrm{Mult}(Y)$ or $T^* \in \mathrm{Mult}(X^*)$ is similar when Theorem I.3.6 is used, and part (c) is a special case of the above. □

The main theorem of this section shows that, under some appropriate restrictions on H, also the converse of Lemma 1.1 holds: All elements of $Z(H)$ are of this particularly simple form.

Theorem 1.2 *Let H be a subspace of $L(X,Y)$ containing the finite rank operators.*

(a) *If $L_T H \subset H$ for each $T \in Z(Y)$ and if $\mathrm{Cun}(X)$ is trivial, then*

$$Z(H) = \{L_T \mid T \in Z(Y)\}.$$

(b) *If $R_T H \subset H$ for all $T \in \mathrm{Cun}(X)$ and if $Z(Y)$ is trivial, then*

$$Z(H) = \{R_T \mid T \in \mathrm{Cun}(X)\}.$$

If the Banach space under consideration is complex, then we have:

(a*) *If $L_T H \subset H$ for each $T \in \mathrm{Mult}(Y)$ and if $\mathrm{Mult}(X^*)$ is trivial, then*

$$\mathrm{Mult}(H) = \{L_T \mid T \in \mathrm{Mult}(Y)\}.$$

Lack of a statement (b*) in the above is due to the absence of a good substitute for $\mathrm{Cun}(X)$, when $Z(X^*)$ is replaced by $\mathrm{Mult}(X^*)$ in the statement of Theorem I.3.14(c) (see the proof of Theorem 1.2(b)).

The basic obstacle for a direct proof of Theorem 1.2 stems from the fact that, in general, only for "small" subspaces such as $H = K(X,Y)$ the extremal structure of B_{H^*} has been determined. Since we will need the description of the extreme functionals on spaces of compact operators in Section VI.3, too, we take a broader perspective and present this representation in the framework of injective tensor products of Banach spaces which we define next.

For $u = \sum_{i=1}^n x_i \otimes y_i \in X \otimes Y$ the injective tensor norm (or ε-norm) is defined by

$$\|u\|_\varepsilon = \sup\Big\{\Big|\sum x^*(x_i)y^*(y_i)\Big| \;\Big|\; x^* \in B_{X^*},\ y^* \in B_{Y^*}\Big\},$$

and the completion of the thus normed space $X \otimes Y$ is denoted by $X\widehat{\otimes}_\varepsilon Y$. This space is called the *injective tensor product* of X and Y, and it is closely related to a space of compact operators, as follows. The linear space $X^* \otimes Y$ can clearly be identified with the space of continuous operators from X to Y of finite rank. Under this identification the ε-norm coincides with the operator norm; hence $A(X,Y)$, the space of norm limits of finite rank operators ("approximable operators"), can canonically be identified with $X^*\widehat{\otimes}_\varepsilon Y$. It is known that whenever X^* or Y has the approximation property, then $A(X,Y) = K(X,Y)$ [422, p. 32f.]. In general, however, the space of compact operators can only be represented as the so-called Schwartz ε-product $X^*\varepsilon Y$, which reduces in the Banach space setting to the space $K_{w^*}(X^{**},Y)$ of compact operators from X^{**} into Y which are weak*-weakly continuous. (The isomorphism is $T \mapsto T^{**}$.)

We also need to know the dual space of $X\widehat{\otimes}_\varepsilon Y$. By construction, $X\widehat{\otimes}_\varepsilon Y$ is isometric to a subspace of $C(B_{X^*} \times B_{Y^*})$ (where the balls are equipped with their weak* topologies). Thus, by the Hahn-Banach and Riesz representation theorems each functional $\varphi \in (X\widehat{\otimes}_\varepsilon Y)^*$ has a representation by a (in fact positive) measure μ on $S = B_{X^*} \times B_{Y^*}$ having the same norm, i.e.

$$\langle \varphi, x \otimes y\rangle = \int_S x^*(x)y^*(y)\, d\mu(x^*,y^*). \tag{1}$$

From the point of view of operator theory, (1) defines an operator $T : X \to Y^*$ by means of

$$\langle Tx, y\rangle = \int_S x^*(x)y^*(y)\, d\mu(x^*,y^*). \tag{2}$$

Operators of this form are called *integral*; they form a linear space $I(X,Y^*)$ on which

$$\|T\|_{int} = \inf\{\|\mu\| \mid (2) \text{ holds}\}$$

defines a complete norm. (Actually, the infimum is attained.) Thus, $(X\widehat{\otimes}_\varepsilon Y)^* = I(X,Y^*)$ for short. For more details we refer to [159, Chapter VIII], [281], and the forthcoming book [153].

Note that $K_{w^*}(X^*,Y)$, too, embeds isometrically into $C(B_{X^*} \times B_{Y^*})$ so that continuous linear functionals may be represented by measures. However, in default of density of finite rank operators $K_{w^*}(X^*,Y)^*$ cannot be thought of as a space of operators.

We now have the following representation of the extreme functionals on spaces of compact operators due to W. Ruess and C. Stegall. In both cases of Theorem 1.3 $p\otimes q$ is supposed to operate on H by $\langle p\otimes q, T\rangle := \langle p, T^*q\rangle$.

Theorem 1.3
(a) *If* $X\widehat{\otimes}_\varepsilon Y \subset H \subset K_{w^*}(X^*, Y)$, *then*

$$\operatorname{ex} B_{H^*} = \{p\otimes q \mid p\in \operatorname{ex} B_{X^*},\ q\in B_{Y^*}\}.$$

(b) *If* $X^*\widehat{\otimes}_\varepsilon Y \subset H \subset K(X,Y)$, *then*

$$\operatorname{ex} B_{H^*} = \{p\otimes q \mid p\in \operatorname{ex} B_{X^{**}},\ q\in B_{Y^*}\}.$$

For the proof of this result we need two lemmas.

Lemma 1.4 *Let* $p\in \operatorname{ex} B_{X^*}$ *and* $y\in S_Y$. *Suppose* μ *is a Radon probability on* $S = B_{X^*}\times B_{Y^*}$ *such that*

$$p(x) = \int_S x\otimes y\,d\mu$$

for all $x\in X$. *Then* $\operatorname{supp}(\mu)$, *the support of* μ, *is contained in*

$$\mathbb{S}\cdot\{p\}\times\{y^*\in B_{Y^*} \mid |\langle y^*, y\rangle| = 1\}.$$

PROOF: Consider the measurable mapping $U: S\to B_{X^*}$, $U(x^*, y^*) = \langle y^*, y\rangle\cdot x^*$. Then we have for the image measure $\nu = U(\mu)$ and $x\in X$

$$\int_{B_{X^*}} x\,d\nu = \int_S x\circ U\,d\mu = \int_S x\otimes y\,d\mu = p(x).$$

Since p is extreme, there is only one probability measure on B_{X^*} representing p. (Here is a quick proof of this fact: Consider the set F of those probability measures representing p. This is a face of the set of all probability measures since p is extreme; hence its extreme points are Dirac measures. But $\delta_q\in F$ if and only if $p = q$ so that $F = \overline{\text{co}}\operatorname{ex} F = \{\delta_p\}$ by the Krein-Milman theorem. For the usual proof see [7, Corollary I.2.4].) It now follows $\nu = \delta_p$. Consequently,

$$\begin{aligned} 1 &= \nu(\{p\}) \\ &= \mu(\{U^{-1}(p)\}) \\ &= \mu(\{(x^*, y^*)\in S \mid \langle y^*, y\rangle\cdot x^* = p\}) \\ &\le \mu(\mathbb{S}\cdot\{p\}\times\{y^*\in B_{Y^*} \mid |\langle y^*, y\rangle| = 1\}). \end{aligned}$$

□

Lemma 1.5 *Let* $T\in I(X, Y^*)$ *with* $\|T\|_{int} = 1$. *If* $T^*(y_0^{**})\in \operatorname{ex} B_{X^*}$ *for some* $y_0^{**}\in S_{Y^{**}}$, *then* $\operatorname{ran}(T^*) = \operatorname{lin}\{T^*(y_0^{**})\}$.

PROOF: We first note that $\|i_{Y^*}T\|_{int} = 1$ [159, Theorem VIII.2.8]. Hence there is a representation of $i_{Y^*}T : X \to Y^{***}$ by means of a probability measure μ on $B_{X^*} \times B_{Y^{***}}$ such that

$$\langle Tx, y^{**}\rangle = \int x^*(x) y^{***}(y^{**})\, d\mu(x^*, y^{***}) = \int x \otimes y^{**}\, d\mu.$$

Now Lemma 1.4 applies to show that $\operatorname{supp}(\mu) \subset \mathbb{S} \cdot \{T^*(y_0^{**})\} \times B_{Y^{***}}$ which yields that the barycentre of μ has the form $(\lambda_0 T^*(y_0^{**}), y_0^{***})$ for some $|\lambda_0| = 1$, $y_0^{***} \in B_{Y^{***}}$. This shows

$$(T^* y^{**})(x) = \int x \otimes y^{**}\, d\mu = \lambda_0\, (T^* y_0^{**})(x) \cdot y_0^{***}(y^{**})$$

for all $y^{**} \in Y^{**}$, $x \in X$. Thus, the proof of the lemma is completed. □

PROOF OF THEOREM 1.3:
It is of course enough to prove (a). Let $\varphi \in \operatorname{ex} B_{H^*}$. Since H embeds isometrically into $C(B_{X^*} \times B_{Y^*}) =: C(S)$, φ has an extension to an extreme functional on $C(S)$. Thus, there exist $p \in B_{X^*}$, $q \in B_{Y^*}$ and $\lambda \in \mathbb{S}$ such that $\varphi = \lambda \cdot \delta_{(p,q)}|_H = (\lambda p) \otimes q$. Necessarily p and q must be extreme if φ is.
We come to the converse and tackle first the case $H = X\widehat{\otimes}_\varepsilon Y$. Suppose $p \in \operatorname{ex} B_{X^*}$ and $q \in B_{Y^*}$, and assume moreover that $\|p \otimes q \pm T\|_{int} \le 1$ for some integral operator T. If $y_0^{**} \in S_{Y^{**}}$ satisfies $y_0^{**}(q) = 1$, then

$$\|p \pm T^* y_0^{**}\| \le \|p \otimes q \pm T\| \le \|p \otimes q \pm T\|_{int} \le 1,$$

whence $T^* y_0^{**} = 0$, and we have for the operator $S = p \otimes q + T$ that $S^* y_0^{**} = p \in \operatorname{ex} B_{X^*}$. Lemma 1.5 entails that $\operatorname{ran} S^* = \operatorname{lin}\{p\}$, therefore

$$S = p \otimes q + T = p \otimes y_0^*$$

for some $y_0^* \in Y^*$. This shows that $T = p \otimes (y_0^* - q)$ and

$$\|q \pm (y_0^* - q)\| = \|p \otimes q \pm T\| \le \|p \otimes q \pm T\|_{int} \le 1.$$

From the extremality of q we now obtain $y_0^* = q$ and $T = 0$. This proves that $p \otimes q \in \operatorname{ex} B_{H^*}$ for $H = X\widehat{\otimes}_\varepsilon Y$.
Finally we suppose that $p \in \operatorname{ex} B_{X^*}$, $q \in B_{Y^*}$ and $X\widehat{\otimes}_\varepsilon Y =: H_0 \subset H \subset K_{w^*}(X^*, Y)$. We already know that $(p \otimes q)|_{H_0} \in \operatorname{ex} B_{H_0^*}$, hence there is an extension of this functional to an extreme functional $\varphi \in \operatorname{ex} B_{H^*}$. By the first part of the proof $\varphi = p_1 \otimes q_1$ for some p_1 and q_1. Clearly, $(p_1 \otimes q_1)|_{H_0} = (p \otimes q)|_{H_0}$ implies that $p = p_1$ and $q = q_1$. Therefore $p \otimes q = \varphi$ and is hence extremal on H. □

It is worthwhile noting that, in general, it is not true that these functionals $p \otimes q$ belong to $\operatorname{ex} B_{H^*}$ for larger subspaces H. Let us interrupt the main line of reasoning to give an example of this.

Example 1.6 *For $X = c_0$ and $Y = C(K)$, K metrizable, the functional $e \otimes \delta_k$, $e \in \operatorname{ex} B_{\ell^\infty}$, is extremal if and only if k is isolated.*

Let us point out why this is so. According to [60], the space

$$\mathfrak{L}_k := L(X,Y)/J_{(k)},$$

where $J_{(k)} = \{T \in L(X,Y) \mid \limsup_{\kappa\to k} \|T^*\delta_\kappa\| = 0\}$, may be written as

$$\mathfrak{L}_k = \mathfrak{L}_k^0 \oplus_1 \mathfrak{L}_k^c \qquad (*)$$

with

$$\mathfrak{L}_k^c = L_k^c/J_{(k)}, \qquad L_k^c = \{T \in L(X,Y) \mid \lim_{\kappa\to k} \|T^*\delta_\kappa - T^*\delta_k\| = 0\}$$

and

$$\mathfrak{L}_k^0 = L_k^0/J_{(k)}, \qquad L_k^0 = \{T \in L(X,Y) \mid \|T^*\delta_k\| = 0\}.$$

Observe that $e \otimes \delta_k \in (L_k^0)^\perp$ and so, functionals of this form belong to $(\mathfrak{L}_k^0)^*$ after canonical identification. Now, if the decomposition in $(*)$ is nontrivial, i.e. if $\mathfrak{L}_k^0 \neq \{0\}$, then $e \otimes \delta_k$ belongs to a nontrivial M-summand of the subspace $(J_{(k)})^\perp$ of $L(X,Y)^*$, which makes it impossible for $e \otimes \delta_k$ to be an extreme functional. But this is exactly what happens when k is *not* isolated:
To find an operator $T \in L_k^0 \setminus J_{(k)}$, select a sequence (k_n) converging to k as well as open and pairwise disjoint neighbourhoods U_n of k_n that do not contain k.
Furthermore, choose continuous functions φ_n of norm ≤ 1 with $\operatorname{supp}\varphi_n \subset U_n$ and $\varphi_n(k_n) = 1$ for all $n \in \mathbb{N}$. Denoting by e_n the elements of the usual Schauder basis of c_0, extend $T(e_n) := \varphi_n$ to a bounded operator (in fact, an isometry) $T : c_0 \to C(K)$. Since

$$T^*\delta_k = w^*\text{-}\lim_{\kappa\to k} T^*\delta_\kappa = 0,$$

this operator does the job. On the other hand, when k is an isolated point, then $e \otimes \delta_k$ is extreme. $(\square)$

Let us come back to the preparation of the proof of Theorem 1.2. To surmount the problem of not having a handsome representation of the elements of $\operatorname{ex} B_{H^*}$ in general, we shall need the following elementary lemma. Let us first agree on some notation.
The symbol H will always denote a space of bounded operators containing the finite rank operators, $p \otimes q$ is a functional of the type defined in Theorem 1.3, and by $\operatorname{supp}\operatorname{ex} B_{X^*}$ we denote the set of "supporting extreme functionals", i.e. those functionals in $\operatorname{ex} B_{X^*}$ attaining their norm. By the Krein-Milman theorem $\operatorname{supp}\operatorname{ex} B_{X^*}$ is weak* dense in $\operatorname{ex} B_{X^*}$. Likewise, for each $x \in X$ there is some $p \in \operatorname{supp}\operatorname{ex} B_{X^*}$ such that $p(x) = \|x\|$.

Lemma 1.7

(a) *Let $p \in \operatorname{ex} B_{E^*}$ for some subspace E of a Banach space X. Then the set N_p of norm preserving extensions of p to X is a weak* closed face of B_{X^*} and, consequently, by the Krein-Milman theorem*

$$N_p = \overline{\operatorname{co}}^{w^*}(\operatorname{ex} B_{X^*} \cap N_p).$$

(b) *For all $T \in L(X,Y)$ we have*

$$\|T\| = \sup\{\langle p, T^*q\rangle \mid p \in \operatorname{supp}\operatorname{ex} B_{X^{**}}, q \in \operatorname{supp}\operatorname{ex} B_{Y^*}\}.$$

(c) *Denote by* $F_{p,q}$ *the weak* closed face of norm preserving extensions of* $p\otimes q \in \operatorname{ex} B_{(X^*\widehat{\otimes}_\varepsilon Y)^*}$ *from* $X^*\widehat{\otimes}_\varepsilon Y$ *to* H *and define*

$$E_0 := \bigcup_{p\otimes q\in B_{(X^*\widehat{\otimes}_\varepsilon Y)^*}} \operatorname{ex} F_{p,q}.$$

Then $E_0 \subset \operatorname{ex} B_{H^*} \subset \overline{E_0}^{w*}$.

PROOF: We content ourselves with a proof of (b) and (c), since (a) is obvious. To prove (b), start with an $x \in B_X$ such that $\|Tx\| > \|T\| - \varepsilon$ for some arbitrary $\varepsilon > 0$ and pick some $q \in \operatorname{supp} \operatorname{ex} B_{Y^*}$ with $\|Tx\| = q(Tx)$. Norm T^*q by $p \in \operatorname{supp} \operatorname{ex} B_{X^{**}}$ to obtain

$$p(T^*q) = \|T^*q\| \geq q(Tx) > \|T\| - \varepsilon.$$

For the proof of (c), note first that the inclusion $E_0 \subset \operatorname{ex} B_{H^*}$ is true since $F_{p,q}$ is a face of B_{H^*} and hence $\operatorname{ex} F_{p,q}$ is contained in $\operatorname{ex} B_{H^*}$. To prove the other inclusion observe that $p\otimes q$ has a natural extension to a functional on the whole of H, which we continue to denote by $p\otimes q$, and that consequently $p\otimes q \in \overline{\operatorname{co}}^{w*} E_0 =: K$. Since by (b) $\|T\| = \sup_{\psi\in K} \psi(T)$ for all $T \in H$, we have $\operatorname{ex} B_{H^*} \subset \overline{E_0}^{w*}$ by the converse of the Krein-Milman theorem. □

The following lemma essentially shows that the functionals $p\otimes q$, though not extreme in general, may nevertheless be treated as if they were, at least as far as the restrictions of operators in $\operatorname{Mult}(H)$ to $X^*\widehat{\otimes}_\varepsilon Y$ are concerned.

Lemma 1.8 *Let* $\Phi \in \operatorname{Mult}(H)$ *and identify* $X^*\widehat{\otimes}_\varepsilon Y$ *with the space of approximable operators from* X *to* Y. *Then the following assertions hold.*

(a) $\Phi|_{X^*\widehat{\otimes}_\varepsilon Y} = 0$ *implies that* $\Phi = 0$.

(b) *For all* $p \in B_{X^{**}}$ *and* $q \in B_{Y^*}$ *there is a number* $a_\Phi(p,q)$ *such that for each operator* $F : X \to Y$ *of finite rank*

$$\langle p, (\Phi(F))^*(q)\rangle = a_\Phi(p,q)\langle p, F^*(q)\rangle.$$

Moreover, Φ *is in* $Z_{0,1}(X)$ *precisely when* $0 \leq a_\Phi(p,q) \leq 1$ *for all* $p\otimes q$.

PROOF: (a) Since $E_0 \subset \operatorname{ex} B_{H^*}$ we have by assumption for each $F \in X^*\otimes Y$ and any $\psi \in \operatorname{ex} F_{p,q}$,

$$0 = \psi(\Phi(F)) = a_\Phi(\psi)\langle p, F^*q\rangle$$

so that $a_\Phi(\psi) = 0$ for all $\psi \in E_0$. Now, the fact that $\operatorname{ex} B_{H^*} \subset \overline{E_0}^{w*}$ in combination with the weak* continuity of the function a_Φ on $\operatorname{ex} B_{H^*}$ (Lemma I.3.2) implies that $\Phi = 0$.

(b) We first show that

$$\langle p, (x^*\otimes y)^*q\rangle = 0 \qquad \forall x^* \in X^*,\ y \in Y$$

implies that

$$\langle p, (\Phi(x^*\otimes y))^*q\rangle = 0 \qquad \forall x^* \in X^*,\ y \in Y.$$

To do this, we use once more that $\operatorname{ex} F_{p,q} \subset \operatorname{ex} B_{H^*}$ and find that for all $\psi \in \operatorname{ex} F_{p,q}$

$$\psi(\Phi(x^* \otimes y)) = a_\Phi(\psi)\psi(x^* \otimes y) = a_\Phi(\psi)\langle p, (x^* \otimes y)^* q\rangle = 0.$$

Our claim then follows since

$$p \otimes q \in \overline{\operatorname{co}}^{w^*} \operatorname{ex} F_{p,q}.$$

To prove the implication claimed note that we just have seen that

$$\ker\left((p \otimes q)|_{X^*\widehat{\otimes}_\varepsilon Y}\right) \subset \ker\left(\Phi^*(p \otimes q)|_{X^*\widehat{\otimes}_\varepsilon Y}\right),$$

which in turn easily implies that, just as desired, for some number $a_\Phi(p,q)$

$$a_\Phi(p,q)(p \otimes q)|_{X^*\widehat{\otimes}_\varepsilon Y} = \Phi^*(p \otimes q)|_{X^*\widehat{\otimes}_\varepsilon Y}.$$

Now let us see what happens when $\Phi \in Z_{0,1}(X)$. Fix a pair (p,q) and pick $x^* \in X^*$ and $y \in Y$ such that $\langle p, (x^* \otimes y)^* q\rangle = 1$. Approximating $p \otimes q$ in the weak* topology by convex combinations $\sum_i t_{\alpha,i}\psi_{\alpha,i}$ with $\psi_{\alpha,i} \in \operatorname{ex} F_{p,q}$ we obtain

$$a_\Phi(p,q) = \langle p, \Phi(x^* \otimes y)^* q\rangle = \lim_\alpha \sum_i t_{\alpha,i} a_\Phi(\psi_{\alpha,i}) \in [0,1].$$

For the converse, suppose that $a_\Phi(p,q) \in [0,1]$ for all p,q. By Lemma 1.7(c), the weak* closure of $\{p \otimes q \mid p \in \operatorname{ex} B_{X^{**}},\ q \in \operatorname{ex} B_{Y^*}\}$ contains $\operatorname{ex} B_{H^*}$ so that Lemma I.3.2 guarantees that

$$0 \le a_\Phi|_{\operatorname{ex} B_{H^*}} \le 1$$

which is what we have claimed. □

We close this set of preparatory lemmata with a result known as the Bishop-Phelps-Bollobás theorem and one of its corollaries. Recall that

$$\Pi(X) = \{(x^*, x) \in B_{X^*} \times B_X \mid x^*(x) = 1\}.$$

Theorem 1.9 *Let X be a Banach space.*

(a) *If $x \in B_X$ and $x^* \in S_{X^*}$ with*

$$|1 - x^*(x)| \le \left(\frac{\varepsilon}{2}\right)^2,$$

there is $(x_\varepsilon^*, x_\varepsilon) \in \Pi(X)$ with

$$\|x - x_\varepsilon\| < \varepsilon \qquad \text{and} \qquad \|x^* - x_\varepsilon^*\| < \varepsilon.$$

(b) *In particular, the functionals on X which attain their norm are norm dense.*

PROOF: [86, p. 7]. □

Actually, part (a) is a special case of a more general result due to Brøndsted and Rockafellar, see [318, p. 165]. Part (b) of Theorem 1.9 is usually called the Bishop-Phelps theorem.

Lemma 1.10 *For each $(x^{**},x^*)\in\Pi(X^*)$ one can find a net (x^*_α,x_α) in $\Pi(X)$ with*

$$w^*\text{-}\lim_\alpha x_\alpha = x^{**} \quad \textit{and} \quad \|\,.\,\|\text{-}\lim_\alpha x^*_\alpha = x^*.$$

PROOF: Choose a net (ξ_α) from B_X with $w^*\text{-}\lim_\alpha \xi_\alpha = x^{**}$ and put

$$\delta_\alpha := |x^{**}(x^*) - \xi_\alpha(x^*)|.$$

By the above, we may find a net $(x^*_\alpha,x_\alpha)\in\Pi(X)$ with

$$\|\xi_\alpha - x_\alpha\| < 2\sqrt{\delta_\alpha} \qquad \text{and} \qquad \|x^*_\alpha - x^*\| < 2\sqrt{\delta_\alpha}.$$

Since $\lim_\alpha \delta_\alpha = 0$, this is a net with the required properties. □

PROOF OF THEOREM 1.2:

One half of the proof of Theorem 1.2 is contained in Lemma 1.1. Let us prove the missing directions.

(a), (a*) Let $\Phi\in\text{Mult}(H)$ be fixed. Define $\Omega_{q,y}: X^*\to X^*$ by

$$\Omega_{q,y}(x^*) = (\Phi(x^*\otimes y))^*(q),$$

where $q\in\text{ex}\,B_{Y^*}$, $y\in Y$ and $q(y)\neq 0$. By Lemma 1.8(b), we have for an arbitrary $p\in\text{ex}\,B_{X^{**}}$

$$\Omega^*_{q,y}(p) = a_\Phi(p,q)q(y)p,$$

so that $\Omega_{q,y}\in\text{Mult}(X^*)$. If, in addition, $\Phi\in Z_{0,1}(H)$, then by Lemma 1.8(b)

$$0\le a_\Phi(p,q) = a_{\Omega_{q,y}}(p)q(y)^{-1}\le 1,$$

whence $q(y)\Omega_{q,y}\in Z_{0,1}(X^*)$. Since always $Z(E)=\overline{\text{lin}}\,Z_{0,1}(E)$ for a Banach space E, we conclude that $\Phi\in Z(H)$ implies $\Omega_{q,y}\in Z(X^*)$. In both cases, there is by assumption on these algebras (recall Theorem I.3.14(b)) a number $\alpha(q,y)$ with

$$\Omega_{q,y} = \alpha(q,y)Id.$$

Comparing equations we find

$$a_\Phi(p,q) = \alpha(q,y)q(y),$$

which in turn implies that $a_\Phi(p,q) =: a(q)$ does not depend on p. Now, take $\xi^*\in X^*$ and $\xi\in X$ with $\xi^*(\xi)=1$ and put

$$Ty := \Phi(\xi^*\otimes y)\xi.$$

By the above, $T^*q = a(q)q$ for all $q\in\text{ex}\,B_{Y^*}$ so that $T\in\text{Mult}(Y)$. Moreover, when Φ is in $Z_{0,1}(H)$ then

$$0\le a_\Phi(p,q) = a(q) = a_T(q)\le 1,$$

and so, $T\in Z_{0,1}(X)$. Therefore, T is in $Z(X)$ whenever $\Phi\in Z(H)$. By Lemma 1.1, we may conclude that $L_T\in\text{Mult}(H)$ or $Z(H)$, respectively.

To finish the first part of the proof, we infer that $\Phi = L_T$. To this end, it is by Lemma 1.8(a) sufficient to show that both operators agree on $X^*\widehat{\otimes}_\varepsilon Y$. In fact, we have for all $(p,q) \in \operatorname{ex} B_{X^{**}} \times \operatorname{ex} B_{Y^*}$ and $u = \sum_{i=1}^n x_i^* \otimes y_i \in X^* \otimes Y$

$$\langle p, ((\Phi - L_T)(u))^* q\rangle = \sum_i \big(\langle p, (\Phi(x_i^* \otimes y_i))^* q\rangle - \langle p, (x_i^* \otimes Ty_i)^* q\rangle\big),$$

and the last expression becomes

$$\sum_i a_\Phi(p,q)\langle p, (x_i^* \otimes y_i)^* q\rangle - \sum_i a(q)p(x_i^*)q(y_i) = 0.$$

This is, by Lemma 1.7(b), enough to prove our claim.
(b) Let $\Phi \in Z(H)$. Similarly to the above, define $\Omega^{x^*,x} : Y \to Y$ by

$$\Omega^{x^*,x}(y) := \Phi(x^* \otimes y)x$$

We claim that $\Omega^{x^*,x} \in Z(Y)$ whenever $(x^*,x) \in \Pi(X)$. To show this, we may restrict ourselves to the case where $\Phi \in Z_{0,1}(X)$. Let $y_1, y_2 \in Y$. By Proposition I.3.9,

$$\|\Phi(x^* \otimes y_1) + (Id - \Phi)(x^* \otimes y_2)\| \le \max\{\|y_1\|, \|y_2\|\}.$$

Since $\|x\| = 1$ and $x^*(x) = 1$,

$$\begin{aligned}\|\Omega^{x^*,x}(y_1) + (Id - \Omega^{x^*,x})(y_2)\| &= \|\Phi(x^* \otimes y_1)x + (Id - \Phi)(x^* \otimes y_2)x\| \\ &\le \max\{\|y_1\|, \|y_2\|\},\end{aligned}$$

and so $\Omega^{x^*,x} \in Z(Y)$, as claimed. Assuming $Z(Y)$ to be trivial, we obtain for all $(x^*,x) \in \Pi(X)$

$$\Omega^{x^*,x} = \beta(x^*,x)Id$$

for some constant $\beta(x^*,x)$. Let $q \in \operatorname{ex} B_{Y^*}$ and $y \in Y$ with $q(y) \neq 0$, put $T^{q,y}(x^*) := (\Phi(x^* \otimes y))^*(q)$ and choose for a fixed pair $(p_0, x_0^*) \in \Pi(X^*)$ with $p_0 \in \operatorname{supp} \operatorname{ex} B_{X^{**}}$ a net $((p_\alpha, x_\alpha^*))$ as indicated in Lemma 1.10. We have

$$\langle p_0, T^{q,y}x_0^*\rangle = \lim_\alpha \langle p_\alpha, T^{q,y}x_\alpha^*\rangle = \lim_\alpha \langle q, \Omega^{p_\alpha, x_\alpha^*}y\rangle = q(y)\lim_\alpha \beta(x_\alpha^*, p_\alpha)$$

so that $\beta(p_0, x_0^*) := \lim_\alpha \beta(x_\alpha^*, p_\alpha)$ exists and does not depend on the particular choice of (p_α, x_α^*). On the other hand,

$$q(y)a_\Phi(p,q) = \langle p, T^{q,y}x^*\rangle = \beta(p,x^*)q(y^*)$$

for all $p \in \operatorname{supp} \operatorname{ex} B_{X^{**}}$ and some norm attaining $x^* \in B_{X^*}$. Consequently, the functions

$$\beta(p,x^*) = a_\Phi(p,q) =: b(p)$$

are independent of their second arguments. It follows that

$$(T^{q,y})^*p = b(p)p$$

for all $p \in \operatorname{supp} \operatorname{ex} B_{X^{**}}$. Since $\operatorname{supp} \operatorname{ex} B_{X^{**}}$ is weak* dense in $\operatorname{ex} B_{X^{**}}$ we deduce from Lemma I.3.2 that $T^{q,y} \in Z(X^*)$. Applying Theorem I.3.14 we find a $T_*^{q,y} \in \operatorname{Cun}(X)$ with $(T_*^{q,y})^* = T^{q,y}$. Fix $(\eta^*, \eta) \in \Pi(Y)$ with $\eta^* \in \operatorname{ex} B_{Y^*}$ and put $T := T_*^{\eta^*,\eta}$. As was shown in Lemma 1.1, R_T belongs to $Z(H)$. Similarly to the proof of (a), we compute

$$\begin{aligned} \langle p, ((\Phi - R_T)(u))^* q\rangle &= \sum_i \left(\langle p, (\Phi(x_i^* \otimes y_i))^* q\rangle - \langle p, (T^* x_i^* \otimes y_i)^* q\rangle\right) \\ &= \sum_i a_\Phi(p,q)\langle p, (x_i^* \otimes y_i)^* q\rangle - \sum_i b(p)p(x_i^*)q(y_i) \\ &= 0, \end{aligned}$$

which is valid for all $(p,q) \in \operatorname{supp} \operatorname{ex} B_{X^{**}} \times \operatorname{ex} B_{Y^{**}}$. But since these functionals form a norming subset of H^* (Lemma 1.7(b)), Φ and R_T coincide on $X^* \widehat{\otimes}_\varepsilon Y$, which in light of Lemma 1.8(a) settles our claim. □

We wish to give some applications of the result proven above. The first one takes advantage of the fact that $Z(X)$ as well as $\operatorname{Mult}(X)$ are dual spaces whenever X is (Theorem I.3.14(c)).

Corollary 1.11 *Suppose that X is a Banach space and that $Z(X^*)$ is trivial. Then $L(X, CK)$ is a dual space if and only if $C(K)$ is. The same result holds true when* $\operatorname{Mult}(X^*)$ *is supposed to be trivial and $C(K)$ is replaced by a function algebra $\mathfrak{A}$ throughout.*

PROOF: If the space Y has a predual Y_* then $L(X,Y)$ always has a representation as a dual space, namely $L(X,Y) = (X \widehat{\otimes}_\pi Y_*)^*$ [159, Cor. VIII.2.2]. On the other hand,

$$Z(L(X, CK)) = Z(CK) = CK$$

is a dual space when $L(X, CK)$ is (Theorem I.3.14(c)). This gives the claim in the centralizer case. The proof of the other case follows the same lines, using $\operatorname{Mult}(\mathfrak{A}) \cong \mathfrak{A}$ (Example I.3.4(c)). □

It is known that $C(K)$ is isometric to a dual Banach space if and only if K is hyperstonean [386, p. 95ff.].

Corollary 1.12

(a) *If there are no nontrivial L-summands in X, then the M-projections on $L(X,Y)$ have the form $T \mapsto P_\infty T$ for some M-projection P_∞ on Y. Consequently, the M-summands of $L(X,Y)$ have the form*

$$\{T \in L(X,Y) \mid \operatorname{ran}(T) \subset J\}$$

for some M-summand J in Y.

(b) *If there are no nontrivial M-summands in Y, then the M-projections on $L(X,Y)$ have the form $T \mapsto TP_1$ for some L-projection P_1 on X. Consequently, the M-summands of $L(X,Y)$ have the form*

$$\{T \in L(X,Y) \mid \operatorname{ran}(T^*) \subset J\}$$

for some M-summand J in X^*.

PROOF: This is an immediate consequence of Theorem 1.2. For part (b) keep Theorem I.1.9 in mind. □

Note that in general, an arbitrary 1-complemented subspace of $L(X,Y)$ does not necessarily have to be of this simple form. Also, the structure of the M-ideals of $L(X,Y)$ is by far not so transparent in the setting of Corollary 1.12. We will, however, present some analogous results for the M-ideals of the operator space $A(X,Y) = X^*\widehat{\otimes}_\varepsilon Y$ in Section VI.3.
It should be noted that Corollary 1.12 still holds for subspaces H of $L(X,Y)$ which are subject to the condition that $L_{P_\infty}(H) \subset H$ resp. $R_{P_1}(H) \subset H$. For instance every closed operator ideal (such as the compact or weakly compact operators) could be considered. In order to simplify the statement of the following corollary, we do not present the most general result possible either.

Corollary 1.13 *For all Banach spaces $X \not\cong \ell^\infty_{\mathbb{R}}(2)$ and any closed two-sided ideal H of $L(X)$ with $X^*\widehat{\otimes}_\varepsilon X \subset H \subset L(X)$ either*

$$Z(H) = \{L_T \mid T \in Z(X)\}$$

or

$$Z(H) = \{R_T \mid T \in \operatorname{Cun}(X)\}$$

holds. Moreover, all M-summands J in H are either of the form

$$J = \{T \in H \mid P_\infty T = T\}$$

or

$$J = \{T \in H \mid TP_1 = T\},$$

respectively, where P_∞ is an M- and P_1 is an L-projection on X.

PROOF: The proof is a direct consequence of Theorem I.3.14(d) and Theorem 1.2. □

We finish this section with two results in the flavour of the classical Banach-Stone theorem.

Corollary 1.14 *For $i = 1,2$ let $H_i \subset L(X_i, \mathfrak{A}_i)$, where the $\mathfrak{A}_i$ are function algebras and the algebras $\operatorname{Mult}(X_i^*)$ are supposed to be trivial. If the spaces H_i contain the finite rank operators and if they are invariant under the operators L_T, $T \in \operatorname{Mult}(\mathfrak{A}_i)$, then*

$$H_1 \cong H_2 \qquad \textit{implies} \qquad \mathfrak{A}_1 \cong \mathfrak{A}_2.$$

PROOF: The proof simply follows from Theorem 1.2(a*), the fact that $W_1 \cong W_2$ implies $\mathrm{Mult}(W_1) \cong \mathrm{Mult}(W_2)$ and

$$\mathfrak{A}_1 \cong \mathrm{Mult}(\mathfrak{A}_1) \cong \mathrm{Mult}(\mathfrak{A}_2) \cong \mathfrak{A}_2,$$

where we have made use of the fact that for a function algebra $\mathfrak{A}$, $\mathrm{Mult}(\mathfrak{A})$ consists of multiplication operators with elements in $\mathfrak{A}$ (Example I.3.4(c)). □

As a special instance of the preceding corollary we obtain the following Banach-Stone theorem for operator spaces.

Corollary 1.15 *If K_1 and K_2 are compact Hausdorff spaces for which $L(CK_1)$ and $L(CK_2)$ are isometrically isomorphic, then K_1 and K_2 are homeomorphic.*

PROOF: By Corollary 1.14 we have $C(K_1) \cong C(K_2)$ so that the classical Banach-Stone theorem (see e.g. [51]) yields the result claimed. □

Corollary 1.16 *Let $H_i \subset L(L^1(\mu_i), X_i)$, $i = 1, 2$, where the μ_i are arbitrary measures, and suppose that the algebras $Z(X_i)$ are trivial. If the spaces H_i contain the finite rank operators, if they are invariant under the operators R_T, $T \in \mathrm{Cun}(L^1(\mu_i))$, and if $H_1 \cong H_2$, then one may conclude that*

$$L^1(\mu_1) \cong L^1(\mu_2).$$

PROOF: We have

$$L^1(\mu_1)^* \cong \mathrm{Cun}(L^1(\mu_1)) \cong \mathrm{Cun}(H_1) \cong \mathrm{Cun}(H_2) \cong \mathrm{Cun}(L^1(\mu_2)) \cong L^1(\mu_2)^*$$

from Theorem 1.2. Since L^1-spaces are unique preduals, i.e.

$$E^* \cong L^1(\mu)^* \quad \Rightarrow \quad E \cong L^1(\mu),$$

(see [386, p. 96f.]), we obtain the desired conclusion. □

We shall have more to say about the relation between M-structure theory and Banach-Stone type results in the Notes and Remarks section.

VI.2 M-ideals in $L(X, C(K))$ for certain X

The present section deals with an attempt to characterise the M-ideals in $L(X, Y)$ in terms of the M-ideals in Y. Unlike the previous section, no satisfactory general condition for this to be possible is known. What will be done in the following is a complete calculation of the M-ideals in $L(X, C(K))$, where X^* is either uniformly convex or else has sufficiently many one-dimensional L-summands. Note that the techniques used below apply independently of the specification of the scalar field.
We start with specifying some subspaces which are always M-ideals of the operator spaces $L(X, C(K))$.

Proposition 2.1 *Suppose D is a closed subset of the compact space K, and let X be a Banach space. Then*

$$J_{(D)} := \{T \in L(X, C(K)) \mid \limsup_{k \to k_0} \|T^*(\delta_k)\| = 0 \ \forall k_0 \in D\}$$

is an M-ideal of $L(X, C(K))$.

The proof of the above is a by now standard application of the n-ball property and will hence be left to our readers; for details we refer to [60] or [626].
Note that in the sense of Corollary V.3.6

$$J_{(D)} = L(X, C(K))_{J_D},$$

where $J_D \subset C(K) \hookrightarrow Z(L(X, C(K)))$ is the M-ideal of continuous functions vanishing on D.
Theorem 2.3 below will show that, for certain X, all M-ideals of $L(X, C(K))$ look like this. We first present an equivalent condition for this to be true. In the following lemma we call an M-ideal in a Banach space *maximal* if there is no nontrivial M-ideal which strictly contains the given one. Also, we employ the notation hT for the operator $x \mapsto h \cdot Tx$, where $T \in L(X, C(K))$, $h \in C(K)$.

Lemma 2.2 *Let X be a Banach space and K a compact Hausdorff space. Then the M-ideals of the form $J_{(D)}$ with D closed exhaust all M-ideals of $L(X, C(K))$ if and only if for all $k \in K$ the closed subspace $J_{(k)} := J_{(\{k\})}$ is a maximal M-ideal.*

PROOF: Only one implication requires a proof:
Suppose that all $J_{(k)}$, $k \in K$, are maximal M-ideals, and denote by J an arbitrary M-ideal of $L(X, C(K))$. Putting

$$D := \{k \in K \mid \lim_{\kappa \to k} \|T^*\delta_\kappa\| = 0 \ \ \forall T \in J\},$$

we must show that $J_{(D)} \subset J$, since the other inclusion is clear. To this end, we will first show that any $T \in J_{(D)}$ can be locally approximated by elements of J, more precisely we claim:

For all $T \in J_{(D)}$ and $k \in K$ there are a neighbourhood U of k and $S \in J$ with

$$\|(T - S)|_U\| := \sup_{\kappa \in U} \|(T - S)^*(\delta_\kappa)\| \leq \varepsilon.$$

In fact, whenever $k \notin D$ – and only this case requires some work – the space $J + J_{(k)}$ strictly contains $J_{(k)}$ by the very definition of D and hence, $J + J_{(k)} = L(X, C(K))$, since the former space is an M-ideal in light of Proposition I.1.11. But then, for any $T \in J_{(D)}$, there is $S \in J \subset J_D$ such that $T - S \in J_{(k)}$ which yields the claim.
To obtain a global approximation of $T \in J_{(D)}$ by an element of J (which will prove the inclusion we have in mind), we fix $\varepsilon > 0$, cover K by finitely many open sets $U_1, \ldots, U_n$ as above and select $S_1, \ldots, S_n \in J$ accordingly with $\|(S_i - T)|_{U_i}\| < \varepsilon$, $i = 1, \ldots, n$.

Of course, all this is possible by compactness and the above claim. If $h_1, \dots, h_n$ is a partition of unity subordinate to $U_1, \dots, U_n$, then

$$\left\| \sum_{i=1}^{n} h_i S_i - T \right\| < \varepsilon,$$

and we are done since $\sum_{i=1}^{n} h_i S_i \in J$ by Lemma 1.1 and Lemma I.3.5(b) (note $M_h \in Z(C(K))$).
The attentive reader will have noticed that we still owe a proof of the not obvious fact that D is closed: Pick k in the complement of D and an operator T with $\liminf_{\kappa \to k} \|T^* \delta_\kappa\| > 0$ (e.g. $T = \delta_k \otimes \mathbf{1}$). As above, $J + J_{(k)} = L(X, C(K))$ and therefore, we may write $T = T_J + T_k$ with $T_J \in J$ and $T_k \in J_{(k)}$. Clearly, $\liminf_{\kappa \to k} \|T_J^*(\delta_\kappa)\| > 0$ and so, $U \cap D = \emptyset$ for a suitable neighbourhood U of k. The proof is finished. □

The following theorem encompasses all what is known on M-ideals in $L(X, C(K))$. Note that we don't put any restriction on the scalar field.

Theorem 2.3 *Let X be Banach space and K a compact Hausdorff space. Suppose further that either*

(a) *X^* is uniformly convex,*

or else that

(b) *$\ker p$ is an M-ideal for each $p \in \operatorname{ex} B_{X^*}$.*

Then a closed subspace J of $L(X, C(K))$ is an M-ideal if and only if J is of the form $J_{(D)}$ for some closed subset D of K.

PROOF: We begin with the assumption that X^* is uniformly convex.
By Lemma 2.2 we must show that, for each $k_0 \in K$, an M-ideal J strictly containing $J_{(k_0)}$ already equals $L(X, C(K))$. Let such a J be given and, by the uniform convexity of X^*, fix for every $\varepsilon > 0$ a number $0 < \delta(\varepsilon) < \varepsilon$ such that

$$\|x^*\| \geq 1 - \delta(\varepsilon),\ \|x^* \pm y^*\| \leq 1 + \delta(\varepsilon) \quad \Longrightarrow \quad \|y^*\| \leq \varepsilon. \tag{$*$}$$

We will first show that for every $\varepsilon > 0$ there is $T_\varepsilon \in J$ with

$$1 - \varepsilon \ \leq\ \liminf_{k \to k_0} \|T_\varepsilon^* \delta_k\| \ \leq\ \limsup_{k \to k_0} \|T_\varepsilon^* \delta_k\| \ \leq\ 1 + \varepsilon.$$

To this end, pick $T \in J \backslash J_{(k_0)}$ with $\limsup_{k \to k_0} \|T^* \delta_k\| > 0$. Choosing $h \in C(K)$ properly and considering hT instead of T we may as well suppose that $\limsup_{k \to k_0} \|T^* \delta_k\| = \|T^*\| = 1$. By the 3-ball property of J, we find $T_\varepsilon \in J$ with

$$\|T_\varepsilon\| \leq 1 + \delta(\varepsilon) \leq 1 + \varepsilon$$

and $\|\pm T + (x_0^* \otimes \mathbf{1} - T_\varepsilon)\| \leq 1 + \delta(\varepsilon)$, where x_0^* is any norm one functional on X. It follows by $(*)$ that $\|x_0^* - T_\varepsilon^* \delta_k\| \leq \varepsilon$ whenever $\|T^* \delta_k\| \leq 1 + \delta(\varepsilon)$. Consequently, by the lower semicontinuity of the map $k \mapsto \|x_0^* - T_\varepsilon^* \delta_k\|$ and our present assumption on T, we have that $\|x_0^* - T_\varepsilon^* \delta_{k_0}\| \le \varepsilon$ and thus $\|T_\varepsilon^* \delta_{k_0}\| \ge 1-\varepsilon$. This entails $\liminf_{k\to k_0} \|T_\varepsilon^* \delta_k\| \ge 1-\varepsilon$. To conclude the proof, fix any $A \in L(X, C(K))$, a positive number η and, by the above (applied to $\varepsilon = \delta(\eta)$), an operator $T_\eta \in J$ with

$$1 - \delta(\eta) \le \liminf_{k\to k_0} \|T_\eta^* \delta_k\| \le \limsup_{k\to k_0} \|T_\eta^* \delta_k\| \le 1 + \delta(\eta).$$

Again, by the 3-ball property of J, there is $S \in J$ with $\|S\| \le 1 + \delta(\eta)$ and $\| \pm T_\eta + (A - S)\| \le 1 + \delta(\eta)$. By $(*)$, we infer that $\|(A-S)|_U\| \le \eta$ in a suitable neighbourhood U of k_0. Now the rest is easy: Fix a continuous map $h : K \to [0,1]$ with $\operatorname{supp} h \subset U$ and $h(k_0) = 1$. An appeal to Lemma I.3.5(b) and Lemma 1.1 together with the fact that $J_{(k_0)} \subset J$ shows $(1-h)A + hS \in J$, and in light of

$$\|A - [(1-h)\,A + hS]\| \le \eta,$$

we see that J is dense in $L(X, C(K))$. Hence $J = L(X, C(K))$.

For the proof of the second part of the present theorem we need the following technical lemma, the proof of which we postpone until the end of this section.

Lemma 2.4 *Let Z be a Banach space, $p \in S_Z$ such that* lin $\{p\}$ *is an L-summand with corresponding L-projection P, and fix $a_1, \dots, a_5 \ge 0$. Then we have:*

(a) $\|\theta p + z\| \le 1 + a_1$ *for every* $\theta \in \mathbb{S}$ *implies* $\|z\| \le a_1$.

(b) $\|z\| \ge 1 - a_1$ *and* $\|Pz\| \le a_2$ *imply* $\|\theta p + z\| \ge 2 - 2a_2 - a_1$ *for all* $\theta \in \mathbb{S}$.

(c) *If for all* $\theta \in \mathbb{S}$

$$\|y\| \le 1 + a_1, \qquad 1 - a_2 \le \|z\| \le 1 + a_3,$$
$$\|\theta p + z\| \ge 2 - a_4, \qquad \|\theta z + (p - y)\| \le 1 + a_5,$$

then $\|p - y\| \le a_1 + 2(a_2 + a_3 + a_4 + a_5)$.

Let us now prove the second part of Theorem 2.3. By Lemma 2.2 it is enough to show:

> *If $k_0 \in K$ and J is an M-ideal in $L(X, C(K))$ properly containing $J_{(k_0)}$ then $J = L(X, C(K))$.*

The assumption $J \ne J_{(k_0)}$ yields an operator $T_1 \in J$ for which

$$\limsup_{k\to k_0} \|T_1^*(\delta_k)\| > 0.$$

However, we will need operators in J which are even worse behaved. Eventually we will prove:

> CLAIM: *There exist a neighbourhood U of k_0, $p \in \operatorname{ex} B_{X^*}$, a number $0 \le q < 1$ and $T_0 \in B_J$ such that*
>
> $$\|p - T_0^*(\delta_k)\| \le q \quad \forall k \in U.$$

Taking this claim for granted, we may complete the proof of the theorem as follows. Let $T \in B_{L(X,C(K))}$. By the 2-ball property (respectively, in the complex case, by Remark I.2.3(c) applied to the compact set of centres $\mathbb{S}T_0$) there is $S \in J$ such that

$$\|\theta T_0 + T - S\| \leq 1 + \varepsilon \quad \forall \theta \in \mathbb{S}$$

where $\varepsilon < 1 - q$. Hence

$$\|\theta p + T^*(\delta_k) - S^*(\delta_k)\| \leq 1 + \varepsilon + q$$

for $k \in K$ and all $\theta \in \mathbb{S}$ so that

$$\|T^*(\delta_k) - S^*(\delta_k)\| \leq \varepsilon + q$$

for these k by Lemma 2.4(a). Finally, we consider a continuous function $h : K \to [0,1]$ which vanishes off U and takes the value 1 at k_0. Then the operator $R \mapsto M_h R$ (where M_h denotes the operator of multiplication by h) belongs to the centralizer of $L(X, C(K))$ (Lemma 1.1). Hence

$$S_0 := M_h S + M_{1-h} T$$

lies in J (by Lemma I.3.5(b) and since $M_{1-h}T \in J_{(k_0)}$), and we have

$$\begin{aligned} \|S_0 - T\| &= \|M_h(S - T)\| \\ &\leq \sup_{k \in K} \|S^*(\delta_k) - T^*(\delta_k)\| \\ &\leq \varepsilon + q. \end{aligned}$$

Since T was arbitrary we may now conclude that the quotient map from $L(X, C(K))$ onto $L(X, C(K))/J$ has norm $\leq \varepsilon + q < 1$. By Riesz' Lemma this means $J = L(X, C(K))$.
Thus it is left to give the proof of the claim. We already know that $\limsup_{k \to k_0} \|T_1^*(\delta_k)\| > 0$ for some $T_1 \in J$. After composition with a suitable multiplication operator we may and shall assume

$$\|T_1\| = \limsup_{k \to k_0} \|T_1^*(\delta_k)\| = 1.$$

Thus, if $A = \{k \mid \|T_1^*(\delta_k)\| > 1 - \eta\}$, then $k_0 \in \overline{A}$. We shall apply this for a positive number η which will later be determined.
We now select 33 linearly independent extreme functionals $p_1, \dots, p_{33}$. By assumption on X each of them spans a one-dimensional L-summand in X^*. P_n is to denote the L-projection onto lin $\{p_n\}$. Then, letting $A_n = \{k \in A \mid \|P_n(T_1^*(\delta_k))\| \leq \frac{1}{33}\}$, we deduce from

$$\sum_{n=1}^{33} \|P_n(T_1^*(\delta_k))\| \leq \|T_1^*(\delta_k)\|$$

the equation $A = \bigcup_{n=1}^{33} A_n$ so that $k_0 \in \overline{A_m}$ for some m. For notational convenience we assume $k_0 \in \overline{A_1}$.
As above, the n-ball property yields some $T_2 \in J$ such that

$$\|T_2\| < 1 + \eta \qquad \text{and} \qquad \|\theta T_1 + p_1 \otimes 1 - T_2\| < 1 + \eta \qquad \forall \theta \in \mathbb{S}$$

(compare Remark I.2.3(b) for the former inequality). Now we apply Lemma 2.4(c) with $z = T_1^*(\delta_k)$, $y = T_2^*(\delta_k)$ (where $k \in A_1$) and $a_1 = \eta$, $a_2 = \eta$, $a_3 = 0$, $a_4 = \frac{2}{33} + \eta$ (this is admissible by Lemma 2.4(b)), $a_5 = \eta$ to obtain

$$\|p_1 - T_2^*(\delta_k)\| \leq 7\eta + \frac{4}{33} \quad \forall k \in A_1.$$

Consequently, we also have, using the weak* lower semicontinuity of the norm,

$$\|p_1 - T_2^*(\delta_{k_0})\| \leq 7\eta + \frac{4}{33}. \tag{$*$}$$

From this we derive the inequality

$$\|P_2(T_2^*(\delta_k))\| \leq 10\eta + \frac{4}{33} \tag{$**$}$$

on some neighbourhood U of k_0: We first note that p_1 belongs to the L-summand complementary to $\mathbb{K} \cdot p_2 = (\ker p_2)^\perp$, therefore

$$1 = \|p_1\| = \|p_1|_{\ker p_2}\|$$

(cf. I.1.12 and I.1.13). Thus, there is $x_0 \in B_X$ satisfying

$$|p_1(x_0)| > 1 - \eta \quad \text{and} \quad p_2(x_0) = 0.$$

Choose a neighbourhood U of k_0 where

$$|T_2(x_0)(k) - T_2(x_0)(k_0)| < \eta$$

so that, as a result of $(*)$,

$$|T_2(x_0)(k)| > 1 - \left(9\eta + \frac{4}{33}\right) \quad \text{for } k \in U.$$

Since $P_2(T_2^*(\delta_k)) \in \text{lin}\,\{p_2\}$ and $p_2(x_0) = 0$ we conclude

$$\begin{aligned} \|P_2(T_2^*(\delta_k))\| &\leq \|T_2^*(\delta_k)\| - \|(Id - P_2)(T_2^*(\delta_k))(x_0)\| \\ &\leq 1 + \eta - |T_2(x_0)(k)| \\ &\leq 10\eta + \frac{4}{33} \end{aligned}$$

for $k \in U$, which proves $(**)$. We may also assume

$$\|T_2^*(\delta_k)\| > 1 - \left(8\eta + \frac{4}{33}\right) \quad \text{for } k \in U \tag{$***$}$$

due to the semicontinuity of $\|\cdot\|$, since $(***)$ holds for k_0 by $(*)$. We now apply the n-ball property once more to obtain an operator $T_3 \in J$ for which

$$\|\theta T_2 + p_2 \otimes 1 - T_3\| < 1 + \eta \quad \forall \theta \in \mathbb{S} \quad \text{and} \quad \|T_3\| < 1 + \eta.$$

A second application of Lemma 2.4(c) with $z = T_2^*(\delta_k)$, $y = T_3^*(\delta_k)$ ($k \in K$) and $a_1 = \eta$, $a_2 = 8\eta + \frac{4}{33}$ (admissible by (* * *)), $a_3 = \eta$, $a_4 = 28\eta + \frac{12}{33}$ (admissible by Lemma 2.4(b) and (**)), $a_5 = \eta$ now shows

$$\|p_2 - T_3^*(\delta_k)\| \leq 77\eta + \frac{32}{33} \quad \text{for} \quad k \in U,$$

and the claim follows easily by a proper choice of η. □

Theorem 2.3 applies in particular to L^1-predual spaces X, e.g. $X = C(K)$ itself, as well as to function algebras (see the remarks preceding the proof of Lemma V.6.7). In fact, we did not need the full strength of the assumptions made; we merely made use of the fact that there are 33 extreme functionals for which the kernels are M-ideals. (The above proof would not have worked with 32 instead!) In other words, Theorem 2.3 remains true if X^* has an L-summand isometric to $\ell^1(33)$. (Of course, this assumption is not very natural, and the number 33 appears here only for technical reasons.)

We conclude this section with the

PROOF OF LEMMA 2.4:
(a) and (b) are obvious consequences of the triangle inequality. For the proof of (c), we write $z = \alpha p + z_1$ and $y = \beta p + y_1$, where z_1, y_1 are in $(Id - P)Z$. We have

$$2 - a_4 \leq \|z + \theta p\| = |\theta + \alpha| + \|z_1\|$$

for every $|\theta| = 1$ so that

$$\begin{aligned} 2 - a_4 &\leq 1 - |\alpha| + \|z_1\| \\ &= 1 - |\alpha| + \|z\| - |\alpha| \\ &\leq 2 - 2|\alpha| + a_3. \end{aligned}$$

It follows that $|\alpha| \leq (1/2)(a_3 + a_4) =: a_6$, and consequently $\|z_1\| = \|z\| - |\alpha| \geq 1 - a_2 - a_6$. On the other hand we have

$$\begin{aligned} 1 + a_5 &\geq \|\theta z + (p - y)\| \\ &= |\theta\alpha + (1 - \beta)| + \|\theta z_1 - y_1\| \\ &\geq |1 - \beta| - a_6 + \|\theta z_1 - y_1\|. \end{aligned}$$

Thus $1 + a_5 + a_6 - |1 - \beta| \geq \|\theta z_1 - y_1\|$ for all $|\theta| = 1$ which yields, since $\|z_1\| \geq 1 - a_2 - a_6$, that $1 + a_5 + a_6 - |1 - \beta| \geq 1 - a_2 - a_6$, i.e. $|1 - \beta| \geq a_2 + a_5 + 2a_6$. This implies that $|\beta| \geq 1 - a_2 - a_5 - 2a_6$, and we get

$$\begin{aligned} \|p - y\| &= |1 - \beta| + \|y_1\| \\ &\leq a_2 + a_5 + 2a_6 + \|y\| - |\beta| \\ &= 2a_2 + 2a_5 + 4a_6 + a_1 \\ &\leq a_1 + 2(a_2 + a_3 + a_4 + a_5), \end{aligned}$$

which gives the claim. □

VI.3 Tensor products

The structure of M-ideals in "small" operator spaces is, due in part to Theorem 1.3, more transparent than the case of $L(X,Y)$. This will be pointed out in this section. For the sake of symmetry, we prefer in our treatment to use the notion of injective tensor products rather than that of spaces of approximable operators.

We will finally pass to the "dual" problem of finding the L-summands in the projective tensor product. Here, the central result of Section VI.1 will play a fundamental rôle.

We first show that an M-ideal in one of the factors of an injective tensor product gives rise to an M-ideal in $X\widehat{\otimes}_\varepsilon Y$ and then that, similarly to the results obtained in Section VI.1, under certain circumstances all the M-ideals arise in this way. (We have recalled the definition of the injective tensor product on p. 265.)

Proposition 3.1 *Let J be an M-ideal in Y. Then, for all Banach spaces X, $X\widehat{\otimes}_\varepsilon J$ is an M-ideal in $X\widehat{\otimes}_\varepsilon Y$.*

PROOF: Denote by $P : Y^* \to J^\perp$ the L-projection corresponding to J and recall from Lemma 1.1 that $Id_{X^{**}} \otimes P^*$ is an M-projection on $X^{**}\widehat{\otimes}_\varepsilon Y^{**}$, which gives rise to an L-projection on $(X^{**}\widehat{\otimes}_\varepsilon Y^{**})^* = I(X^{**}, Y^{***})$, the space of integral operators from X^{**} to Y^{***}. Call this projection Π. Writing $\|\cdot\|_{int}$ for the integral norm of an operator and taking advantage of the fact that $T \in I(X, Y^*)$ iff $T^{**} \in I(X^{**}, Y^{***})$ with equality of integral norms [159, Corollary VIII.2.11], we find

$$\begin{aligned} \|T\|_{int} &= \|\Pi T^{**}\|_{int} + \|T^{**} - \Pi T^{**}\|_{int} \\ &= \|P^{**}T^{**}\|_{int} + \|T^{**} - P^{**}T^{**}\|_{int} \\ &= \|PT\|_{int} + \|T - PT\|_{int}. \end{aligned}$$

(Note that the integral operators form an operator ideal in the sense of Pietsch, and hence, application of P from the left doesn't lead out of this class.) Consequently, left multiplication by P induces an L-projection on $(X\widehat{\otimes}_\varepsilon Y)^* = I(X, Y^*)$. Moreover, $T = PT$ iff $T(X) \subset P(Y^*) = J^\perp$ iff $T \in (X \otimes J)^\perp$. It follows that $X\widehat{\otimes}_\varepsilon J$ is an M-ideal in $X\widehat{\otimes}_\varepsilon Y$. □

Theorem 3.2 *Suppose X has no nontrivial M-ideal. Then every M-ideal Z in $X\widehat{\otimes}_\varepsilon Y$ has the form $Z = X\widehat{\otimes}_\varepsilon J$ with some M-ideal J in Y.*

To prepare the proof we present a result which may be of independent interest.

Proposition 3.3 *Let $p \in \operatorname{ex} B_{X^*}$ and suppose Z is an M-ideal in $X\widehat{\otimes}_\varepsilon Y$. Then the closure of $(p \otimes Id)(Z)$ is an M-ideal in Y.*

PROOF: Let E denote the L-projection from $(X\widehat{\otimes}_\varepsilon Y)^*$ onto $Z^\perp$. Given $y_0^* \in S_{Y^*}$, consider $E(p \otimes y_0^*)$. We shall prove the existence of a (uniquely determined) functional $P(y_0^*) \in Y^*$ such that

$$E(p \otimes y_0^*) = p \otimes P(y_0^*). \tag{$*$}$$

To this end, represent the integral bilinear form $E(p \otimes y_0^*)$ by a positive Radon measure μ_1 on $S := B_{X^*} \times B_{Y^*}$ such that $\|E(p \otimes y_0^*)\| = \mu_1(S)$. Likewise, let $p \otimes y_0^* - E(p \otimes y_0^*)$

be represented by μ_2. Since E is an L-projection, $\mu := \mu_1 + \mu_2$ is a probability measure for which

$$\langle p \otimes y_0^*, x \otimes y \rangle = \int_S x \otimes y \, d\mu$$

for all $x \in X, y \in Y$. Assume for the moment that y_0^* attains its norm on B_Y. In this case

$$p(x) = \int_S x \otimes y_0 \, d\mu$$

for a suitable $y_0 \in S_Y$ and all $x \in X$. Then we have by Lemma 1.4

$$\text{supp}(\mu_i) \subset \text{supp}(\mu) \subset \mathbb{S} \cdot \{p\} \times B_{Y^*}$$

so that there are $\lambda_1 \in \mathbb{S}$, $y_1^* \in B_{Y^*}$ with

$$\begin{aligned}\langle E(p \otimes y_0^*), x \otimes y \rangle &= \int_S x \otimes y \, d\mu_1 \\ &= \langle \lambda_1 p \otimes y_1^* , x \otimes y \rangle \\ &= \langle p \otimes \lambda_1 y_1^* , x \otimes y \rangle\end{aligned}$$

for all $x \in X, y \in Y$. Thus (*) is valid in case y_0^* attains its norm. In the general case, the Bishop-Phelps-Bollobás theorem 1.9 (actually the Bishop-Phelps theorem suffices for this purpose) yields a sequence of norm attaining functionals $y_n^* \in S_{Y^*}$ converging to y_0^* in norm. The validity of (*) for y_n^* gives

$$E(p \otimes y_0^*) = \lim E(p \otimes y_n^*) = \lim p \otimes P(y_n^*)$$

so that $P(y_0^*) = \lim P(y_n^*)$ exists, and (*) is proved in the general case, too. Now define a mapping P from Y^* into itself by (*). Obviously, P is an L-projection. It is left to prove

$$P(Y^*) = (p \otimes Id)(Z)^{\perp}.$$

"$\subset$" is immediate from the definitions of E and P. Conversely, suppose $y^* \in Y^*$ satisfies

$$0 = \langle y^*, (p \otimes Id)(u) \rangle = \langle p \otimes y^*, u \rangle$$

for all $u \in Z$. Then $p \otimes y^* \in Z^{\perp}$ so that

$$p \otimes Py^* = E(p \otimes y^*) = p \otimes y^*.$$

Hence $Py^* = y^*$, and Proposition 3.3 is completely proved. □

PROOF OF THEOREM 3.2:
As in the proof of Proposition 3.3, we denote the L-projection onto $Z^{\perp}$ by E. By Proposition 3.3 we may partition the set ex B_{Y^*} into the two subsets

$$\begin{aligned}C_1 &= \{q \in \text{ex}\, B_{Y^*} \mid E(x^* \otimes q) = x^* \otimes q \text{ for all } x^* \in X^*\} \\ C_2 &= \{q \in \text{ex}\, B_{Y^*} \mid E(x^* \otimes q) = 0 \text{ for all } x^* \in X^*\},\end{aligned}$$

because X was assumed to have no nontrivial M-ideals. Next, we apply Proposition 3.3 once more to obtain a family of L-projections P_p on Y^*, indexed by the extreme functionals $p \in \operatorname{ex} B_{X^*}$, with weak* closed ranges which satisfy

$$E(p \otimes y^*) = p \otimes P_p(y^*)$$

for all $y^* \in Y^*$. Furthermore,

$$P_p(Y^*) = \{y^* \mid p \otimes y^* \in Z^\perp\} =: M_p.$$

Obviously, $C_1 \subset M_p$ for all $p \in \operatorname{ex} B_{X^*}$. On the other hand, let $q \in \operatorname{ex} B_{M_p}$. Then $q \in \operatorname{ex} B_{Y^*}$ (since M_p is an L-summand) and $p \otimes q \in Z^\perp$, that is $q \in C_1$. M_p is weak* closed and thus $M_p = \overline{C_1}^{w*}$ independently of p. In other words,

$$J := \{y \in Y \mid q(y) = 0 \text{ for all } q \in C_1\}$$

is an M-ideal in Y. To prove $X \widehat{\otimes}_\varepsilon J = Z$ it is enough to show $(X \widehat{\otimes}_\varepsilon J)^\perp = Z^\perp$. By the first part of the proof, both spaces are weak* closed L-summands, therefore it is enough to check the coincidence of the extreme points of the unit balls, which must have the form $p \otimes q$ with p and q extremal (Theorem 1.3 and Lemma I.1.5). In fact, $p \otimes q \in Z^\perp$ iff $q \in C_1$ iff $q \in J^\perp$ iff $p \otimes q \in (X \widehat{\otimes}_\varepsilon J)^\perp$. □

The following Corollary is an immediate consequence of Theorem 3.2 and the canonical isometric isomorphism $C(K, X) \cong C(K) \widehat{\otimes}_\varepsilon X$ [159, p. 224f.].

Corollary 3.4 *If K denotes a compact Hausdorff space and X a Banach space without nontrivial M-ideals, then $J \subset C(K, X)$ is an M-ideal if and only if*

$$J = J_D \widehat{\otimes}_\varepsilon X = \{f \in C(K, X) \mid f|_D = 0\}$$

for some closed subset $D \subset K$.

Corollary 3.5 *Let X and Y be Banach spaces without proper M-ideals (i.e. every M-ideal is an M-summand). Then $X \widehat{\otimes}_\varepsilon Y$ fails to have proper M-ideals.*

PROOF: A Banach space without proper M-ideals is isometrically isomorphic to a c_0-sum of Banach spaces without nontrivial M-ideals; this follows as in the proof of Proposition III.2.6. If $X = c_0(X_i)$ and $Y = c_0(Y_j)$ are represented in this way, then

$$X \widehat{\otimes}_\varepsilon Y = c_0(X_i \widehat{\otimes}_\varepsilon Y_j)_{i,j},$$

and $X_i \widehat{\otimes}_\varepsilon Y_j$ has no nontrivial M-ideal by Theorem 3.2. Now an appeal to Proposition I.1.16 shows that each M-ideal in $c_0(X_i \widehat{\otimes}_\varepsilon Y_j)_{i,j}$ must be of the form $c_0(J_{i,j})$, where $J_{i,j}$ is either $\{0\}$ or the space $X_i \widehat{\otimes}_\varepsilon Y_j$. The conclusion now follows. □

In the above proofs the density of the algebraic tensor product (or, in the language of operator spaces, the finite rank operators) was essential. In the case of the whole space of compact operators we have the following results; recall that $K_{w^*}(X^*, Y)$ stands for the space of compact operators which are weak*-weakly continuous and that $K_{w^*}(X^*, Y) = X \widehat{\otimes}_\varepsilon Y$ whenever X or Y has the approximation property.

Proposition 3.6 *If X and Y have no nontrivial M-ideals, then neither has $K_{w^*}(X^*,Y)$.*

PROOF: The proof of Proposition 3.3 shows that the closure of $\{T(p) \mid T \in Z\}$ is an M-ideal in Y if $p \in \operatorname{ex} B_{X^*}$ and Z is an M-ideal in $H := K_{w^*}(X^*,Y)$. As a matter of fact, the essential property to be used is that H may be embedded isometrically into $C(B_{X^*} \times B_{Y^*})$. Analogously, the closure of $\{T^*(q) \mid T \in Z\}$ is an M-ideal in X for $q \in \operatorname{ex} B_{Y^*}$. One can, therefore, partition $\operatorname{ex} B_{Y^*} = C_1 \cup C_2$ in the same way as in the proof of Theorem 3.2. Following this proof one arrives at the conclusion that

$$J = \{y \mid q(y) = 0 \text{ for all } q \in C_1\}$$

is an M-ideal in Y. Hence $J = \{0\}$ or $J = Y$. In the first case it follows that $\operatorname{ex} B_{Y^*} = C_1$ and $Z = \{0\}$. (The L-summand which is complementary to $Z^\perp$ is isometrically isomorphic to Z^*, but has no extreme points because of $C_2 = \emptyset$.) In the second case $\operatorname{ex} B_{Y^*} = C_2$ holds and thus $Z = H$. □

Corollary 3.7 *If X^* and Y have no nontrivial M-ideals, then neither has $K(X,Y)$.*

PROOF: The map $T \mapsto T^{**}$ defines an isometric isomorphism between $K(X,Y)$ and $K_{w^*}(X^{**},Y)$. □

We now turn to the dual situation and determine the L-summands in the projective tensor product of two Banach spaces. Let us recall what that means. Let $u \in X \otimes Y$, the algebraic tensor product. The projective (or π-) norm of u is defined as

$$\|u\|_\pi := \inf \left\{ \sum_{i=1}^n \|x_i\| \, \|y_i\| \;\middle|\; u = \sum_{i=1}^n x_i \otimes y_i \right\}.$$

The completion of $X \otimes Y$ is called the *projective tensor product* and denoted by $X \widehat{\otimes}_\pi Y$. It is known that each $u \in X \widehat{\otimes}_\pi Y$ has, for every $\varepsilon > 0$, a representation as a series $u = \sum_{i=1}^\infty x_i \otimes y_i$ (converging for the π-norm) such that $\|u\|_\pi \le \sum_{i=1}^\infty \|x_i\| \, \|y_i\| + \varepsilon$. The dual space of $X \widehat{\otimes}_\pi Y$ can be identified with $L(X,Y^*)$ under the duality

$$\left\langle T, \sum_{i=1}^\infty x_i \otimes y_i \right\rangle = \sum_{i=1}^\infty (Tx_i)(y_i).$$

The reader who is not acquainted with the fundamental properties of these tensor products should consult e.g. [159, Chapter VIII], or, for full treatment, [281] and [153].

Our most general result on L-summands reads as follows.

Theorem 3.8 *Let K be a closed subspace of $X \widehat{\otimes}_\pi Y$ such that for all $\sum_{i=1}^\infty x_i \otimes y_i \in K$*

$$\sum_{i=1}^\infty x^*(x_i) y^*(y_i) = 0 \qquad \forall x^* \in X^* \; \forall y^* \in Y^*. \tag{$*$}$$

If $\operatorname{Cun}(X)$ is trivial and $(Id \otimes T)K \subset K$ for all $T \in \operatorname{Cun}(Y)$ then

$$\operatorname{Cun}((X \widehat{\otimes}_\pi Y)/K) = \{(Id \otimes T)_K \mid T \in \operatorname{Cun}(Y)\},$$

where $(Id \otimes T)_K$ denotes the natural action of $Id \otimes T$ on $(X \widehat{\otimes}_\pi Y)/K$.
A similar statement holds when $\operatorname{Cun}(Y)$ is trivial.

Concerning the assumption on K in the above theorem we would like to point out that $(*)$ expresses the requirement that $\widehat{u} = 0$ for all $u \in K$, where $\widehat{u}$ is the nuclear operator naturally assigned to the tensor u. Also, note that $u = 0$ iff $\widehat{u} = 0$ whenever X or Y has the approximation property, cf. [159, Theorem VIII.3.4].

PROOF: Consider the natural duality $(X\widehat{\otimes}_\pi Y)^* = L(X, Y^*)$ given by

$$\left\langle S\,, \sum_{i=1}^{\infty} x_i \otimes y_i \right\rangle = \sum_{i=1}^{\infty} Sx_i(y_i).$$

Then, by assumption, the space $K^\perp = ((X\widehat{\otimes}_\pi Y)/K)^*$ contains the finite dimensional operators between X and Y^* and $L_U K^\perp \subset K^\perp$ for all $U \in Z(Y^*) = \{T^* \mid T \in \mathrm{Cun}(Y)\}$ (see Theorem I.3.14(b), also recall the notation $L_U(T) = UT$).
Hence, the conditions imposed on $K^\perp$ are such that Theorem 1.2 may be applied to any Φ^* with $\Phi \in \mathrm{Cun}((X\widehat{\otimes}_\pi Y)/K)$. Consequently, using Theorem 1.2, we find an operator $T \in \mathrm{Cun}(Y)$ with $\Phi^* = L_{T^*}$. This gives

$$\begin{aligned}
\left\langle S\,, \Phi\left(\sum_{i=1}^{\infty} x_i \otimes y_i\right)\right\rangle &= \left\langle \Phi^*(S), \sum_{i=1}^{\infty} x_i \otimes y_i \right\rangle \\
&= \left\langle T^*S\,, \sum_{i=1}^{\infty} x_i \otimes y_i \right\rangle \\
&= \sum_{i=1}^{\infty} \langle Sx_i, Ty_i \rangle \\
&= \left\langle S\,, \sum_{i=1}^{\infty} x_i \otimes Ty_i \right\rangle
\end{aligned}$$

showing that Φ is of the form claimed.
Conversely, in passing to the dual $K^\perp$ of $(X\widehat{\otimes}_\pi Y)/K$, we conclude from Lemma 1.1 that $((Id \otimes T)_K)^* \in Z(K^\perp)$ and hence $(Id \otimes T)_K \in \mathrm{Cun}((X\widehat{\otimes}_\pi Y)/K)$. □

For $K = \{0\}$ the above may be rephrased as follows.

Corollary 3.9 *Let X and Y be Banach spaces. Whenever $X\widehat{\otimes}_\pi Y = L_1 \oplus_1 L_2$, there is a decomposition $X = J_1 \oplus_1 J_2$ such that $L_i = J_i\widehat{\otimes}_\pi Y$, provided that the Banach space Y cannot be decomposed as $Y = Y_1 \oplus_1 Y_2$ in a nontrivial way.*

In the above, the reader should note that J_i is a 1-complemented subspace of X so that $J_i\widehat{\otimes}_\pi Y$ is in fact a subspace of $X\widehat{\otimes}_\pi Y$.
A somewhat more involved application of Theorem 3.8 is given in the next corollary, where $N(X,Y)$ denotes the space of all nuclear operators from X to Y.

Corollary 3.10 *If $\mathrm{Cun}(X^*)$ is trivial then*

$$\mathrm{Cun}(N(X,Y)) = \{L_T \mid T \in \mathrm{Cun}(Y)\}.$$

PROOF: Write
$$N(X,Y) = (X^*\widehat{\otimes}_\pi Y)/K,$$
where K denotes the kernel of the natural surjection $u \mapsto \widehat{u}$ from $X^*\widehat{\otimes}_\pi Y$ onto $N(X,Y)$. If $u = \sum_{i=1}^\infty x_i^* \otimes y_i \in K$ then by definition
$$\widehat{u}(x) = \sum_{i=1}^\infty x_i^*(x) y_i = 0 \qquad \forall x \in X,$$
and taking adjoints, we hence obtain that
$$\sum_{i=1}^\infty x^{**}(x_i^*) y^*(y_i) = 0 \qquad \forall x^{**} \in X^{**}\, \forall y^* \in Y^*.$$
Furthermore, for all $S \in \operatorname{Cun}(Y)$ and all $\sum_{i=1}^\infty x_i^* \otimes y_i \in K$,
$$\left[(Id \otimes S)\left(\sum_{i=1}^\infty x_i^* \otimes y_i\right)\right](x) = \sum_{i=1}^\infty (x_i^* \otimes Sy_i)(x) = 0,$$
and the result now follows from Theorem 3.8. □

Note that Corollary 3.9 implies Corollary 3.10 if Y (or X^*) has the approximation property, since under this assumption $X^*\widehat{\otimes}_\pi Y = N(X,Y)$.

Corollary 3.11 *If X_1 and X_2 have no nontrivial L-summands, then the assumption that $L^1(\mu, X_1) \cong L^1(\nu, X_2)$ implies that $L^1(\mu) \cong L^1(\nu)$.*

PROOF: We have, using Theorem 3.8 and [159, Example VIII.1.10],
$$\begin{aligned}\operatorname{Cun}(L^1(\mu)) &\cong \operatorname{Cun}(L^1(\mu)\widehat{\otimes}_\pi X_1) \cong \operatorname{Cun}(L^1(\mu, X_1)) \\ &\cong \operatorname{Cun}(L^1(\nu, X_2)) \cong \operatorname{Cun}(L^1(\nu)).\end{aligned}$$
and the claim follows as in the proof of Corollary 1.16. □

We finish this section with two counterexamples. The first one shows that, contrary to a rather tempting conjecture, there is at least one case in which there are nontrivial L-summands in $X\widehat{\otimes}_\varepsilon Y$:

Proposition 3.12 $\qquad \ell^2(2,\mathbb{R})\widehat{\otimes}_\varepsilon \ell^2(2,\mathbb{R}) \cong \ell^2(2,\mathbb{R}) \oplus_1 \ell^2(2,\mathbb{R})$.

PROOF: Let $X = \ell^2(2,\mathbb{R})\widehat{\otimes}_\varepsilon \ell^2(2,\mathbb{R}) = L(\ell^2(2,\mathbb{R}))$. It is well-known (see [318, p. 82]) that $\operatorname{ex} B_X = O(2,\mathbb{R})$, the group of orthogonal 2×2-matrices. Let E (respectively $\widehat{E}$) denote the linear span of all orthogonal matrices with determinant $+1$ (respectively -1). Then $X = E \oplus \widehat{E}$ and $E \cong \widehat{E} \cong \ell^2(2,\mathbb{R})$. If P denotes the projection from X onto E, then $P(\operatorname{ex} B_X) \subset \{0,1\} \cdot \operatorname{ex} B_X$ from which one can deduce that P is an L-projection; for details see [401, Theorem 4.6]. □

Our next example is concerned with the question of how the structure topology (see Definition I.3.11) of $X\widehat{\otimes}_\varepsilon Y$ can be derived from the respective topologies of X and Y.

Guided by Theorem 3.2 a first guess might be that the former topology should just coincide with the product of the latter ones. Note that, as an equation of sets, we always have

$$E_{X\widehat{\otimes}_\varepsilon Y} = E_X \times E_Y.$$

(Recall that E_X was defined in Section I.3 as the set of equivalence classes of $\operatorname{ex} B_{X^*}$, where the antipodal points have been glued together.) This natural conjecture, however, is false:

Proposition 3.13 *In general, the structure topology of $X\widehat{\otimes}_\varepsilon Y$ is not the product of the structure topologies of X and Y.*

PROOF: Let $X = Y = \{f \in C_0(\mathbb{R}) \mid nf(n) = f(1)\ \forall n \in \mathbb{N}\}$. By definition, this is a G-space. We now have

$$\operatorname{ex} B_{X^*} = \{\pm\delta_k \mid k \in \mathbb{R} \setminus \{2,3,4,\ldots\}\},$$

cf. p. 89. It is straightforward to check that

$$X\widehat{\otimes}_\varepsilon X = \{f \in C_0(\mathbb{R}^2) \mid mn\, f(m,n) = f(1,1)\ \forall m,n \in \mathbb{N}\}.$$

As an easy application of Proposition II.5.2 we obtain that the space E_X is homeomorphic to $\mathbb{R}/\mathbb{N}$, whereas $E_{X\widehat{\otimes}_\varepsilon X} = \mathbb{R}^2/\mathbb{N}^2$. Thus it is left to prove that $(\mathbb{R}/\mathbb{N})^2$ is not homeomorphic to $\mathbb{R}^2/\mathbb{N}^2$. To this end, denote for $m,n \in \mathbb{N}$ by $D_{m,n}$ the (open) disk with radius $(m+n)^{-1}$ centred at (m,n). Since a neighbourhood of $\mathbb{N}$ always contains a set of the form $\bigcup_{\mu\in\mathbb{N}}(a_\mu, b_\mu)$ with $\mu \in (a_\mu, b_\mu)$, we see that $\bigcup_{m,n=1}^\infty D_{m,n}$ is open in $\mathbb{R}^2/\mathbb{N}^2$, but not in $(\mathbb{R}/\mathbb{N})^2$. □

As a final remark we point out that the space X considered in the above proof provides an example to show that the centralizer $Z(X\widehat{\otimes}_\varepsilon X)$ need not coincide with the norm closure of $Z(X) \otimes Z(X)$. In fact, it follows from the discussion preceding Proposition II.5.8 that

$$Z(X) = \{f \in C^b(\mathbb{R}) \mid f_{|\mathbb{N}} = \text{ const.}\} = C^b(\mathbb{R}/\mathbb{N}),$$

as well as

$$Z(X\widehat{\otimes}_\varepsilon X) = C^b(\mathbb{R}^2/\mathbb{N}^2).$$

Since a function $f \in C^b(\mathbb{R}^2)$ that vanishes on $\mathbb{R}^2 \setminus \bigcup_{m,n=1}^\infty D_{m,n}$ and attains the value 1 on $\mathbb{N}^2$ is not continuous when it is considered as a function on $(\mathbb{R}/\mathbb{N})^2$ (by the same argument as above), we obtain

$$C^b(\mathbb{R}^2/\mathbb{N}^2) \neq C^b((\mathbb{R}/\mathbb{N})^2),$$

hence the result. We remark that also c_0 yields such an example, but we will return to the above space X in the Notes and Remarks section which contains a more detailed discussion of the centralizer of an injective tensor product.

VI.4 M-ideals of compact operators

The final sections of this book are concerned with the phenomenon that for certain Banach spaces $K(X,Y)$ is an M-ideal in $L(X,Y)$; of course, here the case $X = Y$ is of special importance. One of the reasons why one is interested in cases where this happens lies in the fact that every bounded operator from X to Y must then have a best compact approximant (Proposition II.1.1). Also, the uniqueness of Hahn-Banach extensions from $K(X,Y)$ to $L(X,Y)$ in the case of an M-ideal of compact operators deserves special attention as does the fact that the functionals $T \mapsto \langle x^{**}, T^*y^*\rangle$, $x^{**} \in \operatorname{ex} B_{X^{**}}$, $y^* \in \operatorname{ex} B_{Y^*}$, are even extremal on $L(X,Y)$ (and not only on $K(X,Y)$, Theorem 1.3), which holds by Lemma I.1.5.
Another reason is that – as in other circumstances – Hilbert space provides a prominent (and in fact the first known) example of this kind; this follows from Theorem V.4.4 or Example 4.1 below. So part of what follows can be understood as a contribution to the question of how far one can go away from Hilbert space without ruining this property.
We now give the basic example of an M-ideal $K(X,Y)$.

Example 4.1 *Let $1 < p \le q < \infty$. Then $K(\ell^p, \ell^q)$ is an M-ideal in $L(\ell^p, \ell^q)$. If X is any Banach space, then $K(X, c_0)$ is an M-ideal in $L(X, c_0)$.*

PROOF: Let $1 < p \le q < \infty$. Denote the canonical coordinate projection $(a_1, a_2, \ldots) \mapsto (a_1, \ldots, a_n, 0, 0, \ldots)$ on ℓ^p by P_n, and on ℓ^q by Q_n. We wish to verify the 3-ball property (Theorem I.2.2). So let contractive operators S_1, S_2, $S_3 \in K(\ell^p, \ell^q)$ and $T \in L(\ell^p, \ell^q)$ as well as $\varepsilon > 0$ be given. We shall prove that for large n and m

$$\|T + S_i - (Q_nT - TP_m + Q_nTP_m)\| \le 1 + \varepsilon, \qquad i = 1, 2, 3, \tag{$*$}$$

thus establishing that $K(\ell^p, \ell^q)$ is an M-ideal in $L(\ell^p, \ell^q)$, since $Q_nT - TP_m + Q_nTP_m$ is compact.
To prove $(*)$ note first that (P_n) and (Q_n) converge strongly to the respective identity operators, as do their adjoints. Since the convergence is uniform on relatively compact sets such as $S_i(B_{\ell^p})$ by the boundedness of these sequences, we have

$$\begin{aligned}
\lim_{n,m\to\infty} \|Q_nS_iP_m - S_i\| &\le \lim_{n,m\to\infty} \left(\|Q_nS_i - S_i\|\,\|P_m\| + \|S_iP_m - S_i\|\right) \\
&\le \lim_{n,m\to\infty} \left(\|Q_nS_i - S_i\| + \|P_m^*S_i^* - S_i^*\|\right) \\
&= 0.
\end{aligned}$$

Secondly, we have

$$\begin{aligned}
&\|T + Q_nS_iP_m - (Q_nT - TP_m + Q_nTP_m)\| \\
&\quad = \|(Id - Q_n)T(Id - P_m) + Q_nS_iP_m\| \le 1,
\end{aligned}$$

because

$$\begin{aligned}\|[(Id-Q_n)T(Id-P_m)+Q_nS_iP_m]x\| &= \left(\|T(Id-P_m)x\|^q+\|S_iP_mx\|^q\right)^{1/q}\\ &\le \left(\|(Id-P_m)x\|^q+\|P_mx\|^q\right)^{1/q}\\ &\le \left(\|(Id-P_m)x\|^p+\|P_mx\|^p\right)^{1/p}\\ &= \|x\|.\end{aligned}$$

These estimates together yield (∗).
The proof of the second assertion is similar, but easier. In this situation one verifies easily

$$\|T+S_i-Q_nT\|\le 1+\varepsilon, \qquad i=1,2,3,$$

for large n and the coordinate projections Q_n on c_0. □

The above technique and the study of compact operators approximating the identity are fundamental in the theory of M-ideals of compact operators.
Note that for $1<q<p<\infty$ we have $K(\ell^p,\ell^q)=L(\ell^p,\ell^q)$ [421, Th. I.2.7] so that the above assertion extends to this case for trivial reasons. The following stability properties are easily proved with the help of the 3-ball property of Theorem I.2.2; we leave the details to our readers. A far more elaborate result than (a) will be presented in Theorem 4.19 below in the case $X=Y$.

Proposition 4.2
(a) *If $K(X,Y)$ is an M-ideal in $L(X,Y)$ and $E\subset X$ and $F\subset Y$ are 1-complemented subspaces, then $K(E,F)$ is an M-ideal in $L(E,F)$.*
(b) *The class of Banach spaces X and Y for which $K(X,Y)$ is an M-ideal in $L(X,Y)$ is closed with respect to the Banach-Mazur distance.*

A more detailed statement of (b) is this: If X and Y are Banach spaces such that whenever $\varepsilon>0$ there are spaces X_ε and Y_ε with Banach-Mazur distances $d(X,X_\varepsilon)\le 1+\varepsilon$, $d(Y,Y_\varepsilon)\le 1+\varepsilon$ where $K(X_\varepsilon,Y_\varepsilon)$ is an M-ideal in $L(X_\varepsilon,Y_\varepsilon)$, then $K(X,Y)$ is an M-ideal in $L(X,Y)$, too.
The next result shows that one cannot expect the compact operators to form an M-summand; the same argument shows that $K(X,Y)$ is not an L^p-summand in $L(X,Y)$ for $p>1$ unless in the trivial case. Actually, we are going to prove that $K(X,Y)$ is not an M- (or L^p-)summand in lin $K(X,Y)\cup\{T\}$ for any noncompact operator T. But for $p=1$ the situation is somewhat different, see the Notes and Remarks.

Proposition 4.3 *If $K(X,Y)$ is an M-summand in $L(X,Y)$, then $K(X,Y)=L(X,Y)$.*

PROOF: Assume for contradiction that $L(X,Y)=K(X,Y)\oplus_\infty R$ where $R\neq\{0\}$. Let $T\in R$ with $\|T\|=1$. Given $\varepsilon>0$ pick $x_0\in S_X$ such that $\|Tx_0\|\ge 1-\varepsilon$. Fix $x_0^*\in S_{X^*}$ with $x_0^*(x_0)=1$ and consider the compact operator $Sx=x_0^*(x)Tx_0$. Then $\|S+T\|=1$ by assumption, but on the other hand $\|(S+T)x_0\|\ge 2-\varepsilon$. This is absurd. □

We now begin our structural investigations. Before we start, let us make one general remark: All results in the sequel will be formulated in terms of $K(X,Y)$ or $K(X)$. Nevertheless, with the obvious modifications, all results, except otherwise stated, will hold verbatim for the space $A(X,Y)$ or $A(X)$ of norm limits of finite rank operators as well. Recall that $A(X,Y) = K(X,Y)$ if X^* or Y has the approximation property [422, Th. 1.e.4 and 1.e.5].

Our first proposition gives access to the results of Chapter III.

Proposition 4.4 *If $K(X) \subset \mathfrak{L} \subset L(X)$, $Id \in \mathfrak{L}$ and $K(X)$ is an M-ideal in $\mathfrak{L}$, then X is an M-ideal in X^{**}.*

PROOF: The argument consists in quite a direct application of (both directions of) Theorem I.2.2:

Let $x^{**} \in S_{X^{**}}$, $x_1, x_2, x_3 \in B_X$ and let $\varepsilon > 0$. Pick $x^* \in B_{X^*}$ with $x^{**}(x^*) > 1 - \varepsilon$. Then, by assumption, there is $U \in K(X)$ with

$$\max_{1 \le i \le 3} \|Id + x^* \otimes x_i - U\| \le 1 + \varepsilon.$$

(Here $x^* \otimes x_i$ denotes the operator $x \mapsto x^*(x)x_i$.) But this implies

$$\max_{1 \le i \le 3} \|x^{**} + x^{**}(x^*)x_i - U^{**}x^{**}\| \le 1 + \varepsilon,$$

therefore

$$\max_{1 \le i \le 3} \|x^{**} + x_i - U^{**}x^{**}\| \le 1 + 2\varepsilon.$$

Since U is compact, we see that $U^{**}x^{**} \in X$, and hence X is an M-ideal in X^{**}. □

An immediate consequence of Theorems III.3.1, III.3.4, III.3.8 and III.4.6 now is:

Corollary 4.5 *Each Banach space X for which $K(X)$ is an M-ideal in $L(X)$ has property (V) as well as property (u). Furthermore, X^* has the RNP and X is weakly compactly generated. In particular, if X is separable, so is X^*.*

Recall the consequences these properties imply for X (Corollaries III.3.7 and III.4.7). The RNP of X^* was obtained in Theorem III.3.1 as a consequence of the fact that the relative weak and weak* topologies on S_{X^*} coincide (Corollary III.2.15). In the present situation, the latter result can be strengthened.

Proposition 4.6 *If $K(X)$ is an M-ideal in $L(X)$, then the relative norm and weak* topologies on S_{X^*} coincide.*

PROOF: We formulate the proof only in the case of real scalars. Let (x^*_α) in S_{X^*} have the weak* limit $x^* \in S_{X^*}$ and fix $\varepsilon > 0$. By the above, X^* has the RNP and hence there is $e \in S_X$ as well as $t > 0$ such that the slice

$$S(e,t) = \{\xi^* \in X^* \mid 1 \ge \|\xi^*\| \ge \xi^*(e) \ge 1 - t\}$$

has diameter less than ε, see [93, Th. 4.4.1] or [497, Lemma 2.18].

Let $x \in S_X$ with $x^*(x) > 1 - t/10$. Then, for α bigger than a certain α_0, we have $x^*_\alpha(x) > 1 - t/5$. Choose $y^* \in S(e,t)$ with $y^*(e) = 1$ and fix $U \in K(X)$ with

$$\|Id \pm y^* \otimes x - U\| \le 1 + \frac{t}{5}.$$

(Note that this only requires the 2-ball property of $K(X)$ in $L(X)$.) Then, assuming $\alpha \geq \alpha_0$, we obtain

$$\begin{aligned} 1+\frac{t}{5} &\geq \max \|[y^* \otimes x \pm (Id-U)]^*(x^*_\alpha)\| \\ &\geq \max |x^*_\alpha(x)y^*(e) \pm (x^*_\alpha - U^*x^*_\alpha)(e)| \\ &\geq 1-\frac{t}{5}+|(x^*_\alpha - U^*x^*_\alpha)(e)| \end{aligned}$$

and thus, $|(x^*_\alpha - U^*x^*_\alpha)(e)| \leq \frac{2t}{5}$. It follows that

$$\frac{(x^*_\alpha(x)y^* \pm (x^*_\alpha - U^*x^*_\alpha))(e)}{1+t/5} \geq \frac{1-t/5-2t/5}{1+t/5} > 1-t.$$

Using that diam $S(e,t) < \varepsilon$, we get

$$\begin{aligned} 2\|x^*_\alpha - U^*x^*_\alpha\| &\leq \|(x^*_\alpha(x)y^* + x^*_\alpha - U^*x^*_\alpha) - (x^*_\alpha(x)y^* - x^*_\alpha + U^*x^*_\alpha)\| \\ &\leq \varepsilon(1+t/5) \end{aligned}$$

and, consequently, $\|x^*_\alpha - U^*x^*_\alpha\| \leq \varepsilon$, which implies that $\|U^*x^* - x^*\| \leq \varepsilon$, too. Enlarging α_0, if necessary, we may assume that $\|U^*x^*_\alpha - U^*x^*\| \leq \varepsilon$, since U^* is weak* to norm continuous on B_{X^*}. Finally, for sufficiently large α, we find

$$\|x^*_\alpha - x^*\| \leq \|x^*_\alpha - U^*x^*_\alpha\| + \|U^*x^*_\alpha - U^*x^*\| + \|U^*x^* - x^*\| \leq 3\varepsilon,$$

and thus, $x^*_\alpha \to x^*$ in norm. □

Given Theorem 4.17, the previous result is contained in Proposition 4.15, too.
Before we are going to characterise the class of Banach spaces for which $K(X)$ is an M-ideal of $L(X)$, we prove two further consequences this property has. (In fact, in both cases the results hold in the case when $K(X,Y)$ is an M-ideal in $L(X,Y)$.)
The first one provides a somewhat more explicit formula for the essential norm of an operator T, i.e. the number $\|T\|_e = \inf_{K\in K(X)} \|T-K\|$. It will be useful in the following section.

Proposition 4.7 *Let X and Y be Banach spaces and put*

$$\begin{aligned} w(T) &:= \sup\{\limsup_\alpha \|Tx_\alpha\| \mid \|x_\alpha\| = 1,\ x_\alpha \xrightarrow{w} 0\}, \\ w^*(T) &:= \sup\{\limsup_\alpha \|T^*x^*_\alpha\| \mid \|x^*_\alpha\| = 1,\ x^*_\alpha \xrightarrow{w^*} 0\}. \end{aligned}$$

If $K(X,Y)$ is an M-ideal in $L(X,Y)$, then

$$\|T\|_e = \max\{w(T), w^*(T)\},$$

and at least one of the involved suprema is actually attained. If X^* and Y are separable, then sequences suffice in the definitions of w and w^*.

PROOF: Since the adjoint of any compact operator K maps bounded nets converging to zero in the weak* topology to nets which converge to zero in norm, we have $\|T-K\| \geq w^*(T)$ and similarly, $\|T-K\| \geq w(T)$. So, without any assumptions on X or Y, the inequality

$$\|T\|_e \geq \max\{w(T), w^*(T)\}$$

holds. To prove equality, fix $\psi \in \operatorname{ex} B_{K(X,Y)^\perp}$ such that $\psi(T) = \|T\|_e$. Since $K(X,Y)$ is an M-ideal in $L(X,Y)$, we must have that $\psi \in \operatorname{ex} B_{L(X,Y)^*}$ (Lemma I.1.5), and since the norm of each $T \in L(X,Y)$ is the supremum of the numbers $\langle y^* \otimes x, T\rangle = y^*(Tx)$ with $y^* \in B_{Y^*}$ and $x \in B_X$, there are, by the Hahn-Banach and the (converse to the) Krein-Milman theorem, nets (x_α) and (y^*_α), each contained in the respective unit sphere, such that the functionals $y^*_\alpha \otimes x_\alpha$ tend to ψ in the $\sigma(L(X,Y)^*, L(X,Y))$-topology. In passing to appropriate subnets, suppose that $x_\alpha \xrightarrow{w^*} x^{**}$ and $y^*_\alpha \xrightarrow{w^*} y^*$ so that still $x_\alpha \otimes y^*_\alpha \xrightarrow{w^*} \psi$. Now, for any compact operator K mapping X into Y, we have

$$0 = \psi(K) = \lim_\alpha (K^* y^*_\alpha)(x_\alpha) = x^{**}(K^* y^*).$$

It follows that either y^* or x^{**} must be equal to zero. Suppose $x^{**} = 0$. Then $x_\alpha \to 0$ weakly and

$$\|T\|_e = \psi(T) = \lim_\alpha y^*_\alpha(Tx_\alpha) \leq \limsup_\alpha \|Tx_\alpha\| \leq w(T).$$

Should $y^* = 0$, then $y^*_\alpha \to 0$ in the weak* sense, and, as above, $\|T\|_e \leq w(T^*)$. This finishes the proof in the general case; and in the separable case we just observe that we may pass to subsequences (x_{α_n}) and $(y^*_{\alpha_n})$ above. □

The next result improves in our more special situation the fact, proved by Zizler [662], that in general the operators whose adjoints attain their norms on B_{Y^*} are dense in $L(X,Y)$.

Proposition 4.8 *Suppose that $K(X,Y)$ is an M-ideal in $L(X,Y)$.*

(a) *If $T \in L(X,Y)$ has the property that T^* does not attain its norm on B_{Y^*}, then $\|T\| = \|T\|_e$.*

(b) *The set of operators whose adjoints do not attain their norm on B_{Y^*} is nowhere dense in $L(X,Y)$ with respect to the norm topology.*

PROOF: To see that (a) holds, note first that always $\psi(T) = \|T\|$ for some $\psi \in \operatorname{ex} B_{L(X,Y)^*}$. Given the hypothesis of the above proposition, we must have that $\psi \in \operatorname{ex} B_{K(X,Y)^\perp}$. Indeed, otherwise we would have $\psi \in \operatorname{ex} B_{K(X,Y)^*}$ by Lemma I.1.5 (and Remark I.1.13) and thus $\psi(T) = \langle T^{**}x^{**}, y^*\rangle$ for some $x^{**} \in \operatorname{ex} B_{X^{**}}$, $y^* \in \operatorname{ex} B_{Y^*}$ by Theorem 1.3; however, the assumption on T rules out that $\|T^*\| = \|T^* y^*\|$. But this implies

$$\|T\| = \psi(T) = \sup\{\varphi(T) \mid \varphi \in \operatorname{ex} B_{K(X,Y)^\perp}\} = \|T\|_e.$$

We now prove part (b). Since by the above the metric complement

$$K^\theta(X,Y) = \{T \in L(X,Y) \mid \|T\| = \|T\|_e\}$$

contains all operators in question, and is, in addition, norm closed, we are finished once it is shown that $K^\theta(X,Y)$ has empty interior. This follows from Proposition II.1.11 and Corollary II.1.7 since

$$\begin{aligned}\|T\| &= \sup\{|\langle Tx, y^*\rangle| \mid x \in B_X,\ y^* \in B_{Y^*}\}\\ &= \sup\{|\langle \psi, T\rangle| \mid \psi \in B_{K(X,Y)^*}\}\end{aligned}$$

for all $T \in L(X,Y)$. □

The class of Banach spaces for which X is an M-ideal in X^{**} is strictly larger than the class in which $K(X)$ is an M-ideal; this can be seen from Proposition 4.6. We now discuss another reason why this is so. We shall see in a moment that X has the metric compact approximation property if $K(X)$ is an M-ideal in $L(X)$. On the other hand, the famous Enflo-Davie subspace of c_0 fails this property [422, pp. 90ff.], but is M-embedded by Theorem III.1.6.

To prove the announced statement, we need some reformulations of the definition of the λ-compact approximation property. Recall that a Banach space X is said to have this property whenever there is a net of compact operators (K_α) on X, uniformly bounded by λ in norm, which converges strongly to the identity. Parts (ii) and (v) of the following lemma will connect us to Chapter V.

Lemma 4.9 *Let X be a Banach space. Then the following are equivalent.*

(i) *X has the λ-compact approximation property.*

(ii) *$K(X)$ contains a left λ-approximate unit.*

(iii) *There is a net (T_α) in $K(X)$ with $\|T_\alpha\| \le \lambda$ converging to the identity in the weak operator topology.*

Also the following two conditions are equivalent.

(iv) *X^* has the λ-compact approximation property with adjoint operators.*

(v) *$K(X)$ contains a right λ-approximate unit.*

In general, we have (iv) ⇒ (iii), *and if X is M-embedded and $\lambda = 1$, then all the above conditions are equivalent.*

PROOF: (i) ⇒ (ii): Let (T_α) be a net in $K(X)$ with $\|T_\alpha\| \le \lambda$ converging to Id_X pointwise. By the boundedness of this net, $T_\alpha \to Id_X$ uniformly on compact subsets, in particular $T_\alpha K \to K$ for a compact operator K since then $K(B_X)$ is relatively compact.

(ii) ⇒ (iii): Evident.

(iii) ⇒ (i): Since $L(X)$ has the same continuous linear functionals when endowed with the weak operator topology as when bearing the strong operator topology, the same convex sets must be closed with respect to both topologies, see [179, Th. VI.1.4]. Hence, after passing to convex combinations, we may suppose that $T_\alpha \to Id_X$ strongly.

(iv) ⇒ (v): Note that

$$\|KT_\alpha - K\| = \|T_\alpha^* K^* - K^*\| \to 0 \qquad \forall K \in K(X)$$

if (T_α) is a net of compact operators such that $T_\alpha^* \to Id_{X^*}$ strongly and $\|T_\alpha\| \le \lambda$.

(v) $\Rightarrow$ (iv): For a right λ-approximative unit (T_α) and $\|x\| = 1$ we have

$$\|T_\alpha^* x^* - x^*\| = \|(x^* \otimes x)T_\alpha - x^* \otimes x\| \to 0 \quad \forall x^* \in X^*.$$

(iv) $\Rightarrow$ (iii) is trivial, and for the remaining assertion see Proposition III.2.5. □

The trick applied in the proof of the implication (iii) $\Rightarrow$ (i) will be used several times below; it will be referred to as a convex combinations argument.

Proposition 4.10 *Suppose that $\mathfrak{L}$ is a space of bounded operators such that*

$$\mathbb{K}\{Id\} + K(X) \subset \mathfrak{L} \subset L(X).$$

If $K(X)$ is an M-ideal in $\mathfrak{L}$ then there is a net (K_α) in $B_{K(X)}$ such that (K_α^*) and (K_α) converge strongly to the identity on the respective spaces. In particular, X and X^* have the metric compact approximation property. If in addition $\mathfrak{L}$ is an algebra of operators, then $K(X)$ is a two-sided inner M-ideal.

PROOF: By Remark I.1.13, $B_{K(X)}$ is $\sigma(\mathfrak{L}, K(X)^\#)$-dense in $B_\mathfrak{L}$. To prove the result we first observe that the functionals

$$x^{**} \otimes x^* : T \mapsto \langle x^{**}, T^* x^* \rangle$$

belong to $K(X)^\#$. Hence there is a net (L_α) of compact operators such that both $L_\alpha \to Id_X$ and $L_\alpha^* \to Id_{X^*}$ for the respective weak operator topologies. As in the previous proof we obtain the desired net by taking convex combinations.
That $K(X)$ is necessarily an inner M-ideal follows from Proposition 4.4, Lemma 4.9 and Theorem V.3.2. □

Let us present an application of this result.

Proposition 4.11 *Let X be a Banach space.*

(a) *If X is reflexive and $K(X)$ is an M-ideal in $L(X)$, then $K(X)^{**} \cong L(X)$ so that $K(X)$ is M-embedded.*
(b) *If $K(X)$ is M-embedded, then X is reflexive.*

PROOF: (a) It is proved in [224] that the map $V : X \widehat{\otimes}_\pi X^* \to K(X)^*$, defined by $\langle V(u), K \rangle = \sum_{i=1}^\infty x_i^*(Kx_i)$ for $u = \sum_{i=1}^\infty x_i \otimes x_i^*$ (absolutely convergent sum), is a quotient map for reflexive X. Hence, after canonical identifications,

$$K(X)^{**} = (\ker V)^\perp \subset (X \widehat{\otimes}_\pi X^*)^* = L(X, X^{**}) = L(X).$$

Thus, it remains to show that $L(X) \subset (\ker V)^\perp$. So let $T \in L(X)$ and $u \in \ker V$. Indeed we have, with (K_α) as in Proposition 4.10,

$$\begin{aligned} \langle u, T \rangle &= \sum_{i=1}^\infty x_i^*(Tx_i) = \lim_\alpha \sum_{i=1}^\infty x_i^*(K_\alpha T x_i) \\ &= \langle V(u), K_\alpha T \rangle = 0. \end{aligned}$$

(b) This follows immediately from Corollary III.3.7(e) since X^* embeds into $K(X)$. $\square$

We now approach the most fundamental result of this section, Theorem 4.17. We are going to prepare this central result by introducing a pair of new concepts in which the geometric peculiarities of the spaces we are about to investigate are formalised.

Definition 4.12 *We say that a Banach space X has* property (M) *if whenever $u, v \in X$ with $\|u\| = \|v\|$ and (x_α) is a bounded weakly null net in X, then*

$$\limsup_\alpha \|u + x_\alpha\| = \limsup_\alpha \|v + x_\alpha\|.$$

Similarly, X is said to have property (M^*) *if whenever $u^*, v^* \in X^*$ with $\|u^*\| = \|v^*\|$ and whenever (x^*_α) is a bounded weak* null net in X^*, then*

$$\limsup_\alpha \|u^* + x^*_\alpha\| = \limsup_\alpha \|v^* + x^*_\alpha\|.$$

One can see by a gliding hump argument that the spaces ℓ^p for $1 \leq p < \infty$ as well as c_0 all have property (M). More precisely, one has

$$\limsup_\alpha \|u + x_\alpha\| = \Big(\|u\|^p + \limsup_\alpha \|x_\alpha\|^p\Big)^{1/p}$$

in the ℓ^p-case and likewise for c_0. Also, easy examples involving the sequence of Rademacher functions show that $L^p[0,1]$ fails property (M) for $p \neq 2$. The relation between (M) and (M^*) will be explained in a moment.
We now state a simple lemma.

Lemma 4.13 *The following conditions are equivalent:*

(i) *X has property (M).*

(ii) *If $\|u\| = \|v\|$ and (x_α) is a bounded weakly null net such that $\lim_\alpha \|u + x_\alpha\|$ exists, then $\lim_\alpha \|v + x_\alpha\|$ exists, and*

$$\lim_\alpha \|u + x_\alpha\| = \lim_\alpha \|v + x_\alpha\|.$$

(iii) *If $\|u\| \leq \|v\|$ and (x_α) is any bounded weakly null net, then*

$$\limsup_\alpha \|u + x_\alpha\| \leq \limsup_\alpha \|v + x_\alpha\|.$$

(iv) *If (u_α) and (v_α) are relatively norm compact nets with $\|u_\alpha\| \leq \|v_\alpha\|$ for every α and (x_α) is a bounded weakly null net, then*

$$\limsup_\alpha \|u_\alpha + x_\alpha\| \leq \limsup_\alpha \|v_\alpha + x_\alpha\|.$$

Analogous results hold for property (M^*).

PROOF: The equivalence of (i) and (ii) is straightforward, and trivially (iv) $\Rightarrow$ (iii) $\Rightarrow$ (i). To deduce (iii) from (i) choose $\lambda > 1$ so that $\|\lambda u\| = \|v\|$. Observe that $x_\alpha + u$ is a convex combination of $x_\alpha + \lambda u$ and $x_\alpha - \lambda u$ whence

$$\|u + x_\alpha\| \leq \max\{\|\lambda u + x_\alpha\|, \|\lambda u - x_\alpha\|\}.$$

It follows

$$\begin{aligned} \limsup \|u + x_\alpha\| &\leq \max\{\limsup \|v + x_\alpha\|, \limsup \|-v + x_\alpha\|\} \\ &= \limsup \|v + x_\alpha\|, \end{aligned}$$

where we employed property (M) several times. Let us conclude by deducing (iv) from (iii). Indeed, if the conclusion were false we might pass to subnets so that $\limsup_k \|u_{\alpha_k} + x_{\alpha_k}\| > \limsup_k \|v_{\alpha_k} + x_{\alpha_k}\|$ and (u_{α_k}) as well as (v_{α_k}) are convergent. This quickly leads to a contradiction to (iii) since $\|\lim u_{\alpha_k}\| \leq \|\lim v_{\alpha_k}\|$. The proofs for (M^*) are similar. □

We also need:

Lemma 4.14 *Suppose X has property (M) and that $T \in L(X)$ with $\|T\| \leq 1$. If (u_α) and (v_α) are relatively norm compact nets with $\|u_\alpha\| \leq \|v_\alpha\|$ and (x_α) is a bounded weakly null net, then*

$$\limsup_\alpha \|u_\alpha + Tx_\alpha\| \leq \limsup_\alpha \|v_\alpha + x_\alpha\|.$$

PROOF: First suppose $\|T\| = 1$ and (x_α) is any bounded weakly null net. For any $\lambda < 1$ there exists w with $\|w\| = 1$ and $\|Tw\| > \lambda$. Let $w_\alpha = \|v_\alpha\| w$. Hence by Lemma 4.13

$$\begin{aligned} \limsup_\alpha \|\lambda u_\alpha + Tx_\alpha\| &\leq \limsup_\alpha \|Tw_\alpha + Tx_\alpha\| \\ &\leq \limsup_\alpha \|w_\alpha + x_\alpha\| \\ &= \limsup_\alpha \|v_\alpha + x_\alpha\|. \end{aligned}$$

Letting $\lambda \to 1$ we obtain the conclusion for this case. Now suppose $0 \leq \|T\| < 1$. Let $T = \lambda L$ where $\lambda = \|T\|$ and $\|L\| = 1$. Then

$$\begin{aligned} \limsup_\alpha \|u_\alpha + Tx_\alpha\| &\leq \max\{\limsup_\alpha \|u_\alpha + Lx_\alpha\|, \limsup_\alpha \|u_\alpha - Lx_\alpha\|\} \\ &\leq \max\{\limsup_\alpha \|v_\alpha + x_\alpha\|, \limsup_\alpha \|-v_\alpha + x_\alpha\|\} \\ &= \limsup_\alpha \|v_\alpha + x_\alpha\|. \end{aligned}$$

□

The following proposition compares (M) and (M^*). Note that ℓ^1 has (M), but fails (M^*). Also, have a peek at the following Theorem 4.17, (vi) $\Longleftrightarrow$ (vii).

Proposition 4.15 *Let X be a Banach space with property (M^*). Then X has property (M) and X is an M-ideal in X^{**}. Moreover, the relative norm and weak* topologies on S_{X^*} coincide.*

PROOF: First we show that X has property (M). Suppose $\|u\| = \|v\|$ and that (x_α) is a bounded weakly null net such that $\lim_\alpha \|u + x_\alpha\| > \lim_\alpha \|v + x_\alpha\|$ (and both limits exist). Then pick $x_\alpha^* \in B_{X^*}$ so that $x_\alpha^*(u + x_\alpha) = \|u + x_\alpha\|$. By passing to a subnet we may suppose that (x_α^*) converges weak* to some x^*. Now pick v^* with $\|v^*\| = \|x^*\|$ and $v^*(v) = \|x^*\|\|v\|$. Then

$$\begin{aligned} \lim_\alpha \|u + x_\alpha\| &= \lim_\alpha \langle u + x_\alpha, x_\alpha^* \rangle \\ &= \langle u, x^* \rangle + \lim_\alpha \langle x_\alpha, x_\alpha^* - x^* \rangle \\ &\leq \langle v, v^* \rangle + \lim_\alpha \langle x_\alpha, x_\alpha^* - x^* \rangle \\ &= \lim_\alpha \langle v + x_\alpha, v^* + x_\alpha^* - x^* \rangle \\ &\leq \limsup_\alpha \|v^* + x_\alpha^* - x^*\| \|v + x_\alpha\| \\ &\leq \lim_\alpha \|v + x_\alpha\| \end{aligned}$$

since $\limsup_\alpha \|v^* + x_\alpha^* - x^*\| = \limsup_\alpha \|x^* + x_\alpha^* - x^*\| = 1$ by property (M^*).
Next we show that X is an M-ideal in X^{**}. Suppose $\varphi \in X^\perp \subset X^{***}$ and suppose $\psi \in X^*$ (canonically embedded in X^{***}). For $\lambda < 1$ we can pick $x^{**} \in B_{X^{**}}$ so that $\varphi(x^{**}) > \lambda\|\varphi\|$. Pick any x^* in X^* with $\|x^*\| = \|\psi\|$ and $x^{**}(x^*) > \lambda\|\psi\|$. Let (y_d^*) be a net in X^* converging weak* in X^{***} to φ and such that $\|\psi + y_d^*\| \leq \|\psi + \varphi\|$. We can suppose that $x^{**}(y_d^*) > \lambda\|\varphi\|$ for all d. Since $\varphi \in X^\perp$, (y_d^*) converges weak* in X^* to zero. Now

$$\begin{aligned} \|\psi + \varphi\| &\geq \limsup_d \|\psi + y_d^*\| \\ &= \limsup_d \|x^* + y_d^*\| \\ &\geq \limsup_d \langle x^* + y_d^*, x^{**} \rangle \\ &\geq \lambda(\|\psi\| + \|\varphi\|), \end{aligned}$$

and the result follows.
Let now $\|x_\alpha^*\| = \|x_0^*\| = 1$ such that w^*-$\lim x_\alpha^* = x_0^*$. By property (M^*),

$$\limsup \|x^* + (x_\alpha^* - x_0^*)\| = \limsup \|x_0^* + (x_\alpha^* - x_0^*)\| = 1 \qquad \forall x^* \in S_{X^*}.$$

If we pick a weak* strongly exposed point x^* (the existence of those objects is guaranteed by Corollary III.3.2), then the conclusion $\|x_\alpha^* - x_0^*\| \to 0$ follows immediately. □

It follows in particular that a reflexive space has (M) if and only if it has (M^*) if and only if its dual has either of these properties.
We are now ready for the announced characterisation. It will be convenient to use the following notation.

Definition 4.16 *A net of compact operators (K_α) on a Banach space X will be called* a shrinking compact approximation of the identity *provided both $K_\alpha \to Id_X$ and $K_\alpha^* \to Id_{X^*}$ strongly.*

The use of the term "shrinking" in this definition is explained by the corresponding concept in the theory of Schauder bases. Proposition 4.10 says that X admits a shrinking compact approximation of the identity provided $K(X)$ is an M-ideal in $L(X)$.

Theorem 4.17 *Let X be an infinite dimensional Banach space. The following conditions are equivalent:*

(i) *$K(X)$ is an M-ideal in $L(X)$.*

(ii) *There exists a shrinking compact approximation of the identity (K_α) such that*

$$\limsup_\alpha \|SK_\alpha + T(Id - K_\alpha)\| \le \max\{\|S\|, \|T\|\} \qquad \forall S, T \in L(X).$$

(iii) *There exists a shrinking compact approximation of the identity (K_α) such that*

$$\limsup_\alpha \|K_\alpha S + (Id - K_\alpha)T\| \le \max\{\|S\|, \|T\|\} \qquad \forall S, T \in L(X).$$

(iv) *$K(X)$ is an M-ideal in $\mathbb{K}\{Id\} \oplus K(X)$.*

(v) *There exists a shrinking compact approximation of the identity (K_α) with $\|K_\alpha\| \le 1$ such that*

$$\limsup_\alpha \|S + Id - K_\alpha\| \le 1 \qquad \forall S \in B_{K(X)}.$$

(vi) *X has property (M), and there is a shrinking compact approximation of the identity (K_α) such that*

$$\lim_\alpha \|Id - 2K_\alpha\| = 1.$$

(vii) *X has property (M^*), and there is a shrinking compact approximation of the identity (K_α) such that*

$$\lim_\alpha \|Id - 2K_\alpha\| = 1.$$

PROOF: The implications (i) $\iff$ (ii) $\iff$ (iii) $\Rightarrow$ (iv) $\iff$ (v) are clear by Theorem V.3.2, Lemma 4.9 and Proposition 4.10.

(v) $\Rightarrow$ (vi): By equivalence of (iv) and (v), we can apply Theorem V.3.2 with $\mathfrak{A} = \mathbb{K}\{Id\} \oplus K(X)$, $\mathfrak{J} = K(X)$ and obtain a shrinking compact approximation of the identity (K_α) such that

$$\limsup_\alpha \|SK_\alpha + T(Id - K_\alpha)\| \le \max\{\|S\|, \|T\|\} \qquad \forall S, T \in \mathfrak{A}.$$

Now pick $S = -Id$, $T = Id$ to see that $\limsup \|Id - 2K_\alpha\| \le 1$. Another way to derive that conclusion is to apply Theorem V.5.4 and to use a convex combinations argument. It is left to prove that X has property (M). Suppose (x_i) is a bounded weakly null net and that $\|u\| \le \|v\|$. Then there is a rank-one operator S with $\|S\| \le 1$ and $Sv = u$. Fix an index β. Then, since $\|Kx_i\| \to 0$ if K is compact,

$$\begin{aligned} \limsup_i \|u + x_i\| &= \limsup_i \|S(v + x_i) + (Id - K_\beta)x_i\| \\ &\le \|S + Id - K_\beta\| \limsup_i \|v + x_i\| + \|(Id - K_\beta)v\|, \end{aligned}$$

and, taking limits in β, it turns out that X must have property (M), by Lemma 4.13.
(vi) $\Rightarrow$ (ii): We first prove that for each β

$$\limsup_\alpha \|K_\beta + Id - K_\alpha\| \le \|Id - 2K_\beta\|.$$

In fact, $(Id - 2K_\alpha)(Id - 2K_\beta) = Id - 2K_\alpha - 2K_\beta + 4K_\alpha K_\beta$ and hence

$$\begin{aligned} \|Id - K_\alpha - K_\beta + 2K_\alpha K_\beta\| &= \frac{1}{2}(Id + (Id - 2K_\alpha)(Id - 2K_\beta)) \\ &\le \|Id - 2K_\alpha\|\|Id - 2K_\beta\|, \end{aligned}$$

note that $\|Id - 2K_\alpha\| \ge 1$. Thus the assertion follows by taking limits, since $\lim_\alpha \|K_\alpha K_\beta - K_\beta\| = 0$. Let us now prove (ii). We first note that, as the proof of Theorem V.3.2 reveals, it is sufficient to show that

$$\limsup_\alpha \|S + T(Id - K_\alpha)\| \le 1$$

whenever $S \in B_{K(X)}$ and $T \in B_{L(X)}$. Accordingly, suppose $S \in B_{K(X)}$ and $T \in L_{K(X)}$ are given. Fix β. We may pick $x_\alpha \in B_X$ so that

$$\limsup_\alpha \|SK_\beta + T(Id - K_\alpha)\| = \limsup_\alpha \|SK_\beta x_\alpha + T(Id - K_\alpha)x_\alpha\|.$$

The net $(K_\beta x_\alpha)_\alpha$ is contained in a relatively compact set. Hence, noting that, as a result of $\lim_\alpha K_\alpha^* x^* = x^*$ for all $x^* \in X^*$, both $((Id - K_\alpha)x_\alpha)$ and $(T(Id - K_\alpha)x_\alpha)$ are weakly null, we obtain from Lemmas 4.13 and 4.14

$$\begin{aligned} \limsup_\alpha \|SK_\beta x_\alpha + T(Id - K_\alpha)x_\alpha\| &\le \limsup_\alpha \|K_\beta x_\alpha + T(Id - K_\alpha)x_\alpha\| \\ &\le \limsup_\alpha \|(K_\beta + Id - K_\alpha)x_\alpha\| \\ &\le \|Id - 2K_\beta\|. \end{aligned}$$

Thus

$$\limsup_\alpha \|S + T(Id - K_\alpha)\| \le \|S - SK_\beta\| + \|Id - 2K_\beta\|,$$

and choosing β large enough yields the assertion.
(vii) $\Rightarrow$ (vi): This follows immediately from Proposition 4.15.
(v) $\Rightarrow$ (vii): The proof proceeds similar to that of the implication (v) $\Rightarrow$ (vi). This time we suppose that (x_i^*) is a bounded weak* null net and that $\|u^*\| < \|v^*\|$. Then there is a weak* continuous rank-one contraction S^* mapping v^* onto u^*. Now one can follow the above pattern, observing $\lim_i \|K_\beta^* x_i^*\| = 0$. □

We note that in the separable case one can get sequences of compact operators rather than nets, and it is also enough to work with the sequential versions of (M) and (M^*); this results from the above proof. Actually, one can easily check that a space with a separable dual has (M) if and only if it has the sequential version of (M); the corresponding

equivalence holds for (M^*) in the case of separable spaces. (The point is that B_X is weakly metrizable if X^* is separable, and B_{X^*} is weak* metrizable if X is separable.)
We recall that ℓ^p $(1 < p < \infty)$ and c_0 have (M). Furthermore, the coordinate projections P_n of these spaces satisfy condition (vi) of the above characterisation theorem; in fact, the norm condition $\|Id - 2P_n\| = 1$ expresses nothing but the 1-unconditionality of the canonical Schauder basis, see below for the definition. (The case $p = 1$ has to be excluded since the ℓ^1-basis is not shrinking.) Thus we regain Example 4.1. However, we point out that the coordinate projections on ℓ^p, $p < \infty$, do not work in parts (ii) and (iii). In fact, for $S = Id$ and T the shift to the right one obtains

$$\begin{aligned}\limsup_n \|P_nS + (Id - P_n)T\| &\geq \limsup_n \|P_ne_n + (Id - P_n)e_{n+1}\| \\ &\geq \limsup_n \|e_n + e_{n+1}\| = 2^{1/p}.\end{aligned}$$

A similar counterexample, with T the shift to the left, works for (ii). We further mention ℓ^1 as a Banach space having (M) and the metric approximation property for which the compact operators do not form an M-ideal.
Here is a first application of Theorem 4.17.

Corollary 4.18 *If $K(X)$ is an M-ideal in $L(X)$ and $K(Y)$ is an M-ideal in $L(Y)$, then $K(X,Y)$ is an M-ideal in $L(X,Y)$.*

PROOF: Let (K_α) be a shrinking compact approximation of the identity with $\lim_\alpha \|Id - 2K_\alpha\| = 1$. Further let $S \in K(X,Y)$ and $T \in L(X,Y)$ with $\|S\| \leq 1$, $\|T\| \leq 1$. We shall show that

$$\limsup \|S + T(Id - K_\alpha)\| \leq 1 \qquad (*)$$

which proves our claim in view of the 3-ball property, Theorem I.2.2.
The proof of $(*)$ is an adaptation of the argument for the implication (vi) $\Rightarrow$ (ii) of Theorem 4.17. Fix β and pick $x_\alpha \in B_X$ such that

$$\limsup_\alpha \|SK_\beta x_\alpha + T(Id - K_\alpha)x_\alpha\| = \limsup_\alpha \|SK_\beta + T(Id - K_\alpha)\|.$$

Observe, as in the previous proof, that $(Id - K_\alpha)x_\alpha \to 0$ weakly. Now, arguments as in Lemmas 4.13 and 4.14 yield that

$$\limsup_\alpha \|SK_\beta x_\alpha + T(Id - K_\alpha)x_\alpha\| \leq \limsup_\alpha \|K_\beta x_\alpha + (Id - K_\alpha)x_\alpha\|,$$

since X and Y have (M) by Theorem 4.17. The desired inequality $(*)$ can now be derived as above. □

Next we provide a stability result for Banach spaces for which $K(X)$ is an M-ideal in $L(X)$ which is much more subtle than that in Proposition 4.2(a).

Theorem 4.19 *Let X be a Banach space such that $K(X)$ is an M-ideal in $L(X)$. Let E be a subspace of a quotient space of X. Then $K(E)$ is an M-ideal in $L(E)$ if and only if E has the metric compact approximation property.*

PROOF: For the "only if" part see Proposition 4.10. The proof of the "if" part relies on the equivalences of Theorem 4.17. First we note that E is an M-ideal in its bidual since X is (Proposition 4.4, Theorem III.1.6). It follows that if E has the metric compact approximation property it has a shrinking compact approximation of the identity (K_α); use Proposition III.2.5 and a convex combinations argument. We also note that, for $E \subset X/V$, E has property (M) since X/V trivially inherits (M^*) from X and hence has (M) by Proposition 4.15.
It remains to check that E has a shrinking compact approximation of the identity (H_α) such that $\lim_\alpha \|Id - 2H_\alpha\| = 1$. Suppose (L_α) is a shrinking compact approximation of the identity for X such that $\lim_\alpha \|Id - 2L_\alpha\| = 1$. (It is legitimate to suppose that (K_α) and (L_α) are indexed by the same set, since one can always switch to a product index set with the product ordering.) Let $J_E : E \to X/V$ be the inclusion and $Q : X \to X/V$ the quotient map. Let $Y = Q^{-1}(E)$ and denote by $J_Y : Y \to X$ the inclusion and by $Q_Y : Y \to E$ the restriction of Q so that $QJ_Y = J_E Q_Y$. We claim that the net $(QL_\alpha J_Y - J_E K_\alpha Q_Y)$ converges to 0 in $\sigma(K(Y, X/V), K(Y, X/V)^*)$. In fact, the linear span of the functionals $T \mapsto \langle y^{**}, T^* x^*\rangle$, $y^{**} \in Y^{**}$ and $x^* \in (X/V)^*$, is dense in $K(Y, X/V)^*$; this is because $(X/V)^* \subset X^*$ has the RNP (Corollary 4.5), see [224] and [365]. Clearly, $K_\alpha^* J_E^* x^* \to J_E^* x^*$ and $L_\alpha^* Q^* x^* \to Q^* x^*$. Thus

$$\langle y^{**}, (J_Y^* L_\alpha^* Q^* - Q_Y^* K_\alpha^* J_E^*)x^*\rangle \to \langle y^{**}, (J_Y^* Q^* - Q_Y^* J_E^*)x^*\rangle = 0,$$

and since $(QL_\alpha J_Y - J_E K_\alpha Q_Y)$ is bounded, we see that this net converges weakly to 0. Hence there exist $H_\alpha \in \mathrm{co}\{K_\beta \mid \beta \geq \alpha\}$ and $M_\alpha \in \mathrm{co}\{L_\beta \mid \beta \geq \alpha\}$ such that $\lim_\alpha \|QM_\alpha J_Y - J_E H_\alpha Q_Y\| = 0$. Thus $\lim_\alpha \|Q(Id - 2M_\alpha)J_Y - J_E(Id - 2H_\alpha)Q_Y\| = 0$ and so $\lim_\alpha \|Id - 2H_\alpha\| = 1$. □

We remark that the previous theorem could be formulated for quotients of subspaces rather than subspaces of quotients as well, since the two concepts coincide. In fact, one direction was observed in the above proof, and on the other hand, for $V \subset W \subset X$, the quotient W/V is clearly a subspace of the quotient X/V. Hence further iteration of forming subspaces and quotient spaces does not lead to any new results.
Combining the above observations on ℓ^p with Theorem 4.19 and the fact that for separable reflexive spaces the compact approximation property already implies the metric compact approximation property (the proof of the corresponding result on the metric approximation property in [422, p. 40] shows this, too) one obtains the following corollary.

Corollary 4.20 *A subspace X of a quotient of ℓ^p $(1 < p < \infty)$ has the compact approximation property if and only if $K(X)$ is an M-ideal of $L(X)$. If X is a subspace of a quotient of c_0, then the metric compact approximation property of X is equivalent to $K(X)$ being an M-ideal of $L(X)$.*

For further examples, the reader is referred to the following sections.
For the following, we have to recall some definitions: A Banach space is said to have a *(Schauder) basis* if there is a sequence (e_i) such that each $x \in X$ has a unique representation of the form

$$x = \sum_{i=1}^{\infty} \alpha_i e_i$$

for some sequence of scalars (α_i). The basis (e_i) is called *λ-unconditional* if

$$\sup_{|\varepsilon_i|=1} \left\| \sum_{i=1}^{n} \varepsilon_i \alpha_i e_i \right\| \le \lambda \left\| \sum_{i=1}^{n} \alpha_i e_i \right\| \qquad \forall n \in \mathbb{N}.$$

(Of course, the expression "unconditional" in this definition has its origin in the unconditional convergence of the series $\sum_{i=1}^{\infty} \alpha_i e_i$, cf. [422, Section 1.c].) Replacing the one-dimensional "fibers" in these definitions by finite dimensional spaces, one arrives at the concept of a so-called *finite dimensional (Schauder) decomposition* of a Banach space, i.e., a sequence (X_k) of finite dimensional subspaces of X such that each $x \in X$ has a unique representation of the form

$$x = \sum_{i=1}^{\infty} x_i,$$

where $x_i \in X_i$. In the same vein, a finite dimensional decomposition is called λ-unconditional if

$$\sup_{|\varepsilon_i|=1} \left\| \sum_{i=1}^{n} \varepsilon_i x_i \right\| \le \lambda \left\| \sum_{i=1}^{n} x_i \right\| \qquad \forall n \in \mathbb{N}.$$

Here comes the reason for recalling these definitions; this is one of the few instances where we must restrict our attention to the case of approximable operators.

Theorem 4.21 *If X is separable and $A(X)$ is an M-ideal in $L(X)$, then, for every $\varepsilon > 0$, the space X is isometric to a $(1+\varepsilon)$-complemented subspace of a space with a $(1+\varepsilon)$-unconditional finite dimensional Schauder decomposition.*

We found it convenient to formulate a lemma, which contains the core of the proof of this theorem. In the Notes and Remarks section we shall comment on this lemma from the point of view of the so-called u-ideals.

Lemma 4.22 *If X is separable and $A(X)$ is an M-ideal in $L(X)$, then, for every $\varepsilon > 0$, the space X admits a $(1+\varepsilon)$-unconditional expansion of the identity; that is, there are finite rank operators $F_1, F_2, \ldots$ such that*

$$\left\| \sum_{i=1}^{n} \varepsilon_i F_i \right\| \le 1 + \varepsilon \qquad \forall n \in \mathbb{N}, |\varepsilon_i| = 1, \tag{1}$$

$$\sum_{i=1}^{\infty} F_i(x) = x \qquad \forall x \in X. \tag{2}$$

Note that the series (2) converges unconditionally by (1).

PROOF: Since X is separable, so is X^* (Corollary 4.5). Consequently $A(X)$ is a separable M-ideal. In this case Theorem I.2.10 implies the existence of finite rank operators S_n

such that

$$\left\| \sum_{i=1}^{n} \varepsilon_i S_i \right\| \leq 1 + \varepsilon \qquad \forall n \in \mathbb{N}, |\varepsilon_i| = 1,$$

$$\varphi(Id) = \sum_{i=1}^{\infty} \varphi(S_i) \qquad \forall \varphi \in A(X)^*,$$

in particular $x^*(x) = \sum_{i=1}^{\infty} x^*(S_i x)$ for all $x \in X$, $x^* \in X^*$, hence $\lim_{n\to\infty} \sum_{i=1}^{n} S_i = Id$ in the weak operator topology. One can now take convex combinations to obtain a strongly convergent sequence of finite rank operators $T_n = \sum_{k=p_n+1}^{p_{n+1}} \lambda_k^{(n)} \sum_{i=1}^{k} S_i$, so that we get, letting $F_1 = T_1$ and $F_{n+1} = T_{n+1} - T_n$, $x = \sum_{i=1}^{\infty} F_i x$ for all $x \in X$. To see that (1) holds one just has to repeat the calculation performed in the proof of Lemma I.2.9. □

PROOF OF THEOREM 4.21: Let $\varepsilon > 0$ and $1 < (1+\delta)^2 \leq 1 + \varepsilon$. Pick F_n according to Lemma 4.22, with δ in place of ε. Consider the space

$$U = \{(x_n) \mid x_n \in \operatorname{ran}(F_n), \textstyle\sum x_n \text{ converges unconditionally}\},$$

endowed with the norm

$$|(x_n)| = \sup_n \sup_{|\varepsilon_i|=1} \left\| \sum_{i=1}^{n} \varepsilon_i x_i \right\|.$$

It is clear that U has a 1-unconditional finite dimensional decomposition and that the operator $I\colon X \to U$, $I(x) = (F_n x)$ satisfies

$$\|x\| \leq |I(x)| \leq (1+\delta)\|x\|$$

by (1) of Lemma 4.22. Moreover, the operator $Q : U \to U$, $(x_n) \mapsto \left(F_n(\sum_i x_i)\right)$ is a projection onto $\operatorname{ran} I$ with $|Q| \leq 1 + \delta$.

If we renorm U such that the new unit ball becomes the closed convex hull of the old one and $I(B_X)$ and call the new norm $\|\,.\,\|$, then I becomes an isometry, the canonical finite dimensional decomposition of U becomes $(1+\delta)$-unconditional and we get $\|Q\| \leq (1+\delta)^2$. Thus Theorem 4.21 is proven. □

Corollary 4.23 *A Banach space X which satisfies the assumptions of Theorem 4.21 is isometric to a subspace of a space with a $(1+\varepsilon)$-unconditional basis.*

PROOF: This follows by an adaptation of the proof of [422, Th. 1.g.5] □

It is worthwhile remarking that one can even embed X into a space with a shrinking $(1+\varepsilon)$-unconditional basis [397].

We conclude this section with a result that shows that among spaces with a 1-symmetric basis there are only very few examples for which $K(X)$ is an M-ideal of $L(X)$. A basis (e_i) of X is called 1-symmetric if

$$\left\| \sum_{i=1}^{\infty} \varepsilon_i \xi_{\pi(i)} e_{\pi(i)} \right\| = \left\| \sum_{i=1}^{\infty} \xi_i e_i \right\|$$

for all $x = \sum_{i=1}^{\infty} \xi_i e_i \in X$, where the ε_i range over all scalars of modulus one and π over all permutations of N; see [422, Ch. 3] for more details. Clearly the canonical bases in ℓ^p and c_0 are 1-symmetric. We also note that a 1-symmetric basis (e_n) of X is 1-unconditional. If in addition X^* is separable, then $e_n \to 0$ weakly. This well-known result follows for instance from [422, Th. 1.c.9] and will be used below.

Proposition 4.24 *Suppose X is a Banach space with a 1-symmetric basis. Then $K(X)$ is an M-ideal in $L(X)$ if and only if X is isometric to c_0 or ℓ^p for some $1 < p < \infty$.*

PROOF: In light of Example 4.1 we must show that if $K(X)$ is an M-ideal in $L(X)$, then X is isometric to c_0 or ℓ^p for some $1 < p < \infty$. To this end we will use the following result, due to Bohnenblust:

A Banach space X with a normalised basis (e_n) is isometric to either c_0 or ℓ^p for $1 \leq p < \infty$, if (and only if) for any normalised elements $x_1, \ldots, x_n$, with $x_i = \sum_{j=n_i}^{m_i} a_j e_j$ and $n_1 \leq m_1 < n_2 \leq m_2 < n_3 \leq \ldots$,

$$\left\| \sum_{i=1}^{n} \alpha_i e_i \right\| = \left\| \sum_{i=1}^{n} \alpha_i x_i \right\| \tag{$*$}$$

for all $n \in \mathbf{N}$ and all scalars $\alpha_1, \ldots, \alpha_n$.

For a proof see [422, Th. 2.a.9] and put $M = 1$ in that reference. Actually, this result goes back to Kolmogorov and Nagumo, see [448, p. 154].
Now, if $K(X)$ is an M-ideal in $L(X)$, X has (M) by Theorem 4.17, and X^* is separable by Corollary 4.5. Together with the 1-symmetry of the basis, this implies:

If x and x' (resp. y and y') are finitely and disjointly supported elements in X with $\|x\| = \|y\|$ and $\|x'\| = \|y'\|$, then

$$\|x + x'\| = \|y + y'\|. \tag{$**$}$$

Here the support of $x = \sum_{i=1}^{\infty} \xi_i e_i$ is understood as the set $\{i \mid \xi_i \neq 0\}$. To prove $(**)$ we introduce the shift operator $\sigma : \sum_{i=1}^{\infty} \xi_i e_i \mapsto \sum_{i=1}^{\infty} \xi_i e_{i+1}$. By 1-symmetry of the basis, this is an isometry, and moreover we have $\sigma^m(x) \to 0$ weakly for each finitely supported x, since $e_n \to 0$ weakly, as was noted above. Again by 1-symmetry we obtain that $\|x + \sigma^r(x')\|$ does not depend on r, provided r is sufficiently large, so that we get from property (M) for sufficiently large r and s

$$\begin{aligned} \|x + x'\| &= \|x + \sigma^r(x')\| \\ &= \|y + \sigma^r(x')\| \\ &= \|\sigma^s(y) + \sigma^r(x')\| \\ &= \|\sigma^s(y) + \sigma^r(y')\| \\ &= \|y + y'\|. \end{aligned}$$

Clearly, $(**)$ inductively implies $(*)$. Since $K(\ell^1)$ is not an M-ideal in $L(\ell^1)$ by Corollary 4.5, the proof of the proposition is completed. □

VI.5 Banach spaces for which $K(X)$ forms an M-ideal: The (M_p)-spaces

In this section we develop the theory of a certain class of spaces X for which $K(X)$ is an M-ideal in $L(X)$. These spaces can be characterised in terms of an ℓ^p-type version of the approximation property (Theorem 5.3), and their position within the class of all Banach spaces for which $K(X)$ is an M-ideal in $L(X)$ will eventually be elucidated in Theorem 6.6. We also present some applications to the study of smooth points in operator spaces.

We remind the reader that, as in the previous section, similar results can be formulated for norm limits of finite rank operators (instead of compact operators).

Definition 5.1 *Let $1 \leq p \leq \infty$. We say that a Banach space X has property (M_p) or is an (M_p)-space if $K(X \oplus_p X)$ is an M-ideal of $L(X \oplus_p X)$.*

Let us point out immediately that (M_1)-spaces are finite dimensional (and thus not really of interest in the present context). In fact, otherwise there is a net (x^*_α) in the dual unit sphere weak* converging to 0. Thus, for arbitrary $x^* \in S_{X^*}$, we have $(x^*, x^*_\alpha) \to (x^*, 0) \in S_{(X \oplus_1 X)^*}$ weak*, but not in norm. By Proposition 4.6 $K(X \oplus_1 X)$ is not an M-ideal in $L(X \oplus_1 X)$.

Clearly, the basic example of an (M_p)-space is ℓ^p (for $1 < p < \infty$) by Example 4.1 and since $\ell^p \oplus_p \ell^p \cong \ell^p$. Likewise, c_0 has the (M_∞)-property.

For convenience we collect some observations on these spaces.

Proposition 5.2

(a) *If $1 < p \leq \infty$ and X is an (M_p)-space, then $K(X)$ is an M-ideal in $L(X)$.*

(b) *If $1 < p < \infty$ and X is an (M_p)-space, then X is reflexive.*

(c) *If $1 < p < \infty$ and $1/p + 1/p^* = 1$, then X is an (M_p)-space if and only if X^* is an (M_{p^*})-space.*

PROOF: (a) follows from Proposition 4.2.

(b) Otherwise, $X^{**} \oplus_p X^{**}$ would contain a nontrivial L^p-summand and the nontrivial M-ideal $X \oplus_p X$ (Proposition 4.4) which is impossible as shown in [47].

(c) In this case $L(X \oplus_p X)$ and $L(X^* \oplus_{p^*} X^*)$ are isometric with the isomorphism mapping compact operators onto compact operators, since X is reflexive by (b). □

Theorem 5.3 *Let X be a Banach space and suppose $1 < p \leq \infty$.*

(a) *Whenever $1 < p < \infty$, X has (M_p) if and only if there is a net (K_α) in the unit ball of $K(X)$ such that*

$$K_\alpha x \to x \quad \forall x \in X \qquad \text{and} \qquad K^*_\alpha x^* \to x^* \quad \forall x^* \in X^*, \tag{1}$$

and for all $\varepsilon > 0$ there is α_0 such that for $\alpha > \alpha_0$ we have for all $x, y \in X$

$$\|K_\alpha x + (Id - K_\alpha)y\|^p \leq (1+\varepsilon)^p(\|x\|^p + \|y\|^p) \tag{2a}$$

as well as

$$\|K_\alpha x\|^p + \|x - K_\alpha x\|^p \leq (1+\varepsilon)^p \|x\|^p. \tag{2b}$$

(b) *X has (M_∞) if and only if there is a net (K_α) in the unit ball of $K(X)$ satisfying (1) and*

$$\|K_\alpha x + (Id - K_\alpha)y\| \leq (1+\varepsilon)\max\{\|x\|, \|y\|\} \qquad \forall x, y \in X \tag{3}$$

provided α is large enough.

PROOF: (a) Suppose $K(X \oplus_p X)$ is an M-ideal in $L(X \oplus_p X)$. By Theorem 4.17 there is a net of operators (S_α) in the unit ball of $K(X \oplus_p X)$ converging to the identity operator on the square with respect to the strong operator topology as do their adjoints such that

$$\limsup \|S_\alpha A + (Id - S_\alpha)B\| \leq \max\{\|A\|, \|B\|\} \qquad \forall A, B \in L(X \oplus_p X). \tag{4}$$

The S_α may be represented as 2×2 matrices of operators on X, say $S_\alpha = (S_\alpha^{kl})_{k,l=1,2}$. Now (S_α^{12}), (S_α^{21}) and $(S_\alpha^{11} - S_\alpha^{22})$ tend strongly to 0, and the technique employed in the proof of Theorem 4.19 shows that these nets tend to 0 for the weak topology of the Banach space $K(X \oplus_p X)$. Hence a convex combinations argument permits us to assume that S_α is of the particular form

$$S_\alpha = \begin{pmatrix} K_\alpha & 0 \\ 0 & K_\alpha \end{pmatrix}.$$

The (K_α) fulfill (1), and checking (4) on the operators

$$A = \begin{pmatrix} Id & 0 \\ 0 & 0 \end{pmatrix}, \qquad B = \begin{pmatrix} 0 & Id \\ 0 & 0 \end{pmatrix}$$

and, respectively,

$$A = \begin{pmatrix} Id & 0 \\ 0 & 0 \end{pmatrix}, \qquad B = \begin{pmatrix} 0 & 0 \\ Id & 0 \end{pmatrix}$$

will immediately yield (2a) and (2b).

For the converse suppose a net (K_α) given as stated. To prove the M-ideal property we wish to establish the restricted 3-ball property (Theorem I.2.2). That is, given $\varepsilon > 0$, compact contractions K_1, K_2, K_3 and a bounded contraction T on $X \oplus_p X$, we must find a compact operator U such that $\|K_r + T - U\| \leq 1 + \varepsilon$. To this end consider the compact operator

$$U_\alpha = \left(T^{kl} - (Id - K_\alpha)T^{kl}(Id - K_\alpha)\right)_{k,l=1,2}.$$

(Again we represent operators on $X \oplus_p X$ by 2×2-matrices with entries in $L(X)$; note that $\|(T^{kl})\| \leq 1$ translates into

$$\|T^{11}x + T^{12}y\|^p + \|T^{21}x + T^{22}y\|^p \leq \|x\|^p + \|y\|^p$$

for all $x, y \in X$.) We wish to show that $U = U_\alpha$ will do for large enough α. First of all (1) implies

$$\|K_\alpha K_r^{kl} K_\alpha - K_r^{kl}\| \leq \varepsilon$$

for $r = 1,2,3$ and $k,l = 1,2$, provided α is sufficiently large. Thus it is acceptable to pretend that actually $K_\alpha K_r^{kl} K_\alpha = K_r^{kl}$ in the following computation which shows the validity of the 3-ball condition:

$$\begin{aligned}
&\|(K_r + T - U_\alpha)(x,y)\|^p \\
&= \ \big\|[K_\alpha K_r^{11} K_\alpha + (Id - K_\alpha)T^{11}(Id - K_\alpha)]x \\
&\qquad + [K_\alpha K_r^{12} K_\alpha + (Id - K_\alpha)T^{12}(Id - K_\alpha)]y\big\|^p \\
&\quad + \big\|[K_\alpha K_r^{21} K_\alpha + (Id - K_\alpha)T^{21}(Id - K_\alpha)]x \\
&\qquad + [K_\alpha K_r^{22} K_\alpha + (Id - K_\alpha)T^{22}(Id - K_\alpha)]y\big\|^p \\
&\leq \ (1+\varepsilon)^p \big[\|K_r^{11} K_\alpha x + K_r^{12} K_\alpha y\|^p \\
&\qquad + \|T^{11}(Id - K_\alpha)x + T^{12}(Id - K_\alpha)y\|^p \\
&\qquad + \|K_r^{21} K_\alpha x + K_r^{22} K_\alpha y\|^p \\
&\qquad + \|T^{21}(Id - K_\alpha)x + T^{22}(Id - K_\alpha)y\|^p\big] \\
&\leq \ (1+\varepsilon)^p \big[\|K_\alpha x\|^p + \|K_\alpha y\|^p + \|(Id - K_\alpha)x\|^p + \|(Id - K_\alpha)y\|^p\big] \\
&\leq \ (1+\varepsilon)^{2p}(\|x\|^p + \|y\|^p)
\end{aligned}$$

provided α is large enough.

(b) The necessity of the conditions (1) and (3) can be proved in much the same way as in part (a). Although this is basically true for the sufficiency argument, too, we would like to be slightly more precise. Following the above notation we may now suppose that $K_\alpha K_r^{kl} = K_r^{kl}$ for large α, and we then have for $U = \big(K_\alpha T^{kl}\big)_{k,l=1,2}$

$$\begin{aligned}
&\|(K_r + T - U)(x,y)\| \\
&= \ \max\Big\{\|[K_\alpha K_r^{11} + (Id - K_\alpha)T^{11}]x + [K_\alpha K_r^{12} + (Id - K_\alpha)T^{12}]y\|, \\
&\qquad \|[K_\alpha K_r^{21} + (Id - K_\alpha)T^{21}]x + [K_\alpha K_r^{22} + (Id - K_\alpha)T^{22}]y\|\Big\} \\
&\leq \ (1+\varepsilon)\max\Big\{\|K_r^{11}x + K_r^{12}y\| \vee \|T^{11}x + T^{12}y\|, \\
&\qquad \|K_r^{21}x + K_r^{22}y\| \vee \|T^{21}x + T^{22}y\|\Big\} \\
&\leq \ 1+\varepsilon.
\end{aligned}$$

□

By a straightforward verification, one checks that the coordinate projections on $X = \ell^p$ (or more generally, the net of projections belonging canonically to an ℓ^p-sum of finite dimensional spaces) satisfy the above conditions. Therefore the following corollary is valid.

Corollary 5.4 *For every family (X_α) of finite dimensional Banach spaces, $K(\ell^p(X_\alpha))$ and $K(c_0(X_\alpha))$ are M-ideals in the respective spaces of bounded operators.*

The proof of part (b) of Theorem 5.3 yields a bit more than stated above. Namely, if T is an operator into $(X \oplus_\infty X)^{**}$ rather than into $X \oplus_\infty X$ and if we define $U = (K_\alpha^{**}T^{ki})$, then U is an $X \oplus_\infty X$-valued compact operator, and the above proof shows that $K(X \oplus_\infty X)$ is an M-ideal in $L(X \oplus_\infty X, (X \oplus_\infty X)^{**})$. The following proposition will assert this, but its formulation is more symmetric. We denote

$$K^{\mathrm{ad}}(X^*, Y^*) = \{T^* \mid T \in K(Y, X)\}$$

and observe that $L(Y, X^{**}) \cong L(X^*, Y^*)$.

Proposition 5.5 *A Banach space X is an (M_∞)-space if and only if either of the following holds:*

(i) *$K^{\mathrm{ad}}(X^* \oplus_1 X^*)$ is an M-ideal in $L(X^* \oplus_1 X^*)$.*

(ii) *There is a net (S_α) in the unit ball of $K^{\mathrm{ad}}(X^*)$ converging to the identity in the strong operator topology such that*

$$\limsup_{\alpha} \sup_{\|x^*\| \le 1} \left(\|S_\alpha x^*\| + \|(Id - S_\alpha)x^*\| \right) \le 1. \tag{5}$$

PROOF: This was in essence observed in the preceding remarks. It is left to point out that (5) and (3) are easily seen to be equivalent by a simple dualization argument, that, by a convex combinations argument, if $L_\alpha^* = S_\alpha$ and $S_\alpha \to Id_{X^*}$ strongly, then some convex combinations of the L_α converge strongly to Id_X and that both (3) and (5) remain in effect after having taken convex combinations. □

Condition (i) can be thought of as a dual version of the (M_∞)-property.
We now investigate the (M_∞)-spaces more closely. Recall that (iii) below means that $K(Y, X)$ is an M-ideal in $L(Y, X^{**})$ for all Y.

Proposition 5.6 *For a Banach space X the following assertions are equivalent:*

(i) *X is an (M_∞)-space.*

(ii) *For all Banach spaces Y, $K(Y, X)$ is an M-ideal in $L(Y, X)$.*

(iii) *For all Banach spaces Y, $K^{\mathrm{ad}}(X^*, Y^*)$ is an M-ideal in $L(X^*, Y^*)$.*

PROOF: (i) ⇒ (iii): Pick a net $(S_\alpha) = (K_\alpha^*)$ as in Proposition 5.5. We may assume that (K_α) satisfies (1) of Theorem 5.3, due to the by now standard convex combinations argument. Then, if $T \in B_{L(X^*, Y^*)}$, $K_1, K_2, K_3 \in B_{K(Y,X)}$ and $\varepsilon > 0$, we have for sufficiently large α

$$\|K_\alpha K_r - K_r\| \le \frac{\varepsilon}{2}, \qquad r = 1, 2, 3$$

and

$$\sup_{\|x_1\|, \|x_2\| \le 1} \|K_\alpha x_1 + (Id - K_\alpha)x_2\| \le 1 + \frac{\varepsilon}{2}.$$

Putting $U = K_\alpha^{**}T^*|_Y \in K(Y, X)$ (for K_α^{**} is X-valued, since K_α is compact) we obtain for large α

$$\|K_r^* + T - U^*\| \;=\; \|K_r^* + T(Id - K_\alpha^*)\|$$

$$\begin{aligned}
&\leq \|K_\tau^* K_\alpha^* + T(Id - K_\alpha^*)\| + \frac{\varepsilon}{2} \\
&\leq \sup_{\|x^*\|\leq 1} (\|K_\alpha^* x^*\| + \|(Id - K_\alpha^*)x^*\|) + \frac{\varepsilon}{2} \\
&\leq 1 + \varepsilon.
\end{aligned}$$

(iii) $\Rightarrow$ (ii): This is clear since $K(Y,X) \cong K^{ad}(X^*,Y^*)$ and $L(Y,X) \overset{\cong}{\hookrightarrow} L(X^*,Y^*)$ canonically.

(ii) $\Rightarrow$ (i): By assumption, $K(X \oplus_\infty X, X)$ is an M-ideal in $L(X \oplus_\infty X, X)$, and now, the statement that X possesses the (M_∞)-property follows from two general facts: The first is that we have (Lemma 1.1(c))

$$\begin{aligned}
L(X \oplus_\infty X) &= L(X \oplus_\infty X, X) \oplus_\infty L(X \oplus_\infty X, X), \\
K(X \oplus_\infty X) &= K(X \oplus_\infty X, X) \oplus_\infty K(X \oplus_\infty X, X),
\end{aligned}$$

and the second consists in the observation that $J \oplus_\infty J$ remains an M-ideal in $E \oplus_\infty E$ whenever J is an M-ideal in the Banach space E. □

There seem to be only few (M_∞)-spaces; in fact, the closed subspaces and quotients of $c_0(I)$ with the metric compact approximation property (see Corollary 5.12) are basically the only examples we know. In this regard the following result is of interest.

Theorem 5.7 *Let X be a Banach space. Then the following conditions are equivalent:*

(i) *For every $\varepsilon > 0$ the space X is $(1+\varepsilon)$-close with respect to the Banach-Mazur distance to a subspace $Y = Y_\varepsilon$ of c_0 enjoying the metric compact approximation property.*

(ii) *The space X is, for every $\varepsilon > 0$, $(1+\varepsilon)$-close with respect to the Banach-Mazur distance to a space $Y = Y_\varepsilon$ admitting a sequence (K_n) of compact operators converging strongly to the identity on Y with the property that there is a sequence (δ_n) of nonnegative numbers tending to zero with*

$$\|K_n y_1 + (Id - K_n)y_2\| \leq 1 + \delta_n \|y_1 - y_2\| \qquad \forall y_1, y_2 \in B_Y.$$

Note that the approximation condition in (ii) is at least formally stronger than (3) of Theorem 5.3, but we don't have an example to show that the two conditions really differ. For the proof we need two auxiliary results. The first one can be regarded as an isomorphic version of Proposition I.3.9.

Proposition 5.8 *Let T be an operator on a Banach space X and let $\delta > 0$. Then the following statements are equivalent:*

(i) *For all $p \in \operatorname{ex} B_{X^*}$ there is a number α in the unit interval with*

$$\|T^* p - \alpha p\| \leq \delta.$$

(ii) *For all pairs of elements x_1, x_2 in the unit ball of X the following estimate holds:*

$$\|Tx_1 + (Id - T)x_2\| \leq 1 + \delta \|x_1 - x_2\|.$$

(iii) *For each $x^* \in X^*$ there is an element $y^* \in X^*$ such that $\|y^*\| \le \delta\|x^*\|$ and*

$$\|T^*x^* - y^*\| + \|(Id - T^*)x^* + y^*\| = \|x^*\|.$$

PROOF: We first show that (i) implies (ii). To do this, let x_1, $x_2 \in B_X$ and pick $p \in \operatorname{ex} B_{X^*}$ with

$$\|Tx_1 + (Id - T)x_2\| = p(Tx_1 + (Id - T)x_2).$$

By (i), there is $\alpha \in [0, 1]$ such that

$$\begin{aligned} p(Tx_1 + (Id - T)x_2) &\le p(x_2) + \alpha p(x_1 - x_2) + \delta\|x_1 - x_2\| \\ &= \alpha p(x_1) + (1 - \alpha)p(x_2) + \delta\|x_1 - x_2\|, \end{aligned}$$

whence

$$\|Tx_1 + (Id - T)x_2\| \le 1 + \delta\|x_1 - x_2\|$$

as claimed.

Let us now see how one can conclude (iii) from (ii). Endow $X \oplus X$ and $X \oplus X \oplus X$ with norms

$$\|(x_1, x_2)\| = \max\{\|x_1\|, \|x_2\|\} + \delta\|x_1 - x_2\|$$

and, respectively,

$$\|(x_1, x_2, x_3)\| = \max\{\|x_1\|, \|x_2\|\} + \delta\|x_3\|.$$

Using the embedding $(x_1, x_2) \mapsto (x_1, x_2, x_1 - x_2)$ it is easy to check that the dual norm on $(X \oplus X)^* = X^* \oplus X^*$ is given by

$$\|(x_1^*, x_2^*)\| = \inf_{y^* \in Y^*} \max\left\{\|x_1^* - y^*\| + \|x_2^* + y^*\|, \delta^{-1}\|y^*\|\right\}.$$

As a benefit of the weak* lower semicontinuity of the norm and the weak* compactness of the dual unit ball the infimum in the last equation is actually attained. Hence, a pair (x_1^*, x_2^*) belongs to the unit ball of $X^* \oplus X^*$ if and only if

$$\min_{\|y^*\| \le \delta} \|x_1^* - y^*\| + \|x_2^* + y^*\| \le 1.$$

The reason for norming $X \oplus X$ in this way is the fact that the operator

$$\widehat{T}: X \oplus X \to X, \quad (x_1, x_2) \mapsto Tx_1 + (Id - T)x_2$$

now has norm one and that, consequently, in passing to adjoints,

$$\begin{aligned} \|x^*\| &\le \inf_{\|y^*\| \le \delta} (\|T^*x^* - y^*\| + \|(Id - T^*)x^* + y^*\|) \\ &= \min_{\|y^*\| \le \delta} (\|T^*x^* - y^*\| + \|(Id - T^*)x^* + y^*\|) \\ &= \|\widehat{T}^*(x^*)\| \\ &\le \|x^*\| \end{aligned}$$

for any $x^* \in B_{X^*}$. (This also shows that $\|\widehat{T}\| = 1$.) Since the case of an arbitrary $x^* \in X^*$ follows by normalization, we have shown (iii).

To finish the proof we must show that (i) is a consequence of (iii). So let us suppose that (iii) holds and pick $y^* \in X^*$ such that for a fixed functional $p \in \operatorname{ex} B_{X^*}$

$$\|T^*p - y^*\| + \|(Id - T^*)p + y^*\| = 1$$

and $\|y^*\| \le \delta$. We certainly may suppose that the numbers $\|T^*p - y^*\|$ and $\|(Id - T^*)p + y^*\|$ are different from zero. (Otherwise nothing has to be shown.) Writing

$$p = \frac{T^*p - y^*}{\|T^*p - y^*\|}\|T^*p - y^*\| + \frac{(Id - T^*)p + y^*}{\|(Id - T^*)p + y^*\|}\|(Id - T^*)p + y^*\|,$$

we find

$$T^*p - y^* = \|T^*p - y^*\|p,$$

and letting $\alpha := \|T^*p - y^*\|$ it turns out that

$$\|T^*p - \alpha p\| = \|y^*\| \le \delta,$$

which brings the proof to an end. □

Lemma 5.9 *Suppose that K is a compact operator defined on X, which satisfies one of the equivalent conditions of the above proposition for the constant $\delta > 0$. Then, for any $\alpha > \delta$ and $\varepsilon > 0$, there is a finite $(\varepsilon + \delta(\alpha - \delta)^{-1})$-net for the set*

$$\{p \in \operatorname{ex} B_{X^*} \mid \|K^*p\| \ge \alpha\}.$$

PROOF: Write C_δ for the set of all points at a distance less than δ from the set C and denote by $[a, b]$ the convex hull of two real numbers a and b. An appeal to Proposition 5.8 then yields $a(p) \in [0, 1]$ which are "almost eigenvalues" of K corresponding to the "almost eigenvector" $p \in \operatorname{ex} B_{X^*}$. It follows

$$\begin{aligned}
\{p \in \operatorname{ex} B_{X^*} \mid \|K^*p\| \ge \alpha\} &\subset \{p \in \operatorname{ex} B_{X^*} \mid a(p) \ge \alpha - \delta\} \\
&\subset [1, (\alpha - \delta)^{-1}]\,\{a(p)p \mid p \in \operatorname{ex} B_{X^*}, a(p) \ge \alpha - \delta\} \\
&\subset [1, (\alpha - \delta)^{-1}]\,\{K^*p \mid p \in \operatorname{ex} B_{X^*}, a(p) \ge \alpha - \delta\}_\delta \\
&\subset \Big([1, (\alpha - \delta)^{-1}]K^*(B_{X^*})\Big)_{\frac{\delta}{\alpha - \delta}},
\end{aligned}$$

and since the set $K^*(B_{X^*})$ is norm compact by assumption, we are done. □

Let us now come to the

PROOF OF THEOREM 5.7:
To demonstrate that condition (i) implies (ii) it certainly suffices to show that the metric compact approximation property of subspaces of c_0 has the announced asymptotic behaviour. So let X be a subspace of c_0 with the metric compact approximation property and suppose that (S_n) is a sequence of norm-one compact operators converging strongly to the identity. If (P_n) is the sequence of coordinate projections on c_0 then, for every n,

$$\sup_{\|x_1\|, \|x_2\| \le 1} \|P_n x_1 + (Id - P_n)x_2\| \le 1$$

and $JS_n - P_nJ \to 0$ strongly, where J denotes the embedding of X into c_0. Now proceed as in Theorem 4.19 to see that there are sequences (S_n^c) and (P_n^c) of convex combinations of (S_n) and (P_n), respectively, such that $JS_n^c - P_n^cJ \to 0$ in norm. For the obvious choice of the sequence (δ_n) we find for $x_1, x_2 \in B_X$

$$\begin{aligned}\|S_n^c x_1 + (Id - S_n^c)x_2\| &\leq \|P_n^c J(x_1) + (Id - P_n^c)J(x_2)\| + \|JS_n^c - P_n^c J\|\|x_1 - x_2\| \\ &\leq 1 + \delta_n\|x_1 - x_2\|.\end{aligned}$$

This proves the desired implication.

For the implication (ii) $\Rightarrow$ (i), it is sufficient to prove that any Banach space X admitting a sequence (K_n) as stated in the theorem is, for any $\varepsilon > 0$, $(1+\varepsilon)$-close to a subspace of c_0. Invoking Proposition 5.8, we see that there is a sequence (δ_n) of nonnegative numbers tending to zero such that for every $p \in \operatorname{ex} B_{X^*}$ there exists $a_n(p) \in [0,1]$ with

$$\|K_n^* p - a_n(p)p\| \leq \delta_n.$$

Fix a natural number $N \in \mathbb{N}$. Applying Lemma 5.9 (with $\varepsilon = (2N)^{-1}$, $\delta = \delta_n$ and $\alpha = 2(N+1)\delta_n$), we successively define sets $C_n \subset \operatorname{ex} B_{X^*}$ together with functionals $p_n \in \operatorname{ex} B_{X^*}$ as follows:

Let $C_1 := \{p \in \operatorname{ex} B_{X^*} \mid \|K_1^* p\| \geq (2N+1)\delta_1\}$ and $p_1, \ldots, p_{k_1}$ be an N^{-1}-net of C_1. Having performed the nth step, put

$$C_{n+1} := \{p \in \operatorname{ex} B_{X^*} \mid \|K_{n+1}^* p\| \geq (2N+1)\delta_{n+1}\} \setminus \bigcup_{\nu=1}^{n} C_\nu$$

and choose $p_{k_n+1}, \ldots, p_{k_{n+1}}$ such that $\{p_{k_n+1}, \ldots, p_{k_{n+1}}\}$ is an N^{-1}-net for C_{n+1}. Note that whenever $k > k_n$ it follows that

$$\|K_n^* p_k\| < (2N+1)\delta_n. \tag{1}$$

Define $I : X \to \ell^\infty$ by $(Ix)(n) = p_n(x)$ and note that $I(X) \subset c_0$: In fact, whenever $x \in B_X$ and $\eta > 0$ is fixed, we pick $n \in \mathbb{N}$ such that $(4N+2)\delta_n \leq \eta$ and $\|K_n x - x\| \leq \frac{\eta}{2}$. By (1) we then may conclude that for all $k \geq k_n + 1$

$$|I(x)(k)| \leq |K_n^* p_k(x)| + |p_k(x) - K_n^* p_k(x)| \leq (2N+1)\delta_n + \frac{\eta}{2} \leq \eta.$$

Let us estimate some norms. Again, fix $\eta > 0$ and $x \in X$ with norm one, and choose $n \in \mathbb{N}$ such that $\|x\| \leq \|K_n x\| + \frac{\eta}{2}$ and $\delta_n \leq \frac{\eta}{2}$ simultaneously. We have

$$1 = \|x\| \leq \|K_n x\| + \frac{\eta}{2} = p(K_n x) + \frac{\eta}{2}$$

for some $p \in \operatorname{ex} B_{X^*}$, which must be N^{-1}-close to some p_k from the sequence constructed above, because $\lim_n \|K_n^* p\| = 1$ and therefore $p \in C_n$ for some $n \in \mathbb{N}$. Consequently, the right hand side in the above inequality can be estimated from above by

$$a_n(p_k)p_k(x) + N^{-1} + \delta_n + \frac{\eta}{2} \leq \|I(x)\| + N^{-1} + \eta.$$

It follows that $\|I(x)\| \geq (1-\eta-N^{-1})\|x\|$ for all $x \in X$ and hence

$$\|I(x)\| \leq \|x\| \leq \frac{N}{N-1}\|I(x)\|. \tag{2}$$

Let us finally observe that $I(X)$ has the metric compact approximation property. In fact, letting $\widetilde{K}_n : (p_k(x)) \mapsto (p_k(K_n x))$, we have $\lim_n \widetilde{K}_n\xi = \xi$ for all $\xi \in I(X)$ and $\limsup_n \|\widetilde{K}_n\| \leq 1$. For, whenever $\sup_{k\in\mathbb{N}} |p_k(x)| \leq 1$ we have by (2) the inequality $\|x\| \leq N(N-1)^{-1}$ and therefore

$$\begin{aligned}\sup_{k\in\mathbb{N}} |p_k(K_n x)| &\leq |(K_n^* p_k - a_n(p_k)p_k)(x)| + |a_n(p_k)p_k(x)| \\ &\leq \delta_n \frac{N}{N-1} + 1.\end{aligned}$$

Further, $\widetilde{K}_n$ is compact since K_n is. The theorem now follows from (2) and the fact that $N \in \mathbb{N}$ was arbitrary. □

We now show that the (M_p)-spaces contain lots of complemented copies of ℓ^p; the existence of ℓ^p-subspaces will be proved under less stringent assumptions in Theorem 6.4.

Theorem 5.10 *Suppose X is an infinite dimensional Banach space which has property (M_p) for some $1 < p \leq \infty$.*

(a) *Every normalized sequence (x_n) in X tending to 0 weakly has, for every $\lambda > 1$, a subsequence (x_{n_k}) which is λ-equivalent to the unit vector basis in ℓ^p (c_0 if $p = \infty$) such that $\overline{\text{lin}}\,\{x_{n_k} \mid k \in \mathbb{N}\}$ is complemented in X.*

(b) *If $p = \infty$, then every normalized sequence (x_n^*) in X^* tending to 0 in the weak* sense has, for every $\lambda > 1$, a subsequence $(x_{n_k}^*)$ which is λ-equivalent to the unit vector basis in ℓ^1 and such that $\overline{\text{lin}}\,\{x_{n_k}^* \mid k \in \mathbb{N}\}$ is complemented in X^*.*

PROOF: (a) We first treat the case $p < \infty$. Let (ε_n) be a sequence of positive numbers such that $\prod_{n=1}^{\infty}(1+3\varepsilon_n) \leq \lambda$ and $\prod_{n=1}^{\infty}(1-3\varepsilon_n) \geq \lambda^{-1}$. Let $(K_\alpha) \subset K(X)$ be a net as in Theorem 5.3. To begin with, put $y_1 = x_1$. Then

$$\|K_{\alpha_1} y_1 - y_1\| \leq \varepsilon_1$$

for some α_1 so large that (2a) and (2b) are fulfilled with $\varepsilon = \varepsilon_1$. Since K_{α_1} is compact, $\lim_{n\to\infty} K_{\alpha_1} x_n = 0$, and we hence may find $n_2 > 1$, such that for $y_2 = x_{n_2}$

$$\|K_{\alpha_1} y_2\| \leq \varepsilon_1.$$

Next, we choose $\alpha_2 > \alpha_1$ such that for all y in the span of y_1 and y_2

$$\|K_{\alpha_2} y - y\| \leq \varepsilon_2 \|y\|$$

and that, additionally, (2a) and (2b) hold for $\varepsilon = \varepsilon_2$. Then choose $n_3 > n_2$ so that for $y_3 = x_{n_3}$

$$\|K_{\alpha_2} y_3\| \leq \varepsilon_2.$$

Thus, an induction procedure yields an increasing sequence of indices (α_k) and a subsequence (y_k) of the x_n such that K_{α_k} fulfills (2a) and (2b) for $\varepsilon = \varepsilon_k$ as well as

$$\|K_{\alpha_k} y - y\| \le \varepsilon_k \|y\| \qquad \forall y \in \operatorname{lin}\{y_1, \dots, y_k\} \tag{$*$}$$

and

$$\|K_{\alpha_k} y_{k+1}\| \le \varepsilon_k. \tag{$**$}$$

We claim that the y_k are λ-equivalent to the unit vector basis of ℓ^p. To prove an upper ℓ^p-estimate, we now consider the numerical sequence defined by

$$\begin{aligned} C_1 &= 1, \\ C_n &= \prod_{k=1}^{n-1} (1 + 3\varepsilon_k). \end{aligned}$$

Note that $\sup_n C_n \le \lambda$. An induction argument then yields

$$\left\| \sum_{k=1}^{n} a_k y_k \right\| \le C_n \left(\sum_{k=1}^{n} |a_k|^p \right)^{1/p}.$$

Indeed, we find by (2a), $(*)$ and $(**)$

$$\begin{aligned} \left\| \sum_{k=1}^{n+1} a_k y_k \right\| &\le \left\| K_{\alpha_n} \left(\sum_{k=1}^{n} a_k y_k \right) + (Id - K_{\alpha_n})(a_{n+1} y_{n+1}) \right\| \\ &\qquad + \varepsilon_n \left\| \sum_{k=1}^{n} a_k y_k \right\| + \varepsilon_n |a_{n+1}| \\ &\le (1 + \varepsilon_n) \left(\left\| \sum_{k=1}^{n} a_k y_k \right\|^p + |a_{n+1}|^p \right)^{1/p} \\ &\qquad + \varepsilon_n \left\| \sum_{k=1}^{n} a_k y_k \right\| + \varepsilon_n |a_{n+1}|, \end{aligned}$$

and by induction hypothesis and since $C_n \ge 1$ this last expression may be be estimated by

$$\le (1 + \varepsilon_n) C_n \left(\sum_{k=1}^{n+1} |a_k|^p \right)^{1/p} + 2\varepsilon_n C_n \left(\sum_{k=1}^{n+1} |a_k|^p \right)^{1/p} = C_{n+1} \left(\sum_{k=1}^{n+1} |a_k|^p \right)^{1/p}.$$

We end up with the upper ℓ^p-estimate

$$\left\| \sum_{k=1}^{n} a_k y_k \right\| \le \lambda \cdot \left(\sum_{k=1}^{n} |a_k|^p \right)^{1/p}.$$

which holds for all $n \in \mathbb{N}$. For the proof of the lower ℓ^p-estimate we put

$$\begin{aligned} c_1 &= 1, \\ c_n &= \prod_{k=1}^{n-1}(1-3\varepsilon_k) \end{aligned}$$

so that $\inf_n c_n \geq \lambda^{-1}$. As before, we obtain inductively

$$\left\| \sum_{k=1}^{n} a_k y_k \right\| \geq c_n \left(\sum_{k=1}^{n} |a_k|^p \right)^{1/p} :$$

$$\begin{aligned}
&(1+\varepsilon_n)\left\| \sum_{k=1}^{n+1} a_k y_k \right\| \\
\underset{(2b)}{\geq}\quad & \left\{ \left| \left\| K_{\alpha_n}\left(\sum_{k=1}^{n} a_k y_k \right) \right\| - \| K_{\alpha_n}(a_{n+1} y_{n+1}) \| \right|^p \right. \\
& \quad \left. + \left| \| (Id - K_{\alpha_n})(a_{n+1} y_{n+1}) \| - \left\| (Id - K_{\alpha_n}) \left(\sum_{k=1}^{n} a_k y_k \right) \right\| \right|^p \right\}^{1/p} \\
\geq\quad & \left\{ \left\| K_{\alpha_n}\left(\sum_{k=1}^{n} a_k y_k \right) \right\|^p + \| (Id - K_{\alpha_n})(a_{n+1} y_{n+1}) \|^p \right\}^{1/p} \\
& \quad - \left\{ \| K_{\alpha_n}(a_{n+1} y_{n+1}) \|^p + \left\| (Id - K_{\alpha_n}) \left(\sum_{k=1}^{n} a_k y_k \right) \right\|^p \right\}^{1/p} \\
& \quad \text{by the triangle inequality for the } \ell^p\text{-norm in } \mathbb{R}^2 \\
\underset{(*)(**)}{\geq}\quad & (1-2\varepsilon_n)\left(\left\| \sum_{k=1}^{n} a_k y_k \right\|^p + |a_{n+1}|^p \right)^{1/p} \\
\geq\quad & c_n(1-2\varepsilon_n)\left(\sum_{k=1}^{n+1} |a_k|^p \right)^{1/p} \\
& \quad \text{by induction hypothesis and since } c_n \leq 1 \\
\geq\quad & (1+\varepsilon_n) c_{n+1} \left(\sum_{k=1}^{n+1} |a_k|^p \right)^{1/p}.
\end{aligned}$$

This gives

$$\left\| \sum_{k=1}^{n} a_k y_k \right\| \geq \lambda^{-1} \cdot \left(\sum_{k=1}^{n} |a_k|^p \right)^{1/p}.$$

Thus our claim is proved.
We now complete the proof for $1 < p < \infty$. Let $(x_n^*) \subset X^*$ be a sequence biorthogonal to (y_n) with $\sup \|x_n^*\| < \infty$. By passing to a subsequence of (x_n^*) if necessary, we may suppose that $x_n^* \to x^*$ weakly, recall that X is reflexive (Proposition 5.2). Since $(x_n^* - x^*)$ remains biorthogonal to (y_n), we may even assume that $x_n^* \to 0$ weakly. Now X^* has the (M_{p^*})-property (Proposition 5.2 again), and in passing to a further subsequence we may, by the above, think of (x_n^*) as being equivalent to the unit vector basis of ℓ^{p^*}. But then $\overline{\text{lin}}\,\{y_1, y_2, \ldots\}$ is complemented in X according to [448, Theorem 3.7].
The announced result for $p = \infty$ follows by first producing a λ-isometric copy of c_0 as in the first part, which must be complemented by Corollary III.4.7; recall Proposition 4.4.
(b) Again by Corollary III.4.7, we may suppose that X and, consequently, X^* are separable. As in the proof of part (a) we first extract a subsequence (y_n^*) of the x_n^* which is λ-equivalent to the unit vector basis of ℓ^1. More precisely, our construction yields a sequence of compact operators on X such that $K_n^* \to Id_{X^*}$ strongly and, for the ε_n defined in part (a),

$$\begin{aligned} \|K_n^* y_m^* - y_m^*\| &\leq \varepsilon_n \quad \forall n \geq m, \\ \|K_n^* y_m^*\| &\leq \varepsilon_n \quad \forall m > n. \end{aligned} \tag{$*$}$$

We wish to produce a subsequence of the y_n^* and a corresponding biorthogonal sequence in X^{**} equivalent to the unit vector basis of c_0. Once this is achieved, an appeal to [448, Proposition 3.5] yields the assertion of the present theorem.
We start with any bounded sequence (x_n^{**}) in X^{**} biorthogonal to (y_n^*). Now X^* is separable, hence there is a weak* convergent subsequence, say $x_{n_k}^{**} \to x^{**}$. Since $(x_{n_k}^{**} - x^{**})$ remains biorthogonal to $(y_{n_k}^*)$, which remains λ-equivalent to the unit vector basis of ℓ^1, there is no loss of generality in assuming that $x_n^{**} \to 0$ (weak*) from the outset.
Next we observe that $w^*\text{-}\lim_{j\to\infty}(K_j^{**} x_n^{**} - x_n^{**}) = 0$ for each n, and that, as a result of $(*)$,

$$|\langle K_j^{**} x_n^{**} - x_n^{**}, y_m^* \rangle| \leq \varepsilon_j \|x_n^{**}\| \qquad \forall j > n$$

uniformly in m. It follows that there is a sequence (j_n) of natural numbers such that

$$K_{j_n}^{**} x_n^{**} - x_n^{**} \xrightarrow{w^*} 0 \tag{1}$$

and

$$\left\| (K_{j_n}^{**} x_n^{**} - x_n^{**})|_{\text{lin}\,\{y_1^*, y_2^*, \ldots\}} \right\| \leq 2^{-n}. \tag{2}$$

Let $\xi_n = K_{j_n}^{**} x_n^{**}$ and note that $\xi_n \in X$. Moreover, $\xi_n \to 0$ weakly by (1), and $\inf \|\xi_n\| > 0$ by (2). By part (a), we may suppose that (ξ_n) is λ-equivalent to the unit vector basis of c_0. Finally we apply the Hahn-Banach theorem to obtain a sequence (ξ_n^{**}) such that $\|\xi_n - \xi_n^{**}\| \leq 2^{-n}$ and $\langle \xi_n - x_n^{**}, y_m^* \rangle = \langle \xi_n - \xi_n^{**}, y_m^* \rangle$ for all m. Consequently, the sequence (ξ_n^{**}) is biorthogonal to (y_n^*), and a straightforward calculation shows that it is equivalent to the unit vector basis of c_0.
This completes the proof of Theorem 5.10. □

Let us draw some consequences of the results presented so far.

Corollary 5.11 *If X has (M_p) and $Y \subset X$ is infinite dimensional, then, for every $\varepsilon > 0$, Y contains a $(1+\varepsilon)$-copy of ℓ^p which is complemented in X if $1 < p < \infty$ and of c_0 if $p = \infty$.*

PROOF: If $p < \infty$, then Y is reflexive (Proposition 5.2) and hence, by the Eberlein-Smulyan theorem, contains a normalized weakly null sequence. The assertion now follows from Theorem 5.10. If $p = \infty$ and Y were reflexive, then the same argument would produce a c_0-copy inside Y, which is absurd. Hence Y is a nonreflexive M-embedded space, and the result follows from Corollary III.4.7(e). (For $p = \infty$ one might also invoke Rosenthal's ℓ^1-theorem to show that there exists a normalized weakly null sequence and apply Theorem 5.10 directly.) □

Corollary 5.12

(a) *If X has (M_p) and E is a subspace or quotient with the metric compact approximation property, then E has (M_p).*

(b) *If X and Y have (M_p), then so has $X \oplus_p Y$.*

PROOF: (a) This follows from Theorem 4.19 and the definition of the (M_p)-property.
(b) If (K_α) (resp. (L_β)) are chosen according to Theorem 5.3 for X resp. Y, then

$$\begin{pmatrix} K_\alpha & 0 \\ 0 & L_\beta \end{pmatrix}$$

works for $X \oplus_p Y$. □

Corollary 5.13 *An infinite dimensional Banach space has at most one of the properties (M_p).*

PROOF: This is implied by the fact that the ℓ^p-spaces are totally incomparable [421, Theorem I.2.7] and Corollary 5.11. □

Corollary 5.14 *Let X and Y be infinite dimensional Banach spaces and suppose that X belongs to the (M_{p_1})- and Y to the (M_{p_2})-class, where $1 < p_1, p_2 \le \infty$.*

(a) *If $p_1 \le p_2$, then $K(X,Y)$ is an M-ideal in $L(X,Y)$.*

(b) *If $p_1 > p_2$, then $K(X,Y) = L(X,Y)$.*

PROOF: (a) This is a special case of Corollary 4.18. Alternatively, this follows by performing a calculation similar to the one in the proof of Example 4.1; also recall Proposition 5.6 in the case $p_2 = \infty$.
(b) We first deal with the case $p_1 < \infty$. Assume that $T \in L(X,Y)$ is not compact. Then, by the reflexivity of X (Proposition 5.2), there is a sequence (x_n) in X with $x_n \to 0$ weakly, yet $\|Tx_n\| \ge \alpha > 0$ for all $n \in \mathbb{N}$. Applying Theorem 5.10 we are in a position to suppose that both sequences (x_n) and (Tx_n) are equivalent to the respective unit vector bases of ℓ^{p_1} and ℓ^{p_2}, and we derive that the identity operator maps ℓ^{p_1} continuously into ℓ^{p_2}, which is clearly absurd.
The argument for $p_1 = \infty$ is similar, when we pass to the adjoint operator. □

In the following we will show how our knowledge on the classes (M_p) can be used in order to characterise smooth points of some spaces of bounded operators. Note that the following applies to ℓ^p-spaces and, most particularly, to Hilbert spaces.
Recall that a smooth point x of the unit sphere of a Banach space X is defined by the requirement that $x^*(x) = 1$ for a uniquely determined $x^* \in B_{X^*}$. In this case the norm of X is Gâteaux differentiable at x with derivative x^*.

Theorem 5.15 *Suppose that X and Y are infinite dimensional and separable Banach spaces belonging to the classes (M_p) and (M_q) for some $1 < p, q \leq \infty$. Then T is a smooth point of the unit sphere of $L(X,Y)$ if and only if*

(1) *the essential norm of T, i.e. the number $\|T\|_e = \inf_{K \in K(X)} \|T - K\|$, is strictly less than one,*
(2) *there is exactly one y_0^* in the unit ball of Y^* (up to multiplication with scalars of modulus 1) for which $\|T^*(y_0^*)\| = 1$,*
(3) *the point $T^*y_0^*$ is smooth.*

Before we come to the proof of this theorem, let us give an example. Put $X = Y = c_0$ and represent a norm one operator $T \in L(c_0)$ by a matrix (a_{ij}). Since T^* attains its norm at an extreme point of B_{ℓ^1}, there is, for a smooth point T, exactly one row $(a_{i_0 j})_j$ with $\sum_{j=1}^{\infty} |a_{i_0 j}| = 1$. Furthermore, condition (1) of the above statement is equivalent to the fact that there is $\varepsilon > 0$ with $\sum_{j=1}^{\infty} |a_{ij}| < 1 - \varepsilon$ for all $i \neq i_0$. Finally, a point $(\lambda_n) \in \ell^1$ is smooth iff $\lambda_n \neq 0$ for all n. So we have found:

Corollary 5.16 *A norm one operator $T = (a_{ij})$ on c_0 is a smooth point of the unit sphere if and only if there is $i_0 \in \mathbb{N}$ such that $a_{i_0 j} \neq 0$ for all j, $\sum_{j=1}^{\infty} |a_{i_0 j}| = 1$, and, for some $\varepsilon > 0$, $\sum_{j=1}^{\infty} |a_{ij}| < 1 - \varepsilon$ for all $i \neq i_0$.*

We are going to prepare the proof of Theorem 5.15 by two lemmas. In the following, we write $x + J$ for the equivalence class in X/J generated by x.

Lemma 5.17 *Suppose that J is an M-ideal in the Banach space X and that X/J does not have any smooth points. Then an element x in the unit sphere of X is smooth if and only if*

(1) $\|x + J\| < 1$,
(2) *there is precisely one $j^* \in B_{J^*}$ with $j^*(x) = 1$.*

PROOF: Suppose that x is smooth, has norm one, and that $p(x) = 1$ for one (and only one) $p \in \operatorname{ex} B_{X^*}$. By Lemma I.1.5, there are two possible cases: If $p \in \operatorname{ex} B_{J^\perp}$, then $x + J$ would have norm one and would consequently be a smooth point of the unit ball of X/J. Since this is not possible, we anyway end up with the case where $p \in \operatorname{ex} B_{J^*}$. This proves (2), and the above argument also shows that $|p(x)| < 1$ for all $p \in \operatorname{ex} B_{J^\perp}$, and hence $\|x + J\| < 1$.
Conversely, suppose that the above conditions are satisfied for x. Put $F = \{\varphi \in X^* \mid \varphi(x) = \|\varphi\| = 1\}$. Then X is a weak* closed face of B_{X^*} and, again by Lemma I.1.5,

$$\operatorname{ex} F = (F \cap \operatorname{ex} B_{J^\perp}) \cup (F \cap \operatorname{ex} B_{J^*}).$$

Since $F \cap \operatorname{ex} B_{J^\perp} = \emptyset$ as a result of $\|x + J\| < 1$, we must have $\operatorname{ex} F = \{j^*\}$, and the lemma is proven. □

Lemma 5.18 *Suppose that X and Y are separable and belong to classes (M_p) and (M_q) for some $1 < p, q \leq \infty$. Then $L(X,Y)/K(X,Y)$ has no smooth point.*

PROOF: By Corollary 5.14, $K(X,Y)$ is an M-ideal in $L(X,Y)$. As a consequence, $K(X,Y)$ is proximinal in $L(X,Y)$ (Proposition II.1.1), and we may suppose that for any fixed $T \in L(X,Y)$ with $\|T\|_e = 1$ we have $\|T\| = 1$ as well. (Note that the essential norm of T is the norm of the equivalence class $T + K(X,Y)$.) Furthermore, by Proposition 4.7 and separability, we find sequences (x_n) in S_X or (y_n^*) in S_{Y^*} converging to zero in the weak or, respectively, weak* topology, such that $\lim_n \|Tx_n\| = 1$ or $\lim_n \|T^* y_n^*\| = 1$. Since Y^* either belongs to the class (M_{p^*}), where, as usual $1/p + 1/p^* = 1$, or is the dual of a space belonging to (M_∞), the following argument works in either of the above cases, and we may suppose that $\lim_n \|Tx_n\| = 1$. Clearly, (x_n) cannot converge in norm and therefore, by Theorem 5.10, we may think of (x_n) as being equivalent to the standard basis of ℓ^p or c_0 and such that there is a projection P_0 from X onto $\overline{\text{lin}}\,\{x_n \mid n \in \mathbb{N}\}$. Let $P_1 : \overline{\text{lin}}\,\{x_n \mid n \in \mathbb{N}\} \to \overline{\text{lin}}\,\{x_{2n} \mid n \in \mathbb{N}\}$ be the canonical projection and put $P = P_1 P_0$. Then,

$$d(TP, \text{lin}\,\{T(Id - P)\} \cup K(X,Y)) \leq \|TP + T(Id - P)\| = 1.$$

On the other hand, for each scalar λ and any $K \in K(X,Y)$,

$$\|TP - (\lambda T(Id - P) + K)\| \geq \limsup_n \|Tx_{2n} - Kx_{2n}\| = 1$$

and therefore, $d(TP, \text{lin}\,\{T(Id - P)\} \cup K(X,Y)) = 1$. Similarly, we find that $d(T(Id - P), \text{lin}\,\{TP\} \cup K(X,Y)) = 1$. To conclude the proof, choose functionals ψ_1, ψ_2 of unit norm with

$$1 = \psi_1(TP) = \psi_2(T(Id - P))$$

and

$$0 = \psi_2|_{\text{lin}\,\{TP\} \cup K(X,Y)} = \psi_1|_{\text{lin}\,\{T(Id - P)\} \cup K(X,Y)}.$$

Then $\psi_1(T) = \psi_2(T) = 1$ and $\psi_1 \neq \psi_2$ whence $T + K(X,Y)$ cannot be smooth. □

PROOF OF THEOREM 5.15:
By Lemmas 5.17 and 5.18, an operator T is smooth if and only if $\|T\|_e < \|T\|$ and there is exactly one functional ψ in $B_{K(X,Y)^*}$ norming T. Necessarily, ψ is an extreme point of the dual unit ball of $K(X,Y)$. The result then follows from the fact that, by Theorem 1.3, ψ is of the form $x^{**} \otimes y^*$, where $x^{**} \in \text{ex}\, B_{X^{**}}$ and $y^* \in \text{ex}\, B_{Y^*}$. □

We finally use the results of this section to disprove the M-ideal property of $K(X)$ in some instances.

Proposition 5.19

(a) *Let $1 \leq p \leq \infty$. Then $K(X)$ is not an M-ideal in $L(X)$ if $p \neq 2$ and*
  (1) $X = L^p[0,1]$,
  (2) $X = c_p$, *the Schatten class,*
  (3) $X = H^p$, *the p-th Hardy space.*

(b) *$K(X)$ is an M-ideal in $L(X)$ for an L^1-predual space X if and only if $X = c_0(I)$ for some index set I.*

PROOF: (a) The borderline cases $p = 1$ and $p = \infty$ are handled by Proposition 4.4 (except for c_∞). So let $1 < p < \infty$. To prove (1) it is enough to rule out that X have (M_p), since $X \cong X \oplus_p X$. In fact, X has a hilbertian subspace (as a consequence of the Khintchin inequalities, see e.g. [422, Th. 2.b.3]) so that the conclusion of Corollary 5.11 fails.

Unfortunately, in (2) $X \cong X \oplus_p X$ is false. However, c_p has a 1-complemented subspace Y isometric to $\ell^2 \oplus_p \ell^2$ for $1 < p \leq \infty$ [31]. If $K(X)$ were an M-ideal in $L(X)$ then $K(Y)$ would be an M-ideal in $L(Y)$ (Proposition 4.2), and Corollary 5.11 furnishes the contradiction that ℓ^2 should contain a copy of ℓ^p.

The proof of part (3) does not depend on the theory of (M_p)-spaces, but uses Corollary 4.23. Let us suppose that $K(H^p)$ is an M-ideal in $L(H^p)$ for some p. Let $\varepsilon > 0$. Using Corollary 4.23 embed H^p isometrically into a Banach space X with a $(1+\varepsilon)$-unconditional basis (e_n). Denote by (e_n^*) the corresponding coefficient functionals and pick $N \in \mathbb{N}$ such that $\left\|\sum_{k>N}\langle 1, e_k^*\rangle e_k\right\| \leq \varepsilon$. Now consider the functions $f_n(z) = z^n + z^{2n}$ and note that $f_n \to 0$ weakly by the Riemann-Lebesgue lemma. Hence there is $n_0 \in \mathbb{N}$ such that $\left\|\sum_{k=1}^{N}\langle f_n, e_k^*\rangle e_k\right\| \leq \varepsilon$ for $n \geq n_0$. In the following calculation we use the symbol $a \approx b$ to indicate that a and b differ by a term which tends to 0 as $\varepsilon \to 0$. We have for $n \geq n_0$

$$\begin{aligned} \|1 + f_n\| &\approx \left\| \sum_{k=1}^{N} \langle 1, e_k^*\rangle e_k + \sum_{k>N} \langle f_n, e_k^*\rangle e_k \right\| \\ &\approx \left\| \sum_{k=1}^{N} \langle 1, e_k^*\rangle e_k - \sum_{k>N} \langle f_n, e_k^*\rangle e_k \right\| \\ &\approx \|1 - f_n\|, \end{aligned}$$

where we used that (e_n) is $(1+\varepsilon)$-unconditional. Since $f_n(z) = f_1(z^n)$, the f_n are identically distributed and thus $\|1 \pm f_n\|_p = \|1 \pm f_1\|_p$ for all n. Consequently $\|1 + f_1\|_p = \|1 - f_1\|_p$ which enforces $p = 2$.

(b) This is an immediate consequence of Proposition 4.4, Proposition III.2.7 and of course Corollary 5.4. □

A modification of the above argument for H^p shows that $K(C(\mathbb{T})/A)$ is not an M-ideal in $L(C(\mathbb{T})/A)$. Indeed, otherwise, by a strengthening of Theorem 4.21 proved in [397], C/A is isometric to a $(1+\varepsilon)$-complemented subspace of a space with a $(1+\varepsilon)$-unconditional *shrinking* finite dimensional Schauder decomposition, and thus $H_0^1 = (C/A)^*$ is isometric to a subspace of a space with a $(1+\varepsilon)$-unconditional basis, for every $\varepsilon > 0$. A reasoning similar to the above shows that this is impossible. (Note that C/A and H^1 have unconditional bases [641].)

We shall show in the next section that L^p and c_p cannot even be renormed so that the compact operators become an M-ideal. Since L^p is isomorphic to H^p for $1 < p < \infty$ [423, Prop. 2.c.17], this result will contain (3) of part (a) as a special case.

VI.6 Banach spaces for which $K(X)$ forms an M-ideal: ℓ^p-subspaces and renormings

We now return to the investigation of the general case of Banach spaces X for which $K(X)$ is an M-ideal in $L(X)$. The main result of the first part of this section will be that those spaces X necessarily contain good copies of ℓ^p for some $p > 1$ or of c_0 (Theorem 6.4). In the second part we will treat the problem of renorming a Banach space X so that $K(X)$ becomes an M-ideal. For all this the equivalences proved in Theorem 4.17 and in particular the property (M) (Definition 4.12) are of crucial importance.
In this section we will exclusively deal with separable Banach spaces, in which case it suffices to work with sequences in Definition 4.12 and Theorem 4.17 as an inspection of its proof shows.
We first present a new class of Banach spaces for which the compact operators form an M-ideal. These spaces will be useful throughout this section and will eventually be identified as isomorphs of Orlicz sequence spaces (Proposition 6.11). Let N be a norm on $\mathbb{K}^2$ such that

$$N(1,0) = 1 \qquad \text{and} \qquad N(\alpha,\beta) = N(|\alpha|,|\beta|) \quad \forall \alpha,\beta \in \mathbb{K};$$

i.e., N is an absolute norm. Define inductively norms on $\mathbb{K}^d$ by

$$N(\xi_0,\ldots,\xi_d) = N(N(\xi_0,\ldots,\xi_{d-1}),\xi_d)$$

and denote by $\widetilde{\Lambda}(N)$ the vector space of all sequences $\xi = (\xi_0,\xi_1,\ldots)$ such that

$$\|\xi\|_N = \sup_d N(\xi_0,\ldots,\xi_d) < \infty.$$

Equipped with the norm $\|\,.\,\|_N$, $\widetilde{\Lambda}(N)$ is a Banach space. We denote by $\Lambda(N)$ the closed linear span of the unit vectors $e_0, e_1, e_2, \ldots$.

Proposition 6.1

(a) *The unit vectors form a 1-unconditional basis of* $\Lambda(N)$.

(b) *If* (e_{n_k}) *is a subsequence of* (e_n), *then* $\overline{\text{lin}}\,\{e_{n_k} \mid k \in \mathbb{N}\}$ *is canonically isometric with* $\Lambda(N)$.

(c) $\Lambda(N)$ *has property* (M).

(d) $\Lambda(N)$ *is an* M*-ideal in* $\widetilde{\Lambda}(N)$.

(e) *If* $\Lambda(N)^*$ *is separable, then* $K(\Lambda(N))$ *is an* M*-ideal in* $L(\Lambda(N))$.

PROOF: (a), (b) and (c) follow directly from the definition of the norm in $\Lambda(N)$; note that the finitely supported sequences are dense. (d) is an immediate consequence of the 3-ball property (Theorem I.2.2). Finally, (e) is implied by Theorem 4.17(vi): The coordinate projections P_n fulfill $\|Id - 2P_n\| = 1$ by (a), and $P_n^* \to Id_{\Lambda(N)^*}$ strongly (i.e., the unit vector basis is shrinking), since $\Lambda(N)^*$ is separable, see [422, Th. 1.c.9]. □

We now address the problem of ℓ^p-subspaces of Banach spaces for which $K(X)$ is an M-ideal in $L(X)$. Let us introduce some terminology. A *type* on a Banach space X is a function of the form $\tau(x) = \lim_{n\to\infty} \|x + x_n\|$ for some bounded sequence (x_n). We say that (x_n) generates the type τ. If $x_n \to 0$ weakly, τ is called a *weakly null type*, and τ is called nontrivial if (x_n) does not converge strongly. A diagonal argument shows that every bounded sequence in a separable Banach space contains a subsequence generating a type.

In a Banach space with property (M) a weakly null type is a function of $\|x\|$. Consequently, if (x_n) generates a weakly null type τ on such a space and $\|x\| = 1$, then $N(\alpha,\beta) = \lim_{n\to\infty} \|\alpha x + \beta x_n\|$ is a well-defined absolute seminorm on $\mathbb{K}^2$ which does not depend on the particular choice of $x \in S_X$; and it is a norm if τ is nontrivial.

Lemma 6.2 *Suppose X is a separable Banach space with property (M) and (x_n) is a weakly null sequence generating a nontrivial type. Let $N(\alpha,\beta) = \lim_{n\to\infty} \|\alpha x + \beta x_n\|$ for some $x \in S_X$. Then, whenever $\|u\| = 1$ and $\varepsilon > 0$, there is a subsequence (y_n) of (x_n) such that*

$$(1+\varepsilon)^{-1} \left\| \xi_0 u + \sum_{i=1}^{\infty} \xi_i y_i \right\| \le \|\xi\|_N \le (1+\varepsilon) \left\| \xi_0 u + \sum_{i=1}^{\infty} \xi_i y_i \right\|$$

for all finitely nonzero sequences $\xi = (\xi_0, \xi_1, \ldots)$. In particular, there is an absolute norm N such that X contains, for every $\varepsilon > 0$, a subspace which is $(1+\varepsilon)$-isomorphic to $\Lambda(N)$.

PROOF: We just sketch the construction of the y_n which proceeds inductively. It is clearly enough to gain the desired inequalities for $|\xi_i| \le 1$. In the first step we pick an index n_1 such that $\|\xi_0 u + \xi_1 x_{n_1}\|$ is close to $N(\xi_0,\xi_1)$ for all $|\xi_0|, |\xi_1| \le 1$. This is possible by definition of N. (The Arzelà-Ascoli theorem takes care of the uniform, not only pointwise approximation.) Let $y_1 = x_{n_1}$. Then pick an index $n_2 > n_1$ such that $\|\xi_0 u + \xi_1 y_1 + \xi_2 x_{n_2}\|$ is close to $N(N(\xi_0,\xi_1),\xi_2) = N(\xi_0,\xi_1,\xi_2)$. Let $y_2 = x_{n_2}$, etc. The strategy how to prove this lemma should now be clear. □

Lemma 6.3 *Let N be an absolute norm on $\mathbb{K}^2$ with $N(1,0) = 1$. Then there is some p, $1 \le p \le \infty$, such that, for every $\varepsilon > 0$, $\Lambda(N)$ contains a subspace $(1+\varepsilon)$-isomorphic to ℓ^p if $p < \infty$ and to c_0 if $p = \infty$.*

PROOF: The proof of this result relies on a difficult theorem due to Krivine ([383, Th. 0.1] and [392]) which asserts in our case that, given $\delta > 0$ and $n \in \mathbb{N}$, some normalised blocks $b_1, \ldots, b_n$ of the unit vector basis of $\Lambda(N)$ with increasing supports are $(1+\delta)$-equivalent to the standard basis in $\ell^p(n)$. It follows that there is a normalised block basic sequence (u_n) satisfying

$$\lim_{n\to\infty} \|\alpha u_{n-1} + \beta u_n\| = \|(\alpha,\beta)\|_p \quad \forall \alpha,\beta \in \mathbb{K}.$$

But since $\Lambda(N)$ has property (M) by Proposition 6.1, any finitely supported ξ lies eventually "left of" u_{2n}; hence by definition of the norm in $\Lambda(N)$

$$\lim_{n\to\infty} \|\alpha\xi + \beta u_n\| = \|(\alpha,\beta)\|_p \quad \forall \alpha,\beta \in \mathbb{K}$$

whenever $\|\xi\| = 1$. It follows from Lemma 6.2 that some subsequence of (u_{2n}) spans a $(1+\varepsilon)$-copy of ℓ^p resp. c_0. □

We now come to the first major result of this section.

Theorem 6.4 *Let X be a separable Banach space with property (M) and Y a closed infinite dimensional subspace of X. Then there is some p, $1 \leq p \leq \infty$, such that, for every $\varepsilon > 0$, Y contains a subspace $(1+\varepsilon)$-isomorphic to ℓ^p if $p < \infty$ and to c_0 if $p = \infty$. In particular, this conclusion holds if $K(X)$ is an M-ideal in $L(X)$ and Y is an infinite dimensional subspace of a quotient of X, and then necessarily $1 < p \leq \infty$.*

PROOF: Y has property (M), too. If every weakly null sequence in Y converges strongly, i.e., Y has the Schur property, then Y contains a $(1+\varepsilon)$-copy of ℓ^1 by Rosenthal's ℓ^1-theorem and James' distortion theorem [422, Th. 2.e.5 and Prop. 2.e.3]. Otherwise, there is a normalised weakly null sequence in Y, and the desired conclusion follows from Lemma 6.2 and Lemma 6.3.
If $K(X)$ is an M-ideal, then every subspace of a quotient of X has (M), see the proof of Theorem 4.19, and $p = 1$ is ruled out by Corollary 4.5. □

We will later need the following corollary.

Corollary 6.5 *Let X be a separable Banach space with property (M) not containing a copy of ℓ^1. Then there exists p, $1 < p \leq \infty$, and a weakly null sequence (x_n) such that whenever $\|u\| = 1$,*

$$\lim_{n\to\infty} \|\alpha u + \beta x_n\| = \|(\alpha, \beta)\|_p \qquad \forall \alpha, \beta \in \mathbb{K}.$$

PROOF: Choose p according to Theorem 6.4, and let E be a subspace of X isomorphic to ℓ^p (resp. c_0). Applying Theorem 6.4 again we obtain that for each n, E contains a further subspace E_n which is $(1+\varepsilon_n)$-isomorphic to ℓ^p (resp. c_0), where $\varepsilon_n \to 0$ is arbitrary. Consequently each E_n contains a weakly null sequence $(x_{nk})_k$ which is $(1+\varepsilon_n)$-equivalent to the unit vector basis of ℓ^p (resp. c_0) and generates a nontrivial weakly null type. Using the assumption that X has (M) we deduce from this that for every $u \in X$ with $\|u\| = 1$,

$$(1+\varepsilon_n)^{-1}\|(\alpha,\beta)\|_p \leq \lim_{k\to\infty} \|\alpha u + \beta x_{nk}\| = (1+\varepsilon_n)\|(\alpha,\beta)\|_p$$

for all $\alpha, \beta \in \mathbb{K}$. We then obtain (x_n) by passing to a suitable diagonal subsequence (x_{n,k_n}). □

We are now in a position to single out the (M_p)-spaces, studied in the previous section, among those Banach spaces for which $K(X)$ is an M-ideal in $L(X)$.
In [384] the class of separable stable Banach spaces was introduced, see also [243]. A separable Banach space X is called stable if, whenever (x_m) and (y_n) are bounded sequences generating types, then the iterated limits $\lim_n \lim_m \|x_m + y_n\|$ and $\lim_m \lim_n \|x_m + y_n\|$ both exist and coincide. Among the stable Banach spaces are the spaces ℓ^r and L^r for $1 \leq r < \infty$ and their subspaces while c_0 is not stable [384]. Also, the quotient spaces of ℓ^r are stable for $1 < r < \infty$ [520]. We remind the reader that the stability of a Banach space is an isometric notion that will generally be spoilt by passing to an equivalent norm. It is a celebrated result due to Krivine and Maurey [384] that stable Banach spaces contain arbitrarily good copies of ℓ^p for some $p \in [1, \infty)$.

Theorem 6.6 *Let X be a separable Banach space such that $K(X)$ is an M-ideal in $L(X)$. Then the following are equivalent:*

(i) *X has property (M_p) for some $1 < p < \infty$.*
(ii) *X is stable.*
(iii) *There exists p, $1 < p < \infty$, such that for every normalised weakly null sequence (x_n) generating a weakly null type and every $u \in X$ with $\|u\| = 1$,*

$$\lim_{n\to\infty} \|\alpha u + \beta x_n\| = (|\alpha|^p + |\beta|^p)^{1/p}.$$

(iv) *There exists p, $1 < p < \infty$, such that for every weakly null sequence (x_n) and every $u \in X$,*

$$\limsup \|u + x_n\|^p = \|u\|^p + \limsup \|x_n\|^p.$$

PROOF: (i) $\Rightarrow$ (ii): To begin with, we make the following observation. Let (K_i) be as in Theorem 5.3. (By the way, since X is separable, this net may be chosen to be a sequence.) We fix an index i for which (2a) and (2b) from that theorem are fulfilled. Now suppose x_0 and y_0 are given such that

$$\begin{aligned} \|K_i x_0 - x_0\| &\leq \varepsilon, \\ \|K_i y_0\| &\leq \varepsilon. \end{aligned}$$

Then

$$\big|\ \|K_i x_0 + (Id - K_i)y_0\| \ - \ (\|x_0\|^p + \|y_0\|^p)^{1/p}\ \big| \ \leq \ f(\varepsilon) \tag{1}$$

where $f(\varepsilon)$ is a quantity which tends to 0 as ε tends to 0. ($f(\varepsilon)$ depends on M if M is an upper bound for $\|x_0\|$ and $\|y_0\|$.)
To prove this claim, we note for $z_i = K_i x_0 + (Id - K_i)y_0$ from (2b) that

$$\|z_i\| \geq \frac{1}{1+\varepsilon}(\|K_i z_i\|^p + \|(Id - K_i)z_i\|^p)^{1/p}$$

and that

$$\begin{aligned} \|K_i z_i - x_0\| &\leq 4\varepsilon \\ \|(Id - K_i)z_i - y_0\| &\leq 4\varepsilon \end{aligned}$$

by the triangle inequality so that (1) follows as a consequence of this and of (2a).
We now enter the main part of the proof. Thus, let (x_m) and (y_n) be given as above. Since X is reflexive, we may assume that (x_m) and (y_n) are weakly convergent (cf. [384, p. 276]), say

$$x_m \xrightarrow{w} \xi \qquad \text{and} \qquad y_n \xrightarrow{w} \eta.$$

Let $m \in \mathbb{N}$. Given $\varepsilon > 0$, fix i such that (2a) and (2b) are fulfilled for i and such that

$$\|K_i(x_m + \eta) \ - \ (x_m + \eta)\| \leq \varepsilon$$

holds. Then find n_0 with

$$\|K_i(y_n - \eta)\| \leq \varepsilon \qquad \text{for all } n \geq n_0.$$

It follows for these n

$$\big|\ \|x_m + y_n\| \ - \ \|K_i(x_m + \eta) \ + \ (Id - K_i)(y_n - \eta)\|\ \big| \leq 2\varepsilon,$$

hence by (1)

$$\lim_n \|x_m + y_n\| = (\|x_m + \eta\|^p + \beta^p)^{1/p}$$

where $\beta = \lim_n \|y_n - \eta\|$. Likewise,

$$\lim_m \|x_m + \eta\| = (\alpha^p + \|\xi + \eta\|^p)^{1/p}$$

(with $\alpha = \lim_m \|x_m - \xi\|$) so that by symmetry

$$\begin{aligned}\lim_m \lim_n \|x_m + y_n\| &= (\alpha^p + \|\xi + \eta\|^p + \beta^p)^{1/p} \\ &= \lim_n \lim_m \|x_m + y_n\|.\end{aligned}$$

(ii) $\Rightarrow$ (iii): Since X does not contain a copy of ℓ^1, we first mention that X must be reflexive; otherwise, X contains arbitrarily good copies of c_0 (Corollary III.3.7), and the conclusion that c_0 is stable would follow. Corollary 6.5 provides us with a normalised weakly null sequence (y_n) such that, for some $1 < p < \infty$ and all $\|u\| = 1$,

$$\lim_{n\to\infty} \|\alpha u + \beta y_n\| = \|(\alpha, \beta)\|_p \qquad \forall \alpha, \beta \in \mathbb{K}.$$

Now let (x_n) be a normalised weakly null sequence generating a nontrivial type. Since X has (M) by assumption, we get for $\|u\| = 1$

$$\begin{aligned}\lim_{m\to\infty} \|\alpha u + \beta x_m\| &= \lim_{n\to\infty} \lim_{m\to\infty} \|\alpha y_n + \beta x_m\| \\ &= \lim_{m\to\infty} \lim_{n\to\infty} \|\alpha y_n + \beta x_m\| \\ &= (|\alpha|^p + |\beta|^p)^{1/p}.\end{aligned}$$

(iii) $\Leftrightarrow$ (iv): This follows by a simple subsequence argument.
(iii) $\Rightarrow$ (i): We shall verify condition (vi) of Theorem 4.17 for $X \oplus_p X$. If (x_n) and (y_n) generate nontrivial weakly null types on X, then

$$\begin{aligned}\lim_{n\to\infty} \|(x, y) + (x_n, y_n)\|^p &= \lim_{n\to\infty} (\|x + x_n\|^p + \|y + y_n\|^p) \\ &= \|(x, y)\|^p + \lim_{n\to\infty} \|x_n + y_n\|^p\end{aligned}$$

by assumption. Hence $X \oplus_p X$ has (M) by Lemma 4.13(ii). Further, if (K_n) is a shrinking compact approximation of the identity on X with $\lim \|Id - 2K_n\| = 1$, then $K_n \oplus K_n$ serves the same purpose on $X \oplus_p X$. □

To cover the case $p = \infty$ one has to consider the class of weakly stable Banach spaces [32], defined by the requirement that

$$\lim_n \lim_m \|x_m + y_n\| = \lim_m \lim_n \|x_m + y_n\|$$

whenever (x_n) and (y_n) generate types and are weakly convergent. The above proof then carries over to include the case of (M_∞)-spaces.
We now investigate renormings. We first present a sufficient condition for $K(X, Y)$ to be an M-ideal in $L(X, Y)$.

Lemma 6.7 *Let X and Y be separable Banach spaces and let $1 < q \le p < \infty$. Suppose X and Y admit shrinking compact approximations of the identity (K_n) resp. (L_n) such that*

$$\limsup(\|K_n x\|^q + \|(Id - K_n)x\|^q)^{1/q} \le \|x\|, \tag{1}$$

$$\limsup \|L_n y_1 + (Id - L_n)y_2\| \le (\|y_1\|^p + \|y_2\|^p)^{1/p}, \tag{2}$$

where the $\limsup$ *conditions are supposed to hold uniformly on bounded sets. Then $K(X,Y)$ is an M-ideal in $L(X,Y)$.*

PROOF: The proof is the same as in (the easy half of) Theorem 5.3. □

The lemma extends to $p = \infty$ in the obvious way.
To discuss the scope of this lemma, we introduce the temporary notation that a space fulfilling the assumptions made on X resp. Y above has the *lower q-* resp. *upper p-property.* We first observe that a reflexive space X has the lower q-property if and only if X^* has the upper q^*-property where $1/q + 1/q^* = 1$. This follows since

$$(\|K_n x\|^q + \|x - K_n x\|^q)^{1/q} \le (1+\varepsilon)\|x\| \qquad \forall x \in X \tag{3}$$

is equivalent to

$$\|K_n^* x^* + (Id - K_n^*)y^*\| \le (1+\varepsilon)(\|x^*\|^{q^*} + \|y^*\|^{q^*})^{1/q^*} \qquad \forall x^* \in X^*. \tag{4}$$

In fact, (3) means that the operator

$$x \mapsto (K_n x, (Id - K_n)x)$$

from X to $X \oplus_q X$ has norm $\le 1+\varepsilon$, and (4) expresses the fact that its adjoint has norm $\le 1+\varepsilon$.
Thus, in order to find examples where Lemma 6.7 applies we need only worry about the upper p-property; examples with the lower q-property can then be constructed by duality. One can also show, using the convex combinations technique, that the upper p-property is inherited by subspaces and quotients with the compact approximation property. Moreover, a computation reveals that the ℓ^r-sum of spaces $Y_1, Y_2, \dots$ with the upper $p_1, p_2, \dots$-properties enjoys the upper p-property provided $p \le \inf\{r, p_1, p_2, \dots\}$.
An effective way to produce upper p-spaces is to look for reflexive sequence spaces where the unit vectors form a Schauder basis and the inequality

$$\|x_1 + x_2\| \le (\|x_1\|^p + \|x_2\|^p)^{1/p}$$

holds for disjointly supported sequences. It is clear that under this hypothesis the sequence of coordinate projections is an approximation of the identity with the required features. Examples include besides the ℓ^p-spaces the Lorentz spaces $d(w,p)$ and, more generally, the p-convexification of a sequence space whose unit vector basis is 1-unconditional; in particular, the p-convexified Tsirelson space $T^{(p)}$ [115, p. 116] can be considered. (Of course, finite dimensional decompositions can be considered instead of bases.)
It follows for instance that $K(\ell^2, T^{(2)})$ is an M-ideal in $L(\ell^2, T^{(2)})$. On the other hand, the compact operators on $T^{(2)}$ do not form an M-ideal in $L(T^{(2)})$ since $T^{(2)}$ does not contain

any ℓ^p. In this respect the operators on ℓ^2 behave very differently from the operators on the "weak Hilbert space" [502] $T^{(2)}$. We refer to [115] for exhaustive information on Tsirelson-like spaces and for a summary of [502].
Lemma 6.7 also enables us to provide an answer to the question if the L^p-spaces can be renormed so that $K(L^p, L^q)$ becomes an M-ideal in $L(L^p, L^q)$. (Recall from Proposition 5.19 that $K(L^p)$ is not an M-ideal in $L(L^p)$ for the natural norm of $L^p = L^p[0,1]$, and see also Corollary 6.10 below.) We will be able to answer this affirmatively if $1 < p \le 2 \le q < \infty$.

Proposition 6.8 *If $1 < p \le 2 \le q < \infty$, then for some renormings X of $L^p[0,1]$ and Y of $L^q[0,1]$, $K(X,Y)$ is an M-ideal in $L(X,Y)$.*

PROOF: Indeed, under these circumstances the Haar system is an unconditional basis, and we first renorm L^p and L^q so that the Haar basis (h_i) becomes 1-unconditional. Call these renormings X_1 and Y_1. Letting

$$\sum_{i=1}^{\infty} a_i h_i \succeq 0 \qquad \text{if and only if} \qquad a_i \ge 0 \ \forall i$$

induces the structure of a Banach lattice on each of X_1 and Y_1, and since Y_1 is isomorphic to L^q and $2 \le q < \infty$, Y_1 has type 2. Hence, the Banach lattice Y_1 can further be renormed to satisfy

$$\left\| \sum_{i=1}^{\infty} a_i h_i \right\| \le \left(\left\| \sum_{i=1}^{n} a_i h_i \right\|^2 + \left\| \sum_{i=n+1}^{\infty} a_i h_i \right\|^2 \right)^{1/2} \qquad \forall n \in \mathbb{N}$$

(cf. [423, p. 100 and Lemma 1.f.11]). Likewise, since L^p has cotype 2 for $p \le 2$ there is a renorming such that

$$\left(\left\| \sum_{i=1}^{n} a_i h_i \right\|^2 + \left\| \sum_{i=n+1}^{\infty} a_i h_i \right\|^2 \right)^{1/2} \le \left\| \sum_{i=1}^{\infty} a_i h_i \right\| \qquad \forall n \in \mathbb{N}.$$

If these renormings are called X resp. Y, then $K(X,Y)$ is an M-ideal in $L(X,Y)$ by Lemma 6.7. □

We refrain from formulating the abstract lemma behind Proposition 6.8 which involves reflexive Banach spaces with unconditional bases and type 2 resp. cotype 2.
We next wish to prove that unless $p = 2$ there is no renorming X of $L^p[0,1]$ for which $K(X)$ is an M-ideal in $L(X)$. This will turn out to be a corollary to the following result.

Proposition 6.9 *The space $\ell^p(\ell^r)$ for $1 < p, r < \infty$ has a renorming X so that $K(X)$ is an M-ideal in $L(X)$ if and only if $p = r$. In fact, $\ell^p(\ell^r)$ can be renormed to have (M) if and only if $p = r$.*

PROOF: The "if"-part follows from Example 4.1. Now suppose $\ell^p(\ell^r)$ has an equivalent renorming $X = (\ell^p(\ell^r), \|\,.\,\|_M)$ for which $K(X)$ is an M-ideal in $L(X)$. A fortiori, X

has (M) in this case. Let X_k be the space of all sequences $(x_n) \in X$ such that $x_n = 0$ if $n \neq k$. Then $(X_k, \|\,.\,\|_M)$ is isomorphic to ℓ^r and, since $(X, \|\,.\,\|_M)$ has (M), we find by Corollary 6.5 normalised weakly null sequences $(u_{kn})_n \subset X_k$ such that

$$\lim_{n\to\infty} \|x + \beta u_{kn}\|_M = (\|x\|^r + |\beta|^r)^{1/r} \qquad \forall x \in X,\ \beta \in \mathbb{K}.$$

Thus for all $\alpha_1, \dots, \alpha_k$,

$$\lim_{n_1\to\infty} \dots \lim_{n_k\to\infty} \left\| \sum_{i=1}^{k} \alpha_i u_{in_i} \right\|_M = \left(\sum_{i=1}^{k} |\alpha_i|^r \right)^{1/r}.$$

But the left hand side is equivalent to $\left\|\sum_{i=1}^{k} \alpha_i u_{in_i}\right\|_{\ell^p(\ell^r)} = \left(\sum_{i=1}^{k} |\alpha_i|^p\right)^{1/p}$; and we get a contradiction unless $p = r$. □

The following corollary is the isomorphic version of Proposition 5.19. Recall that c_p denotes the Schatten class of operators on Hilbert space of index p.

Corollary 6.10 *Let $1 \le p \le \infty$.*

(a) *$L^p = L^p[0,1]$ has a renorming X such that $K(X)$ is an M-ideal in $L(X)$ if and only if $p = 2$.*
(b) *c_p has a renorming X such that $K(X)$ is an M-ideal in $L(X)$ if and only if $p = 2$.*

PROOF: Both L^2 and c_2 are Hilbert spaces so that the compact operators form an M-ideal. Suppose now that $1 < p < \infty$ and that L^p has a renorming as stated. Since there are isomorphic embeddings (the first one results from the Khintchin inequality)

$$\ell^p(\ell^2) \hookrightarrow \ell^p(L^p) \hookrightarrow L^p(L^p) \cong L^p,$$

we deduce from Theorem 4.17 that $\ell^p(\ell^2)$ has a renorming satisfying (M). Now Proposition 6.9 yields $p = 2$. The argument for c_p is the same, since $\ell^p(\ell^2) \hookrightarrow c_p$ [31].
For c_∞ we observe that $c_0(\ell^2)$ embeds into c_∞, and if c_∞ had a renorming for which the compact operators are an M-ideal, then $c_0(\ell^2)$ could be renormed to have (M). An argument similar to the one in Proposition 6.9 shows that this is impossible.
For L^1, L^∞ and c_1 the same arguments as in Proposition 5.19 apply. □

We finally come to a positive result and resume the discussion of the spaces $\Lambda(N)$ introduced at the beginning of this section. Recall from Section III.1 the definition of the Orlicz sequence spaces h_M and ℓ_M.

Proposition 6.11

(a) *Every space $\widetilde{\Lambda}(N)$ (resp. $\Lambda(N)$) is isomorphic to an Orlicz sequence space ℓ_M (resp. h_M).*
(b) *Every Orlicz sequence space ℓ_M (resp. h_M) is isomorphic to a space $\widetilde{\Lambda}(N)$ (resp. $\Lambda(N)$) for some absolute norm N on $\mathbb{K}^2$.*

PROOF: (a) Let N be an absolute norm on $\mathbb{K}^2$ such that $N(1,0) = 1$. Define $M(t) = N(1,t) - 1$ for $t \geq 0$. Clearly, this is a continuous, increasing, convex function; however, it might be degenerate, meaning $M(t) = 0$ for some $t > 0$. We claim that $\widetilde{\Lambda}(N) = \ell_M$ as sets. A standard closed graph argument then shows that the two spaces are actually (canonically) isomorphic, and since $\Lambda(N)$, resp. h_M, is the closed linear span of the unit vectors, these spaces are canonically isomorphic, too.
We start with proving that $\widetilde{\Lambda}(N) \subset \ell_M$. Suppose that $\|\xi\|_N \leq 1$. Then $N(\xi_0, \dots, \xi_k) \leq 1$ for all k; note that this expression increases with k. For simplicity of notation assume $\xi_0 \neq 0$ (otherwise switch to the smallest index k_0 where $\xi_{k_0} \neq 0$) so that $0 < N(\xi_0, \dots, \xi_k) \leq 1$ for all k. For convenience we put $N(\xi_0) = |\xi_0|$. Then we obtain for all $k \geq 1$

$$\begin{aligned} N(\xi_0, \dots, \xi_k) &= N(N(\xi_0, \dots, \xi_{k-1}), \xi_k) \\ &= N(\xi_0, \dots, \xi_{k-1}) N\Big(1, \frac{\xi_k}{N(\xi_0, \dots, \xi_{k-1})}\Big) \\ &\geq N(\xi_0, \dots, \xi_{k-1}) N(1, \xi_k) \\ &= N(\xi_0, \dots, \xi_{k-1})(1 + M(|\xi_k|)). \end{aligned}$$

Consequently, $N(\xi_0, \dots, \xi_k) \geq \prod_{j=1}^k (1 + M(|\xi_j|))|\xi_0|$ for each k, and $\prod_{j=1}^\infty (1 + M(|\xi_j|)$ converges. Hence $\sum_{j=1}^\infty M(|\xi_j|)$ converges, and $\xi \in \ell_M$.
Let us now suppose for contradiction that $\widetilde{\Lambda}(N) \neq \ell_M$. Then there is a sequence $\xi = (\xi_j)$ with $\sum M(|\xi_j|) < \infty$, yet $\sup_k N(\xi_0, \dots, \xi_k) = \infty$.

CLAIM: $\quad N\Big(1, \dfrac{\xi_k}{N(\xi_0, \dots, \xi_{k-1})}\Big) > 1 + M(|\xi_k|)$ infinitely often.

Indeed, otherwise we would have "$\leq$" for $k \geq K$ instead. It would follow for those k, as in the first part of the proof, that

$$\sup_{k \geq K} N(\xi_0, \dots, \xi_k) \leq \prod_{j=K+1}^{\infty} (1 + M(|\xi_j|)) N(\xi_0, \dots, \xi_K) < \infty,$$

contradicting the choice of ξ. Thus, the claim is proved.
But the claim leads to a contradiction, too, since we have infinitely often

$$N\Big(1, \frac{\xi_k}{N(\xi_0, \dots, \xi_{k-1})}\Big) > 1 + M(|\xi_k|) = N(1, \xi_k) \geq N\Big(1, \frac{\xi_k}{N(\xi_0, \dots, \xi_{k-1})}\Big).$$

This completes the proof of part (a).
(b) Given ℓ_M we wish to find an Orlicz function F such that $\ell_M = \ell_F$ and $F(t) = N(1,t) - 1$, $t \geq 0$, for some absolute norm N. Since $\ell_F = \widetilde{\Lambda}(N)$ by (a), this will yield our assertion.
Without loss of generality we assume that $M(1) = 1$. We define F by $F(t) = M(t)$ for $0 \leq t \leq \frac{1}{2}$ and $F(t) = M(\frac{1}{2}) + 2t - 1$ for $t > \frac{1}{2}$. By convexity of M the right-hand derivative of M at $\frac{1}{2}$ is ≤ 2; therefore F is continuous, increasing and convex. Clearly $\ell_F = \ell_M$. Let us define $N(s,t) = |s|(1 + F(|t/s|))$ for $s \neq 0$ and $N(0,t) = 2|t|$. Note that N is continuous on $\mathbb{K}^2$. To complete the proof of the proposition it is enough to prove

the triangle inequality for N. Let us first suppose that $s_1, s_2 > 0$. Then we have, using the convexity of F,

$$\begin{aligned} N(s_1+s_2, t_1+t_2) &= s_1+s_2+(s_1+s_2)F\Big(\frac{t_1+t_2}{s_1+s_2}\Big) \\ &= s_1+s_2+(s_1+s_2)F\Big(\frac{s_1}{s_1+s_2}\frac{t_1}{s_1}+\frac{s_2}{s_1+s_2}\frac{t_2}{s_2}\Big) \\ &\le s_1+s_2+s_1F\Big(\frac{t_1}{s_1}\Big)+s_2F\Big(\frac{t_2}{s_2}\Big) \\ &= N(s_1,t_1)+N(s_2,t_2). \end{aligned}$$

This inequality extends to the case in which one or both s_i are zero by continuity. It is left to observe that in the general case

$$\begin{aligned} N(s_1+s_2, t_1+t_2) &= N(|s_1+s_2|, |t_1+t_2|) \le N(|s_1|+|s_2|, |t_1|+|t_2|) \\ &\le N(|s_1|,|t_1|)+N(|s_2|,|t_2|) = N(s_1,t_1)+N(s_2,t_2) \end{aligned}$$

due to the fact that $(s,t) \mapsto N(s,t)$ is increasing in each variable on $[0,\infty) \times [0,\infty)$, which is elementary to verify. □

Corollary 6.12

(a) *Every Orlicz sequence space h_M can be renormed to have property (M).*

(b) *An Orlicz sequence space h_M can be renormed to a space X for which $K(X)$ is an M-ideal in $L(X)$ if and only if $(h_M)^*$ is separable.*

PROOF: This follows from Propositions 6.1 and 6.11 and Corollary 4.5. □

Note that $K(h_M)$ is not an M-ideal in $L(h_M)$ for the Luxemburg norm introduced in Section III.1 unless $h_M = \ell^p$, since the unit vector basis is 1-symmetric for this norm, by Proposition 4.24. Also, no renorming of $h_M \neq \ell^p$ such that the compact operators form an M-ideal can be stable; indeed, otherwise some subsequence of the unit vector basis would be equivalent to some ℓ^p-basis by Theorem 6.6 and Lemma 6.2.

Corollary 6.13 (Lindenstrauss-Tzafriri)
Every infinite dimensional subspace of an Orlicz sequence space h_M contains an isomorphic copy of some ℓ^p for $1 \le p < \infty$ or of c_0.

PROOF: This follows from Theorem 6.4 and Corollary 6.12(a). □

Our final result can be thought of as an isomorphic version of Proposition 4.24. To formulate it we need the notion of a subsymmetric basis [422, p. 114]: A basis (e_n) of a Banach space X is called *subsymmetric* if it is unconditional and if for every subsequence (n_k) the basic sequence (e_{n_k}) is equivalent to (e_n). Proposition 6.1 implies that the unit vector basis of $\Lambda(N)$ is subsymmetric.

Theorem 6.14 *Let X be a Banach space with a subsymmetric basis (e_n). Then X can be renormed to a space Y for which $K(Y)$ is an M-ideal in $L(Y)$ if and only if X is isomorphic to an Orlicz sequence space h_M and X^* is separable.*

PROOF: One direction is contained in Corollary 6.12, and it remains to prove that a space which can be renormed as stated is in fact isomorphic to some h_M; X^* is then necessarily separable by Corollary 4.5.
The separability of X^* implies that (e_n) is weakly null, since (e_n) is an unconditional basis; this has already been observed in Proposition 4.24. By Theorem 4.17 the renormed space Y has (M). Lemma 6.2 shows that some subsequence of (e_n) is equivalent to the unit vector basis of some $\Lambda(N)$ and thus of some h_M (Proposition 6.11). Since (e_n) is supposed to be subsymmetric, (e_n) itself is equivalent to the unit vector basis of some h_M. Hence, the claim is proved. □

The proof of this theorem shows a little bit more, namely if $K(X)$ is an M-ideal in $L(X)$, then every subsymmetric basic sequence is actually symmetric.

VI.7 Notes and remarks

GENERAL REMARKS. The main references for Sections VI.1 through VI.3 are [60], [620], [626] and [629]. Section VI.1 is largely based on [629], where Theorem 1.2 and its corollaries can be found. Special cases of these results were previously obtained in [626] ($X = Y = C_{\mathbb{C}}(K)$) and [60] ($Z(X^*)$ trivial, $Y = C(K)$); for the case of arbitrary X and $Y = C(K)$ see [61]. Lemma 1.1(c) is a simple observation which can be traced back as least as far as [401]. Theorem 1.3 is due to Ruess and Stegall [546] in the real and to Lima and Olsen [408] in the complex case. Our proof, building on the Lemmas 1.4 and 1.5 from [620], covers both cases simultaneously. A similar approach is in [604].
In [107] it is proved, by vector measure techniques, that K is hyperstonean if X is reflexive and $L(X, C(K))$ is isometric to a dual space. This can also be deduced from Corollary 1.11, which is easily seen to extend to spaces X such that $Z(X^*)$ is finite dimensional, in particular to reflexive X. The paper [107] contains the additional information that, for reflexive X, the predual of $L(X, C(K))$ is strongly unique. Let us mention that the paper [517] deals with the dual space $L(X, C(K))$ from an M-structure point of view, too.
The results in Section VI.2 are taken from [60]. The main aspect of Theorem 2.3 is that, under the assumptions made, the M-ideals of $L(X, C(K))$ are in one-to-one correspondence with the M-ideals in $C(K)$. A result of this type was first proved by Flinn and Smith [236] for the case in which X is a complex $C(K)$-space. Their arguments rely on numerical range techniques for the Banach algebra $L(C(K))$. The representation of the M-ideals in $L(C(K))$ they give is embellished in [626]. There the subspaces $J_{(D)}$ are introduced, and their relation to the subspaces of J_D-valued operators is investigated. There are also some results concerning the M-ideals in $L(X, C_0(L))$ for a locally compact space L, but here a lot of technical problems arise; see [60] and [236] for details.
Our source for the first half of Section VI.3 is [620], which contains 3.1–3.7 and Proposition 3.12. This paper also discusses M-ideals for the projective hull of the ε-tensor norm and dually L-summands in spaces of absolutely summing operators. Theorem 3.2 is extended to the case of a space X with finitely many M-ideals in [548]. Corollary 3.4 has first appeared in [51]; the M-ideals of $C(K, X)$ are completely described in [51, Prop. 10.1]. Corollary 3.7 has been considered for $X = Y = \ell^p_{\mathbb{C}}$ in [580], and for re-

flexive X and Y in [293]. Theorem 3.8 and its corollaries are taken from [629], but Corollary 3.11 was originally proved in [279] where measure theoretic arguments are employed. Proposition 3.13 is a result from [630]. It was previously proved in [548] that the structure topology is stable by taking products if one of the Banach spaces involved has only finitely many M-ideals. In connection with Section VI.3 we also mention the papers [401] and [519] which deal with intersection properties in tensor products.

The problem of investigating for which Banach spaces $K(X,Y)$ is an M-ideal in $L(X,Y)$ has attracted a number of authors from the beginning of M-ideal theory. The interest in this question originated initially both from the approximation theoretic properties of M-ideals and from the uniqueness conclusion in the Hahn-Banach theorem in the M-ideal setting. Hennefeld [303] was the first to prove that $K(\ell^p)$ is an M-ideal in $L(\ell^p)$, for the explicit purpose of obtaining unique Hahn-Banach extensions. (We recall that Dixmier [165] proved that $K(H)$ is an M-ideal in $L(H)$ for a Hilbert space H.) Saatkamp [551] showed that $K(\ell^p, \ell^q)$ is an M-ideal (meaning, an M-ideal in the corresponding space of bounded operators); these authors directly construct the L-projection in the dual of the bounded operators. Å. Lima was the first to tackle the problem of M-ideals of compact operators by means of the 3-ball property; it is his proof from [402] that we have presented in Example 4.1. Actually the same proof yields the assertion of Corollary 5.4. Several authors observed that $K(X, c_0)$ is an M-ideal [220], [434], [551]; examples in the spirit of Lemma 6.7, which is taken from [624], can be found in [460] and [553]. In the other direction, a number of papers give ad hoc constructions in some spaces of operators to show that the 2-ball property fails and thus that $K(X,Y)$ is not an M-ideal. In this regard we mention [220], [551], [578]. Today these results can be obtained in a more systematic manner. For instance, in order to show that $K(\ell^1, \ell^2)$ is not an M-ideal observe that $K(\ell^1, \ell^2) \cong K(\ell^2, \ell^\infty) \cong C(\beta\mathrm{N}, \ell^2)$, that ℓ^2 has the IP (Definition II.4.1) and that $C(K,X)$ has the IP whenever X has (this is easy). Consequently, should $K(\ell^1, \ell^2)$ be an M-ideal in $L(\ell^1, \ell^2)$, it would even be an M-summand by Theorem II.4.4. This is impossible by Proposition 4.3. In [407] similar reasoning is employed to show that $K(X, C(K))$ is an M-ideal in $L(X, C(K))$ only in the trivial case where X or $C(K)$ is finite dimensional. This paper also contains the result that, for a separable Banach space Y with the MCAP, $K(\ell^1, Y)$ is an M-ideal if and only if Y is an (M_∞)-space.

Lima proved that $K(L^p)$, $p \neq 2$, fails to be an M-ideal in $L(L^p)$, by using properties of the Haar basis. A different proof is due to Li [397], wheras our argument in Proposition 5.19(a), parts (1) and (2), comes from [467]; we remark that part (3) of that proposition is unpublished. However, as was pointed out to us by L. Weis, $K(L^p)$ is an M-ideal in a certain nonunital subalgebra of $L(L^p)$, viz. the algebra consisting of those $T \in L(L^p)$ such that

$$\forall \varepsilon > 0\ \exists A, B \subset [0,1],\ \lambda(A \cup B) < \varepsilon: \quad T - P_A T P_B \text{ is compact.}$$

(Here P_A stands for the operator $f \mapsto \chi_A f$.) This can easily be shown with the help of the 3-ball property.

As we have already pointed out in the Notes and Remarks section to Chapter I, Hennefeld, in [305], introduced the notion of an *HB*-subspace as a more flexible means to deal with uniqueness of Hahn-Banach extensions. Results on *HB*-subspaces of compact operators can be found in [305], [306], [460], [463], [464], [466], [467]; for example, $K(\ell^p, X)$ and $K(X, \ell^p)$ are *HB*-subspaces for $1 < p < \infty$, but in general not M-ideals.

The systematic treatment of M-ideals of compact operators was begun in Hennefeld's paper [305], which contains Proposition 4.24. Clearly his proof differs from ours in details, since property (M) was not at his disposal. Instead, his argument is based on the following lemma: If X has a 1-unconditional basis (e_i) and $K(X)$ is an M-ideal in $L(X)$, then, for every $\varepsilon > 0$ and every finite rank operator S with $\|S\| = 1$, we have $\|S + (Id - P_n)\| \leq 1 + \varepsilon$ infinitely often; here P_n denotes the nth coordinate projection relative to the basis (e_i). The appearance of this lemma more than a decade before similar *necessary* conditions (cf. Theorem 4.17) appeared in [631] and [367] is quite remarkable. Lima [404] contributed Propositions 4.2, 4.4 and 4.5; Proposition 4.3 was observed in [553], however with a less obvious proof. The real breakthrough came with the paper [293] where the MCAP of X and X^* was shown to be necessary for the M-ideal property of $K(X)$, cf. Proposition 4.10. Simpler proofs can be found in [628] and [622], the latter one being presented in the text. Subsequently, it was investigated by several authors to what extent the converse is true. Cho and Johnson [118] obtained Corollary 4.20 (for $p < \infty$). The new technique they introduced into this context was the convex combination argument that allows one to pass from strong operator convergence via Banach space weak convergence to operator norm convergence. This device, which we have used a number of times, seems to be due to Feder [222]; it is successfully applied in [344] too. Cho's and Johnson's proof was simplified in [57] and [621]; the corresponding result for subspaces of c_0 was established in [622] and, independently and with a different proof, in [462].

The next step was performed in [631] where the equivalences of (i) through (iii) of Theorem 4.17 were proved. The remaining equivalences are due to Kalton [367] (see also [366]) who succeeded in singling out a geometrical property of the norm, viz. property (M) resp. property (M^*), which in conjunction with a suitable version of the approximation property characterises Banach spaces for which $K(X)$ forms an M-ideal. The achievement of Kalton's extremely important paper is to devise a characterisation in terms of the Banach space X itself rather than in terms of the operators defined on X. It is this characterisation that permitted him to provide answers to many of the questions in the theory of M-ideals of compact operators left open until then, for instance the general Cho-Johnson type Theorem 4.19. Corollary 4.18 was recently observed by E. Oja [466]. Actually, Kalton only deals with separable Banach spaces, but the extension to the general case presents no major difficulty (see also [466]). Our proofs of these results occasionally differ from his in that we work with the 3-ball property and use the material of Chapter V, whereas Kalton's approach is via the projection in the dual; he also presents an independent argument for the equivalence (i) $\Longleftrightarrow$ (ii) along these lines.

It seems remarkable that a contractive projection on $L(X,Y)^*$ with kernel $K(X,Y)^\perp$ can be constructed more or less explicitly when Y has the MCAP. This was done by J. Johnson [352, Lemma 1] as follows. Pick a net (K_α) in the unit ball of $K(Y)$ such that $K_\alpha \to Id_Y$ strongly, and let $\psi \in K(Y)^{**}$ be a weak* cluster point of (K_α). Let us assume for simplicity that $\psi = w^*\text{-}\lim K_\alpha$. For $\ell \in L(X,Y)^*$ consider $\pi(\ell)$: $T \mapsto \lim\langle \ell, K_\alpha T\rangle = \langle \psi, \varphi\rangle$ where $\varphi \in K(Y)^*$ is the functional $\varphi(K) = \langle \ell, KT\rangle$. Then π is a contractive projection with kernel $K(X,Y)^\perp$ and whose range is isometric to $K(X,Y)^*$. By Proposition I.1.2(b) π must be the L-projection if $K(X,Y)$ is an M-ideal in $L(X,Y)$, thus, to check that $K(X,Y)$ is an M-ideal it is enough to check that this particular projection is an L-projection. This remark applies even in the setting of semi

L-projections defined in the Notes and Remarks to Chapter I. Therefore, if one could prove that X has the MCAP if $K(X)$ is a semi M-ideal in $L(X)$, one would obtain that $K(X)$ is a semi M-ideal in $L(X)$ if and only if it is an M-ideal; as yet, this is open. A similar procedure as above, with $\pi(\ell) : T \mapsto \lim\langle \ell, TK_\alpha \rangle$ works if X^* has the MCAP with adjoint operators. Likewise, one obtains analogous results for the space $A(X,Y)$ of norm limits of finite rank operators if one employs the MAP. But here a slight technical improvement is possible. For, an application of the principle of local reflexivity shows that $\{S^* \mid S \in F(X), \|F\| \leq 1\}$ is strongly dense in $B_{F(X^*)}$. Therefore X^* has the MAP with adjoint operators provided it has the MAP at all. By contrast, there are counterexamples to this last assertion for the MCAP [280]; in fact, only recently was the first example of a Banach space with the MCAP lacking the MAP (even the AP) constructed by Willis [637]. It appears to be still open whether or not such spaces exist among the subspaces of ℓ^p. We also refer to [270], [406] and [471] for related information.
Theorem 4.21, Lemma 4.22 and Corollary 4.23 suggest that the assumption that $K(X)$ is an M-ideal in $L(X)$ is related to unconditionality properties of X; these results are due to Li [397], cf. also [269] and [564]. (Concerning Lemma 4.22, see also the subsection on u-ideals of compact operators below.) The condition $\lim \|Id - 2K_\alpha\| = 1$ from Theorem 4.17 is an indication of this, too. Corollary 6.11(b) supports this point of view from another direction since c_p fails to have local unconditional structure (see [501, Cor. 8.20]). In a similar vein, one can understand the assumption that $K(X,Y)$ is an M-ideal as a richness property; for example it follows as in the proof of Proposition 4.10 that $B_{K(X,Y)}$ is strongly dense in $B_{L(X,Y)}$ which yields the existence of compact operators with some desirable properties. On the other hand, once $K(X,Y) \neq L(X,Y)$ is an M-ideal in $L(X,Y)$ it is necessarily a proper M-ideal (Definition II.3.1) and hence contains c_0 (Theorem II.4.7). But then it follows that $K(X,Y)$ is an uncomplemented subspace of $L(X,Y)$ (see [365], [223], [204] and [345]) which indicates a rich supply of nontrivial (= noncompact) operators. (This can also be derived from Theorem I.2.10 with an appeal to [223].) By the way, up to now *no* example of a complemented subspace $K(X,Y) \neq L(X,Y)$ is known.
Condition (3) in Theorem 5.3 was known to experts as an easy-to-check sufficient condition for $K(Y,X)$ to be an M-ideal in $L(Y,X)$ for all Banach spaces Y, cf. [622]. That (3) is also necessary was proved in [481]. In this paper the idea to consider $X \oplus X$ in order to derive information about X emerged, and it allowed the authors of [481] to provide a proof of Theorem 5.3(b) and Proposition 5.6. The formal definition of the class of (M_p)-spaces was given in [467]. In particular it was shown there that the (M_p)-spaces are stable for $p < \infty$, a fact that might be considered as a first step towards property (M). Our sources for Section VI.5 are, apart from [467] and [481], [632] and [633]; the simple proof that (M_1)-spaces are finite dimensional was pointed out to us by T. S. S. Rao. Theorem 5.7 comes from [632].
Theorem 5.15 and the preparatory Propositions 4.7 and 4.8 are proved in [633]. Let us mention that $K(X,Y)^\theta$ is always nowhere dense (see [553] or [633]); we used a particular case of this in the proof of Proposition 4.8. The quantity $w(T)$ from Proposition 4.7 was introduced in the context of Hilbert space operators in [320] where $w(T) = \|T\|_e$ is proved for $T \in L(H)$. Proposition 4.8 was observed for Hilbert spaces in [321]. The method of proof of Theorem 5.15 is adapted from [379]; this paper, again, deals with Hilbert space operators and uses in addition spectral theoretic arguments. Characterisations of

smooth points in $K(X,Y)$ rather than $L(X,Y)$ can be found in [298] and [547]; these papers also investigate Fréchet smoothness in operator spaces. In another direction one might ask, starting from the main result in [379], i.e. Theorem 5.15 for $X = Y =$ Hilbert space, for characterisations of smooth points in subalgebras of $L(H)$. Such results are proved in [599] and [600].

Section VI.6 is almost entirely based on Kalton's paper [367] and contains the most important consequences of the results presented so far; only the easy implication (i) $\Rightarrow$ (ii) in Theorem 6.6 [467] and Lemma 6.7 together with Proposition 6.8 [624] are not found there. Actually, Kalton's results are more general than the versions stated; for example he shows that a separable order continuous nonatomic Banach lattice can be renormed to have property (M) if and only if it is lattice isomorphic to L^2. Also, instead of the spaces $\Lambda(N)$ one can construct a more general class of spaces enjoying (M) resp. such that the compact operators form an M-ideal. One considers a sequence $\mathcal{N} = (N_1, N_2, \ldots)$ of absolute norms on $\mathbb{K}^2$ with $N_k(1,0) = 1$ and defines inductively $\mathcal{N}(\xi_0, \ldots, \xi_d) = N_d(\mathcal{N}(\xi_0, \ldots, \xi_{d-1}), \xi_d)$. The resulting space, $\Lambda(\mathcal{N})$, fulfills the conclusion of Proposition 6.1 (apart from part (b)). As in Proposition 6.11 one can show that $\Lambda(\mathcal{N})$ is isomorphic to a so-called modular sequence space (see [422, Sect. 4.d], these spaces are also known as Musielak-Orlicz spaces) and vice versa. As a consequence, one obtains that every modular sequence space contains a copy of some ℓ^p. Originally this result is due to Woo [644] who reduced it to the Lindenstrauss-Tzafriri theorem, Corollary 6.13. Lindenstrauss and Tzafriri proved their theorem with the help of a fixed point argument [422, Th. 4.a.9]; this argument even yields $(1+\varepsilon)$-copies of ℓ^p for the Luxemburg norm introduced in Section III.1. This does not imply automatically that one also gets $(1+\varepsilon)$-copies of ℓ^p for the $\Lambda(N)$-norm. Still, this is true as proved in Section VI.6; the nontrivial ingredient of the proof this time being an appeal to Krivine's theorem in Lemma 6.3. Note that isomorphs of ℓ^p not containing $(1+\varepsilon)$-copies of ℓ^p were recently constructed by Odell and Schlumprecht [459]. Yet another approach, via the Krivine-Maurey theorem [384], can be found in [243, p. 167] where Garling proved that h_M is stable (for the Luxemburg norm).

As for Lemma 6.7, we would like to mention that under the assumptions made there, for subspaces $E \subset X$ and $F \subset Y$ such that $K(E,F)$ is strongly dense in $L(E,F)$, the space $K(E,F)$ is an M-ideal in $L(E,F)$ as well. This is shown in [624]. Further, let us record the result from [625] that $K(X,Y)$ is an M-ideal in $L(X,Y)$ if and only if for all $T \in L(X,Y)$, $\|T\| = 1$, there is a net (K_α) in $K(X,Y)$ with $K_\alpha^* y^* \to T^* y^*$ for all $y^* \in Y^*$ satisfying $\limsup \|S + (T - K_\alpha)\| \leq 1$ whenever $S \in K(X,Y)$, $\|S\| = 1$. M-ideals of compact operators are also considered in some papers quoted in the Notes and Remarks to Section V.6 where additional comments can be found. For the general topic of M-ideals in operator spaces we also mention [34], [232], [464], [465], [518].

MORE ON (M_∞)-SPACES. One of the main problems left open in connection with these spaces is to decide whether the class of (M_p)-spaces coincides with the class of subspaces of quotients of $\ell^p(X_\alpha)$, $\dim X_\alpha < \infty$, resp. of $c_0(\Gamma)$ with the MCAP. (Note that a quotient of c_0 is almost isometric to a subspace of c_0, by [14] or [336].) Theorem 5.7 gives a contribution to this question. Another subclass of (separable) (M_∞)-spaces which are known to embed almost isometrically into c_0 are those where the K_α appearing in condition (3) of Theorem 5.3 are projections; then the K_α clearly have finite rank.

Let us sketch the simple argument. Our assumption is that there exists a sequence (F_n) of finite rank projections on X such that

$$F_n x \to x \qquad \forall x \in X,$$

$$\limsup_n \sup_{\|x\|,\|y\| \le 1} \|F_n x + (Id - F_n)y\| \le 1.$$

Let us now observe the following lemma.

LEMMA A. *For all $\varepsilon > 0$ there is some $\delta > 0$ with the following property: If E is a Banach space and the operators $S, T \in L(E)$ satisfy*

$$\left.\begin{array}{rcl} \|Sx + (Id - S)y\| & \le & 1 + \delta \quad \forall x, y \in B_E \\ \|Tx + (Id - T)y\| & \le & 1 + \delta \quad \forall x, y \in B_E \end{array}\right\} \tag{1}$$

then

$$\|ST - TS\| \le \varepsilon. \tag{2}$$

Assuming the lemma to be false we try to achieve a contradiction. So let us suppose that there are some $\varepsilon_0 > 0$, some Banach spaces E_n and operators $S_n, T_n \in L(E_n)$ such that

$$\begin{array}{rcl} \|S_n x_n + (Id - S_n)y_n\| & \le & 1 + \frac{1}{n} \quad \forall x_n, y_n \in B_{E_n}, \\ \|T_n x_n + (Id - T_n)y_n\| & \le & 1 + \frac{1}{n} \quad \forall x_n, y_n \in B_{E_n}, \end{array}$$

but

$$\|S_n T_n - T_n S_n\| > \varepsilon_0.$$

We choose a free ultrafilter $\mathcal{U}$ over the integers and consider the ultraproduct $E := \prod_{\mathcal{U}} E_n$ and the operators $S, T : E \to E$ defined by $[(x_n)] \mapsto [(S_n x_n)]$, and $[(x_n)] \mapsto [(T_n x_n)]$. Then S and T satisfy (1) with $\delta = 0$. On the other hand, there are $\xi_n \in E_n$, $\|\xi_n\| = 1$, such that $\|(S_n T_n - T_n S_n)\xi_n\| > \varepsilon_0$. It follows for $\xi = [(\xi_n)] \in E$ that

$$\|(ST - TS)\xi\| = \lim_{\mathcal{U}} \|(S_n T_n - T_n S_n)\xi_n\| \ge \varepsilon_0,$$

hence $ST \ne TS$, and (2) fails with $\varepsilon = 0$. This contradicts Corollary I.3.9, since the centralizer of E is commutative.
With a similar technique one can show:

LEMMA B. *For all $\varepsilon > 0$ there is some $\delta > 0$ with the following property: If E is a Banach space and $T \in L(E)$ is a projection satisfying*

$$\|Tx + (Id - T)y\| \le 1 + \delta \qquad \forall x, y \in B_E,$$

then

$$\|x\| \le (1 + \varepsilon) \max\{\|Tx\|, \|x - Tx\|\} \qquad \forall x \in E.$$

Thus our projections are almost M-projections (cf. (*) on p. 2), and by Lemma A they almost commute. Let now $0 < \varepsilon < 1/2$. Pick a decreasing sequence of positive numbers (ε_n) such that $\prod_{n=1}^{\infty}(1+\varepsilon_n) \le 1+\varepsilon$, and choose $0 < \delta_n \le \varepsilon_n$ according to Lemmas A and B. After possibly discarding a number of the F_n, we have from our assumption

$$F_n x \to x \qquad \forall x \in X,$$

$$\|F_n x + (Id - F_n)y\| \le 1 + \delta_n \qquad \forall x, y \in B_X.$$

Denote $Y_n := \operatorname{ran}(F_n)$, $Z := (Y_1 \oplus Y_2 \oplus \ldots)_{\ell^\infty}$ and let $I : X \to Z$ be defined by

$$I(x) = (F_1 x, F_2(Id - F_1)x, F_3(Id - F_2)(Id - F_1)x, \ldots).$$

It is clear that I is well-defined and that $\|I\| \le 1 + \varepsilon$. We wish to prove that $\operatorname{ran}(I) \subset Y := (Y_1 \oplus Y_2 \oplus \ldots)_{c_0}$ and that

$$\frac{1}{1+\varepsilon}\|x\| \le \|I(x)\| \le (1+\varepsilon)\|x\| \qquad \forall x \in X. \tag{3}$$

This will show our assertion since Y is almost isometric to a subspace of c_0 and ε is arbitrarily small.
For convenience we let $G_k := Id - F_k$. To prove the former we have to show that $G_n G_{n-1} \cdots G_1 x \to 0$ for each $x \in X$. The idea is to let G_n wander from the left to the right until we end up with something small; the error terms that arise can be estimated by Lemma A. We skip the details. We now turn to the proof of the left hand inequality in (3). From Lemma B we have

$$\|\xi\| \le (1+\varepsilon_n)\max\{\|F_n\xi\|, \|\xi - F_n\xi\|\} \qquad \forall \xi \in X,\ n \in \mathbb{N}. \tag{4}$$

Let $x \in X$. We first apply (4) to $\xi = x$ with $n = 1$. If the left hand term in the maximum in (4) is the bigger one, we get

$$\|x\| \le (1+\varepsilon_1)\|F_1 x\| \le (1+\varepsilon)\|I(x)\|.$$

Otherwise, we have $\|x\| \le (1+\varepsilon_1)\|G_1 x\|$. Then we apply (4) to $\xi = G_1 x$ with $n = 2$. If now the maximum in (4) is attained at the first item, we get

$$\|x\| \le (1+\varepsilon_1)\|G_1 x\| \le (1+\varepsilon_1)(1+\varepsilon_2)\|F_2 G_1 x\| \le (1+\varepsilon)\|I(x)\|.$$

Otherwise, we have $\|G_1 x\| \le (1+\varepsilon_2)\|G_2 G_1 x\|$, and we continue exploiting (4) with $\xi = G_2 G_1 x$, $n = 3$, etc. If, for some n, the first item in the maximum in (4) is the bigger one, we deduce $\|x\| \le (1+\varepsilon)\|I(x)\|$, which is the desired inequality. But should for no n the first item exceed the second, we get for all $n \in \mathbb{N}$

$$\|x\| \le \prod_{k=1}^{n}(1+\varepsilon_k)\|G_n \cdots G_1 x\| \le (1+\varepsilon)\|G_n \cdots G_1 x\|,$$

which, however, tends to 0, as was noticed above. This completes the proof of the inequality $\|x\| \le (1+\varepsilon)\|I(x)\|$ and thus of our claim.

A direct proof of the almost commutativity of almost M-projections is given in [110]; the usage of ultraproduct techniques in the above proof is inspired by [341]. Another piece of information on (M_∞)-spaces is the result from [407] already mentioned that a separable space Y with the MCAP has (M_∞) if and only if $K(\ell^1, Y)$ is an M-ideal in $L(\ell^1, Y)$.

THE CENTRALIZER OF AN INJECTIVE TENSOR PRODUCT. The centralizers of injective tensor products are discussed in various contexts (C^*-algebras, $A(K)$-spaces) for example in [126], [297], [377] and [616]. The most far-reaching results in a general framework seem to have been obtained in [630]. In this paper the centralizer of an injective tensor product is represented as a function space on a suitable product space. More precisely, the following can be shown. We have seen in Example I.3.4(g) that $Z(X)$ can be regarded as a subalgebra of $C^b(Z_X)$, where $Z_X = \overline{\mathrm{ex}}^{w^*} B_{X^*} \setminus \{0\}$, and it follows from Theorem I.3.6 that $Z(X)$ can in fact be identified with the space of bounded continuous functions on the quotient space $\Theta_X = Z_X/\!\sim$ derived from the equivalence relation

$$p \sim q \iff a_T(p) = a_T(q) \quad \forall T \in Z(X).$$

Let us now address the question how $\Theta_{X \widehat{\otimes}_\varepsilon Y}$ relates to the product of Θ_X and Θ_Y. The answer involves the so-called k-product of topological spaces [582]. Let S_1 and S_2 be Hausdorff spaces. Then $S_1 \times_k S_2$ denotes the space $S_1 \times S_2$ endowed with the finest topology which agrees with the product topology on the compact subsets of $S_1 \times S_2$. Here is the main theorem from [630].

THEOREM. *$Z(X \widehat{\otimes}_\varepsilon Y)$ can be identified with $C^b(\Theta_X \times_k \Theta_Y)$.*

The example at the end of Section VI.3 shows that one cannot replace the k-product by the ordinary product here.
Prior to [630] it was known that the algebraic tensor product $Z(X) \otimes Z(Y)$ is dense in $Z(X \widehat{\otimes}_\varepsilon Y)$ for the strong operator topology [635], another proof of this fact was given in [50]. (That $Z(X) \otimes Z(Y) \subset Z(X \widehat{\otimes}_\varepsilon Y)$ follows from the easy part of Theorem 1.3.) The latter paper also contains sufficient conditions for $Z(X) \otimes Z(Y)$ to be norm dense, i.e. $Z(X \widehat{\otimes}_\varepsilon Y) = Z(X) \widehat{\otimes}_\varepsilon Z(Y)$. For example, norm density holds if X and Y are dual spaces; in view of the above theorem this follows once the compactness of Θ_X and Θ_Y is shown, which is done in [630].
A special case of these results is that $Z(C(K, X)) \cong C(K)$ if $Z(X)$ is trivial. For a description of the centralizer of $C(K, X)$ in the general case see [51, Prop. 10.3].

BEST APPROXIMATION BY COMPACT OPERATORS AND THE "BASIC INEQUALITY". The concept of an M-ideal has proved particularly useful for obtaining best approximation results for compact operators, because M-ideals are proximinal (Proposition II.1.1). Thus, M-ideal techniques provide a systematic approach to proximity questions. The close connection between M-ideal theory and best approximation by compact operators was first noticed in [321].
The proximinality of $K(H)$ in $L(H)$, H a Hilbert space, is a result originally due to Gohberg and Krein [276, p. 62]. To obtain the same result now, one just has to take into account that $K(H)$ is an M-ideal in $L(H)$. Since $K(\ell^p)$ is an M-ideal in $L(\ell^p)$ provided $1 < p < \infty$, $K(\ell^p)$ is seen to be proximinal in $L(\ell^p)$ for these p. This corollary extends

to $p = 1$ although $K(\ell^1)$ is not an M-ideal, but not to $p = \infty$ [221]. We remark that a "constructive" proof of the proximinality of $K(\ell^p)$ can be found in [434]. The natural question whether this result extends to L^p was answered in the negative by Benyamini and Lin [74]. Of course, this is stronger than the fact, proved in Proposition 5.19, that $K(L^p)$ is not an M-ideal in $L(L^p)$. Incidentally Benyamini and Lin use the concept of U-proximinality, which we mentioned in the Notes and Remarks to Chapter II, as well: They point out that $K(L^p)$ would even be U-proximinal ("satisfy the successive approximation scheme" in their terminology) if it were proximinal at all, and they show that it is not.

Another approach to proximinality questions was suggested by Axler, Berg, Jewell and Shields [38], [39]; see also [75]. They prove proximinality of $K(X)$ for Banach spaces X such that X^* enjoys the bounded approximation property and X satisfies the so-called "basic inequality", which means the following:

For all $S \in L(X)$, for all bounded nets $(A_\alpha) \subset L(X)$ such that $A_\alpha \to 0$ and $A_\alpha^* \to 0$ strongly and for all $\varepsilon > 0$ there is some index α_0 such that

$$\|S + A_{\alpha_0}\| \le \max\{\|S\|, \|S\|_e + \|A_{\alpha_0}\|\} + \varepsilon. \tag{1}$$

(Here $\|S\|_e$ denotes the essential norm of S, i.e. the norm of the equivalence class $S+K(X)$ in the quotient space $L(X)/K(X)$.) They go on to show that ℓ^p satisfies the basic inequality for $1 < p < \infty$ as does c_0, whereas ℓ^1, ℓ^∞ and the L^p-spaces (for $p \neq 2$) fail the basic inequality. Therefore, they get that $K(\ell^p)$ is proximinal in $L(\ell^p)$ for $1 < p < \infty$. Another result of [38] is the proximinality of $H^\infty + C$ in L^∞, which can also be derived with the help of M-ideal techniques; see Corollary III.1.5. Davidson and Power also obtain theorems on best approximation both by M-ideal methods and by basic inequality techniques [148]; and there are proximinality results on nest algebras (which we discussed in the Notes and Remarks to Chapter V) that can be obtained either using M-ideals or a variant of the basic inequality ([225] and [226]).

These similarities suggest that there might be a close relation between the two methods. This relation is elucidated in [625] where it is proved that, although neither does the basic inequality imply that $K(X)$ is an M-ideal in $L(X)$ nor does the converse hold, the two techniques are basically equivalent. More precisely, one can show:

THEOREM. *For a Banach space X, the following assertions are equivalent:*

(i) *$K(X)$ is an M-ideal in $L(X)$.*

(ii) *For all $T \in L(X)$ there is a net (K_α) in $K(X)$ such that $K_\alpha^* \to T^*$ strongly and, for all $S \in L(X)$,*

$$\limsup \|S + T - K_\alpha\| \le \max\{\|S\|, \|S\|_e + \|T\|\} \tag{2}$$

Here, condition (ii) can be regarded as a "revised basic inequality"; and an inspection of the proofs in [38] and [39] shows that the arguments in these papers are based on (2) rather than the basic inequality (1). Thus it seems all the more surprising that one can actually prove (1) for the ℓ^p-spaces. Let us point out in addition that one can take the coordinate projections in (ii) above for $X = \ell^p$; this contrasts the situation in Theorem 4.17(ii) where they don't work.

For other papers on the problem of best approximation by compact operators we refer e.g. to [22], [41], [73], [218], [219], [221], [320], [388], [618], [619], [654]. Papers dealing with the impact of M-ideals on approximation theory – apart from those already mentioned – include [129], [130], [220], [233], [234], [307], [553], [617].

BANACH-STONE THEOREMS. The classical Banach-Stone theorem asserts that two compact Hausdorff spaces K and L are homeomorphic provided that the sup-normed spaces $C(K)$ and $C(L)$ are isometrically isomorphic [179, p. 442]. A number of authors have investigated the validity of this theorem for spaces of vector-valued continuous functions. An in-depth analysis of this problem is presented in part II of E. Behrends' monograph [51], where the relevance of M-structure concepts for proving vector-valued Banach-Stone theorems was pointed out for the first time. We wish to sketch the main ideas and some of their recent extensions.

It is clear that some geometric condition has to be imposed on a Banach space X in order to render the Banach-Stone theorem for X-valued function spaces valid; the simplest example to demonstrate this is the isometric isomorphism $C(\{1\}, c_0) \cong c_0 \cong c_0 \oplus_\infty c_0 \cong C(\{1,2\}, c_0)$. The example suggests that the richness of the M-structure of c_0 is responsible for the failure of the Banach-Stone theorem in this case. Hence one might expect positive results for Banach spaces whose M-structure is scarce or even lacking. This is indeed so and can be seen as follows. Suppose $\Phi : C(K, X) \to C(L, X)$ is an isometric isomorphism. Then Φ maps M-ideals onto M-ideals. Let us now suppose that X has no nontrivial M-ideals. Then, by Corollary 3.4, $\Phi(J_{\{k\}} \widehat{\otimes}_\varepsilon X) = J_D \widehat{\otimes}_\varepsilon X$ for some closed subset D of L. Next one shows, by a maximality argument or otherwise, that D must be a singleton, say $D = \{l\}$, so that a map $\varphi : K \to L$, $k \mapsto l$ is defined. Finally one concludes by routine arguments that φ is actually a homeomorphism from K onto L. In fact, one can even obtain more information by this method. Namely it turns out that Φ must have the special form $(\Phi f)(l) = \Phi_l(f(\varphi^{-1}(l)))$, where $(\Phi_l)_{l \in L}$ is a strongly continuous family of isometrical isomorphisms on X. Also, if $\Phi : C(K, X) \to C(L, Y)$ is an isometrical isomorphism, with X and Y a priori distinct, then the same arguments apply, and X and Y must be isometric.

A second method to obtain a vector-valued Banach-Stone theorem consists in analysing the centralizer of $C(K, X)$. If $Z(X)$ is trivial, then one knows that $Z(C(K, X)) \cong C(K)$; this was mentioned above. Plainly, if $C(K, X)$ and $C(L, X)$ are isometrically isomorphic, so are their centralizers; hence K and L must be homeomorphic by the scalar Banach-Stone theorem. (Clearly, this is the mechanism on which the proof of Corollary 1.14 relies.) We remark that also in this case deeper consequences can be revealed, see [51, Th. 8.10]. We have just scratched the surface of the topic of $C(K, X)$-isometries; for full treatment we again refer to [51].

In another direction, the original Banach-Stone theorem has been extended by Amir [15] and Cambern [104] to include small-bound isomorphisms; see also [173]. Their result is that K and L are homeomorphic if $d(C(K), C(L)) < 2$ (d denotes the Banach-Mazur distance). The constant 2 is known to be optimal here [135]. (At this stage the Milutin theorem is worth recalling, which states that $C(K)$ and $C(L)$ are isomorphic whenever K and L are uncountable compact metric spaces; see [421, p. 174].) Let us agree to say that a Banach space X has the isomorphic Banach-Stone property if there exists some $\delta > 0$ such that, whenever $d(C(K, X), C(L, X)) < 1 + \delta$, K and L are homeomorphic.

The isomorphic Banach-Stone property is tackled in [338] and [105], and more recently in [59] and [65] where the above-sketched techniques are adapted to the nonisometric setting. For instance, it is shown in [65] that an isomorphism $\Phi : C(K,X) \to C(L,X)$ maps the M-ideal $J_{\{k\}}\widehat{\otimes}_\varepsilon X$ onto a subspace close enough to an M-ideal $J_{\{l\}}\widehat{\otimes}_\varepsilon X$ to recover l uniquely from k, provided $\|\Phi\|\,\|\Phi^{-1}\|$ is sufficiently small and X has in some sense no isomorphic M-structure. More precisely, in that paper it is required that the two-dimensional space $\ell^1(2)$ is not finitely representable in X^*, i.e.,

$$\lambda(X^*) := \inf\{d(E, \ell^1(2)) \mid E \subset X^*, \ \dim E = 2\} > 1.$$

In the real case this condition means precisely that X is uniformly nonsquare in the sense of [335]. Clearly, uniformly convex spaces X fulfill $\lambda(X^*) > 1$ and hence have the isomorphic Banach-Stone property. This was originally proved in [105].
Another approach to the problem, suggested in [59] and further elaborated in [109], is to investigate the so-called ε-mulipliers, which are defined by the requirement that the condition of Theorem I.3.6(ii) be fulfilled up to ε, and the class $\mathfrak{C}$ of Banach spaces on which ε-multipliers are close to multiples of the identity operator. It can be shown that any Banach space in $\mathfrak{C}$ enjoys the isomorphic Banach-Stone property, and a sufficient condition to belong to $\mathfrak{C}$ is that some ultrapower of X has a trivial multiplier algebra. In particular, L^1-spaces have the isomorphic Banach-Stone property; note that $\lambda(L^\infty) = 1$ so that the previous method does not apply.
Another recent direction of research inquires into Banach-Stone theorems for spaces of weak* continuous functions $C(K, (X^*, w^*))$. In this case the operator space $L(X, C(K))$ can be represented as $C(K, (X^*, w^*))$ [179, p. 490], and so the Corollaries 1.14 and 1.15 contribute to this circle of ideas. Particular results, apart from these corollaries, can be found in the papers [61], [106], [108], [109] and their references.
We also refer to [339] and [342] for information on Banach-Stone theorems, and to [110], [278] and [279] for similar results on Bochner L^p-spaces, where the L^p-structure rather than the M-structure is essential.

DAUGAVET'S EQUATION. We have pointed out in Proposition 4.3 that, for an infinite dimensional Banach space X, $K(X)$ is not an L^p- or M-summand in any subspace of $L(X)$ properly containing it, for $p > 1$. On the other hand, Theorem 4.17 asserts that $K(X)$ is an M-ideal in $L(X)$ once it is an M-ideal in $K(X) \oplus \mathbb{K}\{Id\}$. Therefore the following result, due to Daugavet [145], is surprising:

$K(C[0,1])$ is an L-summand in $K(C[0,1]) \oplus \mathbb{K}\{Id\}$.

In a plain formula this result can be rephrased as

$$\|Id + T\| = 1 + \|T\| \qquad \forall T \in K(C[0,1]).$$

The above equation is nowadays referred to as Daugavet's equation. Today it is known that every weakly compact operator on X satisfies Daugavet's equation if $X = C(K)$, K without isolated points, or $X = L^1(\mu)$, μ without atoms. This theorem appears in the literature several times; instead of giving a complete account of the relevant references, we mention the papers [3] and [561] where this is done. Other papers on this subject are [1], [323] and [643].

Another result along these lines, proved, rediscovered and reproved in [178], [322], [2] and [561], states that, if X is a real $C(K)$- or L^1-space, then

$$\max \|Id \pm T\| = 1 + \|T\| \qquad \forall T \in L(X); \qquad (*)$$

in other words, T or $-T$ satisfies Daugavet's equation.
We would like to offer one more argument for this which attempts to make the geometric background clear. Let E be a real Banach space. For $x \in E$ we let $I(x)$ be the intersection of all closed balls containing both x and $-x$. It is proved in [52, Ex. 2.3(e)] that $p(I(x)) \subset [-p(x), p(x)]$ for every extreme functional $p \in \operatorname{ex} B_{E^*}$. Hence, if $\|x\| = 1$ and $I(x) = B_E$, then $|p(x)| = 1$, from which we deduce

$$1 + p(y) = \max p(\pm x + y) \leq \max \|x \pm y\|.$$

Thus $\max \|x \pm y\| = 1 + \|y\|$ for all $y \in E$. It is left to observe that $I(x) = B_E$ holds for $E = L(C(K))$ and $x = Id$, whence $(*)$ for operators on $C(K)$-spaces. The case of L^1-spaces follows by duality. Likewise it follows for complex scalars that $\max\{\|Id + \theta T\| \mid |\theta| = 1\} = 1 + \|T\|$.

u-IDEALS OF COMPACT OPERATORS. In the Notes and Remarks to Chapter IV we presented a number of results on u-ideals from the important paper [264]. Now we resume the discussion of this notion and turn our interest to u-ideals of compact operators. It was in this connection that u-ideals were first introduced by Casazza and Kalton in [114]. Recall that a u-ideal is a closed subspace J of a Banach space E such that there is a projection P on E^* with kernel $J^\perp$ satisfying $\|Id - 2P\| = 1$. In case the scalars are complex and P satisfies $\|Id - (1 + \lambda)P\| = 1$ for all $|\lambda| = 1$ (which amounts to saying that P is a hermitian projection), then J is called an h-ideal.
The study of u-ideals of compact operators is intimately related to the investigation of an unconditional version of the metric compact approximation property. Let us say that a separable Banach space X has the unconditional MCAP if there exists a sequence of compact operators (K_n) converging strongly to Id_X such that $\lim \|Id_X - 2K_n\| = 1$. (Note that this really implies the MCAP.) In the case of complex scalars we define the $\mathbb{C}$-unconditional MCAP by requiring that $\lim \|Id_X - (1 + \lambda)K_n\| = 1$ for all $|\lambda| = 1$. Likewise, the unconditional MAP and the $\mathbb{C}$-unconditional MAP are introduced. These properties can be characterised in terms of unconditional expansions. By [264, Prop. 8.2] (see also [114, Th. 3.8]) X has the unconditional MCAP if and only if for all $\varepsilon > 0$ there is a sequence (A_n) of compact operators such that $\sum_{n=1}^{\infty} A_n x = x$ for all $x \in X$ and $\|\sum_{n=1}^{m} \varepsilon_n A_n\| \leq 1 + \varepsilon$ for all $m \in \mathbb{N}$ and $|\varepsilon_n| = 1$. Let us observe that $\sum A_n x$ is unconditionally convergent, which explains why this version of the MCAP is called unconditional. Similar results hold for the remaining unconditional approximation properties. Thus we see that the conclusion of Theorem 4.21 holds under much weaker assumptions than those made there; note that X has the unconditional MCAP if $K(X)$ is an M-ideal in $L(X)$ by (a variant of) Lemma 4.22. It is clear that a Banach space with a 1-unconditional basis enjoys the unconditional MAP; on the other hand, the argument in [397] shows that $L^p[0,1]$, $1 < p < \infty$, $p \neq 2$, fails the unconditional MAP for its canonical norm, but clearly can be renormed to have it, because the Haar basis is unconditional. Unconditional versions of the AP have been employed by Kalton [365] and Feder [222],

[223], among others, to show that the compact operators between certain Banach spaces form an uncomplemented subspace in the space of all bounded linear operators.
Suppose now that a separable Banach space X has the unconditional MCAP (resp., if the scalars are complex, the $\mathbb{C}$-unconditional MCAP). Then one can show that X is a u-ideal in X^{**} and that $K(X)$ is a u-ideal in $L(X)$ (resp., they are h-ideals). The converse holds for reflexive X or, in the complex case which is better understood, if X^* is separable [264, Th. 8.3]. Note that $K(X)^{**} \cong L(X)$ for reflexive X with the CAP so that the preceding result actually deals with the embedding of $K(X)$ in its bidual in this case. The general problem of when $K(X)$ is a u- or h-ideal in $K(X)^{**}$ is more subtle. Here one can show for complex spaces with separable duals satisfying the $\mathbb{C}$-unconditional MCAP that $K(X)$ has property (u), actually $\varkappa_h(K(X)) = 1$ (see p. 213 for this notation), however $K(X)$ need not be an h-ideal in $K(X)^{**}$. The definite criterion to check this involves M-ideals in a decisive manner, since [264, Th. 8.6] asserts that, for a complex Banach space with a separable dual, $K(X)$ is an h-ideal in $K(X)^{**}$ if and only if X has the $\mathbb{C}$-unconditional MCAP and X is an M-ideal in X^{**}. Let us mention that the Orlicz sequence spaces and the preduals of the Lorentz sequence spaces discussed in Example III.1.4 provide examples where these conditions are fulfilled.

Bibliography

[1] Y. Abramovich. *New classes of spaces on which compact operators satisfy the Daugavet equation.* To appear in J. Operator Theory.

[2] Y. Abramovich. *A generalization of a theorem of J. Holub.* Proc. Amer. Math. Soc. **108** (1990) 937–939.

[3] Y. Abramovich, C. D. Aliprantis, and O. Burkinshaw. *The Daugavet equation in uniformly convex Banach spaces.* J. Funct. Anal. **97** (1991) 215–230.

[4] I. Aharoni and J. Lindenstrauss. *Uniform equivalence between Banach spaces.* Bull. Amer. Math. Soc. **84** (1978) 281–283.

[5] A. B. Aleksandrov. *Essays on non locally convex Hardy classes.* In: V. P. Havin and N. K. Nikol'skii, editors, *Complex Analysis and Spectral Theory. Seminar, Leningrad 1979/80,* Lecture Notes in Math. 864, pages 1–89. Springer, Berlin-Heidelberg-New York, 1981.

[6] P. Alexandroff and H. Hopf. *Topologie I.* Springer, Berlin-Heidelberg-New York, 1974.

[7] E. M. Alfsen. *Compact Convex Sets and Boundary Integrals.* Springer, Berlin-Heidelberg-New York, 1971.

[8] E. M. Alfsen. *M-structure and intersection properties of balls in Banach spaces.* Israel J. Math. **13** (1972) 235–245.

[9] E. M. Alfsen and T. B. Andersen. *Split faces of compact convex sets.* Proc. London Math. Soc. **21** (1970) 415–442.

[10] E. M. Alfsen and T. B. Andersen. *On the concept of centre in $A(K)$.* J. London Math. Soc. **4** (1972) 411–417.

[11] E. M. Alfsen and E. G. Effros. *Structure in real Banach spaces. Part I and II.* Ann. of Math. **96** (1972) 98–173.

[12] E. M. Alfsen and B. Hirsberg. *On dominated extensions in linear subspaces of $C_{\mathbb{C}}(X)$.* Pacific J. Math. **36** (1971) 567–584.

[13] E. M. Alfsen, F. W. Shultz, and E. Størmer. *A Gelfand-Neumark theorem for Jordan algebras.* Adv. Math. **28** (1978) 11–56.

[14] D. ALSPACH. *Quotients of c_0 are almost isometric to subspaces of c_0.* Proc. Amer. Math. Soc. **76** (1979) 285–288.

[15] D. AMIR. *On isomorphisms of continuous function spaces.* Israel J. Math. **3** (1965) 205–210.

[16] D. AMIR AND J. LINDENSTRAUSS. *The structure of weakly compact sets in Banach spaces.* Ann. of Math. **88** (1968) 35–46.

[17] T. B. ANDERSEN. *Linear extensions, projections, and split faces.* J. Funct. Anal. **17** (1974) 161–173.

[18] T. ANDO. *Linear functionals on Orlicz spaces.* Nieuw Archief v. Wiskunde (3) **8** (1960) 1–16.

[19] T. ANDO. *Closed range theorems for convex sets and linear liftings.* Pacific J. Math. **44** (1973) 393–410.

[20] T. ANDO. *A theorem on non-empty intersection of convex sets and its applications.* J. Approx. Th. **13** (1975) 158–166.

[21] T. ANDO. *On the predual of H^∞.* Commentationes Mathematicae. Special volume in honour of W. Orlicz. (1978) 33–40.

[22] K. T. ANDREWS AND J. D. WARD. *Proximity in operator algebras on L^1.* J. Operator Theory **17** (1987) 213–221.

[23] C. APARICIO, F. OCAÑA, R. PAYÁ, AND A. RODRÍGUEZ-PALACIOS. *A non-smooth extension of Fréchet differentiability of the norm with applications to numerical ranges.* Glasgow Math. J. **28** (1986) 121–137.

[24] J. ARAZY. *More on convergence in unitary matrix spaces.* Proc. Amer. Math. Soc. **83** (1981) 44–48.

[25] J. ARAZY. *An application of infinite dimensional holomorphy to the geometry of Banach spaces.* In: J. Lindenstrauss and V. D. Milman, editors, *Geometrical Aspects of Functional Analysis, Israel Seminar 1985/86,* Lecture Notes in Math. 1267, pages 122–150, 1987.

[26] J. ARAZY AND S. D. FISHER. *Some aspects of the minimal Möbius invariant space of analytic functions on the unit disk.* In: M. Cwikel and J. Peetre, editors, *Interpolation Spaces and Allied Topics in Analysis. Proc. Conf. Lund 1983,* Lecture Notes in Math. 1070, pages 24–44. Springer, Berlin-Heidelberg-New York, 1984.

[27] J. ARAZY, S. D. FISHER, AND J. PEETRE. *Möbius invariant function spaces.* J. Reine Angew. Math. **363** (1985) 110–145.

[28] J. ARAZY AND Y. FRIEDMAN. *Contractive projections in C_1 and C_∞.* Mem. Amer. Math. Soc. **200** (1978) .

[29] J. ARAZY AND Y. FRIEDMAN. *Contractive projections in C_p.* C. R. Acad. Sc. Paris, Sér. A **312** (1990) 439–444.

[30] J. ARAZY AND Y. FRIEDMAN. *Contractive projections in C_p.* Mem. Amer. Math. Soc. **459** (1992) .

[31] J. ARAZY AND J. LINDENSTRAUSS. *On some linear topological properties of the space c_p of operators on Hilbert space.* Compositio Math. **30** (1975) 81–111.

[32] S. ARGYROS, S. NEGREPONTIS, AND TH. ZACHARIADES. *Weakly stable Banach spaces.* Israel J. Math. **57** (1987) 68–88.

[33] A. ARIAS. *Some properties of noncommutative H^1-spaces.* Proc. Amer. Math. Soc. **112** (1991) 465–472.

[34] J. ARIAS DE REYNA, J. DIESTEL, V. LOMONOSOV, AND L. RODRIGUEZ-PIAZZA. *Some observations about the space of weakly continuous functions from a compact space into a Banach space.* Preprint, 1992.

[35] L. ASIMOW AND A. J. ELLIS. *Convexity Theory and its Applications in Functional Analysis.* Academic Press, London, 1980.

[36] S. AXLER. *The Bergman space, the Bloch space, and commutators of multiplication operators.* Duke Math. J. **53** (1986) 315–332.

[37] S. AXLER. *Bergman spaces and their operators.* In: J. B. Conway and B. B. Morrell, editors, *Surveys of Some Recent Results in Operator Theory, Vol. I.* Pitman Research Notes in Mathematics 171, pages 1–50. Longman, 1988.

[38] S. AXLER, I. D. BERG, N. JEWELL, AND A. SHIELDS. *Approximation by compact operators and the space $H^\infty + C$.* Ann. of Math. **109** (1979) 601–612.

[39] S. AXLER, N. JEWELL, AND A. SHIELDS. *The essential numerical range of an operator and its adjoint.* Trans. Amer. Math. Soc. **261** (1980) 159–167.

[40] G. F. BACHELIS AND S. E. EBENSTEIN. *On $\Lambda(p)$ sets.* Pacific J. Math. **54** (1974) 35–38.

[41] H. BANG AND E. ODELL. *On the best compact approximation problem for operators between L_p-spaces.* J. Approx. Th. **51** (1987) 274–287.

[42] T. BARTON AND G. GODEFROY. *Remarks on the predual of a JB^*-triple.* J. London Math. Soc. **34** (1986) 300–304.

[43] T. BARTON AND R. M. TIMONEY. *Weak*-continuity of Jordan triple products and its applications.* Math. Scand. **59** (1986) 177–191.

[44] H. BAUER. *Šilovscher Rand und Dirichletsches Problem.* Ann. Inst. Fourier **11** (1961) 89–136.

[45] B. BEAUZAMY AND J.-T. LAPRESTÉ. *Modèles étalés des epaces de Banach.* Travaux en Cours. Hermann, Paris, 1984.

[46] J. B. BEDNAR AND H. E. LACEY. *Concerning Banach spaces whose duals are abstract L-spaces.* Pacific J. Math. **41** (1972) 13–24.

[47] E. BEHRENDS. *L^p-Struktur in Banachräumen.* Studia Math. **55** (1976) 71–85.

[48] E. BEHRENDS. *L^p-Struktur in Banachräumen II.* Studia Math. **62** (1978) 47–63.

[49] E. BEHRENDS. *M-structure in tensor products of Banach spaces.* In: S. Machado, editor, *Functional Analysis, Holomorphy, and Approximation Theory. Proc. Conf. Rio de Janeiro 1978,* Lecture Notes in Math. 843, pages 41–54. Springer, Berlin-Heidelberg-New York, 1978.

[50] E. BEHRENDS. *The centralizer of tensor products of Banach spaces (a function space representation).* Pacific J. Math. **81** (1979) 291–301.

[51] E. BEHRENDS. *M-Structure and the Banach-Stone Theorem.* Lecture Notes in Math. 736. Springer, Berlin-Heidelberg-New York, 1979.

[52] E. BEHRENDS. *Operators which respect norm intervals.* Studia Math. **70** (1982) 203–220.

[53] E. BEHRENDS. *Multiplier representations and an application to the problem whether $A\otimes_\varepsilon X$ determines A and/or X.* Math. Scand. **52** (1983) 117–144.

[54] E. BEHRENDS. *The continuity space of a Banach space.* In: H. Baumgärtel, G. Laßner, A. Pietsch, and A. Uhlmann, editors, *Proc. of the Second Intern. Conf. on Operator Algebras, Ideals, and Their Appl. in Theor. Physics, Leipzig, 1983,* pages 139–144. Teubner, 1984.

[55] E. BEHRENDS. *M-complements of M-ideals.* Rev. Roum. Math. Pures Appl. **29** (1984) 537–541.

[56] E. BEHRENDS. *Normal operators and multipliers on complex Banach spaces and a symmetry property of L^1-predual spaces.* Israel J. Math. **47** (1984) 23–28.

[57] E. BEHRENDS. *Sur les M-idéaux des espaces d'opérateurs compacts.* In: B. Beauzamy, B. Maurey, and G. Pisier, editors, *Séminaire d'Analyse Fonctionnelle 1984/85. Université Paris VI/VII*, pages 11–18, 1985.

[58] E. BEHRENDS. *A generalization of the principle of local reflexivity.* Rev. Roum. Math. Pures Appl. **31** (1986) 293–296.

[59] E. BEHRENDS. *Isomorphic Banach-Stone theorems and isomorphisms which are close to isometries.* Pacific J. Math. **133** (1988) 229–250.

[60] E. BEHRENDS. *On the geometry of spaces of C_0K-valued operators.* Studia Math. **90** (1988) 135–151.

[61] E. BEHRENDS. *C_0K-valued local operators.* Rev. Roum. Math. Pures Appl. **35** (1990) 109–123.

[62] E. BEHRENDS. *On the principle of local reflexivity.* Studia Math. **100** (1991) 109–128.

[63] E. BEHRENDS. *Points of symmetry of convex sets in the two-dimensional complex space – a counterexample to D. Yost's problem.* Math. Ann. **290** (1991) 463–471.

[64] E. BEHRENDS. *Compact convex sets in the two-dimensional complex linear space with the Yost property.* In: K. D. Bierstedt, J. Bonet, J. Horváth, and M. Maestre, editors, *Progress in Functional Analysis. Proc. International Functional Analysis Meeting, Peñíscola 1990*, pages 383––392. North-Holland, 1992.

[65] E. BEHRENDS AND M. CAMBERN. *An isomorphic Banach-Stone theorem.* Studia Math. **90** (1988) 15–26.

[66] E. BEHRENDS, R. DANCKWERTS, R. EVANS, S. GÖBEL, P. GREIM, K. MEYFARTH, AND W. MÜLLER. *L^p-Structure in Real Banach Spaces.* Lecture Notes in Math. 613. Springer, Berlin-Heidelberg-New York, 1977.

[67] E. BEHRENDS AND P. HARMAND. *Banach spaces which are proper M-ideals.* Studia Math. **81** (1985) 159–169.

[68] E. BEHRENDS AND G. WODINSKI. *A note on hermitian operators on complex Banach spaces.* Bull. Pol. Acad. Sci. **35** (1987) 801–803.

[69] S. F. BELLENOT. *Local reflexivity of normed spaces, operators, and Fréchet spaces.* J. Funct. Anal. **59** (1984) 1–11.

[70] C. B. BENNETT AND R. S. SHARPLEY. *Interpolation of Operators.* Academic Press, 1988.

[71] Y. BENYAMINI. *Constants of simultaneous extension of continuous functions.* Israel J. Math. **16** (1971) 258–262.

[72] Y. BENYAMINI. *Near isometries in the class of L^1-preduals.* Israel J. Math. **20** (1975) 275–281.

[73] Y. BENYAMINI. *Asymptotic centers and best compact approximation of operators into $C(K)$.* Constr. Approx. **1** (1985) 217–229.

[74] Y. BENYAMINI AND P. K. LIN. *An operator on L^p without best compact approximation.* Israel J. Math. **51** (1985) 298–304.

[75] I. D. BERG. *On norm approximation of functions of operators in the Calkin algebra.* Proc. Roy. Ir. Acad., Sect. A **78** (1978) 143–148.

[76] M. C. F. BERGLUND. *Ideal C^*-algebras.* Duke Math. J. **40** (1973) 241–257.

[77] M. BESBES. *Points fixes des contractions définies sur un convexe L^0-fermé de L^1.* C. R. Acad. Sc. Paris, Sér. A **311** (1990) 243–246.

[78] K. D. BIERSTEDT AND W. H. SUMMERS. *Biduals of weighted Banach spaces of analytic functions.* J. Austral. Math. Soc. **54** (1993) 70–79.

[79] E. BISHOP. *A general Rudin-Carleson theorem.* Proc. Amer. Math. Soc. **13** (1962) 140–143.

[80] D. P. BLECHER AND V. I. PAULSEN. *Tensor products of operator spaces.* J. Funct. Anal. **99** (1991) 262–292.

[81] N. BOBOC AND A. CORNEA. *Convex cones of lower semicontinuous functions.* Rev. Roum. Math. Pures Appl. **12** (1967) 471–525.

[82] J. BOCLÉ. *Sur la théorie ergodique.* Ann. Inst. Fourier **10** (1960) 1–45.

[83] F. BOMBAL. *On (V^*) sets and Pełczyński's property (V^*).* Glasgow Math. J. **32** (1990) 109–120.

[84] F. F. BONSALL AND J. DUNCAN. *Numerical Ranges of Operators on Normed Spaces and of Elements of Normed Algebras.* London Mathematical Society Lecture Note Series 2. Cambridge University Press, 1971.

[85] F. F. BONSALL AND J. DUNCAN. *Complete Normed Algebras.* Springer, Berlin-Heidelberg-New York, 1973.

[86] F. F. BONSALL AND J. DUNCAN. *Numerical Ranges II.* London Mathematical Society Lecture Note Series 10. Cambridge University Press, 1973.

[87] N. BOURBAKI. *Topologie générale.* Hermann, Paris, second edition, 1958.

[88] N. BOURBAKI. *Algèbre commutative.* Hermann, Paris, 1961.

[89] J. BOURGAIN. *Nouvelles propriétés des espaces L^1/H_0^1 et H^∞.* In: *Séminaire d'analyse fonctionnelle,* Exposé III. École Polytechnique, 1980–1981.

[90] J. BOURGAIN. *New Banach space properties of the disc algebra and H^∞.* Acta Math. **152** (1984) 1–48.

[91] J. BOURGAIN AND F. DELBAEN. *A class of special $\mathcal{L}_\infty$-spaces.* Acta Math. **145** (1980) 155–176.

[92] J. BOURGAIN, D. H. FREMLIN, AND M. TALAGRAND. *Pointwise compact sets of Baire-measurable functions.* Amer. J. Math. **100** (1978) 845–886.

[93] R. D. BOURGIN. *Geometric Aspects of Convex Sets with the Radon-Nikodým Property.* Lecture Notes in Math. 993. Springer, Berlin-Heidelberg-New York, 1983.

[94] E. BRIEM AND M. RAO. *Normpreserving extensions in subspaces of $C(X)$.* Pacific J. Math. **42** (1972) 581–587.

[95] A. BROWDER. *Introduction to Function Algebras.* W. A. Benjamin Inc., New York, Amsterdam, 1969.

[96] A. L. BROWN. *On the canonical projection of the third dual of a Banach space onto the first dual.* Bull. Austral. Math. Soc. **15** (1976) 351–354.

[97] L. BROWN AND T. ITO. *Classes of Banach spaces with unique isometric preduals.* Pacific J. Math. **90** (1980) 261–283.

[98] R. G. M. BRUMMELHUIS. *A note on Riesz sets and lacunary sets.* J. Austral. Math. Soc. Ser. A **48** (1990) 57–65.

[99] A. V. BUHVALOV AND G. J. LOZANOVSKII. *On sets closed in measure in spaces of measurable functions.* Trans. Moscow Math. Soc. **2** (1978) 127–148.

[100] J. BUNCE. *The intersection of closed ideals in a simplex space need not be an ideal.* J. London Math. Soc. **1** (1969) 67–68.

[101] L. J. BUNCE AND C.-H. CHU. *Dual spaces of JB^*-triples and the Radon-Nikodym property.* Math. Z. **208** (1991) 327–334.

[102] L. J. BUNCE AND C.-H. CHU. *Compact operators, multipliers and Radon-Nikodym property in JB^*-triples.* Pacific J. Math. **153** (1992) 249–265.

[103] J. C. CABELLO-PIÑAR, J. F. MENA-JURADO, R. PAYÁ, AND A. RODRÍGUEZ-PALACIOS. *Banach spaces which are absolute subspaces in their biduals.* Quart. J. Math. Oxford (2) **42** (1991) 175–182.

[104] M. CAMBERN. *On isomorphisms with small bound.* Proc. Amer. Math. Soc. **18** (1967) 1062–1066.

[105] M. CAMBERN. *Isomorphisms of spaces of norm-continuous functions.* Pacific J. Math. **116** (1985) 243–254.

[106] M. CAMBERN AND P. GREIM. *Mappings of continuous functions on hyperstonean spaces.* Acta Univ. Carol. Math. Phys. **28** (1987) 31–40.

[107] M. CAMBERN AND P. GREIM. *Uniqueness of preduals for spaces of continuous vector functions.* Can. Math. Bull. **32** (1989) 98–104.

[108] M. CAMBERN AND K. JAROSZ. *Isometries of spaces of weak* continuous functions.* Proc. Amer. Math. Soc. **106** (1989) 707–712.

[109] M. CAMBERN AND K. JAROSZ. *Ultraproducts, ε-multipliers, and isomorphisms.* Proc. Amer. Math. Soc. **105** (1989) 929–937.

[110] M. CAMBERN, K. JAROSZ, AND G. WODINSKI. *Almost-L^p-projections and L^p isomorphisms.* Proc. Royal Soc. Edinburgh, Ser. A **113** (1989) 13–25.

[111] N. L. CAROTHERS AND S. J. DILWORTH. *Subspaces of $L_{p,q}$.* Proc. Amer. Math. Soc. **104** (1988) 537–545.

[112] H. CARTAN AND S. EILENBERG. *Homological Algebra.* Princeton University Press, 1956.

[113] P. CARTIER. *Homologie cyclique: Rapport sur des travaux recents de Connes, Karoubi, Loday, Quillen, ...* Séminaire Bourbaki, Exposé 621. Astérisque **121–122** (1985) 123–146.

[114] P. G. CASAZZA AND N. J. KALTON. *Notes on approximation properties in separable Banach spaces.* In: P. F. X. Müller and W. Schachermayer, editors, *Geometry of Banach Spaces, Proc. Conf. Strobl 1989,* London Mathematical Society Lecture Note Series 158, pages 49–63. Cambridge University Press, 1990.

[115] P. G. CASAZZA AND T. J. SHURA. *Tsirelson's Space.* Lecture Notes in Math. 1363. Springer, Berlin-Heidelberg-New York, 1989.

[116] P. CEMBRANOS. *On Banach spaces of vector valued continuous functions.* Bull. Austral. Math. Soc. **28** (1983) 175–186.

[117] P. CEMBRANOS, N. J. KALTON, E. SAAB, AND P. SAAB. *Pełczyński's property (V) on $C(\Omega, E)$ spaces.* Math. Ann. **271** (1985) 91–97.

[118] C.-M. CHO AND W. B. JOHNSON. *A characterization of subspaces X of ℓ_p for which $K(X)$ is an M-ideal in $L(X)$.* Proc. Amer. Math. Soc. **93** (1985) 466–470.

[119] C.-M. Cho and W. B. Johnson. *M-ideals and ideals in $L(X)$.* J. Operator Theory **16** (1986) 245–260.

[120] M.-D. Choi and E. G. Effros. *The completely positive lifting problem for C^*-algebras.* Ann. of Math. **104** (1976) 585–609.

[121] M.-D. Choi and E. G. Effros. *Lifting problems and the cohomology of C^*-algebras.* Canadian J. Math. **29** (1977) 1092–1111.

[122] M.-D. Choi and E. G. Effros. *Nuclear C^*-algebras and injectivity: the general case.* Indiana Univ. Math. J. **26** (1977) 443–446.

[123] M.-D. Choi and E. G. Effros. *Nuclear C^*-algebras and the approximation property.* Amer. J. Math. **100** (1978) 61–79.

[124] E. Christensen, E. G. Effros, and A. M. Sinclair. *Completely bounded multilinear maps and C^*-algebraic cohomology.* Invent. Math. **90** (1987) 279–296.

[125] E. Christensen and A. M. Sinclair. *On the vanishing of $H^n(\mathfrak{A},\mathfrak{A}^*)$ for certain C^*-algebras.* Pacific J. Math. **137** (1989) 55–63.

[126] C.-H. Chu. *On standard elements and tensor products of compact convex sets.* J. London Math. Soc. **14** (1976) 71–78.

[127] C.-H. Chu. *A note on scattered C^*-algebras and the Radon-Nikodým property.* J. London Math. Soc. **24** (1981) 533–536.

[128] C.-H. Chu and B. Iochum. *On the Radon-Nikodým property in Jordan triples.* Proc. Amer. Math. Soc. **99** (1987) 462–464.

[129] C. K. Chui, P. W. Smith, R. R. Smith, and J. D. Ward. *L-ideals and numerical range preservation.* Illinois J. Math. **21** (1977) 365–373.

[130] C. K. Chui, P. W. Smith, and J. D. Ward. *Approximation by M-ideals in the disc algebra.* In: E. W. Cheney, editor, *Approximation Theory III. Proc. Conf. Honoring G. G. Lorentz, Austin 1980*, pages 305–309. Academic Press, 1980.

[131] J. A. Cima. *The basic properties of Bloch functions.* Internat. J. Math. & Math. Sci. **2** (1979) 369–413.

[132] J. A. Cima and W. R. Wogen. *Extreme points of the unit ball of the Bloch space B_0.* Michigan Math. J. **25** (1978) 213–222.

[133] J. A. Cima and W. R. Wogen. *On isometries of the Bloch space.* Illinois J. Math. **24** (1980) 313–316.

[134] P. Civin and B. Yood. *The second conjugate space of a Banach algebra as an algebra.* Pacific J. Math. **11** (1961) 847–870.

[135] H. B. Cohen. *A bound-two isomorphism between $C(X)$ Banach spaces.* Proc. Amer. Math. Soc. **50** (1975) 215–217.

[136] A. Connes. *Non-commutative differential geometry.* Publ. I.H.E.S. **65** (1985) 41–144.

[137] H. H. Corson and J. Lindenstrauss. *On simultaneous extension of continuous functions.* Bull. Amer. Math. Soc. **71** (1965) 542–545.

[138] F. Cunningham. *L-structure in L-spaces.* Trans. Amer. Math. Soc. **95** (1960) 274–299.

[139] F. Cunningham. *M-structure in Banach spaces.* Proc. Cambridge Phil. Soc. **63** (1967) 613–629.

[140] F. Cunningham. *Square Banach spaces.* Proc. Cambridge Phil. Soc. **66** (1969) 553–558.

[141] F. CUNNINGHAM. *Corrigendum to 'Square Banach spaces'.* Proc. Cambridge Phil. Soc. **76** (1974) 143–144.

[142] F. CUNNINGHAM, E. G. EFFROS, AND N. M. ROY. *M-structure in dual Banach spaces.* Israel J. Math. **14** (1973) 304–308.

[143] F. CUNNINGHAM AND N. M. ROY. *Extreme functionals on an upper semicontinuous function space.* Proc. Amer. Math. Soc. **42** (1974) 461–465.

[144] F. K. DASHIELL AND J. LINDENSTRAUSS. *Some examples concerning strictly convex norms on $C(K)$ spaces.* Israel J. Math. **16** (1973) 329–342.

[145] I. K. DAUGAVET. *On a property of compact operators in the space C.* Uspekhi Mat. Nauk **18** (1963) 157–158. (Russian).

[146] J. DAUNS AND K. H. HOFMANN. *Representations of rings by continuous sections.* Mem. Amer. Math. Soc. **83** (1968) .

[147] K. R. DAVIDSON. *Nest Algebras.* Pitman Research Notes in Mathematics 191. Longman, 1988.

[148] K. R. DAVIDSON AND S. C. POWER. *Best approximation in C^*-algebras.* J. Reine Angew. Math. **368** (1986) 43–62.

[149] A. M. DAVIE. *Linear extension operators for spaces and algebras of functions.* American J. Math. **94** (1972) 156–172.

[150] W. J. DAVIS AND J. LINDENSTRAUSS. *On total nonnorming subspaces.* Proc. Amer. Math. Soc. **31** (1972) 109–111.

[151] M. M. DAY. *Normed Linear Spaces.* Springer, Berlin-Heidelberg-New York, third edition, 1973.

[152] D. W. DEAN. *The equation $L(E, X^{**}) = L(E, X)^{**}$ and the principle of local reflexivity.* Proc. Amer. Math. Soc. **40** (1973) 146–148.

[153] A. DEFANT AND K. FLORET. *Tensor Norms and Operator Ideals.* North-Holland, Amsterdam, 1993.

[154] F. DEUTSCH. *Cardinality of extreme points of the unit sphere with applications to nonexistence theorems in approximation theory.* In: G. G. Lorentz, editor, *Approximation Theory. Proc. Int. Symp. Austin 1973*, pages 319–323. Academic Press, 1973.

[155] J. DIESTEL. *Geometry of Banach Spaces – Selected Topics.* Lecture Notes in Math. 485. Springer, Berlin-Heidelberg-New York, 1975.

[156] J. DIESTEL. *A survey of results related to the Dunford-Pettis property.* In: W. H. Graves, editor, *Proc. of the Conf. on Integration, Topology, and Geometry in Linear Spaces*, pages 15–60. Contemp. Math. **2**, 1980.

[157] J. DIESTEL. *An introduction to 2-summing operators. Measure theory in action.* In: G. A. Goldin and R. F. Wheeler, editors, *Measure theory and its applications. Proc. of the 1980 Conf. Northern Illinois University, DeKalb, Illinois*, pages 1–49, 1981.

[158] J. DIESTEL. *Sequences and Series in Banach Spaces.* Springer, Berlin-Heidelberg-New York, 1984.

[159] J. DIESTEL AND J. J. UHL. *Vector Measures.* Mathematical Surveys 15. American Mathematical Society, Providence, Rhode Island, 1977.

[160] J. DIEUDONNÉ. *A History of Algebraic and Differential Topology 1900–1960.* Birkhäuser, 1989.

[161] S. DINEEN. *The second dual of a JB^* triple system.* In: J. Mujica, editor, *Complex Analysis, Functional Analysis and Approximation Theory,* pages 67–69. North-Holland, 1986.

[162] S. DINEEN, M. KLIMEK, AND R. M. TIMONEY. *Biholomorphic mappings and Banach function modules.* J. Reine Angew. Math. **387** (1988) 122–147.

[163] S. DINEEN AND R. M. TIMONEY. *The centroid of a JB^*-triple system.* Math. Scand. **62** (1988) 327–342.

[164] J. DIXMIER. *Sur un théorème de Banach.* Duke Math. J. **15** (1948) 1057–1071.

[165] J. DIXMIER. *Les fonctionnelles linéaires sur l'ensemble des opérateurs bornés d'un espace de Hilbert.* Ann. of Math. **51** (1950) 387–408.

[166] J. DIXMIER. *Les algèbres d'opérateurs dans l'espace hilbertien.* Gauthier-Villars, Paris, second edition, 1969.

[167] J. DIXMIER. *Les C^*-algèbres et leurs représentations.* Gauthier-Villars, Paris, second edition, 1969.

[168] P. DOMAŃSKI. *Principle of local reflexivity for operators and quojections.* Arch. Math. **54** (1990) 567–575.

[169] R. DOSS. *Elementary proof of the Rudin-Carleson and the F. and M. Riesz theorems.* Proc. Amer. Math. Soc. **82** (1981) 599–602.

[170] R. G. DOUGLAS. *Banach Algebra Techniques in Operator Theory.* Academic Press, New York, 1972.

[171] P. N. DOWLING. *Radon-Nikodym properties associated with subsets of countable discrete abelian groups.* Trans. Amer. Math. Soc. **327** (1991) 879–890.

[172] P. N. DOWLING AND G. A. EDGAR. *Some characterizations of the analytic Radon-Nikodým property in Banach spaces.* J. Funct. Anal. **80** (1988) 349–357.

[173] L. DREWNOWSKI. *A remark on the Amir-Cambern theorem.* Functiones et Appr. **16** (1988) 181–190.

[174] J. DUGUNDJI. *An extension of Tietze's theorem.* Pacific J. Math. **1** (1951) 353–367.

[175] D. VAN DULST. *Reflexive and Superreflexive Banach Spaces.* Math. Centre Tracts 102. Amsterdam, 1978.

[176] D. VAN DULST. *Characterizations of Banach spaces not containing ℓ^1.* CWI Tracts 59. Centrum voor Wiskunde en Informatica, Amsterdam, 1989.

[177] J. DUNCAN AND S. A. R. HOSSEINIUN. *The second dual of a Banach algebra.* Proc. Royal Soc. Edinburgh **84** (1979) 309–325.

[178] J. DUNCAN, C. M. MCGREGOR, J. D. PRYCE, AND A. J. WHITE. *The numerical index of a normed space.* J. London Math. Soc. **2** (1970) 481–488.

[179] N. DUNFORD AND J. T. SCHWARTZ. *Linear Operators. Part 1: General Theory.* Interscience Publishers, New York, 1958.

[180] N. DUNFORD AND J. T. SCHWARTZ. *Linear Operators. Part 3: Spectral Operators.* Interscience Publishers, New York, 1971.

[181] P. L. DUREN. *Theory of H^p-spaces.* Academic Press, New York, 1970.

[182] G. A. EDGAR. *Measurability in a Banach space.* Indiana Univ. Math. J. **26** (1977) 663–677.

[183] G. A. EDGAR. *Measurability in a Banach space, II.* Indiana Univ. Math. J. **28** (1979) 559–579.

[184] G. A. EDGAR. *An ordering for the Banach spaces.* Pacific J. Math. **108** (1983) 83–98.

[185] C. M. EDWARDS. *On Jordan W^*-algebras.* Bull. Sci. Math. **104** (1980) 393–403.

[186] R. E. EDWARDS. *Fourier Series. A Modern Introduction. Volume 2.* Springer, Berlin-Heidelberg-New York, second edition, 1982.

[187] E. G. EFFROS. *Order ideals in a C^*-algebra and its dual.* Duke Math. J. **30** (1963) 391–412.

[188] E. G. EFFROS. *Structure in simplexes.* Acta Math. **117** (1967) 103–121.

[189] E. G. EFFROS. *Structure in simplexes II.* J. Funct. Anal. **1** (1967) 379–391.

[190] E. G. EFFROS. *On a class of real Banach spaces.* Israel J. Math. **9** (1971) 430–458.

[191] E. G. EFFROS. *On a class of complex Banach spaces.* Illinois J. Math. **18** (1974) 48–59.

[192] E. G. EFFROS. *Advances in quantized functional analysis.* In: A. M. Gleason, editor, *Proc. Intern. Congr. Math., Berkeley 1986*, pages 906–916. Amer. Math. Soc., Providence, R. I., 1987.

[193] E. G. EFFROS AND U. HAAGERUP. *Lifting problems and local reflexivity for C^*-algebras.* Duke Math. J. **52** (1985) 103–128.

[194] E. G. EFFROS AND J. L. KAZDAN. *Applications of Choquet simplexes to elliptic and parabolic boundary value problems.* J. Diff. Equations **8** (1970) 95–134.

[195] E. G. EFFROS AND Z.-J. RUAN. *On approximation properties for operator spaces.* Intern. J. of Math. **1** (1990) 163–187.

[196] E. G. EFFROS AND Z.-J. RUAN. *On non-self-adjoint operator algebras.* Proc. Amer. Math. Soc. **110** (1990) 915–922.

[197] E. G. EFFROS AND Z.-J. RUAN. *The Grothendieck-Pietsch and Dvoretzky-Rogers theorems for operator spaces.* Preprint, 1991.

[198] E. G. EFFROS AND Z.-J. RUAN. *Mapping spaces and liftings for operator spaces.* Preprint, 1991.

[199] E. G. EFFROS AND Z.-J. RUAN. *A new approach to operator spaces.* Can. Math. Bull. **34** (1991) 329–337.

[200] E. G. EFFROS AND Z.-J. RUAN. *On the abstract characterization of operator spaces.* Preprint, 1992.

[201] G. A. ELLIOTT. *On approximately finite-dimensional von Neumann algebras, II.* Can. Math. Bull. **21** (1978) 415–418.

[202] G. A. ELLIOTT AND D. OLESEN. *A simple proof of the Dauns-Hofmann theorem.* Math. Scand. **34** (1974) 231–234.

[203] A. J. ELLIS, T.S.S.R.K. RAO, A. K. ROY, AND U. UTTERSRUD. *Facial characterizations of complex Lindenstrauss spaces.* Trans. Amer. Math. Soc. **268** (1981) 173–186.

[204] G. EMMANUELE. *A remark on the containment of c_0 in spaces of compact operators.* Math. Proc. Cambridge Phil. Soc. **111** (1992) 331–335.

[205] R. ENGELKING. *General Topology.* PWN, Warszawa, 1977.

[206] R. EVANS. *A characterization of M-summands.* Proc. Cambridge Phil. Soc. **76** (1974) 157–159.

[207] R. EVANS. *Boolean algebras of projections.* Per. Math. Hung. **9** (1978) 293–295.

[208] R. EVANS. *Resolving Banach spaces.* Studia Math. **69** (1980) 91–107.

[209] R. EVANS. *Embedding $C(K)$ in $B(X)$.* Math. Scand. **48** (1981) 119–136.

[210] R. EXEL. *The F. and M. Riesz theorem for C^*-algebras.* J. Operator Theory **23** (1990) 351–368.

[211] M. FABIAN. *On projectional resolution of the identity on the duals of certain Banach spaces.* Bull. Austral. Math. Soc. **35** (1987) 363–371.

[212] M. FABIAN. *Each weakly countably determined Asplund space admits a Fréchet differentiable norm.* Bull. Austral. Math. Soc. **36** (1987) 367–374.

[213] M. FABIAN AND G. GODEFROY. *The dual of every Asplund space admits a projectional resolution of the identity.* Studia Math. **91** (1988) 141–151.

[214] M. FABIAN, L. ZAJÍČEK, AND V. ZIZLER. *On residuality of the set of rotund norms on a Banach space.* Math. Ann. **258** (1982) 349–351.

[215] H. FAKHOURY. *Préduaux de L-espace: Notion de centre.* J. Funct. Anal. **9** (1972) 189–207.

[216] H. FAKHOURY. *Sélections linéaires associées au théorème de Hahn-Banach.* J. Funct. Anal. **11** (1972) 436–452.

[217] H. FAKHOURY. *Existence d'une projection continue de meilleure approximation dans certains espaces de Banach.* J. Math. pures et appl. **53** (1974) 1–16.

[218] H. FAKHOURY. *Approximation par des opérateurs compacts ou faiblement compact à valeurs dans $C(X)$.* Ann. Inst. Fourier **27.4** (1977) 147–167.

[219] H. FAKHOURY. *Approximation des bornés d'un espace de Banach par des compacts et applications à l'approximation des opérateurs bornés.* J. Approx. Th. **26** (1979) 79–100.

[220] H. FAKHOURY. *Sur les M-idéaux dans certains espaces d'opérateurs et l'approximation par des opérateurs compacts.* Can. Math. Bull. **23** (1980) 401–411.

[221] M. FEDER. *On a certain subset of $L_1(0,1)$ and non-existence of best approximation in some spaces of operators.* J. Approx. Th. **29** (1980) 170–177.

[222] M. FEDER. *On subspaces of spaces with an unconditional basis and spaces of operators.* Illinois J. Math. **24** (1980) 196–205.

[223] M. FEDER. *On the non-existence of a projection onto the space of compact operators.* Can. Math. Bull. **25** (1982) 78–81.

[224] M. FEDER AND P. SAPHAR. *Spaces of compact operators and their dual spaces.* Israel J. Math. **21** (1975) 38–49.

[225] T. G. FEEMAN. *Best approximation and quasitriangular algebras.* Trans. Amer. Math. Soc. **288** (1985) 179–187.

[226] T. G. FEEMAN. *M-ideals and quasi-triangular algebras.* Illinois J. Math. **31** (1987) 89–98.

[227] T. G. FEEMAN. *A note on M-summands in dual spaces.* Can. Math. Bull. **30** (1987) 393–398.

[228] A. FERNÁNDEZ-LÓPEZ, E. GARCÍA-RUS, AND A. RODRÍGUEZ-PALACIOS. *A Zelmanov prime theorem for JB^*-algebras.* J. London Math. Soc. **46** (1992) 319–335.

[229] C. FINET. *Lacunary sets for groups and hypergroups.* J. Austral. Math. Soc. Ser. A **54** (1993) 39–60.

[230] C. Finet and V. Tardivel-Nachef. *Lacunary sets on transformation groups.* Preprint, 1992.

[231] C. Finet and V. Tardivel-Nachef. *A variant of a Yamaguchi's result.* Hokkaido Math. J. **21** (1992) 483–489.

[232] D. J. Fleming and D. M. Giarusso. *M-ideals in $L(\ell^1, E)$.* Can. Math. Bull. **29** (1986) 3–10.

[233] R. J. Fleming and J. E. Jamison. *Banach spaces with a basic inequality property and the best compact approximation property.* In: P. Nevai and A. Pinkus, editors, *Progress in Approximation Theory,* pages 347–362. Academic Press, 1991.

[234] R. J. Fleming and J. E. Jamison. *M-ideals and a basic inequality in Banach spaces.* In: P. Nevai and A. Pinkus, editors, *Progress in Approximation Theory,* pages 363–378. Academic Press, 1991.

[235] P. H. Flinn. *A characterization of M-ideals in $B(\ell_p)$ for $1 < p < \infty$.* Pacific J. Math. **98** (1982) 73–80.

[236] P. H. Flinn and R. R. Smith. *M-structure in the Banach algebra of operators on $C_0(\Omega)$.* Trans. Amer. Math. Soc. **281** (1984) 233–242.

[237] Y. Friedman and B. Russo. *Solution of the contractive projection problem.* J. Funct. Anal. **60** (1985) 56–79.

[238] Y. Friedman and B. Russo. *The Gelfand-Naimark theorem for JB^*-triples.* Duke Math. J. **53** (1986) 139–148.

[239] T. W. Gamelin. *Restrictions of subspaces of $C(X)$.* Trans. Amer. Math. Soc. **112** (1964) 278–286.

[240] T. W. Gamelin. *Uniform Algebras.* Prentice-Hall, Englewood Cliffs, 1969.

[241] T. W. Gamelin, D. E. Marshall, R. Younis, and W. R. Zame. *Function theory and M-ideals.* Ark. Mat. **23** (1985) 261–279.

[242] D. J. H. Garling. *On symmetric sequence spaces.* Proc. London Math. Soc. **16** (1966) 85–106.

[243] D. J. H. Garling. *Stable Banach spaces, random measures and Orlicz function spaces.* In: H. Heyer, editor, *Probability Measures on Groups. Proc. Conf. Oberwolfach 1981,* Lecture Notes in Math. 928, pages 121–175. Springer, Berlin-Heidelberg-New York, 1982.

[244] J. B. Garnett. *Bounded Analytic Functions.* Academic Press, New York, 1981.

[245] G. Gierz. *Bundles of Topological Vector Spaces and Their Duality.* Lecture Notes in Math. 955. Springer, Berlin-Heidelberg-New York, 1982.

[246] G. Gierz. *Integral representations of linear functionals on function modules.* Rocky Mountain J. Math. **17** (1987) 545–554.

[247] J. R. Giles, D. A. Gregory, and B. Sims. *Geometrical implications of upper semi-continuity of the duality mapping on a Banach space.* Pacific J. Math. **79** (1978) 99–109.

[248] A. Gleit and R. McGuigan. *A note on polyhedral Banach spaces.* Proc. Amer. Math. Soc. **33** (1972) 398–404.

[249] I. Glicksberg. *Measures orthogonal to algebras and sets of antisymmetry.* Trans. Amer. Math. Soc. **105** (1962) 415–435.

[250] G. Godefroy. *Espaces de Banach: existence et unicité de certains préduaux.* Ann. Inst. Fourier **28.3** (1978) 87–105.

[251] G. GODEFROY. *Nouvelles classes d'espaces de Banach à prédual unique.* In: *Séminaire d'analyse fonctionnelle,* Exposé VI. École Polytechnique, 1980–1981.

[252] G. GODEFROY. *Points de Namioka, espaces normants, applications à la théorie isométrique de la dualité.* Israel J. Math. **38** (1981) 209–220.

[253] G. GODEFROY. *Projections bicontractantes du bidual E'' d'un Banach E sur l'espace E. Applications.* Unpublished manuscript, 1981.

[254] G. GODEFROY. *Parties admissibles d'un espace de Banach. Applications.* Ann. scient. Éc. Norm. Sup. **16** (1983) 109–122.

[255] G. GODEFROY. *Symétries isométriques. Applications à la dualité des espaces réticulés.* Israel J. Math. **44** (1983) 61–74.

[256] G. GODEFROY. *Quelques remarques sur l'unicité des préduaux.* Quart. J. Math. Oxford (2) **35** (1984) 147–152.

[257] G. GODEFROY. *Sous-espaces bien disposés de L^1 - Applications.* Trans. Amer. Math. Soc. **286** (1984) 227–249.

[258] G. GODEFROY. *On Riesz subsets of abelian discrete groups.* Israel J. Math. **61** (1988) 301–331.

[259] G. GODEFROY. *Existence and uniqueness of isometric preduals: a survey.* In: Bor-Luh Lin, editor, *Banach Space Theory. Proc. of the Iowa Workshop on Banach Space Theory 1987*, pages 131–193. Contemp. Math. **85**, 1989.

[260] G. GODEFROY. *Metric characterizations of first Baire class linear forms and octahedral norms.* Studia Math. **95** (1989) 1–15.

[261] G. GODEFROY AND B. IOCHUM. *Arens-regularity of Banach algebras and the geometry of Banach spaces.* J. Funct. Anal. **80** (1988) 47–59.

[262] G. GODEFROY AND N. J. KALTON. *The ball topology and its applications.* In: Bor-Luh Lin, editor, *Banach Space Theory. Proc. of the Iowa Workshop on Banach Space Theory 1987*, pages 195–237. Contemp. Math. **85**, 1989.

[263] G. GODEFROY, N. J. KALTON, AND P. D. SAPHAR. *Idéaux inconditionnels dans les espaces de Banach.* C. R. Acad. Sc. Paris, Sér. A **313** (1991) 845–849.

[264] G. GODEFROY, N. J. KALTON, AND P. D. SAPHAR. *Unconditional ideals in Banach spaces.* Studia Math. (1993) (to appear).

[265] G. GODEFROY AND D. LI. *Banach spaces which are M-ideals in their bidual have property (u).* Ann. Inst. Fourier **39.2** (1989) 361–371.

[266] G. GODEFROY AND D. LI. *Some natural families of M-ideals.* Math. Scand. **66** (1990) 249–263.

[267] G. GODEFROY AND F. LUST-PIQUARD. *Some applications of geometry of Banach spaces to harmonic analysis.* Colloq. Math. **60/61** (1990) 443–456.

[268] G. GODEFROY AND P. SAAB. *Quelques espaces de Banach ayant les propriétés (V) ou (V^*) de A. Pełczyński.* C. R. Acad. Sc. Paris, Sér. A **303** (1986) 503–506.

[269] G. GODEFROY AND P. SAAB. *Weakly unconditionally convergent series in M-ideals.* Math. Scand. **64** (1989) 307–318.

[270] G. GODEFROY AND P. SAPHAR. *Duality in spaces of operators and smooth norms on Banach spaces.* Illinois J. Math. **32** (1988) 672–695.

[271] G. GODEFROY AND M. TALAGRAND. *Classes d'espaces de Banach à prédual unique.* C. R. Acad. Sc. Paris, Sér. A **292** (1981) 323–325.

[272] G. GODEFROY, S. TROYANSKI, J. H. M. WHITFIELD, AND V. ZIZLER. *Smoothness in weakly compactly generated Banach spaces.* J. Funct. Anal. **52** (1983) 344–353.

[273] G. GODEFROY, S. TROYANSKI, J. H. M. WHITFIELD, AND V. ZIZLER. *Locally uniformly rotund renorming and injections into* $c_0(\Gamma)$. Can. Math. Bull. **27** (1984) 494–500.

[274] G. GODINI. *Best approximation and intersection of balls.* In: A. Pietsch, N. Popa, and I. Singer, editors, *Banach Space Theory and its Applications. Proc. First Romanian-GDR Seminar, Bucharest 1981,* Lecture Notes in Math. 991, pages 44–54. Springer, Berlin-Heidelberg-New York, 1983.

[275] K. GOEBEL AND W. A. KIRK. *Topics in Metric Fixed Point Theory.* Cambridge Studies in Advanced Mathematics 28. Cambridge University Press, 1990.

[276] I. C. GOHBERG AND M. G. KREĬN. *Introduction to the Theory of Linear Nonselfadjoint Operators.* Translations of Mathematical Monographs, Vol. 18. American Mathematical Society, Providence, Rhode Island, 1969.

[277] I. C. GOHBERG, A. S. MARKUS, AND I. A. FELDMAN. *Normally solvable oerators and ideals associated with them.* Amer. Math. Soc. Transl. **61** (1967 (1960)) 63–84.

[278] P. GREIM. *Isometries and* L^p*-structure of separably valued Bochner* L^p*-spaces.* In: J. M. Belley, J. Dubois, and P. Morales, editors, *Measure Theory and Its Applications. Proc. Conf. Sherbrooke 1982.* Lecture Notes in Math. 1033, pages 209–218. Springer, Berlin-Heidelberg-New York, 1983.

[279] P. GREIM. *Banach spaces with the* L^1*-Banach-Stone property.* Trans. Amer. Math. Soc. **287** (1985) 819–828.

[280] N. GRØNBÆK AND G. WILLIS. *Approximate identities in Banach algebras of compact operators.* Can. Math. Bull. (1993) (to appear).

[281] A. GROTHENDIECK. *Produits tensoriels topologiques et espaces nucléaires.* Mem. Amer. Math. Soc. **16** (1955) .

[282] A. GROTHENDIECK. *Une caractérisation vectorielle-métrique des espaces* L^1. Canadian J. Math. **7** (1955) 552–561.

[283] S. GUERRE AND Y. RAYNAUD. *Sur les isométries de* $L^p(X)$ *et le théorème vectoriel ergodique.* Canadian J. Math. **40** (1988) 360–391.

[284] P. R. HALMOS. *Lectures on Boolean Algebras.* Van Nostrand, 1963.

[285] H. HANCHE-OLSEN. *Split faces and ideal structure of operator algebras.* Math. Scand. **48** (1981) 137–144.

[286] H. HANCHE-OLSEN AND E. STØRMER. *Jordan Operator Algebras.* Pitman, Boston, London, Melbourne, 1984.

[287] O. HANNER. *Intersections of translates of convex bodies.* Math. Scand. **4** (1956) 65–87.

[288] R. W. HANSELL. *On Borel mappings and Baire functions.* Trans. Amer. Math. Soc. **194** (1974) 195–211.

[289] R. W. HANSELL. *First class functions with values in nonseparable spaces.* In: Th. M. Rassias, editor, *Constantin Carathéodory: An International Tribute,* pages 461–475. World Scientific Publ. Co., 1991.

[290] A. B. HANSEN AND Å. LIMA. *The structure of finite dimensional Banach spaces with the 3.2. intersection property.* Acta Math. **146** (1981) 1–23.

[291] K.E. HARE. *An elementary proof of a result on* $\Lambda(p)$ *sets.* Proc. Amer. Math. Soc. **104** (1988) 829–834.

[292] P. HARMAND. *M-Ideale und die Einbettung eines Banachraumes in seinen Bidualraum.* Dissertation, FU Berlin, 1983.

[293] P. HARMAND AND Å. LIMA. *Banach spaces which are M-ideals in their biduals.* Trans. Amer. Math. Soc. **283** (1984) 253–264.

[294] P. HARMAND AND T. S. S. R. K. RAO. *An intersection property of balls and relations with M-ideals.* Math. Z. **197** (1988) 277–290.

[295] F. HAUSDORFF. *Set Theory.* Chelsea Publishing, New York, third edition, 1978.

[296] R. HAYDON, E. ODELL, AND H. P. ROSENTHAL. *On certain classes of Baire-1 functions with applications to Banach space theory.* In: E. Odell and H. P. Rosenthal, editors, *Functional Analysis. Proceedings of the Seminar at the University of Texas at Austin,* Lecture Notes in Math. 1470, pages 1–35. Springer, Berlin-Heidelberg-New York, 1991.

[297] R. HAYDON AND S. WASSERMANN. *A commutation result for tensor products of C^*-algebras.* Bull. London Math. Soc. **5** (1973) 283–287.

[298] S. HEINRICH. *The differentiability of the norm in spaces of operators.* Funct. Anal. Appl. **9** (1975) 360–362.

[299] S. HEINRICH. *Ultraproducts in Banach space theory.* J. Reine Angew. Math. **313** (1980) 72–104.

[300] S. HEINRICH. *Ultraproducts of L^1-predual spaces.* Fund. Math. **113** (1981) 221–234.

[301] S. HEINRICH, C. W. HENSON, AND L. C. MOORE. *Elementary equivalence of L_1-preduals.* In: A. Pietsch, N. Popa, and I. Singer, editors, *Banach Space Theory and its Applications. Proc. First Romanian-GDR Seminar, Bucharest 1981,* Lecture Notes in Math. 991, pages 79–90. Springer, Berlin-Heidelberg-New York, 1983.

[302] A. YA. HELEMSKII. *The Homology of Banach and Topological Algebras.* Mathematics and Its Applications (Soviet Series) 41. Kluwer Academic Publishers, Dordrecht, 1989.

[303] J. HENNEFELD. *A decomposition for $B(X)^*$ and unique Hahn-Banach extensions.* Pacific J. Math. **46** (1973) 197–199.

[304] J. HENNEFELD. *Compact extremal operators.* Illinois J. Math. **21** (1977) 61–65.

[305] J. HENNEFELD. *M-ideals, HB-subspaces, and compact operators.* Indiana Univ. Math. J. **28** (1979) 927–934.

[306] J. HENNEFELD. *A note on M-ideals in $B(X)$.* Proc. Amer. Math. Soc. **78** (1980) 89–92.

[307] J. HENNEFELD. *M-ideals, related spaces, and some approximation properties.* In: A. Pietsch, N. Popa, and I. Singer, editors, *Banach Space Theory and its Applications. Proc. First Romanian-GDR Seminar, Bucharest 1981,* Lecture Notes in Math. 991, pages 96–102. Springer, Berlin-Heidelberg-New York, 1983.

[308] R. H. HERMAN. *On the uniqueness of the ideals of compact and strictly singular operators.* Studia Math. **29** (1968) 161–165.

[309] E. HEWITT AND K. A. ROSS. *Abstract Harmonic Analysis I.* Springer, Berlin-Heidelberg-New York, second edition, 1979.

[310] E. HEWITT AND K. A. ROSS. *Abstract Harmonic Analysis II.* Springer, Berlin-Heidelberg-New York, 1970.

[311] P. J. HILTON AND U. STAMMBACH. *A Course in Homological Algebra.* Springer, Berlin-Heidelberg-New York, 1970.

[312] B. HIRSBERG. *M-ideals in complex function spaces and algebras.* Israel J. Math. **12** (1972) 133–146.

[313] B. HIRSBERG AND A. J. LAZAR. *Complex Lindenstrauss spaces with extreme points.* Trans. Amer. Math. Soc. **186** (1973) 141–150.

[314] A. HITCHCOCK. *Dial M for Murder.* Starring Grace Kelly and Ray Milland. Warner Brothers, 1954.

[315] G. HOCHSCHILD. *On the cohomology groups of an associative algebra.* Ann. of Math. **46** (1945) 58–76.

[316] G. HOCHSCHILD. *Cohomology and representations of associative algebras.* Duke Math. J. **14** (1947) 921–948.

[317] K. HOFFMAN. *Banach Spaces of Analytic Functions.* Prentice-Hall, Englewood Cliffs, 1962.

[318] R. B. HOLMES. *Geometric Functional Analysis and its Applications.* Springer, Berlin-Heidelberg-New York, 1975.

[319] R. B. HOLMES. *M-ideals in approximation theory.* In: G. G. Lorentz, C. K. Chui, and L. L. Schumaker, editors, *Approximation Theory II, Proc. Int. Symp.*, pages 391–396, Austin, 1976.

[320] R. B. HOLMES AND B. R. KRIPKE. *Best approximation by compact operators.* Indiana Univ. Math. J. **21** (1971) 255–263.

[321] R. B. HOLMES, B. SCRANTON, AND J. WARD. *Approximation from the space of compact operators and other M-ideals.* Duke Math. J. **42** (1975) 259–269.

[322] J. R. HOLUB. *A property of weakly compact operators on $C[0,1]$.* Proc. Amer. Math. Soc. **97** (1986) 396–398.

[323] J. R. HOLUB. *Daugavet's equation and operators on $C[0,1]$.* Math. Nachr. **141** (1989) 177–181.

[324] G. HORN. *Characterization of the predual and ideal structure of a JBW^*-triple.* Math. Scand. **61** (1987) 117–133.

[325] G. HORN. *Classification of JBW^*-triples of type I.* Math. Z. **196** (1987) 271–291.

[326] K. HRBACEK AND T. JECH. *Introduction to Set Theory.* Marcel Dekker, New York and Basel, 1978.

[327] O. HUSTAD. *A note on complex $\mathcal{P}_1$-spaces.* Israel J. Math. **16** (1973) 117–119.

[328] O. HUSTAD. *Intersection properties of balls in complex Banach spaces whose duals are L_1 spaces.* Acta Math. **132** (1974) 283–312.

[329] K. IZUCHI. *Extreme points of unit balls of quotients of L^∞ by Douglas algebras.* Illinois J. Math. **30** (1986) 138–147.

[330] K. IZUCHI. *QC-level sets and quotients of Douglas algebras.* J. Funct. Anal. **65** (1986) 293–308.

[331] K. IZUCHI. *Countably generated Douglas algebras.* Trans. Amer. Math. Soc. **299** (1987) 171–192.

[332] K. IZUCHI AND Y. IZUCHI. *Annihilating measures for Douglas algebras.* Yokohama Math. J. **32** (1984) 135–151.

[333] K. IZUCHI AND Y. IZUCHI. *Extreme and exposed points in quotients of Douglas algebras by H^∞ or $H^\infty + C$.* Yokohama Math. J. **32** (1984) 45–54.

[334] K. IZUCHI AND R. YOUNIS. *On the quotient space of two Douglas algebras.* Math. Japonica **31** (1986) 399–406.

[335] R. C. JAMES. *Uniformly non-square Banach spaces.* Ann. of Math. **80** (1964) 542–550.

[336] R. C. JAMES. *Renorming the Banach space* c_0. Proc. Amer. Math. Soc. **80** (1980) 631–634. *Erratum.* Ibid. **83** (1981) 442.

[337] G. J. O. JAMESON. *Topology and Normed Spaces.* Chapman and Hall, London, 1974.

[338] K. JAROSZ. *A generalization of the Banach-Stone theorem.* Studia Math. **73** (1982) 33–39.

[339] K. JAROSZ. *Perturbations of Banach Algebras.* Lecture Notes in Math. 1120. Springer, Berlin-Heidelberg-New York, 1985.

[340] K. JAROSZ. *Multipliers in complex Banach spaces and structure of the unit balls.* Studia Math. **87** (1987) 197–213.

[341] K. JAROSZ. *Ultraproducts and small bound perturbations.* Pacific J. Math. **148** (1991) 81–88.

[342] K. JAROSZ AND V. D. PATHAK. *Isometries and small bound isomorphisms of function spaces.* In: K. Jarosz, editor, *Function Spaces. Proc. Conf. Function Spaces, Edwardsville 1990,* Lecture Notes in Pure and Applied Mathematics 136, pages 241–271. Marcel Dekker, 1992.

[343] J. E. JAYNE AND C. A. ROGERS. *Borel selectors for upper semi-continuous set-valued maps.* Acta Math. **155** (1985) 41–79.

[344] K. JOHN. *On the compact non-nuclear operator problem.* Math. Ann. **287** (1990) 509–514.

[345] K. JOHN. *The uncomplemented subspace* $K(X,Y)$. Czech. Math. J. **42** (1992) 167–173.

[346] K. JOHN AND V. ZIZLER. *Smoothness and its equivalents in weakly compactly generated Banach spaces.* J. Funct. Anal. **15** (1974) 1–11.

[347] B. E. JOHNSON. *The Wedderburn decomposition of Banach algebras with finite-dimensional radical.* Amer. J. Math. **90** (1968) 866–876.

[348] B. E. JOHNSON. *Cohomology in Banach Algebras.* Mem. Amer. Math. Soc. **127** (1972) .

[349] B. E. JOHNSON. *Introduction to cohomology in Banach algebras.* In: J. H. Williamson, editor, *Algebras in Analysis*, pages 84–100. Academic Press, London, 1975.

[350] B. E. JOHNSON. *Perturbations of Banach algebras.* J. London Math. Soc. **34** (1977) 439–458.

[351] B. E. JOHNSON, R. V. KADISON, AND J. R. RINGROSE. *Cohomology of operator algebras III: Reduction to normal cohomology.* Bull. Soc. Math. France **100** (1972) 73–96.

[352] J. JOHNSON. *Remarks on Banach spaces of compact operators.* J. Funct. Anal. **32** (1979) 304–311.

[353] J. JOHNSON AND J. WOLFE. *On the norm of the canonical projection of* E^{***} *onto* $E^{\perp}$. Proc. Amer. Math. Soc. **75** (1979) 50–52.

[354] W. B. JOHNSON. *A complementary universal conjugate Banach space and its relation to the approximation problem.* Israel J. Math. **13** (1972) 301–310.

[355] W. B. JOHNSON. *Complementably universal separable Banach spaces: An application of counterexamples to the approximation problem.* In: R. Bartle, editor, *Studies in Functional Analysis*, pages 81–114. MAA Studies in Mathematics, Vol. 21, 1980.

[356] W. B. JOHNSON AND J. LINDENSTRAUSS. *Some remarks on weakly compactly generated Banach spaces.* Israel J. Math. **17** (1974) 219–230. *Corrigendum.* Ibid. **32** (1979) 382–383.

[357] R. V. KADISON. *Isometries of operator algebras.* Ann. of Math. **54** (1951) 325–338.

[358] R. V. KADISON. *Derivations of operator algebras.* Ann. of Math. **83** (1966) 280–293.

[359] R. V. KADISON AND J. R. RINGROSE. *Cohomology of operator algebras I. Type I von Neumann algebras.* Acta Math. **126** (1971) 227–243.

[360] R. V. KADISON AND J. R. RINGROSE. *Cohomology of operator algebras II. Extended cobounding and the hyperfinite case.* Ark. Mat. **9** (1971) 55–63.

[361] R. V. KADISON AND J. R. RINGROSE. *Fundamentals of the Theory of Operator Algebras. Vol. I: Elementary Theory.* Academic Press, 1983.

[362] R. V. KADISON AND J. R. RINGROSE. *Fundamentals of the Theory of Operator Algebras. Vol. II: Advanced Theory.* Academic Press, 1986.

[363] J.-P. KAHANE. *Best approximation in $L^1(T)$.* Bull. Amer. Math. Soc. **80** (1974) 788–804.

[364] S. KAKUTANI. *Concrete representation of abstract M-spaces. (A characterization of the space of continuous functions.).* Ann. of Math. **42** (1941) 994–1024.

[365] N. J. KALTON. *Spaces of compact operators.* Math. Ann. **208** (1974) 267–278.

[366] N. J. KALTON. *Banach spaces for which the ideal of compact operators is an M-ideal.* C. R. Acad. Sc. Paris, Sér. A **313** (1991) 509–513.

[367] N. J. KALTON. *M-ideals of compact operators.* Illinois J. Math. **37** (1993) 147–169.

[368] N. J. KALTON, E. SAAB, AND P. SAAB. *$L^p(X)$ $(1 \le p < \infty)$ has the property (u) whenever X does.* Bull. Sc. math., 2e série **115** (1991) 369–377.

[369] L. V. KANTOROVICH AND G. P. AKILOV. *Functional Analysis.* Pergamon Press, Oxford, second edition, 1982.

[370] D. KASTLER. *Cyclic Cohomology within the Differential Envelope.* Travaux en cours 30. Hermann, Paris, 1988.

[371] W. KAUP. *Algebraic characterization of symmetric complex Banach manifolds.* Math. Ann. **228** (1977) 39–64.

[372] W. KAUP. *A Riemann mapping theorem for bounded symmetric domains in complex Banach spaces.* Math. Z. **183** (1983) 503–529.

[373] W. KAUP. *Contractive projections on Jordan C^*-algebras and generalizations.* Math. Scand. **54** (1984) 95–100.

[374] W. KAUP AND H. UPMEIER. *Banach spaces with biholomorphically equivalent unit balls are isomorphic.* Proc. Amer. Math. Soc. **58** (1976) 129–133.

[375] A. S. KECHRIS AND A. LOUVEAU. *A classification of Baire class 1 functions.* Trans. Amer. Math. Soc. **318** (1990) 209–236.

[376] E. KIRCHBERG. *C^*-nuclearity implies CPAP.* Math. Nachrichten **76** (1977) 203–212.

[377] J. W. KITCHEN AND D. A. ROBBINS. *Tensor products of Banach bundles.* Pacific J. Math. **94** (1981) 151–169.

[378] J. W. KITCHEN AND D. A. ROBBINS. *Gelfand representations of Banach modules.* Dissertationes Mathematicae **202** (1982) .

[379] F. KITTANEH AND R. YOUNIS. *Smooth points of certain operator spaces.* Integr. Equat. Oper. Th. **13** (1990) 849–855.

[380] V. L. KLEE. *Some new results on smoothness and rotundity in normed linear spaces.* Math. Ann. **139** (1959) 51–63.

[381] H. KNAUST AND E. ODELL. *On c_0 sequences in Banach spaces.* Israel J. Math. **67** (1989) 153–169.

[382] J. KOMLÓS. *A generalization of a problem of Steinhaus.* Acta Math. Acad. Sci. Hung. **18** (1967) 217–229.

[383] J.-L. KRIVINE. *Sous-espaces de dimension finie des espaces de Banach réticulés.* Ann. of Math. **104** (1976) 1–29.

[384] J.-L. KRIVINE AND B. MAUREY. *Espaces de Banach stables.* Israel J. Math. **39** (1981) 273–295.

[385] K. KURATOWSKI. *Topology I.* Academic Press, New York, 1966.

[386] H. E. LACEY. *The Isometric Theory of Classical Banach Spaces.* Springer, Berlin-Heidelberg-New York, 1974.

[387] K.-S. LAU. *The dual ball of a Lindenstrauss space.* Math. Scand. **33** (1973) 323–337.

[388] K.-S. LAU. *On a sufficient condition for proximity.* Trans. Amer. Math. Soc. **251** (1979) 343–356.

[389] A. J. LAZAR. *Spaces of affine continuous functions on simplexes.* Trans. Amer. Math. Soc. **134** (1968) 503–525.

[390] A. J. LAZAR AND J. LINDENSTRAUSS. *Banach spaces whose duals are L_1 spaces and their representing matrices.* Acta Math. **126** (1971) 165–193.

[391] A. J. LAZAR, D. E. WULBERT, AND P. D. MORRIS. *Continuous selections for metric projections.* J. Funct. Anal. **3** (1969) 193–216.

[392] H. LEMBERG. *Nouvelle démonstration d'un théorème de J. L. Krivine sur la finie représentation de ℓ_p dans un espace de Banach.* Israel J. Math. **39** (1981) 341–348.

[393] V. L. LEVIN. *The Lebesgue decomposition for functionals on the vector-function space L_X^∞.* Funct. Anal. Appl. **8** (1974) 314–317.

[394] D. LEWIS AND C. STEGALL. *Banach spaces whose duals are isomorphic to $\ell_1(\Gamma)$.* J. Funct. Anal. **12** (1973) 177–187.

[395] D. LI. *Espaces L-facteurs de leurs biduaux: bonne disposition, meilleure approximation et propriété de Radon-Nikodym.* Quart. J. Math. Oxford (2) **38** (1987) 229–243.

[396] D. LI. *Lifting properties for some quotients of L^1-spaces and other spaces L-summand in their bidual.* Math. Z. **199** (1988) 321–329.

[397] D. LI. *Quantitative unconditionality of Banach spaces E for which $K(E)$ is an M-ideal in $L(E)$.* Studia Math. **96** (1990) 39–50.

[398] Å. LIMA. *An application of a theorem of Hirsberg and Lazar.* Math. Scand. **38** (1976) 325–340.

[399] Å. LIMA. *Complex Banach spaces whose duals are L_1-spaces.* Israel J. Math. **24** (1976) 59–72.

[400] Å. LIMA. *Intersection properties of balls and subspaces in Banach spaces.* Trans. Amer. Math. Soc. **227** (1977) 1–62.

[401] Å. LIMA. *Intersection properties of balls in spaces of compact operators.* Ann. Inst. Fourier **28.3** (1978) 35–65.

[402] Å. LIMA. *M-ideals of compact operators in classical Banach spaces.* Math. Scand. **44** (1979) 207–217.

[403] A. LIMA. *Banach spaces with the 4.3. intersection property.* Proc. Amer. Math. Soc. **80** (1980) 431–434.

[404] A. LIMA. *On M-ideals and best approximation.* Indiana Univ. Math. J. **31** (1982) 27–36.

[405] A. LIMA. *Uniqueness of Hahn-Banach extensions and liftings of linear dependences.* Math. Scand. **53** (1983) 97–113.

[406] A. LIMA. *The metric approximation property, norm-one projections and intersection properties of balls.* Israel J. Math. (1992) (to appear).

[407] A. LIMA, E. OJA, T. S. S. R. K. RAO, AND D. WERNER. *Geometry of operator spaces.* Preprint, 1992.

[408] A. LIMA AND G. OLSEN. *Extreme points in duals of complex operator spaces.* Proc. Amer. Math. Soc. **94** (1985) 437–440.

[409] A. LIMA, G. OLSEN, AND U. UTTERSRUD. *Intersections of M-ideals and G-spaces.* Pacific J. Math. **104** (1983) 175–177.

[410] A. LIMA AND A. K. ROY. *Characterizations of complex L^1-preduals.* Quart. J. Math. Oxford (2) **35** (1984) 439–453.

[411] A. LIMA AND A. K. ROY. *Some results on intersection properties of balls in complex Banach spaces.* Studia Math. **83** (1986) 37–45.

[412] A. LIMA AND U. UTTERSRUD. *Centers of symmetry in finite intersections of balls in Banach spaces.* Israel J. Math. **44** (1983) 189–200.

[413] A. LIMA AND D. YOST. *Absolutely Chebyshev subspaces.* In: S. Fitzpatrick and J. Giles, editors, *Workshop/Miniconference Funct. Analysis/Optimization. Canberra 1988*, pages 116–127. Proc. Cent. Math. Anal. Austral. Nat. Univ. **20**, 1988.

[414] J. LINDENSTRAUSS. *Extension of compact operators.* Mem. Amer. Math. Soc. **48** (1964) .

[415] J. LINDENSTRAUSS. *On nonlinear projections in Banach spaces.* Michigan Math. J. **11** (1964) 263–287.

[416] J. LINDENSTRAUSS. *On reflexive spaces having the metric approximation property.* Israel J. Math. **3** (1965) 199–204.

[417] J. LINDENSTRAUSS. *On nonseparable reflexive Banach spaces.* Bull. Amer. Math. Soc. **72** (1966) 967–970.

[418] J. LINDENSTRAUSS. *Weakly compact sets — their topological properties and the Banach spaces they generate.* In: R. D. Anderson, editor, *Symposium on infinite dimensional topology*, pages 235–273. Ann. of Math. Studies **69**, 1972.

[419] J. LINDENSTRAUSS AND A. PEŁCZYŃSKI. *Absolutely summing operators in $\mathcal{L}_p$-spaces and their applications.* Studia Math. **29** (1968) 275–326.

[420] J. LINDENSTRAUSS AND H. P. ROSENTHAL. *The $\mathcal{L}_p$ spaces.* Israel J. Math. **7** (1969) 325–349.

[421] J. LINDENSTRAUSS AND L. TZAFRIRI. *Classical Banach Spaces.* Lecture Notes in Math. 338. Springer, Berlin-Heidelberg-New York, 1973.

[422] J. LINDENSTRAUSS AND L. TZAFRIRI. *Classical Banach Spaces I.* Springer, Berlin-Heidelberg-New York, 1977.

[423] J. LINDENSTRAUSS AND L. TZAFRIRI. *Classical Banach Spaces II.* Springer, Berlin-Heidelberg-New York, 1979.

[424] J. LINDENSTRAUSS AND D. E. WULBERT. *On the classification of the Banach spaces whose duals are L_1 spaces.* J. Funct. Anal. **4** (1969) 332–349.

[425] J. M. LÓPEZ AND K. A. ROSS. *Sidon Sets.* Marcel Dekker, New York, 1975.

[426] G. G. LORENTZ. *Some new functional spaces.* Ann. of Math. **51** (1950) 37–55.

[427] D. H. LUECKING. *The compact Hankel operators form an M-ideal in the space of Hankel operators.* Proc. Amer. Math. Soc. **79** (1980) 222–224.

[428] D. H. LUECKING AND R. M. YOUNIS. *Quotients of L^∞ by Douglas algebras and best approximation.* Trans. Amer. Math. Soc. **276** (1983) 699–706.

[429] J. LUKEŠ, J. MALÝ, AND L. ZAJÍČEK. *Fine Topology Methods in Real Analysis and Potential Theory.* Lecture Notes in Math. 1189. Springer, Berlin-Heidelberg-New York, 1986.

[430] W. LUSKY. *On separable Lindenstrauss spaces.* J. Funct. Anal. **26** (1977) 103–120.

[431] W. LUSKY. *On the structure of $Hv_0(D)$ and $hv_0(D)$.* Math. Nachr. (1993) (to appear).

[432] F. LUST. *Ensembles de Rosenthal et ensembles de Riesz.* C. R. Acad. Sc. Paris, Sér. A **282** (1976) 833–835.

[433] F. LUST-PIQUARD. *Propriétés géométriques des sous-espaces invariants par translation de $L^1(G)$ et $C(G)$.* In: *Séminaire sur la géométrie des espaces de Banach*, Exposé XXVI. École Polytechnique, 1977–1978.

[434] J. MACH AND J. D. WARD. *Approximation by compact operators on certain Banach spaces.* J. Approx. Th. **23** (1978) 274–286.

[435] J. MARTÍNEZ-MORENO, J. F. MENA-JURADO, R. PAYÁ, AND A. RODRÍGUEZ-PALACIOS. *An approach to numerical ranges without Banach algebra theory.* Illinois J. Math. **29** (1985) 609–626.

[436] B. MAUREY. *Types and ℓ_1-subspaces.* In: E. Odell and H. P. Rosenthal, editors, *Texas Functional Analysis Seminar*, pages 123–137. The University of Texas at Austin, Longhorn Notes, 1982–1983.

[437] R. D. MCWILLIAMS. *A note on weak sequential convergence.* Pacific J. Math. **12** (1962) 333–335.

[438] R. D. MCWILLIAMS. *Iterated w^*-sequential closure of a Banach space in its second conjugate.* Proc. Amer. Math. Soc. **16** (1965) 1195–1199.

[439] J. F. MENA-JURADO, R. PAYÁ, AND A. RODRÍGUEZ-PALACIOS. *Semisummands and semiideals in Banach spaces.* Israel J. Math. **51** (1985) 33–67.

[440] J. F. MENA-JURADO, R. PAYÁ, AND A. RODRÍGUEZ-PALACIOS. *Absolute subspaces of Banach spaces.* Quart. J. Math. Oxford (2) **40** (1989) 43–64.

[441] J. F. MENA-JURADO, R. PAYÁ, A. RODRÍGUEZ-PALACIOS, AND D. YOST. *Absolutely proximinal subspaces of Banach spaces.* J. Approx. Th. **65** (1991) 46–72.

[442] Y. MEYER. *Spectres des mesures et mesures absolument continues.* Studia Math. **30** (1968) 87–99.

[443] E. MICHAEL. *Continuous selections I.* Ann. of Math. **63** (1956) 361–382.

[444] E. MICHAEL AND I. NAMIOKA. *Barely continuous functions.* Bull. Acad. Polon. Sci. **24** (1976) 889–892.

[445] E. MICHAEL AND A. PEŁCZYŃSKI. *A linear extension theorem.* Illinois J. Math. **11** (1967) 563–579.

[446] E. Michael and A. Pełczyński. *Peaked partition subspaces of $C(X)$.* Illinois J. Math. **11** (1967) 555–562.

[447] I. M. Miheev. *Trigonometric series with gaps.* Anal. Math. **9** (1983) 43–55.

[448] V. D. Milman. *The geometric theory of Banach spaces I. The theory of basic and minimal systems.* Russian Math. Surveys **25.3** (1970) 111–170.

[449] I. Namioka and R. F. Wheeler. *Gul'ko's proof of the Amir-Lindenstrauss theorem.* In: R. G. Bartle et al., editor, *Geometry of normed linear spaces, Proc. Conf. Hon. M. M. Day, Urbana-Champaign / Ill. 1983*, pages 113–120. Contemp. Math. **52**, 1986.

[450] C. Nara. *Uniqueness of the predual of the Bloch space and its strongly exposed points.* Illinois J. Math. **34** (1990) 98–107.

[451] S. Negrepontis. *Banach spaces and topology.* In: K. Kunen and J. E. Vaughan, editors, *Handbook of set-theoretic topology*, pages 1045–1142. Elsevier Science Publishers B. V., 1984.

[452] C. Niculescu. *Weak compactness in Banach lattices.* J. Operator Theory **6** (1981) 217–231.

[453] C. Niculescu. *A commutative extremal extension of Hilbert-Schmidt theorem.* Rev. Roum. Math. Pures Appl. **34** (1988) 247–261.

[454] C. Niculescu. *Lectures on Alfsen-Effros theory.* Preprint, 1990.

[455] C. Niculescu and D. T. Vuza. *Alfsen-Effros type order relations defined by vector norms.* Rev. Roum. Math. Pures Appl. **33** (1988) 751–766.

[456] N. J. Nielsen and G. H. Olsen. *Complex preduals of L_1 and subspaces of $\ell_\infty^n(\mathbb{C})$.* Math. Scand. **40** (1977) 271–287.

[457] E. Odell. *On quotients of Banach spaces having shrinking unconditional bases.* Preprint, 1990.

[458] E. Odell and H. P. Rosenthal. *A double dual characterization of separable Banach spaces containing l_1.* Israel J. Math. **20** (1975) 375–384.

[459] E. Odell and T. Schlumprecht. *The distortion problem.* Preprint, 1992.

[460] E. Oja. *On the uniqueness of the norm-preserving extension of a linear functional in the Hahn-Banach theorem.* Izv. Akad. Nauk Est. SSR **33** (1984) 424–438. (Russian).

[461] E. Oja. *Strong uniqueness of the extension of linear functionals according to the Hahn-Banach theorem.* Math. Notes **43** (1988) 134–139.

[462] E. Oja. *Dual de l'espace des opérateurs linéaires continus.* C. R. Acad. Sc. Paris, Sér. A **309** (1989) 983–986.

[463] E. Oja. *Isometric properties of the subspace of compact operators in a space of continuous linear operators.* Math. Notes **45** (1989) 472–475.

[464] E. Oja. *On M-ideals of compact operators and Lorentz sequence spaces.* Proc. Estonian Acad. Sci. Phys. Math. **40** (1991) 31–36.

[465] E. Oja. *Remarks on the dual of the space of continuous linear operators.* Acta et Comment. Universitatis Tartuensis **928** (1991) 89–96.

[466] E. Oja. *A note on M-ideals of compact operators.* Acta et Comment. Universitatis Tartuensis (1993) (to appear).

[467] E. Oja and D. Werner. *Remarks on M-ideals of compact operators on $X \oplus_p X$.* Math. Nachr. **152** (1991) 101–111.

[468] G. H. OLSEN. *On the classification of complex Lindenstrauss spaces.* Math. Scand. **35** (1974) 237–258.

[469] G. H. OLSEN. *Edwards' separation theorem for complex Lindenstrauss spaces with application to selection and embedding theorems.* Math. Scand. **38** (1976) 97–105.

[470] J. ORIHUELA AND M. VALDIVIA. *Projective generators and resolutions of identity in Banach spaces.* Rev. Mat. Univ. Complutense Madr. **2** (1989) 179–199.

[471] T. W. PALMER. *The bidual of the compact operators.* Trans. Amer. Math. Soc. **288** (1985) 827–839.

[472] S. H. PARK. *A relationship between the $1\frac{1}{2}$-ball property and the strong $1\frac{1}{2}$-ball property.* J. Approx. Th. **57** (1989) 332–346.

[473] S. H. PARK. *Uniform Hausdorff strong uniqueness.* J. Approx. Th. **58** (1989) 78–89.

[474] J. R. PARTINGTON. *Hermitian operators for absolute norms and absolute direct sums.* Linear Algebra and its Appl. **23** (1979) 275–280.

[475] J. R. PARTINGTON. *An Introduction to Hankel Operators.* Cambridge University Press, Cambridge, 1988.

[476] V. I. PAULSEN. *Completely Bounded Maps and Dilations.* Pitman Research Notes in Mathematics 146. Longman, 1986.

[477] R. PAYÁ. *Numerical range of operators and structure in Banach spaces.* Quart. J. Math. Oxford (2) **33** (1982) 357–364.

[478] R. PAYÁ, J. PÉREZ, AND A. RODRÍGUEZ-PALACIOS. *Noncommutative Jordan C^*-algebras.* Manuscr. Math. **37** (1982) 87–120.

[479] R. PAYÁ, J. PÉREZ, AND A. RODRÍGUEZ-PALACIOS. *Type I factor representations of non-commutative JB^*-algebras.* Proc. London Math. Soc. **48** (1984) 428–444.

[480] R. PAYÁ AND A. RODRÍGUEZ-PALACIOS. *Banach spaces which are semi-L-summands in their biduals.* Math. Ann. **289** (1991) 529–542.

[481] R. PAYÁ AND W. WERNER. *An approximation property related to M-ideals of compact operators.* Proc. Amer. Math. Soc. **111** (1991) 993–1001.

[482] R. PAYÁ AND D. YOST. *The two-ball property: transitivity and examples.* Mathematika **35** (1988) 190–197.

[483] G. K. PEDERSEN. *Spectral formulas in quotient C^*-algebras.* Math. Z. **148** (1976) 299–300.

[484] A. PEŁCZYŃSKI. *Some aspects of the present theory of Banach spaces.* In: Stefan Banach, *Œuvres, Vol. II*, pages 221–302, PWN, Warszawa, 1979.

[485] A. PEŁCZYŃSKI. *A connection between weakly unconditional convergence and weak completeness of Banach spaces.* Bull. Acad. Polon. Sci. **6** (1958) 251–253.

[486] A. PEŁCZYŃSKI. *Banach spaces on which every unconditionally converging operator is weakly compact.* Bull. Acad. Polon. Sci. **10** (1962) 641–648.

[487] A. PEŁCZYŃSKI. *On simultaneous extension of continuous functions.* Studia Math. **24** (1964) 285–304.

[488] A. PEŁCZYŃSKI. *Supplement to my paper 'On simultaneous extension of continuous functions'.* Studia Math. **25** (1964) 157–161.

[489] A. PEŁCZYŃSKI. *Linear extensions, linear averagings, and their applications to linear topological classification of spaces of continuous functions.* Dissertationes Mathematicae **58** (1968) .

[490] A. PEŁCZYŃSKI. *Banach spaces of analytic functions and absolutely summing operators.* CBMS No. 30. American Mathematical Society, Providence, Rhode Island, 1977.

[491] A. PEŁCZYŃSKI AND P. WOJTASZCZYK. *Banach spaces with finite dimensional expansions of identity and universal bases of finite dimensional subspaces.* Studia Math. **40** (1971) 91–108.

[492] F. PERDRIZET. *Espaces de Banach ordonnés et idéaux.* J. Math. pures et appl. **49** (1970) 61–98.

[493] H. PFITZNER. *L-summands in their biduals have property (V^*).* Studia Math. (1993) (to appear).

[494] H. PFITZNER. *Weak compactness in C^*-algebras is determined commutatively.* Math. Ann. (1993) (to appear).

[495] R. R. PHELPS. *Uniqueness of Hahn-Banach extensions and unique best approximation.* Trans. Amer. Math. Soc. **95** (1960) 238–255.

[496] R. R. PHELPS. *The Choquet representation in the complex case.* Bull. Amer. Math. Soc. **83** (1977) 299–312.

[497] R. R. PHELPS. *Convex Functions, Monotone Operators and Differentiability.* Lecture Notes in Math. 1364. Springer, Berlin-Heidelberg-New York, 1989.

[498] A. PIETSCH. *Operator Ideals.* North Holland, Amsterdam-New York-Oxford, 1980.

[499] L. PIGNO AND S. SAEKI. *On the theorem of F. and M. Riesz.* Colloq. Math. **60/61** (1990) 315–320.

[500] G. PISIER. *Counterexamples to a conjecture of Grothendieck.* Acta Math. **151** (1983) 181–208.

[501] G. PISIER. *Factorisation of linear operators and geometry of Banach spaces.* CBMS No. 60. American Mathematical Society, Providence, Rhode Island, 1986.

[502] G. PISIER. *Weak Hilbert spaces.* Proc. London Math. Soc. **56** (1988) 547–579.

[503] G. PISIER. *Completely bounded maps between sets of Banach space operators.* Indiana Univ. Math. J. **39** (1990) 249–277.

[504] H. POINCARÉ. *Analysis Situs.* In: *Œuvres, Vol. VI*, pages 193–288. Gauthier-Villars, 1953.

[505] Y. T. POON AND Z.-J. RUAN. *M-ideals and quotients of subdiagonal algebras.* J. Funct. Anal. **105** (1992) 144–170.

[506] S. C. POWER. *Hankel operators on Hilbert space.* Bull. London Math. Soc. **12** (1980) 422–442.

[507] S. C. POWER. *Hankel Operators on Hilbert Space.* Pitman, Boston, London, Melbourne, 1982.

[508] S. C. POWER. *Analysis in nest algebras.* In: J. B. Conway and B. B. Morell, editors, *Surveys of Some Recent Results in Operator Theory, Vol. II.* Pitman Research Notes in Mathematics 192, pages 189–234. Longman, 1988.

[509] R. T. POWERS. *Representations of uniformly hyperfinite algebras and their associated von Neumann rings.* Ann. of Math. **86** (1967) 138–171.

[510] R. T. PROSSER. *On the ideal structure of operator algebras.* Mem. Amer. Math. Soc. **45** (1963) .

[511] K. PRZESŁAWSKI AND D. YOST. *Continuity properties of selectors and Michael's theorem.* Michigan Math. J. **36** (1989) 113–134.

[512] D. Przeworska-Rolewicz and S. Rolewicz. *Equations in Linear Spaces.* PWN, Warszawa, 1968.

[513] F. Räbiger. *Beiträge zur Strukturtheorie der Grothendieckräume.* In: Sitzungsberichte der Heidelberger Akademie der Wissenschaften, Mathematisch-naturwissenschaftliche Klasse, Jahrgang 1985, 4. Abhandlung. Springer, Berlin-Heidelberg-New York, 1985.

[514] I. Raeburn and J. L. Taylor. *Hochschild cohomology and perturbations of Banach algebras.* J. Funct. Anal. **25** (1977) 258–267.

[515] M. Rao. *On simultaneous extensions.* Inventiones Math. **13** (1971) 284–294.

[516] T. S. S. R. K. Rao. *Characterizations of some classes of L^1-preduals by the Alfsen–Effros structure topology.* Israel J. Math. **42** (1982) 20–32.

[517] T. S. S. R. K. Rao. *$L(X, C(K))$ as a dual space.* Proc. Amer. Math. Soc. **110** (1990) 727–729.

[518] T. S. S. R. K. Rao. *$C(K, X)$ as an M-ideal in $WC(K, X)$.* Proc. Indian Acad. Sci. **102** (1992) 217–224.

[519] T. S. S. R. K. Rao, A. K. Roy, and K. Sundaresan. *Intersection properties of balls in tensor products of some Banach spaces.* Math. Scand. **65** (1989) 103–118.

[520] Y. Raynaud. *Deux nouveaux exemples d'espaces de Banach stables.* C. R. Acad. Sc. Paris, Sér. A **292** (1981) 715–717.

[521] S. Reisner. *Minimal volume-product in Banach spaces with a 1-unconditional basis.* J. London Math. Soc. **36** (1987) 126–136.

[522] H. Reiter. *Contributions to harmonic analysis: VI.* Ann. of Math. **77** (1963) 552–562.

[523] J. R. Ringrose. *Cohomology of operator algebras.* In: *Lectures on Operator Algebras,* Lecture Notes in Math. 247, pages 359–435. Springer, Berlin-Heidelberg-New York, 1972.

[524] A. Rodríguez-Palacios. *Non-associative normed algebras spanned by hermitian elements.* Proc. London Math. Soc. **47** (1983) 258–274.

[525] A. Rodríguez-Palacios. *A note on Arens regularity.* Quart. J. Math. Oxford (2) **38** (1987) 91–93.

[526] A. Rodríguez-Palacios. *Infinite-dimensional sets of constant width and their applications.* Extracta Math. **5** (1990) 41–52.

[527] A. Rodríguez-Palacios. *Jordan structures in analysis.* In: Proc. Conf. Jordan Algebras, Oberwolfach 1992, (to appear).

[528] C. A. Rogers. *Functions of the first Baire class.* J. London Math. Soc. **37** (1988) 535–544.

[529] H. P. Rosenthal. *On relatively disjoint families of measures, with some applications to Banach space theory.* Studia Math. **37** (1970) 13–36, 311–313.

[530] H. P. Rosenthal. *On subspaces of L^p.* Ann. of Math. **97** (1973) 344–373.

[531] H. P. Rosenthal. *A characterization of Banach spaces containing c_0.* Preprint, 1992.

[532] A. K. Roy. *Convex functions on the dual ball of a complex Lindenstrauss space.* J. London Math. Soc. **20** (1979) 529–540.

[533] A. K. Roy. *A short proof of a theorem on complex Lindenstrauss spaces.* Israel J. Math. **38** (1981) 41–45.

[534] N. M. Roy. *A characterization of square Banach spaces.* Israel J. Math. **17** (1974) 142–148.

[535] N. M. Roy. *Contractive projections in square Banach spaces.* Proc. Amer. Math. Soc. **59** (1976) 291–296.

[536] N. M. Roy. *An M-ideal characterization of G-spaces.* Pacific J. Math. **92** (1981) 151–160.

[537] N. M. Roy. *Extreme points and $\ell_1(\Gamma)$-spaces.* Proc. Amer. Math. Soc. **86** (1982) 216–218.

[538] Z.-J. Ruan. *Subspaces of C^*-algebras.* J. Funct. Anal. **76** (1988) 217–230.

[539] Z.-J. Ruan. *On the predual of dual algebras.* J. Operator Theory (to appear) .

[540] L. A. Rubel and A. L. Shields. *The second duals of certain spaces of analytic functions.* J. Austral. Math. Soc. **11** (1970) 276–280.

[541] W. Rudin. *Trigonometric series with gaps.* J. Math. and Mech. **9** (1960) 203–227.

[542] W. Rudin. *Fourier Analysis on Groups.* Interscience Publishers, New York-London, 1962.

[543] W. Rudin. *Functional Analysis.* McGraw-Hill, New York, 1973.

[544] W. Rudin. *Spaces of type $H^\infty + C$.* Ann. Inst. Fourier **25.1** (1975) 99–125.

[545] W. Rudin. *Real and Complex Analysis.* McGraw-Hill, New York, third edition, 1986.

[546] W. Ruess and Ch. Stegall. *Extreme points in duals of operator spaces.* Math. Ann. **261** (1982) 535–546.

[547] W. Ruess and Ch. Stegall. *Exposed and denting points in duals of operator spaces.* Israel J. Math. **53** (1986) 163–190.

[548] W. Ruess and D. Werner. *Structural properties of operator spaces.* Acta Univ. Carol. Math. Phys. **28** (1987) 127–136.

[549] E. Saab and P. Saab. *On Pełczyński's properties (V) and (V^*).* Pacific J. Math. **125** (1986) 205–210.

[550] E. Saab and P. Saab. *On stability problems of some properties in Banach spaces.* In: K. Jarosz, editor, *Function Spaces. Proc. Conf. Function Spaces, Edwardsville 1990*, Lecture Notes in Pure and Applied Mathematics 136, pages 367–394. Marcel Dekker, 1992.

[551] K. Saatkamp. *M-ideals of compact operators.* Math. Z. **158** (1978) 253–263.

[552] K. Saatkamp. *Schnitteigenschaften und beste Approximation.* Dissertation, Universität Bonn, 1979.

[553] K. Saatkamp. *Best approximation in the space of bounded operators and its applications.* Math. Ann. **250** (1980) 35–54.

[554] S. Sakai. *Derivations of W^*-algebras.* Ann. of Math. **83** (1966) 273–279.

[555] S. Sakai. *C^*-Algebras and W^*-Algebras.* Springer, Berlin-Heidelberg-New York, 1971.

[556] D. Sarason. *Algebras of functions on the unit circle.* Bull. Amer. Math. Soc. **79** (1973) 286–299.

[557] D. Sarason. *Algebras between L^∞ and H^∞.* In: O. B. Bekken, B. K. Øksendal, and A. Stray, editors, *Spaces of Analytic Functions. Proc Conf. Kristiansand 1975*, Lecture Notes in Math. 512, pages 117–130. Springer, Berlin-Heidelberg-New York, 1976.

[558] D. Sarason. *Function Theory on the Unit Circle.* Virginia Poly. Inst. and State Univ., Blacksburg, Virginia, 1979.

[559] H. H. Schaefer. *Topological Vector Spaces.* Springer, Berlin-Heidelberg-New York, third edition, 1971.

[560] H. H. Schaefer. *Banach Lattices and Positive Operators.* Springer, Berlin-Heidelberg-New York, 1974.

[561] K. D. SCHMIDT. *Daugavet's equation and orthomorphisms.* Proc. Amer. Math. Soc. **108** (1990) 905–911.

[562] J. SCHREIER. *Ein Gegenbeispiel zur schwachen Konvergenz.* Studia Math. **2** (1930) 58–62.

[563] A. K. SEDA. *Integral representation of linear functionals on spaces of sections.* Proc. Amer. Math. Soc. **91** (1984) 549–555.

[564] A. SERSOURI. *Propriété (u) dans les espaces d'opérateurs.* Bull. Pol. Acad. Sci. **36** (1988) 655–659.

[565] J. H. SHAPIRO. *Subspaces of $L^p(G)$ spanned by characters: $0 < p < 1$.* Israel J. Math. **29** (1978) 248–264.

[566] A. L. SHIELDS AND D. L. WILLIAMS. *Bounded projections, duality, and multipliers in spaces of analytic functions.* Trans. Amer. Math. Soc. **162** (1971) 287–302.

[567] A. L. SHIELDS AND D. L. WILLIAMS. *Bounded projections, duality, and multipliers in spaces of harmonic functions.* J. Reine Angew. Math. **299/300** (1978) 256–279.

[568] T. J. SHURA AND D. TRAUTMAN. *The λ-property in Schreier's space S and the Lorentz space $d(a,1)$.* Glasgow Math. J. **32** (1990) 277–284.

[569] B. SIMON. *Trace Ideals and Their Applications.* London Mathematical Society Lecture Note Series 35. Cambridge University Press, 1979.

[570] B. SIMS. *"Ultra"-Techniques in Banach Space Theory.* Queen's Papers in Pure and Applied Mathematics No. 60. Queen's University, Kingston, Ontario, 1982.

[571] B. SIMS AND D. YOST. *Banach spaces with many projections.* In: B. Jefferies, A. McIntosh, and W. Ricker, editors, *Operator theory and partial differential equations, Miniconf. North Ryde/Austral. 1986*, pages 335–342. Proc. Cent. Math. Anal. Austral. Nat. Univ. **14**, 1986.

[572] B. SIMS AND D. YOST. *Linear Hahn-Banach extension operators.* Proc. Edinburgh Math. Soc. **32** (1989) 53–57.

[573] I. SINGER. *Bases in Banach Spaces I.* Springer, Berlin-Heidelberg-New York, 1970.

[574] M. A. SMITH AND F. SULLIVAN. *Extremely smooth Banach spaces.* In: J. Baker, C. Cleaver, and J. Diestel, editors, *Banach Spaces of Analytic Functions. Proc. Conf. Kent, Ohio 1976.* Lecture Notes in Math. 604, pages 125–137. Springer, Berlin-Heidelberg-New York, 1977.

[575] R. R. SMITH. *An addendum to 'M-ideal structure in Banach algebras'.* J. Funct. Anal. **32** (1979) 269–271.

[576] R. R. SMITH. *On Banach algebra elements of thin numerical range.* Math. Proc. Cambridge Phil. Soc. **86** (1979) 71–83.

[577] R. R. SMITH. *The numerical range in the second dual of a Banach algebra.* Math. Proc. Cambridge Phil. Soc. **89** (1981) 301–307.

[578] R. R. SMITH AND J. D. WARD. *M-ideal structure in Banach algebras.* J. Funct. Anal. **27** (1978) 337–349.

[579] R. R. SMITH AND J. D. WARD. *Applications of convexity and M-ideal theory to quotient Banach algebras.* Quart. J. Math. Oxford (2) **30** (1979) 365–384.

[580] R. R. SMITH AND J. D. WARD. *M-ideals in $B(\ell_p)$.* Pacific J. Math. **81** (1979) 227–237.

[581] A. R. SOUROUR. *Isometries of norm ideals of compact operators.* J. Funct. Anal. **43** (1981) 69–77.

[582] N. E. STEENROD. *A convenient category of topological spaces.* Michigan Math. J. **14** (1967) 133–152.

[583] CH. STEGALL. *A proof of the theorem of Amir and Lindenstrauss.* Israel J. Math. **68** (1989) 185–192.

[584] M. H. STONE. *Applications of the theory of Boolean rings to general topology.* Trans. Amer. Math. Soc. **41** (1937) 375–481.

[585] E. STØRMER. *On partially ordered vector spaces and their duals with applications to simplexes and C^*-algebras.* Proc. London Math. Soc. **18** (1968) 245–265.

[586] E. L. STOUT. *The Theory of Uniform Algebras.* Bogden & Quigley, Tarrytown-on-Hudson, 1971.

[587] I. SUCIU. *Function Algebras.* Noordhoff, Leyden, 1975.

[588] F. SULLIVAN. *Geometric properties determined by the higher duals of a Banach space.* Illinois J. Math. **21** (1977) 315–331.

[589] C. SUNDBERG. *$H^\infty + BUC$ does not have the best approximation property.* Ark. Mat. **22** (1984) 287–292.

[590] V. S. SUNDER. *On ideals and duals of C^*-algebras.* Indian J. Pure Appl. Math. **15** (1984) 601–608.

[591] D. G. TACON. *The conjugate of a smooth Banach space.* Bull. Austral. Math. Soc. **2** (1970) 415–425.

[592] M. TAKESAKI. *On the conjugate space of an operator algebra.* Tôhoku Math. J. **10** (1958) 194–203.

[593] M. TAKESAKI. *Theory of Operator Algebras I.* Springer, Berlin-Heidelberg-New York, 1979.

[594] M. TALAGRAND. *La structure des espaces de Banach réticulés ayant la propriété de Radon-Nikodym.* Israel J. Math. **44** (1983) 213–220.

[595] M. TALAGRAND. *A new type of affine Borel function.* Math. Scand. **54** (1984) 183–188.

[596] M. TALAGRAND. *Pettis integral and measure theory.* Mem. Amer. Math. Soc. **307** (1984)

[597] M. TALAGRAND. *Weak Cauchy sequences in $L^1(E)$.* Amer. J. Math. **106** (1984) 703–724.

[598] M. TALAGRAND. *Renormages de quelques $C(K)$.* Israel J. Math. **54** (1986) 327–334.

[599] K. TAYLOR AND W. WERNER. *Differentiability of the norm in von Neumann algebras.* Proc. Amer. Math. Soc. (1993) (to appear).

[600] K. TAYLOR AND W. WERNER. *Differentiability of the norm in C^*-algebras.* In: Proc. Conf. Functional Analysis, Essen 1991, (to appear).

[601] P. D. TAYLOR. *A characterization of G-spaces.* Israel J. Math. **10** (1971) 131–134.

[602] S. L. TROYANSKI. *An example of a smooth space whose dual is not strictly normed.* Studia Math. **35** (1970) 305–309. (Russian).

[603] S. L. TROYANSKI. *On locally uniformly convex and differentiable norms in certain non-separable Banach spaces.* Studia Math. **37** (1971) 173–180.

[604] I. I. TSEITLIN. *The extreme points of the unit ball of certain spaces of operators.* Math. Notes **20** (1976) 848–852.

[605] J. W. TUKEY. *Some notes on the separation of convex sets.* Portugaliae Math. **3** (1942) 95–102.

[606] H. UPMEIER. *Symmetric Banach Manifolds and Jordan C^*-Algebras.* North-Holland, Amsterdam-New York-Oxford, 1985.

[607] H. UPMEIER. *Jordan Algebras in Analysis, Operator Theory, and Quantum Mechanics.* CBMS No. 67. American Mathematical Society, Providence, Rhode Island, 1987.

[608] U. UTTERSRUD. *On M-ideals and the Alfsen-Effros structure topology.* Math. Scand. **43** (1978) 369–381.

[609] U. UTTERSRUD. *Geometrical properties of subclasses of complex L_1-preduals.* Israel J. Math. **72** (1990) 353–371.

[610] M. VALDIVIA. *Resolutions of the identity in certain Banach spaces.* Collect. Math. **39** (1988) 127–140.

[611] M. VALDIVIA. *Resolution of identity in certain metrizable locally convex spaces.* Rev. Real Acad. Cienc. Exact. Fís. Natur. Madrid **83** (1989) 75–96.

[612] M. VALDIVIA. *Projective resolution of identity in $C(K)$ spaces.* Arch. Math. **54** (1990) 493–498.

[613] L. VAŠÁK. *On one generalization of weakly compactly generated Banach spaces.* Studia Math. **70** (1981) 11–19.

[614] J. VESTERSTRØM. *Positive linear extension operators for spaces of affine functions.* Israel J. Math. **16** (1973) 203–211.

[615] J.-P. VIGUÉ. *Domaines bornés symétriques.* In: E. Vesentini, editor, *Geometry Seminar "Luigi Bianchi", Lect. Sc. Norm. Super., Pisa 1982,* Lecture Notes in Math. 1022, pages 125–177, 1983.

[616] G. F. VINCENT-SMITH. *The centre of the tensor product of $A(K)$-spaces and C^*-algebras.* Quart. J. Math. Oxford (2) **28** (1977) 87–91.

[617] J. WADA. *Sets of best approximations to elements in certain function spaces.* In: F. Greenleaf and D. Gulick, editors, *Proc. Conf. Banach Algebras and Several Complex Variables. New Haven, Conn. 1983,* pages 267–272. Contemp. Math. **32**, 1984.

[618] L. WEIS. *Integral operators and changes of density.* Indiana Univ. Math. J. **31** (1982) 83–96.

[619] L. WEIS. *Approximation by weakly compact operators on L_1.* Math. Nachr. **119** (1984) 321–326.

[620] D. WERNER. *M-structure in tensor products of Banach spaces.* Math. Scand. **61** (1987) 149–164.

[621] D. WERNER. *De nouveau: M-idéaux des espaces d'opérateurs compacts.* Séminaire d'Initiation à l'Analyse, année 1988/89. Université Paris VI **28** (1989) Exposé 17.

[622] D. WERNER. *Remarks on M-ideals of compact operators.* Quart. J. Math. Oxford (2) **41** (1990) 501–507.

[623] D. WERNER. *Contributions to the theory of M-ideals in Banach spaces.* Habilitationsschrift, FU Berlin, 1991.

[624] D. WERNER. *New classes of Banach spaces which are M-ideals in their biduals.* Math. Proc. Cambridge Phil. Soc. **111** (1992) 337–354.

[625] D. WERNER. *M-ideals and the 'basic inequality'.* J. Approx. Th. (1993) (to appear).

[626] D. Werner and W. Werner. *On the M-structure of the operator space $L(CK)$*. Studia Math. **87** (1987) 133–138.

[627] E. Werner. *Une preuve du théorème de Odell-Rosenthal par les renormages.* Unpublished manuscript, 1989.

[628] W. Werner. *On the M-structure of spaces of bounded operators and Banach algebras.* Dissertation, FU Berlin, 1988.

[629] W. Werner. *Some results concerning the M-structure of operator spaces.* Math. Ann. **282** (1988) 545–553.

[630] W. Werner. *The centraliser of the injective tensor product.* Bull. Austral. Math. Soc. **44** (1991) 357–365.

[631] W. Werner. *Inner M-ideals in Banach algebras.* Math. Ann. **291** (1991) 205–223.

[632] W. Werner. *The asymptotic behaviour of the metric approximation property on subspaces of c_0.* Arch. Math. **59** (1992) 186–191.

[633] W. Werner. *Smooth points in some spaces of bounded operators.* Integr. Equat. Oper. Th. **15** (1992) 496–502.

[634] R. Whitley. *Projecting m onto c_0.* Amer. Math. Monthly **73** (1966) 285–286.

[635] A. W. Wickstead. *The centraliser of $E \otimes_\lambda F$.* Pacific J. Math. **65** (1976) 563–571.

[636] A. Wilansky. *Semi-Fredholm maps of FK spaces.* Math. Z. **144** (1975) 9–12.

[637] G. Willis. *The compact approximation property does not imply the approximation property.* Studia Math. **103** (1992) 99–108.

[638] W. Wils. *The ideal center of partially ordered vector spaces.* Acta Math. **127** (1971) 41–77.

[639] G. Wodinski. *Multiplikatoren in komplexen Banachräumen.* Dissertation, FU Berlin, 1986.

[640] P. Wojtaszczyk. *On projections in spaces of bounded analytic functions with applications.* Studia Math. **65** (1979) 147–173.

[641] P. Wojtaszczyk. *The Franklin system is an unconditional basis in H_1.* Ark. Mat. **20** (1982) 293–300.

[642] P. Wojtaszczyk. *Banach Spaces for Analysts.* Cambridge Studies in Advanced Mathematics 25. Cambridge University Press, 1991.

[643] P. Wojtaszczyk. *Some remarks on the Daugavet equation.* Proc. Amer. Math. Soc. **115** (1992) 1047–1052.

[644] J. Y. T. Woo. *On modular sequence spaces.* Studia Math. **48** (1973) 271–289.

[645] J. D. M. Wright. *Jordan C^*-algebras.* Michigan Math. J. **24** (1977) 291–302.

[646] A. C. Yorke. *Geometry in quotient reflexive spaces.* J. Austral. Math. Soc. Ser. A **35** (1983) 263–268.

[647] K. Yosida and E. Hewitt. *Finitely additive measures.* Trans. Amer. Math. Soc. **72** (1952) 46–66.

[648] D. Yost. *Best approximation and intersections of balls in Banach spaces.* Bull. Austral. Math. Soc. **20** (1979) 285–300.

[649] D. Yost. *M-ideals, the strong 2-ball property and some renorming theorems.* Proc. Amer. Math. Soc. **81** (1981) 299–303.

[650] D. YOST. *The n-ball properties in real and complex Banach spaces.* Math. Scand. **50** (1982) 100–110.

[651] D. YOST. *Intersecting balls in spaces of vector valued functions.* In: A. Pietsch, N. Popa, and I. Singer, editors, *Banach Space Theory and its Applications. Proc. First Romanian-GDR Seminar, Bucharest 1981,* Lecture Notes in Math. 991, pages 296–302. Springer, Berlin-Heidelberg-New York, 1983.

[652] D. YOST. *Best approximation operators in functional analysis.* In: N. S. Trudinger and G. Williams, editors, *Nonlinear analysis. Miniconf. Canberra 1984*, pages 249–270. Proc. Cent. Math. Anal. Austral. Nat. Univ. **8**, 1984.

[653] D. YOST. *Semi-M-ideals in complex Banach spaces.* Rev. Roum. Math. Pures Appl. **29** (1984) 619–623.

[654] D. YOST. *Approximation by compact operators between $C(X)$ spaces.* J. Approx. Th. **49** (1987) 99–109.

[655] D. YOST. *Banach spaces isomorphic to proper M-ideals.* Colloq. Math. **56** (1988) 99–106.

[656] D. YOST. *There can be no Lipschitz version of Michael's selection theorem.* In: S. T. L. Choy, J. P. Jesudason, and P. Y. Lee, editors, *Analysis Conference, Singapore 1986*, pages 295–299. North Holland, 1988.

[657] D. YOST. *Irreducible convex sets.* Mathematika **38** (1991) 134–155.

[658] R. YOUNIS. *Best approximation in certain Douglas algebras.* Proc. Amer. Math. Soc. **80** (1980) 639–642.

[659] R. YOUNIS. *Properties of certain algebras between L^∞ and H^∞.* J. Funct. Anal. **44** (1981) 381–387.

[660] R. YOUNIS. *M-ideals of L^∞/H^∞ and support sets.* Illinois J. Math. **29** (1985) 96–102.

[661] M. ZIPPIN. *The separable extension problem.* Israel J. Math. **26** (1977) 372–387.

[662] V. ZIZLER. *On some extremal problems in Banach spaces.* Math. Scand. **32** (1973) 214–224.

[663] V. ZIZLER. *Locally uniformly rotund renorming and decompositions of Banach spaces.* Bull. Austral. Math. Soc. **29** (1984) 259–265.

Symbols

$X, Y, \ldots$	Banach spaces
$\mathfrak{A}, \mathfrak{B}, \ldots$	Banach algebras
X^*	dual space of X
X_*	predual space of X
$X_{\mathbb{R}}$	complex Banach space X considered as a real space
$J^{\#}$	cf. Remark I.1.13
$X^{\sharp}$	cf. Definition IV.3.8
X_s	*the* complementary L-summand of an L-embedded space X
J^{θ}	metric complement of J
$Y^{\perp}$	annihilator of subspace $Y \subset X$
$Y_{\perp}$	elements being annihilated by $Y \subset X^*$
J_D	M-ideal of continuous functions vanishing on the set D
$X \cong Y$	X is isometric to Y
$X \simeq Y$	X is isomorphic to Y
$d(X, Y)$	Banach-Mazur distance between X and Y
B_X	closed unit ball of X
S_X	unit sphere of X
$B_X(x, r)$	$\{y \in X \mid \|y - x\| \le r\}$
i_X	natural embedding of X into X^{**}
π_{X^*}	natural projection from X^{***} onto X^*
$P_J(x)$	set of best approximations to x from J
$\|T\|_e$	essential norm of T
$\mathbb{K}$	$\mathbb{R}$ or $\mathbb{C}$
$\mathbb{T}$	complex numbers of modulus 1
$\mathbb{D}$	open unit disk in the complex plane
$\mathbb{S}$	$\{z \in \mathbb{K} \mid \lvert z\rvert = 1\}$

A	disk algebra
H^p, H_0^p	Hardy spaces
h_M, ℓ_M	Orlicz sequence spaces
H_M, L_M	Orlicz function spaces
$\Lambda(N)$, $\widetilde{\Lambda}(N)$	cf. Section VI.6
$d(w,1)$	Lorentz sequence space
$L^{p,1}, L^{p,\infty}$	Lorentz function spaces
$\ell^p(n)$	n-dimensional ℓ^p-space
$\ell^p(X)$	vector valued ℓ^p-space
$L^0(\mu)$	space of (equivalence classes of) measurable functions
L^1_Λ, M_Λ	space of Λ-spectral L^1-functions resp. measures
$C_0(S)$	space of continuous functions on S vanishing at infinity
$C(S)$	space of continuous functions on S (S compact Hausdorff)
$M(S)$	space of regular Borel measures on S
$\mathrm{Mult}(X)$	multiplier algebra of X
$Z(X)$	centralizer of X
$\mathrm{Cun}(X)$	Cunningham algebra of X
$\mathrm{Mult}_{\mathrm{inn}}(\mathfrak{A})$	inner elements of $\mathrm{Mult}(\mathfrak{A})$
$Z_{\mathrm{inn}}(\mathfrak{A})$	inner elements of $Z(\mathfrak{A})$
$a_T(\cdot)$	symbol of $T \in \mathrm{Mult}(X)$
$\overline{M}^{w*}$	weak* closure of M
$\overline{M}^{w}$	weak closure of M
$\mathrm{int}\, M$	interior of M
$\mathrm{ex}\, M$	extreme points of M
$\mathrm{sexp}\, M$	strongly exposed points of M
$\mathrm{co}\, M$	convex hull of M
$\mathrm{lin}\, M$	linear hull of M
E_X	$\mathrm{ex}\, B_{X^*}/\!\sim$ where $p \sim q$ if they are linearly dependent
Z_X	$\overline{\mathrm{ex}}^{w*} B_{X^*} \setminus \{0\}$
$\ker T$	kernel of an operator T
$\mathrm{ran}\, T$	range of an operator T
$L(X,Y)$	space of bounded operators from X to Y
$K(X,Y)$	space of compact operators from X to Y
$K_{w^*}(X^*,Y)$	space of weak*-weakly continuous compact operators from X^* to Y
$A(X,Y)$	space of approximable operators (i.e. norm limits of finite rank operators) from X to Y
$F(X,Y)$	space of finite rank operators from X to Y
$N(X,Y)$	space of nuclear operators from X to Y
$L(X), \ldots$	corresponding space of operators from X to X
$X \widehat{\otimes}_\varepsilon Y$	completed injective tensor product
$X \widehat{\otimes}_\pi Y$	completed projective tensor product

$X \oplus_1 Y$	direct sum of X and Y equipped with the sum-norm
$X \oplus_p Y$	direct sum of X and Y equipped with the ℓ^p-norm
$X \oplus_\infty Y$	direct sum of X and Y equipped with the max-norm
$\sigma(X, J^*)$	cf. Remark I.1.13
s_{op}	strong operator topology
w_{op}	weak operator topology
$\mathbb{H}(\mathfrak{A})$	hermitian elements of $\mathfrak{A}$
e	unit of a unital Banach algebra
L_a, R_a	multiplication operator $x \mapsto ax$ resp. $x \mapsto xa$
M_a	multiplication operator in the commutative case
$\mathsf{S}_{\mathfrak{A}}$	state space of $\mathfrak{A}$
$\mathsf{M}_{\mathfrak{A}}$	maximal ideal space of a function algebra $\mathfrak{A}$
$\mathfrak{A}_{\mathfrak{J}}$	cf. Proposition V.3.5
$\Pi(X)$	$\{(x^*, x) \in B_{X^*} \times B_X \mid x^*(x) = 1\}$
$V(T)$	spatial numerical range of $T \in L(X)$
$v(a, \mathfrak{A})$	numerical range of $a \in \mathfrak{A}$
Γ	dual group of a compact abelian group G
$\widehat{f}(\cdot)$	Fourier transform of $f \in L^1(G)$
φ_γ	character $f \mapsto \widehat{f}(\gamma)$ on $L^1(G)$
f_x	translate of $f \in L^1(G)$, $f_x(y) = f(xy)$
$\langle \cdot, \cdot \rangle$	natural duality
χ_A	indicator function of a set A
$\{f \leq a\}$	$\{x \mid f(x) \leq a\}$
$\complement A$	complement of A
dens X	density character of X
P/X	cf. Lemma I.1.15
d_μ	metric of convergence in measure
$d(x, Y)$	distance of $x \in X$ to $Y \subset X$
αL	one-point compactification of the locally compact space L
δ_x	Dirac measure

Index

Printing: Weihert-Druck GmbH, Darmstadt
Binding: Buchbinderei Schäffer, Grünstadt

Lecture Notes in Mathematics

For information about Vols. 1–1370 please contact your bookseller or Springer-Verlag

Vol. 1412: V.V. Kalashnikov, V.M. Zolotarev (Eds.), Stability Problems for Stochastic Models. Proceedings, 1987. X, 380 pages. 1989.

Vol. 1413: S. Wright, Uniqueness of the Injective III_1 Factor. III, 108 pages. 1989.

Vol. 1414: E. Ramirez de Arellano (Ed.), Algebraic Geometry and Complex Analysis. Proceedings, 1987. VI, 180 pages. 1989.

Vol. 1415: M. Langevin, M. Waldschmidt (Eds.), Cinquante Ans de Polynômes. Fifty Years of Polynomials. Proceedings, 1988. IX, 235 pages. 1990.

Vol. 1416: C. Albert (Ed.), Géométrie Symplectique et Mécanique. Proceedings, 1988. V, 289 pages. 1990.

Vol. 1417: A.J. Sommese, A. Biancofiore, E.L. Livorni (Eds.), Algebraic Geometry. Proceedings, 1988. V, 320 pages. 1990.

Vol. 1418: M. Mimura (Ed.), Homotopy Theory and Related Topics. Proceedings, 1988. V, 241 pages. 1990.

Vol. 1419: P.S. Bullen, P.Y. Lee, J.L. Mawhin, P. Muldowney, W.F. Pfeffer (Eds.), New Integrals. Proceedings, 1988. V, 202 pages. 1990.

Vol. 1420: M. Galbiati, A. Tognoli (Eds.), Real Analytic Geometry. Proceedings, 1988. IV, 366 pages. 1990.

Vol. 1421: H.A. Biagioni, A Nonlinear Theory of Generalized Functions, XII, 214 pages. 1990.

Vol. 1422: V. Villani (Ed.), Complex Geometry and Analysis. Proceedings, 1988. V, 109 pages. 1990.

Vol. 1423: S.O. Kochman, Stable Homotopy Groups of Spheres: A Computer-Assisted Approach. VIII, 330 pages. 1990.

Vol. 1424: F.E. Burstall, J.H. Rawnsley, Twistor Theory for Riemannian Symmetric Spaces. III, 112 pages. 1990.

Vol. 1425: R.A. Piccinini (Ed.), Groups of Self-Equivalences and Related Topics. Proceedings, 1988. V, 214 pages. 1990.

Vol. 1426: J. Azéma, P.A. Meyer, M. Yor (Eds.), Séminaire de Probabilités XXIV, 1988/89. V, 490 pages. 1990.

Vol. 1427: A. Ancona, D. Geman, N. Ikeda, École d'Eté de Probabilités de Saint Flour XVIII, 1988. Ed.: P.L. Hennequin. VII, 330 pages. 1990.

Vol. 1428: K. Erdmann, Blocks of Tame Representation Type and Related Algebras. XV. 312 pages. 1990.

Vol. 1429: S. Homer, A. Nerode, R.A. Platek, G.E. Sacks, A. Scedrov, Logic and Computer Science. Seminar, 1988. Editor: P. Odifreddi. V, 162 pages. 1990.

Vol. 1430: W. Bruns, A. Simis (Eds.), Commutative Algebra. Proceedings. 1988. V, 160 pages. 1990.

Vol. 1431: J.G. Heywood, K. Masuda, R. Rautmann, V.A. Solonnikov (Eds.), The Navier-Stokes Equations – Theory and Numerical Methods. Proceedings, 1988. VII, 238 pages. 1990.

Vol. 1432: K. Ambos-Spies, G.H. Müller, G.E. Sacks (Eds.), Recursion Theory Week. Proceedings, 1989. VI, 393 pages. 1990.

Vol. 1433: S. Lang, W. Cherry, Topics in Nevanlinna Theory. II, 174 pages. 1990.

Vol. 1434: K. Nagasaka, E. Fouvry (Eds.), Analytic Number Theory. Proceedings, 1988. VI, 218 pages. 1990.

Vol. 1435: St. Ruscheweyh, E.B. Saff, L.C. Salinas, R.S. Varga (Eds.), Computational Methods and Function Theory. Proceedings, 1989. VI, 211 pages. 1990.

Vol. 1436: S. Xambó-Descamps (Ed.), Enumerative Geometry. Proceedings, 1987. V, 303 pages. 1990.

Vol. 1437: H. Inassaridze (Ed.), K-theory and Homological Algebra. Seminar, 1987–88. V, 313 pages. 1990.

Vol. 1438: P.G. Lemarié (Ed.) Les Ondelettes en 1989. Seminar. IV, 212 pages. 1990.

Vol. 1439: E. Bujalance, J.J. Etayo, J.M. Gamboa, G. Gromadzki. Automorphism Groups of Compact Bordered Klein Surfaces: A Combinatorial Approach. XIII, 201 pages. 1990.

Vol. 1440: P. Latiolais (Ed.), Topology and Combinatorial Groups Theory. Seminar, 1985–1988. VI, 207 pages. 1990.

Vol. 1441: M. Coornaert, T. Delzant, A. Papadopoulos. Géométrie et théorie des groupes. X, 165 pages. 1990.

Vol. 1442: L. Accardi, M. von Waldenfels (Eds.), Quantum Probability and Applications V. Proceedings, 1988. VI, 413 pages. 1990.

Vol. 1443: K.H. Dovermann, R. Schultz, Equivariant Surgery Theories and Their Periodicity Properties. VI, 227 pages. 1990.

Vol. 1444: H. Korezlioglu, A.S. Ustunel (Eds.), Stochastic Analysis and Related Topics VI. Proceedings, 1988. V, 268 pages. 1990.

Vol. 1445: F. Schulz, Regularity Theory for Quasilinear Elliptic Systems and – Monge Ampère Equations in Two Dimensions. XV, 123 pages. 1990.

Vol. 1446: Methods of Nonconvex Analysis. Seminar, 1989. Editor: A. Cellina. V, 206 pages. 1990.

Vol. 1447: J.-G. Labesse, J. Schwermer (Eds), Cohomology of Arithmetic Groups and Automorphic Forms. Proceedings, 1989. V, 358 pages. 1990.

Vol. 1448: S.K. Jain, S.R. López-Permouth (Eds.), Non-Commutative Ring Theory. Proceedings, 1989. V, 166 pages. 1990.

Vol. 1449: W. Odyniec, G. Lewicki, Minimal Projections in Banach Spaces. VIII, 168 pages. 1990.

Vol. 1450: H. Fujita, T. Ikebe, S.T. Kuroda (Eds.), Functional-Analytic Methods for Partial Differential Equations. Proceedings, 1989. VII, 252 pages. 1990.

Vol. 1451: L. Alvarez-Gaumé, E. Arbarello, C. De Concini, N.J. Hitchin, Global Geometry and Mathematical Physics. Montecatini Terme 1988. Seminar. Editors: M. Francaviglia, F. Gherardelli. IX, 197 pages. 1990.

Vol. 1452: E. Hlawka, R.F. Tichy (Eds.), Number-Theoretic Analysis. Seminar, 1988–89. V, 220 pages. 1990.

Vol. 1453: Yu.G. Borisovich, Yu.E. Gliklikh (Eds.), Global Analysis – Studies and Applications IV. V, 320 pages. 1990.

Vol. 1454: F. Baldassari, S. Bosch, B. Dwork (Eds.), p-adic Analysis. Proceedings, 1989. V, 382 pages. 1990.

Vol. 1455: J.-P. Françoise, R. Roussarie (Eds.), Bifurcations of Planar Vector Fields. Proceedings, 1989. VI, 396 pages. 1990.

Vol. 1456: L.G. Kovács (Ed.), Groups – Canberra 1989. Proceedings. XII, 198 pages. 1990.

Vol. 1457: O. Axelsson, L.Yu. Kolotilina (Eds.), Preconditioned Conjugate Gradient Methods. Proceedings, 1989. V, 196 pages. 1990.

Vol. 1458: R. Schaaf, Global Solution Branches of Two Point Boundary Value Problems. XIX, 141 pages. 1990.

Vol. 1459: D. Tiba, Optimal Control of Nonsmooth Distributed Parameter Systems. VII, 159 pages. 1990.

Vol. 1460: G. Toscani, V. Boffi, S. Rionero (Eds.), Mathematical Aspects of Fluid Plasma Dynamics. Proceedings, 1988. V, 221 pages. 1991.

Vol. 1461: R. Gorenflo, S. Vessella, Abel Integral Equations. VII, 215 pages. 1991.

Vol. 1462: D. Mond, J. Montaldi (Eds.), Singularity Theory and its Applications. Warwick 1989, Part I. VIII, 405 pages. 1991.

Vol. 1463: R. Roberts, I. Stewart (Eds.), Singularity Theory and its Applications. Warwick 1989, Part II. VIII, 322 pages. 1991.

Vol. 1464: D. L. Burkholder, E. Pardoux, A. Sznitman, Ecole d'Eté de Probabilités de Saint- Flour XIX-1989. Editor: P. L. Hennequin. VI, 256 pages. 1991.

Vol. 1465: G. David, Wavelets and Singular Integrals on Curves and Surfaces. X, 107 pages. 1991.

Vol. 1466: W. Banaszczyk, Additive Subgroups of Topological Vector Spaces. VII, 178 pages. 1991.

Vol. 1467: W. M. Schmidt, Diophantine Approximations and Diophantine Equations. VIII, 217 pages. 1991.

Vol. 1468: J. Noguchi, T. Ohsawa (Eds.), Prospects in Complex Geometry. Proceedings, 1989. VII, 421 pages. 1991.

Vol. 1469: J. Lindenstrauss, V. D. Milman (Eds.), Geometric Aspects of Functional Analysis. Seminar 1989-90. XI, 191 pages. 1991.

Vol. 1470: E. Odell, H. Rosenthal (Eds.), Functional Analysis. Proceedings, 1987-89. VII, 199 pages. 1991.

Vol. 1471: A. A. Panchishkin, Non-Archimedean L-Functions of Siegel and Hilbert Modular Forms. VII, 157 pages. 1991.

Vol. 1472: T. T. Nielsen, Bose Algebras: The Complex and Real Wave Representations. V, 132 pages. 1991.

Vol. 1473: Y. Hino, S. Murakami, T. Naito, Functional Differential Equations with Infinite Delay. X, 317 pages. 1991.

Vol. 1474: S. Jackowski, B. Oliver, K. Pawałowski (Eds.), Algebraic Topology, Poznań 1989. Proceedings. VIII, 397 pages. 1991.

Vol. 1475: S. Busenberg, M. Martelli (Eds.), Delay Differential Equations and Dynamical Systems. Proceedings, 1990. VIII, 249 pages. 1991.

Vol. 1476: M. Bekkali, Topics in Set Theory. VII, 120 pages. 1991.

Vol. 1477: R. Jajte, Strong Limit Theorems in Noncommutative L_2-Spaces. X, 113 pages. 1991.

Vol. 1478: M.-P. Malliavin (Ed.), Topics in Invariant Theory. Seminar 1989-1990. VI, 272 pages. 1991.

Vol. 1479: S. Bloch, I. Dolgachev, W. Fulton (Eds.), Algebraic Geometry. Proceedings, 1989. VII, 300 pages. 1991.

Vol. 1480: F. Dumortier, R. Roussarie, J. Sotomayor, H. Żołądek, Bifurcations of Planar Vector Fields: Nilpotent Singularities and Abelian Integrals. VIII, 226 pages. 1991.

Vol. 1481: D. Ferus, U. Pinkall, U. Simon, B. Wegner (Eds.), Global Differential Geometry and Global Analysis. Proceedings, 1991. VIII, 283 pages. 1991.

Vol. 1482: J. Chabrowski, The Dirichlet Problem with L^2-Boundary Data for Elliptic Linear Equations. VI, 173 pages. 1991.

Vol. 1483: E. Reithmeier, Periodic Solutions of Nonlinear Dynamical Systems. VI, 171 pages. 1991.

Vol. 1484: H. Delfs, Homology of Locally Semialgebraic Spaces. IX, 136 pages. 1991.

Vol. 1485: J. Azéma, P. A. Meyer, M. Yor (Eds.), Séminaire de Probabilités XXV. VIII, 440 pages. 1991.

Vol. 1486: L. Arnold, H. Crauel, J.-P. Eckmann (Eds.), Lyapunov Exponents. Proceedings, 1990. VIII, 365 pages. 1991.

Vol. 1487: E. Freitag, Singular Modular Forms and Theta Relations. VI, 172 pages. 1991.

Vol. 1488: A. Carboni, M. C. Pedicchio, G. Rosolini (Eds.), Category Theory. Proceedings, 1990. VII, 494 pages. 1991.

Vol. 1489: A. Mielke, Hamiltonian and Lagrangian Flows on Center Manifolds. X, 140 pages. 1991.

Vol. 1490: K. Metsch, Linear Spaces with Few Lines. XIII, 196 pages. 1991.

Vol. 1491: E. Lluis-Puebla, J.-L. Loday, H. Gillet, C. Soulé, V. Snaith, Higher Algebraic K-Theory: an overview. IX, 164 pages. 1992.

Vol. 1492: K. R. Wicks, Fractals and Hyperspaces. VIII, 168 pages. 1991.

Vol. 1493: E. Benoît (Ed.), Dynamic Bifurcations. Proceedings, Luminy 1990. VII, 219 pages. 1991.

Vol. 1494: M.-T. Cheng, X.-W. Zhou, D.-G. Deng (Eds.), Harmonic Analysis. Proceedings, 1988. IX, 226 pages. 1991.

Vol. 1495: J. M. Bony, G. Grubb, L. Hörmander, H. Komatsu, J. Sjöstrand, Microlocal Analysis and Applications. Montecatini Terme, 1989. Editors: L. Cattabriga, L. Rodino. VII, 349 pages. 1991.

Vol. 1496: C. Foias, B. Francis, J. W. Helton, H. Kwakernaak, J. B. Pearson, H_∞-Control Theory. Como, 1990. Editors: E. Mosca, L. Pandolfi. VII, 336 pages. 1991.

Vol. 1497: G. T. Herman, A. K. Louis, F. Natterer (Eds.), Mathematical Methods in Tomography. Proceedings 1990. X, 268 pages. 1991.

Vol. 1498: R. Lang, Spectral Theory of Random Schrödinger Operators. X, 125 pages. 1991.

Vol. 1499: K. Taira, Boundary Value Problems and Markov Processes. IX, 132 pages. 1991.

Vol. 1500: J.-P. Serre, Lie Algebras and Lie Groups. VII, 168 pages. 1992.

Vol. 1501: A. De Masi, E. Presutti, Mathematical Methods for Hydrodynamic Limits. IX, 196 pages. 1991.

Vol. 1502: C. Simpson, Asymptotic Behavior of Monodromy. V, 139 pages. 1991.

Vol. 1503: S. Shokranian, The Selberg-Arthur Trace Formula (Lectures by J. Arthur). VII, 97 pages. 1991.

Vol. 1504: J. Cheeger, M. Gromov, C. Okonek, P. Pansu, Geometric Topology: Recent Developments. Editors: P. de Bartolomeis, F. Tricerri. VII, 197 pages. 1991.

Vol. 1505: K. Kajitani, T. Nishitani, The Hyperbolic Cauchy Problem. VII, 168 pages. 1991.

Vol. 1506: A. Buium, Differential Algebraic Groups of Finite Dimension. XV, 145 pages. 1992.

Vol. 1507: K. Hulek, T. Peternell, M. Schneider, F.-O. Schreyer (Eds.), Complex Algebraic Varieties. Proceedings, 1990. VII, 179 pages. 1992.

Vol. 1508: M. Vuorinen (Ed.), Quasiconformal Space Mappings. A Collection of Surveys 1960-1990. IX, 148 pages. 1992.

Vol. 1509: J. Aguadé, M. Castellet, F. R. Cohen (Eds.), Algebraic Topology - Homotopy and Group Cohomology. Proceedings, 1990. X, 330 pages. 1992.

Vol. 1510: P. P. Kulish (Ed.), Quantum Groups. Proceedings, 1990. XII, 398 pages. 1992.

Vol. 1511: B. S. Yadav, D. Singh (Eds.), Functional Analysis and Operator Theory. Proceedings, 1990. VIII, 223 pages. 1992.

Vol. 1512: L. M. Adleman, M.-D. A. Huang, Primality Testing and Abelian Varieties Over Finite Fields. VII, 142 pages. 1992.

Vol. 1513: L. S. Block, W. A. Coppel, Dynamics in One Dimension. VIII, 249 pages. 1992.

Vol. 1514: U. Krengel, K. Richter, V. Warstat (Eds.), Ergodic Theory and Related Topics III, Proceedings, 1990. VIII, 236 pages. 1992.

Vol. 1515: E. Ballico, F. Catanese, C. Ciliberto (Eds.), Classification of Irregular Varieties. Proceedings, 1990. VII, 149 pages. 1992.

Vol. 1516: R. A. Lorentz, Multivariate Birkhoff Interpolation. IX, 192 pages. 1992.

Vol. 1517: K. Keimel, W. Roth, Ordered Cones and Approximation. VI, 134 pages. 1992.

Vol. 1518: H. Stichtenoth, M. A. Tsfasman (Eds.), Coding Theory and Algebraic Geometry. Proceedings, 1991. VIII, 223 pages. 1992.

Vol. 1519: M. W. Short, The Primitive Soluble Permutation Groups of Degree less than 256. IX, 145 pages. 1992.

Vol. 1520: Yu. G. Borisovich, Yu. E. Gliklikh (Eds.), Global Analysis – Studies and Applications V. VII, 284 pages. 1992.

Vol. 1521: S. Busenberg, B. Forte, H. K. Kuiken, Mathematical Modelling of Industrial Process. Bari, 1990. Editors: V. Capasso, A. Fasano. VII, 162 pages. 1992.

Vol. 1522: J.-M. Delort, F. B. I. Transformation. VII, 101 pages. 1992.

Vol. 1523: W. Xue, Rings with Morita Duality. X, 168 pages. 1992.

Vol. 1524: M. Coste, L. Mahé, M.-F. Roy (Eds.), Real Algebraic Geometry. Proceedings, 1991. VIII, 418 pages. 1992.

Vol. 1525: C. Casacuberta, M. Castellet (Eds.), Mathematical Research Today and Tomorrow. VII, 112 pages. 1992.

Vol. 1526: J. Azéma, P. A. Meyer, M. Yor (Eds.), Séminaire de Probabilités XXVI. X, 633 pages. 1992.

Vol. 1527: M. I. Freidlin, J.-F. Le Gall, Ecole d'Eté de Probabilités de Saint-Flour XX – 1990. Editor: P. L. Hennequin. VIII, 244 pages. 1992.

Vol. 1528: G. Isac, Complementarity Problems. VI, 297 pages. 1992.

Vol. 1529: J. van Neerven, The Adjoint of a Semigroup of Linear Operators. X, 195 pages. 1992.

Vol. 1530: J. G. Heywood, K. Masuda, R. Rautmann, S. A. Solonnikov (Eds.), The Navier-Stokes Equations II – Theory and Numerical Methods. IX, 322 pages. 1992.

Vol. 1531: M. Stoer, Design of Survivable Networks. IV, 206 pages. 1992.

Vol. 1532: J. F. Colombeau, Multiplication of Distributions. X, 184 pages. 1992.

Vol. 1533: P. Jipsen, H. Rose, Varieties of Lattices. X, 162 pages. 1992.

Vol. 1534: C. Greither, Cyclic Galois Extensions of Commutative Rings. X, 145 pages. 1992.

Vol. 1535: A. B. Evans, Orthomorphism Graphs of Groups. VIII, 114 pages. 1992.

Vol. 1536: M. K. Kwong, A. Zettl, Norm Inequalities for Derivatives and Differences. VII, 150 pages. 1992.

Vol. 1537: P. Fitzpatrick, M. Martelli, J. Mawhin, R. Nussbaum, Topological Methods for Ordinary Differential Equations. Montecatini Terme, 1991. Editors: M. Furi, P. Zecca. VII, 218 pages. 1993.

Vol. 1538: P.-A. Meyer, Quantum Probability for Probabilists. X, 287 pages. 1993.

Vol. 1539: M. Coornaert, A. Papadopoulos, Symbolic Dynamics and Hyperbolic Groups. VIII, 138 pages. 1993.

Vol. 1540: H. Komatsu (Ed.), Functional Analysis and Related Topics, 1991. Proceedings. XXI, 413 pages. 1993.

Vol. 1541: D. A. Dawson, B. Maisonneuve, J. Spencer, Ecole d' Eté de Probabilités de Saint-Flour XXI - 1991. Editor: P. L. Hennequin. VIII, 356 pages. 1993.

Vol. 1542: J.Fröhlich, Th.Kerler, Quantum Groups, Quantum Categories and Quantum Field Theory. VII, 431 pages. 1993.

Vol. 1543: A. L. Dontchev, T. Zolezzi, Well-Posed Optimization Problems. XII, 421 pages. 1993.

Vol. 1544: M.Schürmann, White Noise on Bialgebras. VII, 146 pages. 1993.

Vol. 1545: J. Morgan, K. O'Grady, Differential Topology of Complex Surfaces. VIII, 224 pages. 1993.

Vol. 1546: V. V. Kalashnikov, V. M. Zolotarev (Eds.), Stability Problems for Stochastic Models. Proceedings, 1991. VIII, 229 pages. 1993.

Vol. 1547: P. Harmand, D. Werner, W. Werner, M-ideals in Banach Spaces and Banach Algebras. VIII, 387 pages. 1993.

Vol. 1548: T. Urabe, Dynkin Graphs and Quadrilateral Singularities. VI, 233 pages. 1993.

Vol. 1549: G. Vainikko, Multidimensional Weakly Singular Integral Equations. XI, 159 pages. 1993.

Vol. 1551: L. Arkeryd, P. L. Lions, P.A. Markowich, S.R. S. Varadhan. Nonequilibrium Problems in Many-Particle Systems. Montecatini, 1992. Editors: C. Cerugnani, M. Pulvirenti. VII, 158 pages 1993.

Vol. 1552: J. Hilgert, K.-H. Neeb, Lie Semigroups and their Applications. XII, 315 pages. 1993.